International Conference on Composite Material, Polymer Science and Engineering (CMPSE 2017)

MATEC Web of Conferences Volume 130 (2017)

Toyama City, Japan
24 - 25 June 2017

Editors:

Katsuyuki Kida

ISBN: 978-1-5108-5050-7

Printed from e-media with permission by:

Curran Associates, Inc.
57 Morehouse Lane
Red Hook, NY 12571

Some format issues inherent in the e-media version may also appear in this print version.

Printed by Curran Associates, Inc. (2017)

For additional information, please contact EDP Sciences – Web of Conferences at the address below.

EDP Sciences – Web of Conferences
17, Avenue du Hoggar
Parc d'Activité de Courtabœuf
BP 112
F-91944 Les Ulis Cedex A
France

Phone: +33 (0) 1 69 18 75 75
Fax: +33 (0) 1 69 28 84 91

contact-edps@webofconferences.org

Additional copies of this publication are available from:

Curran Associates, Inc.
57 Morehouse Lane
Red Hook, NY 12571 USA
Phone: 845-758-0400
Fax: 845-758-2633
Email: curran@proceedings.com
Web: www.proceedings.com

TABLE OF CONTENTS

The International Conference on Composite Material, Polymer Science and Engineering (CMPSE2017)

June 24-25, 2017

Japan

Edited by

Prof. Katsuyuki Kida, Department of Mechanical and Intellectual

Systems Engineering, University of Toyama, Japan

Assistant professor Koshiro Mizobe, University of Toyama, Japan

CMPSE2017 Committee

Conference Chair

Professor Katsuyuki Kida, University of Toyama, Japan

Scientific Editor

Assistant professor Koshiro Mizobe, University of Toyama, Japan

Program Committee

International Academic Committee Chairs （Malaysia）

Abdul Talib bin Din, Universiti Teknikal Malaysia Melaka (UTeM), Malaysia

Zakiah Ahmad, University Teknologi Mara, Malaysia

International Academic Committee Chair （India）

Bhingole Pramod P., Institute of Infrastructure, Technology, Research and Management (IITRAM), INDIA

International Academic Committee Chair （Japan）

Koshiro Mizobe, University of Toyama, Japan

Takahiro Matsueda, Tokyo Metropolitan University, Japan

Hitonobu Koike, Kyushu Sangyo University, Japan

Kazuaki Nakane, Osaka University, Japan

International Academic Committee Chair （Russia）

Dmitry A. Chinakhov, Yurga Institute of Technology of National Research Tomsk Polytechnic University, Russia

Sponsor

University of Toyama, Japan

Preface

Dear Distinguished Authors and Guests,

This proceeding book consists of 50 papers from 'The 1st International Conference on Composite Material, Polymer Science and Engineering (CMPSE), which was held on June 24th and 25th, 2017 in Toyama, Japan. This conference focused on recent scientific research and engineering technology in the field of composite materials, polymer science and engineering. The main topics of the 1st CMPSE were 'Metal composite', 'Ceramic composite', 'Polymer composite', and other materials related issues, which cover the whole area ranging from material process (chemical synthesis) to practical engineering.

All papers were accepted through a peer-review process by the editorial committee and international reviewers, and were categorized into 9 chapters.

On behalf of the organizing committee, the editor would like to acknowledge the technical comments from the reviewers and thank all of the authors for their contributions to this proceeding book. And thanks are expressed to Koshiro Mizobe from the University of Toyama who, as the conference secretary, did a thorough job of planning and organizing the conference. Also thanks to Kazuaki Nakane from Osaka University, Hitonobu Koike from Kyushu Sangyo University and Takahiro Matsueda from Tokyo Metropolitan University for their editorial work. With our warmest regards,

Editor of CMPSE2017

Statement of Peer review

In submitting conference proceedings to MATEC Web of Conferences, I certify to the Publisher that I adhere to the **Policy on Publishing Integrity** of the journal in order to safeguard good scientific practice in publishing.

1. All articles have been subjected to peer review administered by the proceedings editors.
2. Reviews have been conducted by expert referees, who have been requested to provide unbiased and constructive comments aimed, whenever possible, at improving the work.
3. Proceedings editors have taken all reasonable steps to ensure the quality of the materials they publish and their decision to accept or reject a paper for publication has been based only on the merits of the work and the relevance to the journal.

Title, date and place of the conference

The International Conference on Composite Material, Polymer Science and Engineering (CMPSE2017)
June 24-25, 2017
University of Toyama, Toyama City, Japan

Proceedings editor(s):

Katsuyuki Kida, University of Toyama, Japan

Date and editor's signature

September 30, 2017

17, avenue du Hoggar PA de Courtabœuf – BP 112 - 91944 Les Ulis Cedex A (France)

DOI: 10.1051/matecconf/201713001001

Synthesis and Evaluation of Olivine Material Coated with Carbon

Masakazu Togo[1, a] and Atsushi Nakahira[1]

[1]Department of Materials Science, Osaka Prefecture University, 1-1 Gakuen-cho, Nakaku, Sakai, Osaka, JAPAN

Abstract. Lithium iron phosphate, $LiFePO_4$, was synthesized by hydrothermal process and subsequently coated with carbon by the thermal decomposition of acetylene gas. The products were characterized by XRD, SEM, and TG-DTA. As-synthesized $LiFePO_4$ was submicron sized plate-like particle. After heating at various temperature in nitrogen atmosphere, the particle size and the crystalline sizes were grown with increasing the heating temperature. Above 700°C, the grain growth was remarkably. Carbon coating temperature was set at 500-600°C because of fine particle and good crystallinity. As a carbon raw material, acetylene gas was flowing to the as-synthesized $LiFePO_4$ in nitrogen atmosphere, and the $LiFePO_4$/C composite was obtained. TG curves showed weight loss above 500°C, which was thought to be associated with carbon layer composition.

1 Introduction

$LiCoO_2$ has been used as a cathode material for lithium ion batteries. Usage of cobalt prohibits to reduce the material costs because of scarcity. The amount of needed lithium ion battery increases significantly for the application of EV and energy storage system. Then, the development of Co-free cathode material is quietly important. Lithium iron phosphate, $LiFePO_4$, is the material, in which iron having high number of Clark is rich in resources, and is inexpensive. Padhi et al. reported that polyanion material like as $LiFePO_4$ and Li_2FeSiO_4, showed the charge-discharge property as cathode for LIB. [1] $LiFePO_4$ shows 3.5 V v.s. Li of reversible potential and 170 $mAhg^{-1}$ of theoretical capacity. These values are the energy density comparable to $LiCoO_2$.

$LiFePO_4$ has orthorhombic crystal olivine type structure. The crystal structure of $LiFePO_4$ is showed in Figure 1. FeO_6 octahedra and PO_4 tetrahedra have the structure that a ridge shared one side, and Li arranges it in a b-axis. PO_4 tetrahedra is strongly connected by covalent bond, and it restrains the oxygen decomposition. The structure stability of produce for $LiFePO_4$ the safety. On the other hand, ionic conduction for $LiFePO_4$ is difficult because a Li ion in structure is localized electronically. The discharge reaction of $LiFePO_4$ is showed in the following equation.

$$LiFePO_4 \rightarrow Li_{1-x}FePO_4 + xLi^+ + xe^- \qquad (1)$$

In the discharge reaction, $FePO_4$ phase is formed by deintercalation of the lithium ion. With the biphase reaction of $LiFePO_4$ and $FePO_4$, equilibrium potential is seen in around 3.5V. Both structure is similar, and cycle capability with charge-discharge reaction is good because the symmetricalness of both does not change. As a model of these reaction, domino-cascade model is proposed. [2] In a border between $LiFePO_4$ and $FePO_4$ with the disorder of the atom placement, electronic conductivity is relatively high, and it is thought that Li^+ is deintercalated from there. The conductivity of 10^{-8}-10^{-9} S/cm for $LiFePO_4$, however, is very lower than that of 10^{-2} S/cm for $LiCoO_2$. [3] Therefore, the problem for commercial use is this low conductivity, and the improvement of conductivity can lead to the good rate capability. In order to improve the conductivity, many researchers have been tried. First, focused on the material composition, it reported that the multi cation doped $LiFePO_4$ showed the good rate capability. It was because that the doped cations, such as Mg^{2+}, Al^{3+}, Zr^{4+}, Nb^{5+} and W^{6+}, reduce the volumetric resistance. [4] On the other hand, the microstructure for the electrode materials is also important. If the electrode material makes a fine particle, its reactivity increases due to the improvement of the supecific surface area. In addition, the diffusion length of the ion is shortened by microparticulation, and this is quietly advantageous. There were the reports focused on the morphology with high performance. [5] In order to obtain the finer particle, a various synthesis method has been examined. A. Yamada reported that the optimization of solid state reaction conditions produced a fine particle, and it showed a good performance. [6] However, the diffusion necessary for a complete reaction needs the condition at high temperature for long-time. As the synthesis method at lower temperature, many methods such as hydrothermal method [7], sol-gel method [8, 9], and splay pyrolysis method [10] are reported. In this study, using the hydrothermal process, the synthesis of $LiFePO_4$ composed of a fine particle was attempted.

In order to cope with high rate charge-discharge reaction, $LiFePO_4$ needs not only the microparticulation but also

[a] Corresponding author: st110029@edu.osakafu-u.ac.jp

 DOI: 10.1051/matecconf/201713001001

the composition with conductive assistant. Coating on $LiFePO_4$ particle surfaces is suitable to make a little conductive assistant in the composition methods. The conduction layer has two kinds of an electronic conduction and the ionic conduction. B. Kang reported the ionic conduction layer. [11] The amorphous $Li_4P_2O_7$ phase as the ionic conduction layer was precipitated by making non-stoichiometric composition. This amorphous phase prohibit the grain growth, and finer particle would be obtained. As the result, the obtained active material showed good property with 130 mAh/g of the discharging capacity at high rate. On the contrary, as the electronic conduction layer, carbon coating has been examined. There are various kinds of raw carbon material and that composition methods. Then, the $LiFePO_4$/C composite produced by heating the mixture of the active material and ractose in inert atmosphere [12], and depositing carbon from heat decomposition of propylene gas. [13] The products by wet method are easy to be affected to the crystalinity and the grain growth among heating process. In this research, $LiFePO_4$ was synthesized by hydrothermal method, and the products was coated with carbon by the thermal decomposition of acetylene gas. The influence of carbon coating process to $LiFePO_4$ was examined.

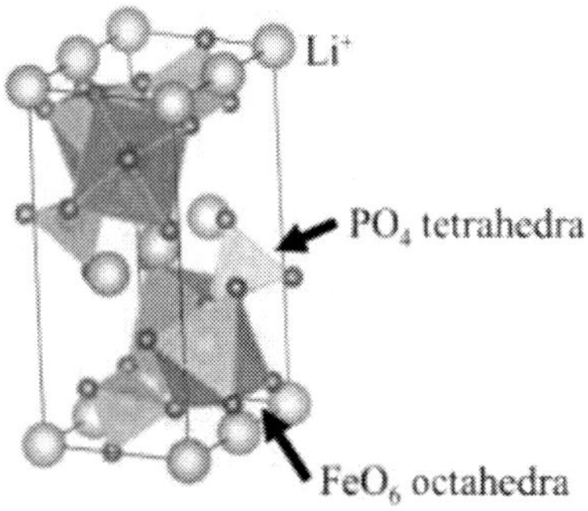

Figure 1. Olivine-type structure of $LiFePO_4$.

2 Experimental procedure

2.1 Synthesis

2.1.1 Synthesis of $LiFePO_4$ by hydrothermal process

$LiOH.H_2O$, $(NH_4)_2HPO_4$ and $FeSO_4.7H_2O$ (Wako Chemical) were used as starting materials. Each starting materials were weighted to become a molar ratio of Li: P: Fe = 2: 1: 1. After measurement, they were put into distilled water in Teflon vessel and vigorously mixed for a few minutes. Teflon vessel with mixed solution was encapsulated into autoclave and was hydrothermally treated at 200°C for 24 hours. Obtained slurry was filtrated and then dried at 50°C for overnight.

2.1.2 Heat treatment of $LiFePO_4$

$LiFePO_4$ synthesized by hydrothermal treatment was heated in nitrogen atmosphere. 0.5g of $LiFePO_4$ was measured, and it was put in carbon crucible. Carbon crucible was moved in a furnace. After the furnace was evacuated slowly, nitrogen gas was carried out with flow 5L/min for 30min. Then, the flow quantity of nitrogen gas was controlled 0.2L/min. The furnace was raised up to target temperature (T=400 - 800°C) with heat rate of 4°C/min and kept it for 2 hours.

2.1.3 $LiFePO_4$/C composites by thermal decomposition

Rotary kiln (Takasago Industry Co., Ltd.) showed in Figure 2 was used for the manufacture of the $LiFePO_4$/C composite. 5g of $LiFePO_4$ was measured, and it was put in carbon capsule for product. Capsule was rolled at the rate of 1rpm. In order to reduce the oxygen concentration in the capsule, N_2 gas was carried out with flow of 5L/min for 15min. Then, 0.2 L/min of acetylene gas was flowing to the capsule through a pipe of the alumina. The capsule was heated externally at 550°C and 600°C for 1 hour, respectively.

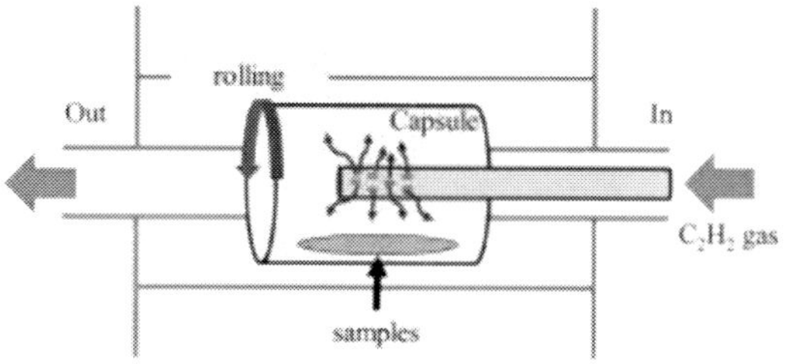

Figure 2. Schematic rotary kiln

2.2 Evaluation

The crystal phase of samples was indexed by XRD (UltimaIV, Rigaku Co., Japan). Scans were performed at 2θ = 10-40°with scan rate of 4°/min using Cu-Kα radiation. In order to estimate the lattice parameter of samples, internal reference method was used using pure Si powder. The morphology was observed by FE-SEM (s-4500, Hitachi, Japan) with applied voltage of 15 kV. Thermal gravimetric analysis was carried out under air atmosphere between room temperature and 700°C, and the flow rate of the synthetic air was 2 mL/min.

3 Results and discussions

3.1. Synthesis of $LiFePO_4$ by hydrothermal process

$LiFePO_4$ was synthesized by hydrothermal process at 200°C for 24h. XRD patterns of the products were showed in Figure 3. The diffraction peaks attributed to olivine-type structure (JCPDS#83-2092) were shown, and the other peaks considered the secondary phase like Li_3PO_4 and Fe_2P were not confirmed. So, the products by hydrothermal method was identified as olivine-type structure. The lattice parameter estimated by inter reference method was a=10.336Å, b=6.002Å, c=4.695Å. This result was good agreement with JCPDS value (a=10.3340Å, b=6.0100Å, c=4.6930Å). In the case of hydrothermal process, it was reported that the lattice parameter of $LiFePO_4$ depended on the reaction condition, especially reaction temperature. [14] At higher reaction temperatures, the unit cell volume was closer to the value of 290.5$Å^3$. The unit cell volume of the products by hydrothermal process was 291.3$Å^3$. As this difference was thought to be case by the crystal purity, the products by hydrothermal process in this study would have the relatively high crystallinity. In addition, the crystalline from (311) plane was 35 nm. Therefore, using of this as-synthesized $LiFePO_4$, the heat-treatments for $LiFePO_4$ and the subsequent synthesis of $LiFePO_4$/C composite was attempted.

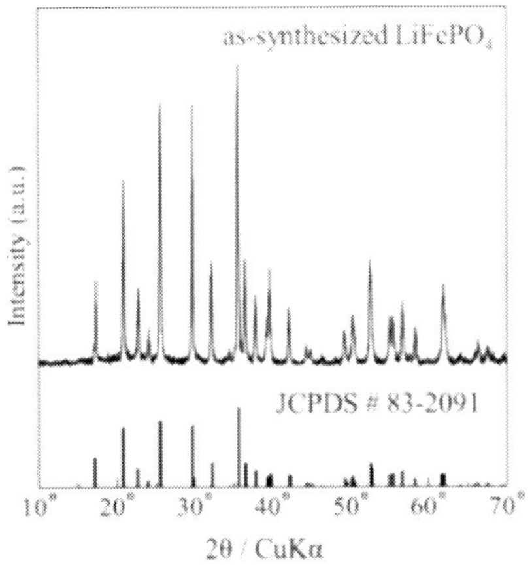

Figure 3 XRD patterns of the product synthesized by hydrothermal process.

3.2 Heat treatments of $LiFePO_4$

The as-synthesized $LiFePO_4$ was heated at 400-800°C in nitrogen atmosphere. Heating process is affected to the crystalline and the grain growth, so that is important role to influence the last products. The influence of heating process was examined. In Figure 4, the crystal phase of the products heated at various temperature in N_2 atmosphere was identified by XRD analysis. After heating, no diffraction peaks of the secondary phases except the peaks attributed olivine-type phase was appeared. In addition, the peak intensity increased with the higher temperature. Therefore, it was thought that the decomposition of $LiFePO_4$ or the phase transformation due to heating process was not occurred, after heat-treatments, but the crystallinity was well done. Next, the crystalline sizes were estimated by Scherar equation with the following, and the result was plotting in Figure 5.

$$D = K\lambda / \beta \cos\theta \quad (2)$$

The peaks belonged to nine lattice plane, (200), (101), (210), (201), (020), (301), (311), (121) and (410), were used for analysis. In (311) plane, the crystalline size was 41nm at 400°C larger than that of as-synthesized $LiFePO_4$. The crystalline size tended to be increased linearly with heating temperature. At 800°C, 84nm of the crystalline size was estimated. Similarly, the same tendency was seen with the other diffraction peaks. In addition, the remarkable difference was appeared in the range of 600°C to 700°C.

In Figure 5, SEM images of the heated products was shown. The particle size of the as-synthesized $LiFePO_4$ was about 0.5μm and the morphology was plate-like particle. Finer particles were also observed in a part. There was no effect of particle size on longer reaction time, so no growth particle was generated during heat-treatments. Next, the size of samples at 400°C was about 0.5μm similar to no-heated samples and the shape became the round particle without finer particles. In addition, the particle size increased remarkably with heated temperature above 600°C. At 800°C, the coarse particle was observed, and the particle size was up to 1μm. From these observation, it was thought that the grain growth occurred above 600°C.

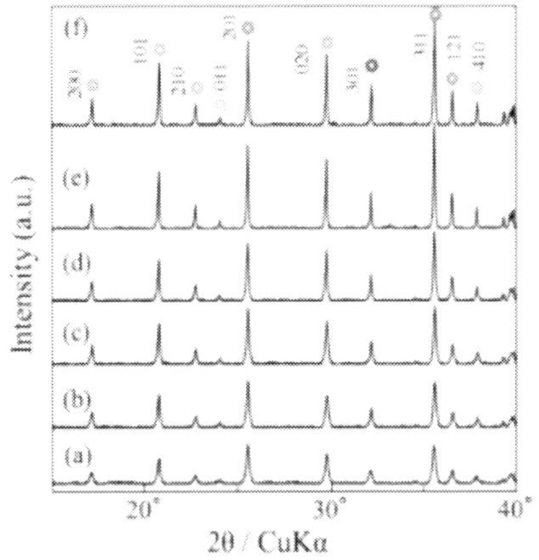

Figure 4 XRD patterns of the olivine products annealed in N_2 atmosphere at 400°C, 500°C, 600°C, 700°C, 800°C

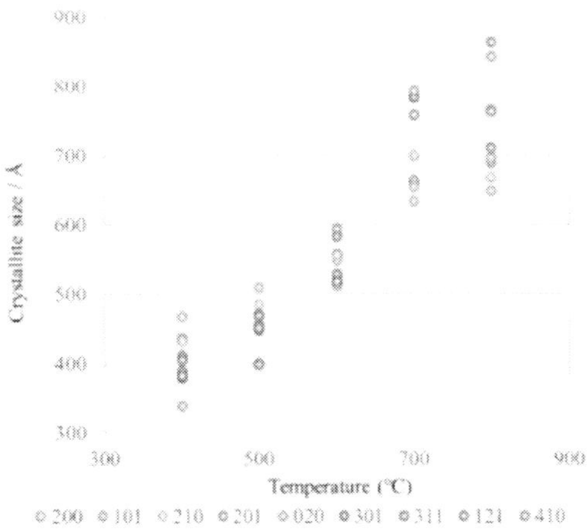

Figure 4 Crystalline sizes estimated by Scherar equation at each annealed temperature.

 DOI: 10.1051/matecconf/201713001001

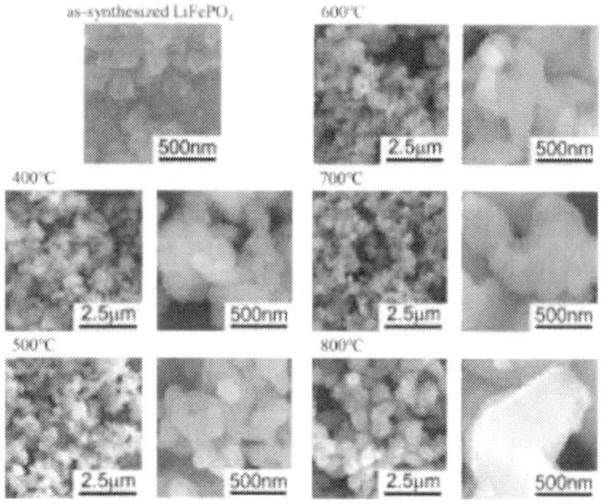

Figure 5 SEM images of the olivine products annealed in N_2 atmosphere at 400℃, 500℃, 600℃, 700℃, 800℃

As a result, heating process led to the improvement of the crystallinity. Below 600°C, amorphous phase or amorphous layer on particle surface was crystalized and the grain growth was not remarkable. Above 600°C, the grain growth was occurred, and it led to crystalline improvement.

3.3 $LiFePO_4$/C composites by thermal decomposition

The $LiFePO_4$/C composite was synthesized by the thermal decomposition of acetylene gas. With flowing of acetylene gas, the as-synthesized $LiFePO_4$ was heated at various temperature for 1 hour. The reaction temperature was settled at 550°C and 600°C which the crystallinity was improved and no grain growth occurred. XRD patterns of the products thermal-decomposed at 550°C and 600°C with flow of acetylene gas were showed in Figure 6. The diffraction peaks were attributed to olivine-type structure, and the secondary phase were not appeared. During this thermal decomposition, no oxidation was occurred for obtained olivine. SEM images were shown in Figure 7. The particle size was about 0.6μm. At both 550°C and 600°C, no grain growth during thermal decomposition was occurred remarkably. But, whereas the round particle was observed for samples at 550°C, the square shape was appeared partially at 600°C. The thermal decomposition of acetylene gas was exothermal reaction of 228kJ/mol. The morphology change of samples that morphology of squared shapes was similar to that at higher reaction temperature was thought to be related with exothermal reaction.

Thermal stability characterization of the obtained $LiFePO_4$/C composite was examined in air condition. TG-DTA curves were draw in Figure 8. In air, $LiFePO_4$ was oxidized with following the reaction. [15]

$$LiFePO_4 + 1/4O_2 \rightarrow 1/3Li_3Fe_2(PO_4) + 1/6\, Fe_2O_3 \quad (3)$$

Estimated total weight gain is 4.2%. In case of the as-synthesized $LiFePO_4$, after initial weight loss associated to loss of water under 250°C, TG curves showed weight gain of 2.8%. From DTA curves, the exothermic peak was observed at around 300°C, and the oxidation would be occurred. On the other hand, initial weight loss was not appeared for $LiFePO_4$/C products. Then, weight gain associated to the oxidation started at around 300°C and weight loss started at around 500°C. This weight loss was thought to be caused by the burning of carbon phase. Therefore, the $LiFePO_4$/C composite were successfully synthesized. In addition, the oxidation starting temperature was different from the $LiFePO_4$/C heated at 550°C in that heated at 600°C. Thus, it was thought that the oxidation reaction was dependent on coated carbon layer on $LiFePO_4$/C composites.

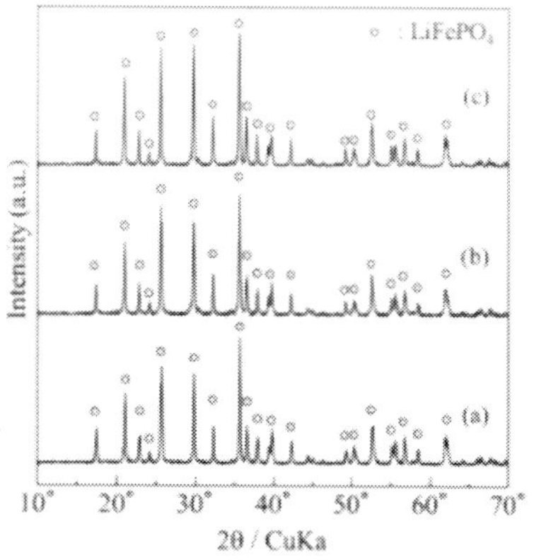

Figure 6 XRD patterns of $LiFePO_4$/C composite samples annealed in N_2 atmosphere. (a) as-synthesized $LiFePO_4$, (b) annealed at 550°C with flow of acetylene gas and (c) annealed at 600 °C with flow of acetylene gas.

Figure 7 SEM images of $LiFePO_4$/C composite samples annealed in N_2 atmosphere at 550 ℃, 600 ℃

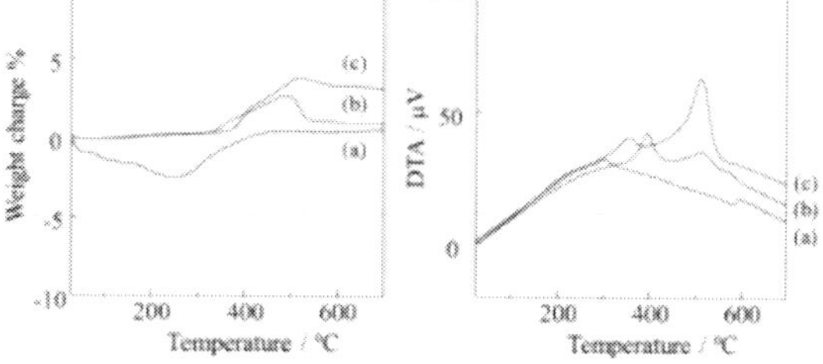

Figure 8 TG-DTA analysis of the $LiFePO_4$/C composite samples: (a) as-synthesized $LiFePO_4$, (b) annealed at 550°C and (c) annealed at 600°C.

4 Conclusion

In this study, $LiFePO_4$ was synthesized by hydrothermal method, and the products was coated with carbon by the thermal decompositionof acetylene gas. The influence of carbon coating process to $LiFePO_4$ was examined. The obtained $LiFePO_4$ samples was crystaline phase and lattice parameter was a=10.336Å, b=6.002 Å, c=4.695Å. Heating process was influenced to the microstructure and the crystallinity. Below 600°C, amorphous phase or amorphous layer on particle surface was crystalized. Above 600°C, the grain growth was occured, and it led to crystalline improvement. Carbon coating temperature settled at 500-600°C because of fine particle and good crystalinity. TG-DTA curves of the $LiFePO_4/C$ products showed weight loss above 500°C, and it thought to be associated with carbon layer composition.

References

1. A. K. Padhi, K. S. Nanjundaswamy and J. B. Goodenough, J. Electrochem. Soc., **144**, 1609 (1997)
2. C. Delmas, M. Maccario, L. Croguennec, F. Le Cras and F. Weil, Nature materials, **71**, 665 (2008)
3. D. Morgan, A. Van der Ven and G. Ceder, Electrochem. and Solid-State Lett., 7(2) A30-A32 (2004)
4. S. -Y. Chung, J. T. Blocking and Y. -M. Chiang, Nature Mater., **2**, 123 (2002)
5. H. Deng et al. , *Journal of Power Sources*, **220**, 342-347 (2012)
6. A. Yamada, S. C. Chung and K. Hinokuma, J. Electrochem. Soc., **148**, A224 (2001)
7. J. Ni, M. Morishita, Y. Kawabe, M. Watada, N. Takeshi and T. Sakai, Journal of Power Sources, 195, 2877-2882 (2010)
8. Y. Xu, Y. Lu, L. Yan, Z. Tang and R. Yang, J. Power Sources, **160**, 570 (2006)
9. Z. Zhang, H. Feng, X. Wu, L. Wang, A. Zhang, T. Xia, H. Dong and M. Liu, Electrochim. Acta, **54**, 3206 (2009)
10. M. Konarova and J. Taniguchi, J. Power Sources, **195**, 3661 (2010)
11. B. Kang and G. Ceder, Nature, **458**, 190 (2009)
12. K. Zaghib, A. Mauger, F. Gendron and C. M. Julien, Chem. Mater., **20**, 462 (2008)
13. I. Belharouak, C. Johnson and K. Amine, Electrochem. Commun., **7**, 983 (2005)
14. O. Xiuqin, P. Lin, G. Haichen, W. Yichen and L. Jianwei, J. Mater. Chem., **22**, 9064 (2012)

High-Porosity Carbon Processed from Mixtures of Furfuryl Resin-Poly(ethylene glycol)-Benzenesulfonyl Chloride and Its Application for Catalyzing Tri-iodide Reduction

Shunjian Xu[a]

Xinyu Institute of New Energy, Xinyu University, Xinyu 338004, China

Abstract. High-porosity carbon (HPC) was first prepared from a novel mixture system of furfuryl resin as carbon-yielding precursor, poly(ethylene glycol) with low molecular weight as pore-forming agent and benzenesulfonyl chloride (BSC) as cured catalyst, and then employed as catalyst for counter electrode in dye-sensitized solar cells (DSCs). The pore structures of HPC are tailored by changing BSC amount from 2% to 14%. Especially, when BSC amount ascends from 4% to 8%, the apparent porosity of HPC rises from 59.58% to 68.02% and the median pore size from 0.429 to 3.164 μm, except that some globular carbons initially occur on the carbon skeleton. Since the HPC derived from 4% BSC has high porosity and the most quantity of small-sized pores to enlarge its surface area allowing for the most catalytic active sites for tri-iodide reduction, the resulted counter electrode exhibits the best catalytic activity and the highest efficiency for use in DSCs.

1 Introduction

Porous carbons have been widely employed in many fields, such as energy storage, catalysis, purification and gas separation [1,2]. More recently the porous carbon has been developed as an alternative catalyst to replace the platinum for counter electrode in dye-sensitized solar cells (DSCs), due to its high corrosion inert, low cost and admirable catalytic activity [3,4]. In such application, the pore structure of porous carbon, such as apparent porosity, surface area and pore size, is very critical in determining the catalytic activity of the electrode for the tri-iodide (I_3^-) reduction and the final photovoltaic performances of the device.

The polymerization-induced phase separation and pyrolysis (PIPSP) is an impactful route to tailor the pore structure of porous carbon [5-7]. In this process, carbon-yielding precursor (CYP) is first mixed with pore-forming agent (PFA) and optional additives (such as starch and water). The polymerization of CYP results in the phase separation and a rigid body. The rigid body is then pyrolyzed to form porous carbon with removing fluid PFA. The pore structure of the porous carbon can be controlled by changing the composition of CYP-PFA system. Therefore, to understand the relationship between the composition of CYP-PFA system and the pore structure of the porous carbon is indispensable for the modification and application of the porous carbon. To date, many CYP-PFA systems, such as phenol resin (PF)-ethylene glycol (EG) [5], poly(furfuryl alcohol)-polyethylene glycol (PEG) with large molecular weight [6]and (PF+furfuryl alcohol)-EG [7], have been developed to fabricate the porous carbons. However, few reports have addressed the potential synthesis of the porous carbon based on a system of furfuryl resin (FR), PEG with low molecular weight and benzenesulfonyl chloride (BSC) as well as the application of porous carbon with high porosity as catalyst in DSCs.

In this study, the novel system containing FR, low-molecular-weight PEG and BSC was first developed to fabricate high-porosity carbon (HPC), with emphasis on controlling the pore structure of the HPC through changing BSC amount. The maximum apparent porosity of the HPC reaches 68.02%. The HPC was then employed as catalyst for counter electrode in DSCs.

[a] Corresponding author: xushunjian@126.com

2 Experimental

2.1 Preparation and characterizations of porous carbon

To synthesize the HPC, FR (Industrial level), PEG (CP, Average molecular weight: 190−210), BSC (CP, 98.5%) and deionized water were mixed by stirring mechanically at room temperature, then poured into mold and cured at 40 °C for 3 h. The sample was then postcured at 100 °C for 40 h. After 40 h, the rigid body was removed from the mold. The sample was finally pyrolyzed by slowly ramping to 800 °C in a flowing N_2, followed by cooling naturally to room temperature. The weight ratio of FR and PEG was 2:3. The weight fraction of BSC was 2%, 4%, 8% and 14%, respectively. And the resulting HPC was designated as HPC-1, HPC-2, HPC-3 and HPC-4, accordingly.

The crystal structures of HPC were identified by X-ray diffraction (XRD; D8 ADVANCE, Bruker). The morphologies of HPC were characterized by scanning electron microscopy (SEM; EVO MA10, Zeiss). The median pore size, apparent porosity, bulk density and pore size distributions of HPC were characterized by mercury porosimetry (Autopore 9510, Micromeritics).

2.2 Configuration and measurements of porous carbon based cells

The HPC counter electrode was prepared by coating a blended paste on FTO, and then heated to 120 °C for 2 h. The blended paste was obtained by grinding 150 mg HPC powder, 15 mg carboxymethyl cellulose (CMC) and 2.5 ml alcohol aqueous solution. The DSCs was assembled with a N719 dye-sensitized TiO_2 photoanode and a HPC counter electrode sandwiching an electrolyte. The electrolyte was a blended solvent of acetonitrile and EG containing 0.5 M KI, 0.05 M I_2 and 0.5 M tributylphosphate. The photoanode was prepared as following [3]: TiO_2 film was firstly coated on FTO substrate by printing a paste containing anatase TiO_2 particles. After sintering at 450 °C for 30 min, the film was treated with 40 mM $TiCl_4$ aqueous solution, and then further heat-treated at 450 °C for 30 min. The photoanode was finally obtained by immersing the TiO_2 film into a N719 ethanol solution (0.5 mM) at room temperature for 24 h.

Cyclic voltammetry (CV) experiments were performed using a three-electrode system. Electrochemical impedance spectroscopy (EIS) measurements were conducted at 10 mV of amplitude. Photocurrent–voltage (*J–V*) curves of DSCs were measured under the illumination of a Xenon light source (CHF−XM500, Trusttech). The light intensity was 100 mW·cm^{-2}.

3 Results and discussion

Fig. 1 illustrates the micrographs of HPC-1 to HPC-4. It is obvious that the morphologies of HPC show differences. Both HPC-1 and HPC-2 have interconnected porous structure with smooth carbon skeleton, and the carbon skeleton in HPC-2 is slightly finer than that in HPC-1. In contrast, with the increase of BSC amount to 8% or 14% (HPC-3 or HPC-4), some globular carbons stacked on the carbon skeleton are detected, except that the carbon skeleton becomes coarse. Accompanied by the evident differences of carbon skeletons, the pore characteristics also vary obviously, which can be confirmed by the pore size distributions and pore characteristic parameters of HPC, as shown in Fig. 2(a) and Table 1, respectively. From HPC-2 to HPC-3, the range of main pore size distribution changes from 0.2 ~ 0.6 μm to 2 ~ 4 μm, and the median pore size (Φ_m) ascents from 0.429 to 3.164 μm. Furthermore, the apparent porosity (P_a) rises from 59.58% to 68.02%. Table 1 also displays that the Φ_m of HPC-2 is slightly smaller than that of HPC-1, and the P_a of both HPC are almost identical, indicating higher surface area for HPC-2. In addition, the bulk density (ρ_b) of HPC decreases from 0.601 to 0.472 g·cm^{-3} with increasing BSC amount.

The images of rigid body reveal that the morphology of HPC is inherited from that of the rigid body, as shown in the inserts in Figs. 1(b) and (c). The pyrolysis mainly brings about the contraction of the pore size and wall thickness. It indicates that the different morphologies of HPC with various BSC amount are formed during curing.

The XRD patterns of HPC are demonstrated in Fig. 2(b). The position and intensity of both (002) and (100) peaks are almost unchanged with increasing BSC amount, suggesting similar crystallinity for all HPC [8].The influence of BSC amount on the gel time of FR is demonstrated in Fig. 2(c). It is obvious that the gel time falls rapidly with increasing the BSC amount from 2% to 8%, and then declines slightly with further increasing the BSC amount.

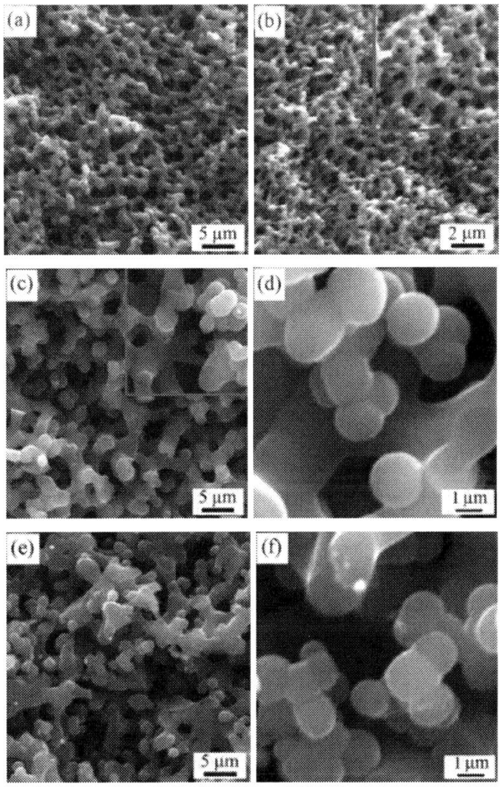

Figure 1. Variation in morphologies of (a) HPC-1; (b) HPC-2; (c) and (d) HPC-3; (e) and (f) HPC-4. The inserts show the images of the corresponding rigid bodies (before pyrolysis).

The formation and different morphologies of the HPC should be preliminarily explained as follows. The BSC as a cured catalyst accelerates FR polymerization (Fig. 2c) and further changes the phase separation kinetics in FR-PEG-BSC system during curing, resulting in the patterned variation of FR-rich and PEG-rich phases in rigid body. After pyrolysis, the HPC inherits the morphologies of rigid body with the transformation of

FR-rich phase into carbon skeleton and the remove of PEG-rich phase to form pores (Figs. 1 and 2b). Under high BSC amount (reached 8% or 14%), the fast polymerization of FR promotes the growing-up of FR-rich and EG-rich phases. Meanwhile, the rapid growing-up of EG-rich phase would interrupts the interconnection of partial FR-rich phase, and the opened FR-rich phase tends to form globularity under surface tension. Therefore, the HPC (HPC-3 or HPC-4) has large pores and coarse carbon skeleton, except that some globular carbons occur.

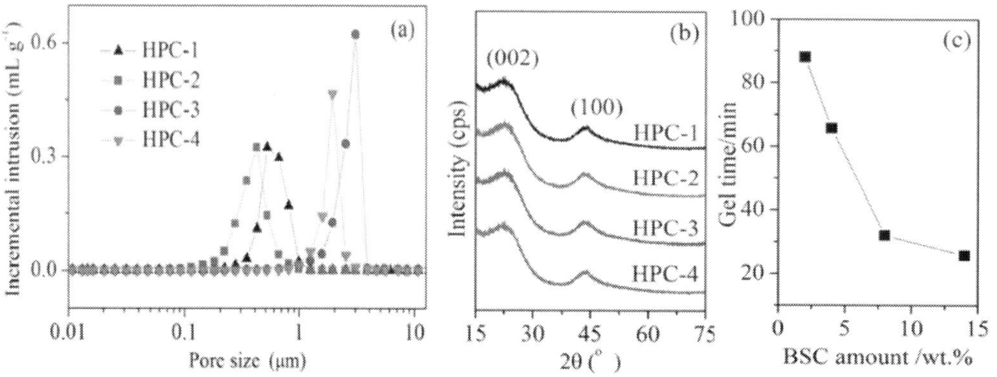

Figure 2. (a) Pore size distributions of HPC-1 to HPC-4; (b) XRD patterns of HPC-1 to HPC-4; (c) Relationship between the BSC amount and the gel time of FR.

Table 1. Pore characteristic parameters of HPC and their photovoltaic performances for use in DSCs.

Catalyst	Φ_m (μm)	P_a (%)	ρ_b (g·cm^{-3})	J_{sc} (mA·cm^{-2})	V_{oc} (V)	FF	η (%)
HPC-1	0.658	59.82	0.601	11.58	0.676	0.385	3.01
HPC-2	0.429	59.58	0.593	12.57	0.690	0.461	4.00
HPC-3	3.164	68.02	0.473	11.85	0.672	0.357	2.84
HPC-4	2.601	63.70	0.472	8.63	0.696	0.336	2.02

*Φ_m: Median pore size; P_a: Apparent porosity; ρ_b: Bulk density; J_{sc}: Short-circuit current density; V_{oc}: Open-circuit voltage; FF: Fill factor; η: Efficiency.

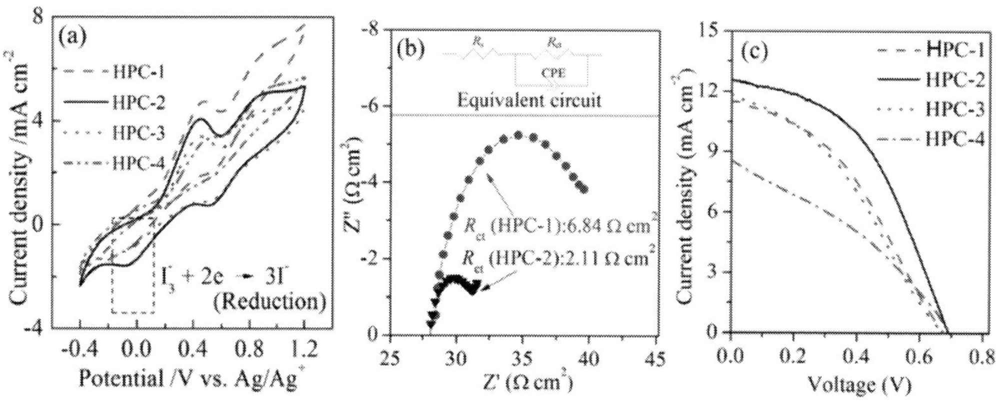

Figure 3. Catalytic activities of HPC and their photovoltaic performances for use in DSCs. (a) CVs of HPC as counter electrode; (b) EIS of HPC as counter electrode; (c) *J-V* curves of HPC based DSCs.

The CV curves of HPC as counter electrode are shown in Fig. 3(a). In the peak represented the reduction of I_3^- to I^-, the current density of HPC-2 is higher than that of HPC-1 or HPC-3, and is far higher than that of HPC-4. It suggests that HPC-2 has the best catalytic activity [9, 10].The EIS displayed in Fig. 3(b) confirms that the catalytic activity of HPC-2 is better than that of HPC-1. The values of charge transfer resistance (R_{ct}) are 2.11 and 6.84 Ω cm^2 corresponding to HPC-2 and HPC-1, respectively. The excellent catalytic activity of HPC-2 is mainly attributed to its distinctive pore structure since the difference of crystal structure among HPC can be ignorable. Compared with HPC-1, HPC-3 or HPC-4, HPC-2 possesses more pores with small size to enlarge its surface area allowing for more catalytic active sites for the I_3^- reduction. As a result, HPC-2 has better catalytic activity. The *J-V* curves of HPC based DSCs are illustrated in Fig. 3(c). The high catalytic activity of HPC brings about the high efficiency (η) for use in DSCs. The

η of DSCs with HPC-2, HPC-1, HPC-3 and HPC-4 is 4%, 3.01%, 2.84% and 2.02%, respectively, as shown in Table 1.

For comparison, two conventional carbon materials, graphite and activated carbon (AC), were individually used as catalysts to prepare the counter electrodes, and the η values of the corresponding DSCs were 0.96% and 2.71%, respectively [11]. It is clear that the HPC-1, HPC-2 or HPC-3 is superior to the graphite or AC for catalyzing the I_3^- reduction.

4 Summary

HPC is fabricated from a novel FR-PEG-BSC system, and its pore structures are tailored by changing BSC amount. Especially, when BSC amount ascends from 4% to 8%, the apparent porosity of HPC rises from 59.58% to 68.02% and the median pore size from 0.429 to 3.164 μm, except that some globular carbons initially occur on carbon skeleton. The CV and EIS show that HPC derived from 4% BSC as counter electrode exhibits the best catalytic activity, which is mainly attributed to the enough catalytic active sites on its large skeleton surface. Importantly, the high catalytic activity of HPC brings about the high efficiency for use in DSCs.

Acknowledgements

This work was financially supported by National Natural Science Foundation of China (51462035), Training Programme Foundation for Young Scientist of Jiangxi (20133BCB23035), Natural Science Foundation of Jiangxi (20161BAB206106) and Educational Commission of Jiangxi (KJLD13100).

References

1. Z. Li, W. Yang, X.W. Xu, Y.S. Tang, Z.W. Zeng, F. Yang, L.Q. Zhang, G.Q. Ning, C.M. Xu, Y.F. Li, J. Power Sources **327**, 289 (2016)
2. H. Hadoun, Z. Sadaoui, N. Souami, D. Sahel, I. Toumert, Appl. Surf. Sci. **280**, 1 (2013)
3. S.J. Xu, G.W. Liu, J. Li, G.J. Qiao, Electrochim. Acta **76**, 504 (2012)
4. S.Q. Fan, B. Fang, J.H. Kim, B. Jeong, C. Kim, J.S. Yu, J.J. Co, Langmuir **26**, 13644 (2010)
5. S.J. Xu, J. Li, G.J. Qiao, H.J. Wang, T.J. Lu, Carbon **47**, 2103 (2009)
6. M.S. Strano, H. Agarwal, J. Pedrick, D. Redman, H.C. Foley, Carbon **41**, 2501 (2003)
7. Y. Zhang, Z. Yuan, Y. Zhou, Mater. Lett. **109**, 124 (2013)
8. A. Braun, M. Bärtsch, B. Schnyder, R. Kötz, O. Haas, H.G. Haubold, G. Goerigk, J. Non-Cryst. Solids **260**, 1 (1999)
9. W.Y. Cheng, C.C. Wang, S.Y. Lu, Carbon **54**, 291 (2013)
10. G.T. Yue, J.H. Wu, Y.M. Xiao, M.L. Huang, J.M. Lin, L.Q. Fan, Z. Lan, Electrochim. Acta **92**, 64 (2013)
11. Y.F. Luo, X.Y. Liu, S.J. Xu, Z. Wei, New Chem. Mater **40**, 58 (2012)

The Influence of Annealing Process on Crystallinity and Structural Properties of Cotton/Spandex Fabric

Md. Nahid Pervez[1,2,3], Md. Eman Talukder[2], Manoj Kanti Datta[2], Md Shipan Mia[2], Ashaduz Zaman[2], Meer Md Rasel Khan[2], Yingjie Cai[1,4] and Lina Lin[1,5, a]

[1]Hubei Provincial Engineering Laboratory for Clean Production and High Value Utilization of Bio-based Textile Materials, Wuhan Textile University, Wuhan 430073, China
[2]School of Chemistry and Chemical Engineering, Wuhan Textile University, Wuhan 430073, China
[3]Research Institute of Flexible Materials, School of Textiles & Design, Heriot-Watt University, Galashiels TD1 3HF, UK
[4]Engineering Research Centre for Clean Production of Textile Dyeing and Printing, Ministry of Education, Wuhan Textile University, Wuhan 430073, China
[5]College of Textile Science and Engineering, Wuhan Textile University, Wuhan 430200, China

Abstract. This research aims to elucidate the crystallinity, structural and mechanical properties of cotton/spandex knitted fabric after conducting heat setting i.e., annealing process with optimized condition for the first time. Experimental results disclosed that, crystallinity index (CI) (%) of heat-treated cotton/spandex were increased with increased temperature and also mechanical properties i.e., tensile strength was improved as the temperature rose from 120 to 140°C and decreased with augmenting temperature ranges, which is in good agreement with the findings of CI % through X-ray diffraction (XRD) patterns. The attendance of cellulose dehydration was further confirmed by Fourier transform infrared (FT-IR) spectra in order to observe the structural integrity.

1 Introduction

Heat-setting (annealing) procedure, seen as an essential issue, is undertaken during the initial phase of the textile industry, considering the fact that it proffers the appropriate product's output to make it appropriate for additional steps and for the customer to utilise it [1, 2]. Heat-setting, based on the report of Karmaker can be defined as a treatment that helps in imparting shape retention and crease resistance, as well as resilience and elasticity to the fibres, and also in causing change in strength and stretchability, as well as in softness dyeability, together with in crystalline structure, and in some occasions, on the material colour [3, 4]. There is a direct connection between the structural and chemical modifications taking place in the fibre and these changes. The ultimate loss of fabric dimensions is the result of under-setting, while power can be lowered as a result of over heat-setting, leading to spandex and companion fibres discoloration. Potential distortion or fabric deformation from residual uneven tension can be reduced by using the relaxation treatment leading to the power development and fabric recovery. The relaxation of the fabric should be performed prior to heat-setting so that marks and puckering during dyeing could be averted in order to see to it that the final garment's good dimensional stability is attained. Heat setting, apart from seeing to it that dimensional stability is imparted on the yarn and the fabric, is also important in some other areas, such as the tendency for twisted or doubled filaments to possess the strength to curl or snarl, making such to have a harmful effect on additional processing. Heat setting, in every one of these operation, sometimes, accomplishes the basic objective in swelling agents and tension manifestation, specifically in taking the product toward equilibrium in order to see to it that resistance is attained against thermal shrinkage and dimensional changes, as well as in curling and creasing. It is possible for improvements to be attained in physical properties, specifically in mechanical properties [5]. In their work with polyester fibres [6,7], Gupta and Kumar made a comparison of slack and taut heat set yarns with regards to a non-heat set sample based in changes in their structures and mechanical properties. The range of temperatures was between 100 and 220°C while the period of exposure ranged between 1 and 60 minutes. It was discovered that there was increase in the level of crystallinity, or crystalline fraction as the heating temperature increased. It was also observed that changes in crystalline and amorphous phases using x-ray method was more, compared to density measurements. A very fascinating study undertaken by Kitano et al. [8] obtained the high crystalline percentage of 95% with PET fibre at high temperature. By measuring tensile properties, Venkatesh et al. [9] conducted an investigation on the mechanical properties of synthetic fibre changes as a result of heat setting. The structural behaviour of polymers is what determines their mechanical properties.

[a] Corresponding author: linalin@wtu.edu.cn

The complex nature of polymers' structures is what made them to become a subject of much fascination. It is quite clear that there may be fundamental differences with regards to polymeric materials in many essential features based on how the preparation of the polymer was made. The possibility of obtaining technologically significant variations when it comes to structure and mechanical properties through varied crystallization and processing condition is well known [8]. The heat-setting procedure, generally, is undertaken above the transition temperature of the glass and lowers than the melting temperature for an appropriate time. Other important factors determining the heat-setting process's effectiveness and efficiency are the fabric width and overfeed. Type of impurities added during manufacturing of yarn and fabric, padding solution's wetting ability and the moisture in heat-setting chambers are among other factors [10]. These parameters are what the heat-setting process strongly depends on while just a small variation with regards to these limitations will have a huge effect on the properties of the fabric. So there must be careful selection, optimisation, and monitoring of these parameters.

In this paper, we made an introduction of heat treatment into the cotton/spandex blended fabrics in the form of its expertise for the structural behaviour accomplishment to be advanced and for the rising needs of industry to be met. Nonetheless, the investigation, so far conducted, with regards to heat treatment effect on crystallinity and cotton/spandex blended fabric structural properties, has been quite limited.

2 Experimental

2.1 Materials

For experimental work knitted single jersey fabrics with (1-5%) spandex and 95% cotton have been used. The fabric samples were knitted on Mayer and Cie (Germany) single knit machine by using 30/1 Ne cotton yarn and 24 denier spandex filament yarn and produced fabric had 74 course/inch, 38 wales/inch and fabric density of 276 g sq. m^{-1}.

2.2 Method

The stenter machine (BRUCKNER, Germany) was used to carry out the heat setting process. At first, samples were placed between parallel bars at 10 cm length and stretched out to different degrees on an aluminium frame with pre-determined tension. There were six levels of temperature and time such as (120, 130, 140, 150, 160, 170 °C) and (15, 30, 45, 60, 75, 90 s) respectively, while all other machine variables were held constant. Then, the process conditions were denoted as C1, C2, C3, C4, C5 and C6. After heat setting, the samples were cooled before being removed from the aluminium frame. Then the samples were conditioned under standard conditions of 65±2 % relative humidity and 21±1°C for 24 h.

2.3 Measurement and characterization

The tensile strength of samples was calculated according to our previous report [11]. The crystallinity behaviour of the samples was determined by using X-ray diffractometer (Shimadzu XRD-6000). The functional groups of treated samples were determined using a Fourier transform infrared (FTIR) spectroscopy (Bruker, United Kingdom) at the wavenumber range 500–4000 cm^{-1} [12, 13].

3 Results and discussion

3.1 Tensile strength

The study of physical properties of fabrics is important due to apparel usually undergoes small extension and relaxation during wear and tensile strength is one of them [14]. Figure 1 shows the tensile strength values of the samples before and after under different heat-setting conditions. It can be seen that the tensile strength is increased after heat setting except for the sample number A1, as while it is decreased after heat setting. The highest tensile strength is shown in case of sample A3 and process condition C3 about 5 % improvement compared to untreated sample and also there is little improvement about 3% observed from sample A4 and A5. This result was attributed due to the fiber adhesion and fabric becomes stiffer during the heat setting [15] and also may be affiliated to the rearrangement of fibres after heat treatment which improved the tensile strength of the blended fabric [16]. Finally from these results it can be conclude that: (1) heat treatment at 140 °C results to increase tensile strength and (2) degradation in fabric starts at about 160°C which results to decrease in tensile strength. Sample A3 was chosen for further study.

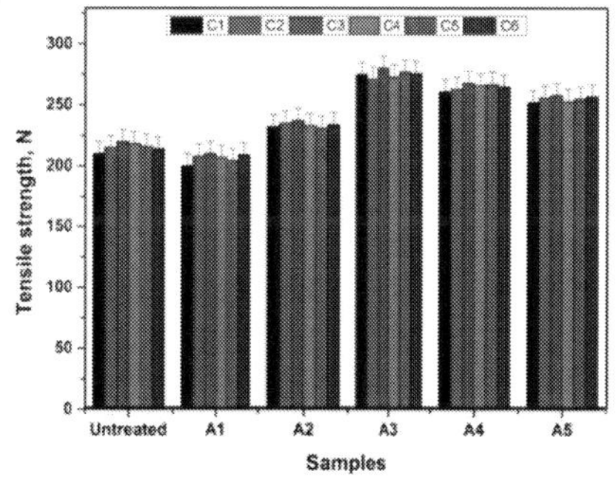

Figure 1. Tensile strength values.

3.2 Crystallinity study

The X-ray diffraction spectra of untreated and treated samples (A3) at various heat-setting conditions is presented in Figure 2. The results show that the untreated sample, which is the same as a semi crystalline polymer structure, presented some large and wide reflections due to its internal structure, and can be assigned according to the literature [17-19], with one significant peak appearing

DOI: 10.1051/matecconf/201713002001

at 16.1 ° with little crystallinity. As can be seen in Figure 2, when the heat-setting condition surpassed C1, two peaks appeared that were (002) and (100). As the temperature increased, (002) peaks shifted from 19.58 ° to 22.58 °, then the (100) diffraction peaks enhance. This result indicates that blended fabric is rearranged from disorderly to orderly and the crystalline structure improves remarkably with increasing temperature.

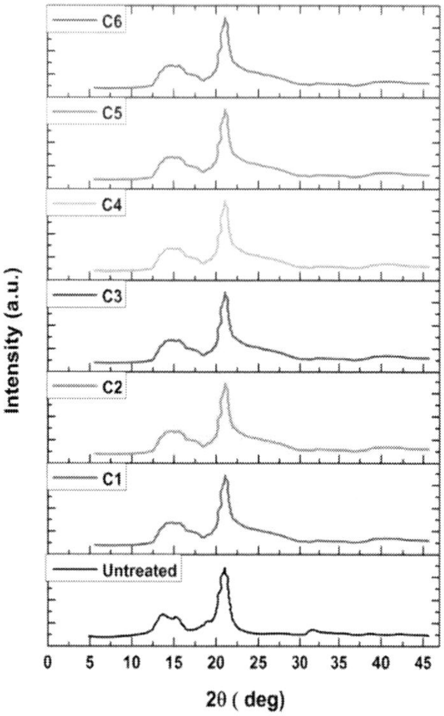

Figure 2. XRD spectra of sample (A3) at different heat setting condition.

From general literature, it can be demonstrated that the peak at 22 ° ≤ 2θ ≤ 23 ° represents the characteristic height of cellulose I, which corresponds to the (002) crystallographic planes [20]. The crystallinity percentage for the untreated sample was found to be almost 3%. The treated samples showed higher intensity for these peaks compared to the untreated samples, and this is due to the higher duration of heat setting temperature and the presence of spandex fibre acting as a nucleating agent. When annealing temperature at 140 °C, the crystallinity percentage for blended fabric was observed up to 72%, which is attributed due to; (1) at hygrothermal condition, cellulose molecules in the quasicrystalline region were easy to soften and began to rearrange or reorient [21]; and (2) during heat setting, the presence of free hydroxyl groups within the cellulose macromolecules are supposed to be involved in a number of intramolecular and intermolecular hydrogen bonds, which may result in various ordered crystalline arrangements [22]. In addition, the decomposition of the amorphous region of cellulose is also responsible for observing this phenomenon. This is supported through Basch and Lewin [23], who suggested that thermal decomposition of the cellulose initiated in the much less-ordered areas through demonstrating a drop inside the areas of polymerisation. It has been pronounced that the facilitated thermal decomposition of amorphous, as well as crystalline, regions changed following the dehydration reactions and other types of rearrangements of cellulose [24]. As far as the heat-setting process on the cotton/spandex is concerned, it may be stated that generally heat setting leads to a restructuring or recrystallisation with a higher degree of crystallinity and smaller crystallites with advanced orientation. In this recrystallisation process, at temperatures above 130 °C, a considerable portion of the amorphous region is also transformed into crystalline regions [25]. The increase of the CI indicates the improvement in the cellulose structure and finally contributes to enhancing the tensile strength of the treated fibre, the results of which were verified by Figure 1. Moreover, at a later stage of the heat setting process, the amount of crystallinity percentage decreased due to continued degradation both in crystalline and non-crystalline regions. The higher numbers of chain scission reactions accelerated the amorphous character of cellulose, which reduced the total amount of cellulose crystalline regions [26]. Additionally, no free hydroxyl groups were involved in the reaction to generate hydrogen bonds.

3.2 FT-IR analysis

The FTIR spectra of untreated and treated samples (A3) at various heat-setting conditions in 500-4000 cm^{-1} region are presented in Figure 3. The cellulose characteristic bands, C-H stretching (3000-2700 cm^{-1}) and C-H wagging (1429 cm^{-1}) bands, began to broaden at 120°C in untreated, whereas these bands remained stable at temperatures up to 170°C in blended fabric. Another substantial distinction was the development of the bands at 1645 cm^{-1} and 1605 cm^{-1}, corresponding to C=O stretching of unsaturated aldehydes and C=C stretching of alkenes, respectively, which indicate the dehydration of cellulose [27,28]. These bands were intensified at 120-140°C for C1-C3 condition, and at 140-170°C for C4-C6 condition. The facilitated dehydration at lower temperatures was also supported by the early disappearance of C-O-C stretching (1152 cm^{-1}) and C-O stretching (1020 cm^{-1}) bands. The shift of the C=C stretching band toward 1585 cm^{-1}, signifying conjugation of C=C bonds, occurred at 120°C, but no such shift was observed until the temperature reached 140°C. Additional bands then appeared in 2000 cm^{-1} for C4-C6 treatment conditions that denoted its progressive dehydration of the cellulose in the range of 150-180 °C, followed by the decomposition of the dehydrocellulose that could be formed through etherification between superimposed cellulose chains (C-2 and C-4) [29]. Since the shifts cause the bands to be closer to that of neat samples, it indicates that a reduction in the interaction of -OH groups with water molecules and dehydration of cellulose has occurred. However, at higher temperatures, the frequencies of the bands decrease slightly with increasing temperature.

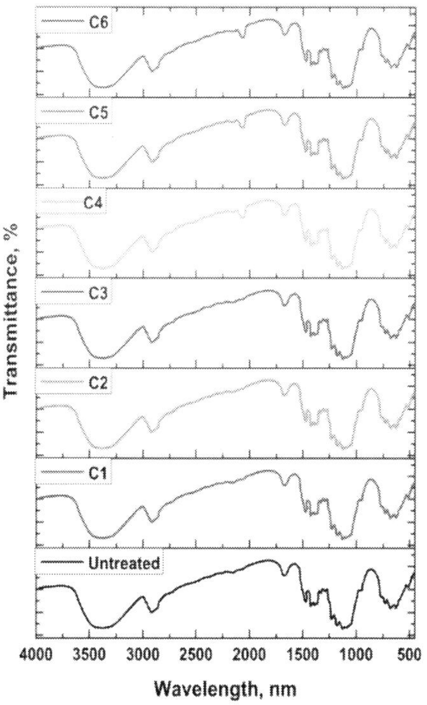

Figure 3. FT-IR spectra of sample (A3) at different heat setting condition.

4 Conclusion

Based on a series of experimental studies in this work, an analytical probe was performed to observe the annealing effect on crystalline, structural and mechanical properties in cotton/spandex knitted fabric. The samples were annealed in 120 to 170°C range and then subjected to regarding tests at room temperature. The tensile properties of cotton/spandex were significant improved by annealing treatment in the range of 120-140°C. It was noted from XRD diffraction spectrum, heat treatment had no bright aftereffect on the position of the cellulose crystalline region. However, the crystallinity showed a gradual increase with heat- treated temperature increasing that was attributed to the heterogeneous nucleation effect.

Acknowledgement

This work was supported by the China National Textile and Apparel Council (2013 "Textile Vision" Applied Basic Research) a grant from the Hubei Province Science and Technology Support Program (Project 2013BAA043), and the Collaborative Innovation Plan of Hubei Province for Key Technology of Eco-Ramie Industry (E JIAO KE HAN 2014 No.8).

References

1. J. Gacén, D. Cayuela, J. Maillo, I. Gacén, J. Text. Inst. **93**, 29 (2002)
2. M.A. Jahid, M.N. Pervez, American Journal of Materials Science and Technology **5**, 57 (2015)
3. S.R. Karmakar, *Chemical Technology in the Pre-treatment Processes of Textiles* (Elsevier Science, UK, 1999)
4. S. Mukhopadhyay, *Advances in Fibre Science* (Textile Institute, Manchester, 1992)
5. V.B. Gupta, J. Appl. Polym. Sci. **83**, 586 (2002)
6. V.B. Gupta, S. Kumar, J. Appl. Polym. Sci. **26**, 1865 (1981)
7. V.B. Gupta, S. Kumar, J. Appl. Polym. Sci. **26**, 1877 (1981)
8. Y. Kitano, Y. Kinoshita, T. Ashida, Polymer **36**, 1947 (1995)
9. G.M. Venkatesh, P.J. Bose, R.V. Shah, N.E. Sweltz, J. Appl. Polym. Sci. **22**, 2357 (1978)
10. S.M. Desai, Colourage **58**, 79 (2011)
11. M. Kamruzzaman, N. Pervez, E. Talukder, A. Kamal, Y.–J. Cai, Advances in Engineering Research, **103**, 391 (2016)
12. M.N. Pervez, A. Khan, I.A. Khan, American Journal of Polymer Science & Engineering **3**, 90 (2015)
13. A. Khan, N. Pervez, I.A. Khan, S. Ahmad, R. Masood, T. Hussain, F. Telegin, Asian. J. Chem. **28**, 1191 (2016)
14. C.W. Kan, Y.L. Lam, Materials **6**, 314 (2013).
15. N. Furvik, A. Bernskiöld, N. Gralen, J. Text. Inst. Trans. **46**, T662 (1955)
16. L. Wang, G. Cai, W. Yu, Synthetic Fiber in China, **31**, 21 (2008)
17. N. Dencheva, Z. Denchev, M. J. Oliveira, S.S. Funari, J. Appl. Polym. Sci. **103**, 2242 (2007)
18. Y.L. Lu, Y. Zhang, G.B. Zhang, M.S. Yang, S.K. Yan, D.Y. Shen, Polymer, **45**, 8999 (2004)
19. E. Kaya, L.J. Mathias, J. Appl. Polym. Sci. **123**, 92 (2012)
20. B.A. Ass, M.N. Belgacem, E. Frollini, Carbohyd. Polym. **63**, 19 (2006)
21. M.T.R. Bhuiyan, N. Hirai, N. Sobue. J. Wood Sci. **46** 431 (2000)
22. M. Poletto, A.J. Zattera, M.M. Forte, R.M Santana, Bioresour. Technol. **109**, 148 (2012)
23. A. Basch, M. Lewin, J. Polym. Sci., Polym. Chem. Ed. **11**, 3071 (1973)
24. G. Dobele, G. Rossinskaja, G. Telysheva, D. Meier, O. Faix, J. Anal. Appl. Pyr. **49**, 307 (1999)
25. S. Sarkeshick, H. Tavanai, M. Zarrebini, M. Morshed, J. Text. Inst. **100**, 128 (2009)
26. L. Kong, Z. Zhao, Z. He, S. Yi, Results Phys. **7**, 914 (2017)
27. M. Tsuboi, J. Polym. Sci. **25**, 159 (1957)
28. M.N. Pervez, F. Shafiq, M.M. Jilani, Z. Sarwar, Y.J. Cai, Materials Science Forum, **893**, 71 (2017)
29. S. Nam, B.D. Condon, Y. Liu, Q. He, Polym. Degrad. Stabil. **138**, 133 (2017)

Discussion on Avoidance of Staining Problem in Cotton/CDPET Dyeing

Md. Nahid Pervez[1,2,3], Umarsharif Y. Inamdar[2], Md. Eman Talukder[2], Md. Ziaur Rahman[2], Mst. Munera Khatun[2], Mst. Zakia Sultana[2], Yingjie Cai[1,4] and Lina Lin[1,5, a]

[1]*Hubei Provincial Engineering Laboratory for Clean Production and High Value Utilization of Bio-based Textile Materials, Wuhan Textile University, Wuhan 430073, China*
[2]*School of Chemistry and Chemical Engineering, Wuhan Textile University, Wuhan 430073, China*
[3]*Research Institute of Flexible Materials, School of Textiles & Design, Heriot-Watt University, Galashiels TD1 3HF, UK*
[4]*Engineering Research Centre for Clean Production of Textile Dyeing and Printing, Ministry of Education, Wuhan Textile University, Wuhan 430073, China*
[5]*College of Textile Science and Engineering, Wuhan Textile University, Wuhan 430200, China*

Abstract. In this research, a novel approach has been proposed to remove stain problem from blended fabrics as an alternative option with traditional reduction clearing process. For this purpose, cationic dyeable polyester and cotton blended fabrics were dyed as traditional technique based on formulations approach and the colour strength (K/S), dye uptake (%) and the whiteness index properties were evaluated. The results of color strength indicated that dye was more uniformly absorbed by the blended fabric compared to the untreated fabric. In addition, when untreated fabric was dyed in the presence of a leveling agent (Lyogen DFTN), the K/S and levelness of dye uptake (%) values were enhanced. Finally, very less staining was observed that proved by CIE Whiteness Index values and indicates that formulation (F-IV) has more beneficial to reduce staining from Cotton/CDPET blended fabric (65/35).

1 Introduction

Consumers have greatly hailed the chemical and natural fibres' blended products due to the fact of the wearing and wrinkle resistance nature of chemical fibres and the moisture absorption and high permeability on the part of natural fibres [1]. There is an ascendant marketplace percentage for polyester/cotton (P/C) blends in the textile industry of 58.45% share in worldwide marketplace as a result of their cost of aesthetic and affable achievement of the consumer [2]. Nonetheless, when disperse dyes under dyeing conditions' conventional high-temperature polyester, such as 130 °C and pH 4–5, are used in the dyeing of blends, the result is a poor wet fastness such as washing and perspiration. The hydrophobic polyester fibre can easily be accessed by disperse dyes since polyester fibre has glass transition temperature (Tg) that is relatively low compared to cotton. Thus, there is heavy staining of the polyester component, resulting in the blend having wet fastness properties that are very low [3]. Contrary to conventional polyester (PET) fibres, cationic dyes can be used in dyeing cationic dyeable polyester (CDP) with a lot of ideal characteristics exhibited. Cationic dyeable polyester (CDP) can be described as a distinct polyester fibre which underwent transformation during the process of polymerisation. The fabrication of this was attained through the adding of dimethyl ester sodium salt of 5-sulfoisophthalic acid (DMS salt), in order that anionic sites, not possessed by normal polyester fibres, could be generated by imparting PET cationic dyeability for the solution of dull colour problem [4]. Figure 1 shows the structure of cationic dyeable polyester. There is much less number of sections containing anionic sites compared to the number of normal ester sections.

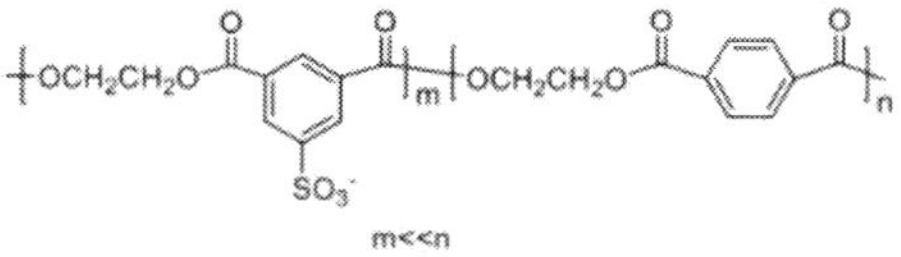

Figure 1. Structure of cationic dyeable polyester.

Identical properties that have normal PET fibres are possessed by modified polyester. Cationic dyes can be used to dye anionic groups. Again, the development of this fibre was made for the lowering of polymer's glass transition temperature, which usually lowered by 10% compared to normal polyesters, so that additional segmental mobility and open polymer structure could be obtained [5], leading to lower temperature dyeing. Again, the hydrolysing of this modified polymer takes place more readily as it is highly sensitive to heat setting. A comprehensive research has been undertaken on CDP [6-8] as a result of the brilliant colour that the dyed fabric displayed and higher colour yield, as well as better wet and sublimation fastnesses and the rest. The application of cationic dyes has been successfully performed at

[a] Corresponding author: linalin@wtu.edu.cn

100°C and normal pressure on easy cationic dyeable polyester (ECDP) while the dyeing of blends made up of comprising ECDP and cellulosic or protein fibres could be accomplished without damaging the fibres [9]. Abolishing of disperse dye on the PET surface and making adjustment to achieve the proper wetness fastness levels are important as well. There is a general acceptance of the technical importance of reduction clearing in order that the accuracy of the colour and the fastness properties could be enhanced, specifically to wet treatments [10].

A new approach has been undertaken in this research to see to it that the staining of the dyes at the time of dyeing behaviour is overcome with the use of levelling agent. This is followed by reduction clearing when subjected to whiteness index as response limitation on pilot scale experiment.

2 Experimental

2.1 Materials

Cotton/cationic dyeable polyester (CT/CDPET) blended woven fabric was collected from local textile mill. The commercial dyes of Cationic Red 4G and Reactive Red HE8B were purchased from DyStar Group, India. Lyogen DFTN (levelling agent), Ladiquest DWA (soaping agent) and Ekaline F (washing off agent) were purchased from Clariant, India.

2.2 Method

The bleaching solution was prepared by adding 1 gL^{-1} of wetting agent, 1 gL^{-1} of peroxide stabiliser, 6 gL^{-1} of hydrogen peroxide (30%), and 2 gL^{-1} of NaOH. The blended fabric was immersed in the bleaching solution at a liquor-to-goods ratio of 20:1 at 98°C for 60 minutes, and then the bleached fabric was washed with hot water and rinsed thoroughly with cold water and dried under ambient conditions. Briefly, Cotton/CDPET blended fabrics (20/80, 35/65, 65/35, 80/20) were placed in a Cationic Red 4G dyebath at a liquor-to-goods ratio of 20:1 and pH 4–5 at room temperature. The cationic dyed fabrics were then put into a Reactive Red HE8B dyebath at a liquor-to-goods ratio of 20:1 at room temperature, 50 gL^{-1} sodium sulphate and 15 gL^{-1} soda ash were used for exhaustion and fixation of the reactive dyebath. After fixation, the dyed fabrics were then rinsed at 35°C and the wash liquor pH was measured, followed by neutralisation with 1% of acetic acid at room temperature for 10 minutes and washed with 1 gL^{-1} of detergent at a liquor-to-goods ratio of 50:1 at 80°C for 10 minutes. Finally, the fabrics were rinsed thoroughly with cold water and air dried. The dyed fabrics were firstly rinsed with warm water and then a reduction clearing process was applied. The fabrics were treated at 70°C for 20 minutes using at a liquor-to-goods ratio of 20:1. The reduction clearing bath comprised 3 gL^{-1} sodium hydrosulphite ($Na_2S_2O_4$) and 3 gL^{-1} sodium hydroxide (NaOH). Reduction cleared samples were then rinsed with water and dried at room temperature. According to Table 1, the experimental process was followed throughout the whole study.

Table 1. Experimental process formulations.

Formulation I	Formulation II	Formulation III	Formulation IV	Formulation V
Bleaching ↓ Dyeing ↓ Reduction clearing ↓ Hot wash ↓ Washing off ↓ Dry	Bleaching ↓ Dyeing ↓ Reduction clearing (Reduced 50% concentration) ↓ Hot wash ↓ Washing off ↓ Dry	Bleaching ↓ Dyeing + Lyogen DFTN (1% o.m.f.) ↓ Reduction clearing ↓ Hot wash ↓ Dry	Bleaching ↓ Dyeing + Lyogen DFTN (3% o.m.f.) ↓ Reduction clearing ↓ Hot wash ↓ Dry	Bleaching ↓ Dyeing + Lyogen DFTN (5% o.m.f.) ↓ Reduction clearing ↓ Hot wash ↓ Dry

2.3 Measurements and standards

Color strength (K/S) and fastness values were determined according to our previous work [11]. The dye uptake percentage (D) of fabrics was calculated using [12] and the whiteness index (WI), denoted as CIE units, was measured for fabrics as per AATCC standard test method using international data colour.

3 Results and discussion

3.1 Colour Strength and Dye uptake

It is well-known that the colour strength parameter (K/S) has been widely used for colour measurement of dyed substrates [13]. Colour strength values (K/S) of dyed blended fabrics were shown in Figure 2. It is clear that the colour strength was found to be higher in the formulation (F-IV) range with 65/35 fabric and then decreased thereafter, and it should be noted that a very small amount of dye might have been lost from the blended (65/35) fabric surface. The colour strength (K/S) was better while the reflectance on the wavelength of maximum absorption (λ_{max}) decreased and suggested a greater uniform dye distribution [14, 15]. The high K/S values obtained from the fabrics with (65/35, 80/20) blended in formulation (F-III, F-IV) suggest that the reflectance's of these fabrics were lower than the others (20/80, 35/65) blended in formulation (F-I, F-II and F-V). This must be a result of different light absorption characteristics. The levelling agent decreases the initial dyeing rate in the dyeing process to ensure level dyeing. The effect of the levelling agent amount on the colour yield of formulations (F-III, F-IV and F-V) was given higher with an increased amount of levelling agent, as shown in Figure 2, which was attributed to the interaction between the dye and the levelling agent. Lyogen DFTN, the levelling agent used in this study, was non-ionic aliphatic polyglycol ether and shows high affinity to dyes. The colour strength was very low when a large amount of levelling agent was added. It seems that the pressure of the levelling agent with dye was so sturdy that it became difficult for the dye molecule to be separated from the levelling agent and migrate into the fibre when there is an excessive amount of levelling agent present. Therefore, a minimum amount of levelling agent ought to be used for suitable colour strength as long as uniform dyeing may be obtained. The colour strength of the 65/35 blended fabric was higher in formulation (F-IV) cases, indicating that it was easier for the dye to migrate into the less compact 65/35 fabric than others.

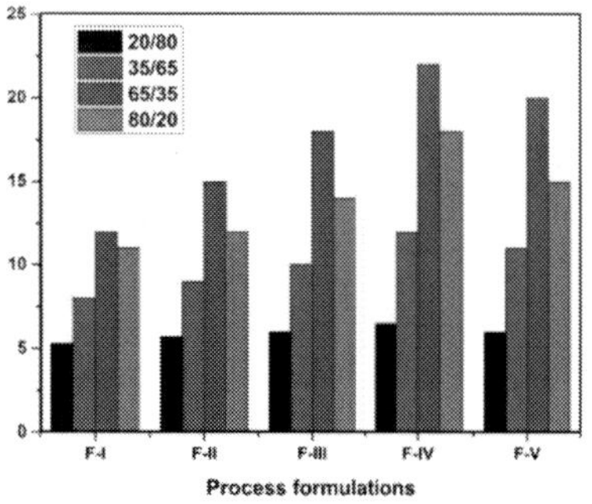

Figure 2. Color strength values.

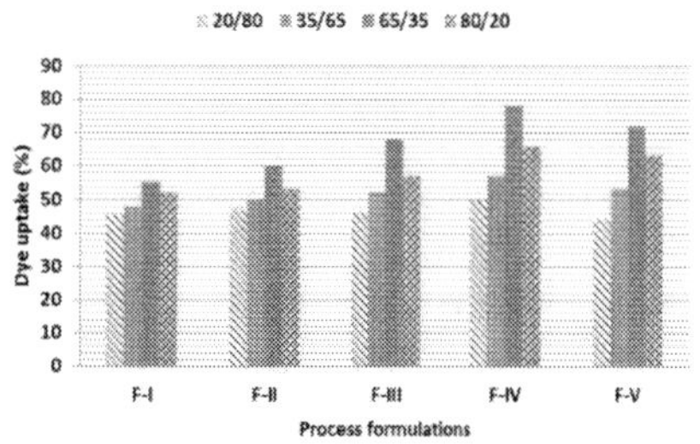

Figure 3. Dye uptake (%) of blended fabrics.

The dye uptake (%) of blended fabrics was shown in Figure 3. It has been shown that the dye uptake (%) was found to be higher in the formulation (F-IV) range with 65/35 blended fabric among all the samples. This result was attributed to the hydroxyl group of the cellulose carrying a slight negative charge, which is undesirable for dye fibre interaction because of the anionic nature of reactive dye used in this study. In addition, by using electrolyte in a dyeing bath, slight positive charge was acquired on the fabric surface. This will promote more dye fibre interaction and resulted in more dye uptake percentage [16-18]. Levelling agents have been evolved with reactive dyes to overcome the skitteriness of the fibre. Lyogen DFTN considerably enhanced the rate and extent of the dye uptake percentage. The glass transition temperature (Tg) of polyester material increases the chain mobility and leads to an extra shape ultimately being opened that can also enhance the uptake percentage by less penetration of the dye molecule [19].

Table 2 and 3 show the wash and rubbing fastness properties of the blended fabrics. The results indicate that the levelling agent involved in dyeing has slightly improved the fastness properties of the dyed blended fabrics. It was attributed to the interaction between the dye and levelling agent [20]. The attraction force of the levelling agent with the dye was so strong that it was difficult for the dye molecule to be separated from the levelling agent and migrate into the fibre in the case of high concentration of levelling agent. Therefore, the minimum amount of levelling agent should be used for good colour fastness yield as long as level dyeing can be obtained.

Table 2. Wash fastness values.

Process Formulations	20/80	35/65	65/35	80/20
F-I	3-4	3-4	4	3-4
F-II	3-4	3-4	4	3-4
F-III	4	4	4-5	3-4
F-IV	4-5	4	4-5	4
F-V	4	3-4	4	4

Table 3. Dry rubbing fastness values.

Process Formulations	20/80	35/65	65/35	80/20
F-I	4-5	4	4-5	3-4
F-II	4-5	4-5	4	4
F-III	4	4-5	4-5	3-4
F-IV	4	4	4	3-4
F-V	4	4	4	4

3.2 Whiteness index

Fabric whiteness maintenance is one of the most important cleaning performance indications in the textile industry. Whiteness can be defined as a facet of colour corresponding to a high luminosity and a lack of hue and greyness. In order to evaluate whiteness quantitatively, the CIE (Commission International de l'Eclairage) advocated a whiteness index based solely on studies by using Ganz [20]. This formula is used to calculate whiteness, W, and is based on illuminant D65 and 10° standard observer. It must be mentioned that the CIE Whiteness Index is not generally a valid assessment of whiteness for values below 40 [21]. Figure 4 shows the effects of various process formulations on CIE whiteness change in blended, bleached and dyed fabric. It is evident that the degree of change in whiteness is subject to the type of formulations. Although bleaching conditions were standardised, they were not representative of factory practice. The higher values of whiteness were obtained in the case of formulation (F-IV) with 65/35 fabric among all samples after bleaching conditions. It is apparent that the mentioned formulations' (F-I, F-II, F-III and F-V) results were not satisfactory, and there is a noticeable difference in the whiteness index of dyed fabric in comparison with bleached fabric, and washing off agent was not able to remove stains completely from the blended fabric. Therefore, these formulations failed to reduce the staining of fabric and were not found to be more effective in erasing staining of blended fabric. The concentration of the reduction clearing treatment was then reduced to 50% and the fabric was subjected to washing off agents in order to remove the slight tint from the fabric. However, the fabric became more stained. This result predicts that the tinting of the fabric may be due to the washing off agents. It can be observed that in formulation (F-IV), there is a very slight difference in the whiteness index of dyed fabric compared to bleached fabric, i.e. much less staining was observed, which was proved by the CIE Whiteness Index and indicates that formulation (F-IV) is beneficial for reducing staining from Cotton/ CDPET blended fabric (65/35).

Less staining observed may be due to the Lyogen acting as a levelling agent along with diffusion acceleration, ensuring that maximum exhaustion of dye goes into the fibre rather than remaining in the bath. However, there could also be unfixed dye that has been removed to a greater extent by reduction cleaning. This indicates that the industrial problem can be solved by using Lyogen DFTN as a levelling agent in the dyeing bath, followed by reduction clearing. The dye has been fixed instead of remaining on surface.

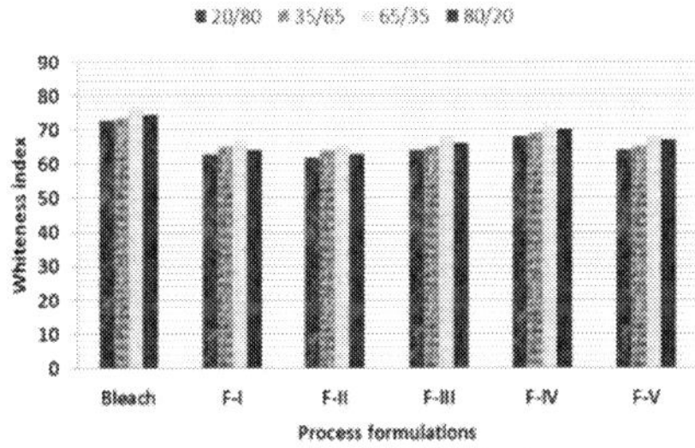

Figure 4. Whiteness index of blended fabrics.

4 Conclusion

In this study, various attempts have been made to overcome the staining of the cotton/CDPET blended fabric based on different formulations and desired result obtained in case of dyeing performed with a leveling agent (commercial name: Lyogen DFTN) followed by reduction clearing. Specifically, the results were found to be the best of the formulation (F-IV) for 65/35 blended fabric, among all the formulations performed. In addition, the obtained results confirmed that the Reactive dye do not contribute in staining problem. Our results are consistent with the proposal that both the reduction cleaning and addition of Lyogen DFTN to the dyebath enhance the absorption of dyes in a similar manner by modifying the surface of the blended fabric and played a role to avoid staining problem.

Acknowledgement

This work was supported by the China National Textile and Apparel Council (2013 "Textile Vision" Applied Basic Research) a grant from the Hubei Province Science and Technology Support Program (Project 2013BAA043), and the Collaborative Innovation Plan of Hubei Province for Key Technology of Eco-Ramie Industry (E JIAO KE HAN 2014 No.8).

References

1. A.A. Mousa, Y.A. Youssef, R. Farouk, T.M. Ayiesh, M.H. Arief, Color. Technol. **127**, 304 (2011)
2. C. Meena, et al.,
3. C.R. Meena, A. Nathany, R.V. Adivarekar, N. Sekar, European International Journal of Science and Technology **2**, 6 (2013)
4. J.H. Choi, A.D. Towns, Color. Technol. **117**, 127 (2001)
5. S.K. Pal, R.S. Gandhi, V.K. Kothari, J. Appl. Polym. Sci. **61**, 401 (1996)
6. E.M. Aizenshtein, L.V. Ignatovskaya, L.A. Anan'eva, O.T. Okuneva, V.L. Molokov, R.V. Simonova, Fibre Chem. **32,** 187 (2000)
7. K.V. Datye, B.H. Palan, J. Appl. Polym. Sci. **38**, 1447 (1989)
8. R. Yang, M. Dai, J. Text. Inst. **89**, 173 (1998)
9. S. Pal, R. Gandhi, V. Kothari, J. Appl. Polym. Sci. **61**, 40 (1996)
10. G.L. Zhao, R. Wu, J. Curiskis, A. Deboos, Text. Res. J. **74**, 27 (2004)
11. W. Tiedemann, J. Schad, Melliand Int. Text. Rep., **79**, E230 (1998)
12. M.N. Pervez, M.A. Rahman, International Journal of Scientific and Engineering Research **6**, 660 (2015)
13. M. Senthilkumar, N. Selvakumar, R. Shamey, Dyes Pigments **90**, 225 (2011)
14. S. Acharya, N. Abidi, R. Rajbhandari, F. Meulewaeter, Cellulose, **21**, 4693 (2014)
15. M. Akgun, B. Becerir, H.R. Alpay, Fiber. Polym. **8**, 495 (2007)
16. M. Naebe, et al., Text. Res. J. **80**, 312 (2010)
17. M. Naebe, P.G. Cookson, J. Rippon, Text. Res. J. **80**, 312 (2010)
18. W.M. Raslan, A. Bendak, E.M. Khalil, T. Fawzi, Color. Technol. **126**, 231 (2010)
19. J. Dave, R. Kumar, H. Srivastava, J. Appl. Polym. Sci. **33**, 455 (1987)
20. S.P. Mccarthy, L. Rebenfeld, H.D. Weigmann, Text. Res. J. **51**, 323 (1981)
21. J. Dumbleton, J. Bell, T. Murayama, J. Appl. Polym. Sci. **12**, 2491 (1968)
22. J.J. Lee, W.S. Shim, I.S. Kim, J.P. Kim, Fiber. Polym. **6**, 244 (2005)
23. E. Ganz, Appl. Opt. **18**, 1073 (1979)
24. R. Levene, R. Prozan, Color. Technol. **108**, 375 (1992)

Effect of Ring Machine Front Roller Rubber Cot Hardness on Properties of Coarser Yarn

Abdul Rehman Akbar[1, 2, 3], Muhammad Zeshan[3], Mohammad Neaz[2], Yingjie Cai[1, 4], and Lina Lin[1, 2, 4, a]

[1]*Hubei Provincial Engineering Laboratory for Clean Production and High Value Utilization of Bio-based Textile Materials, Wuhan Textile University, Wuhan 430200, China*
[2]*College of Textile Science and Engineering, Wuhan Textile University, Wuhan 430200, China*
[3]*Department of Yarn Manufacturing, National Textile University, Faisalabad 37250, Pakistan*
[4]*Engineering Research Centre for Clean Production of Textile Dyeing and Printing, Ministry of Education, Wuhan Textile University, Wuhan 430200, China*

Abstract. The hardness of the rubber cot is one of the most important factors that influences the properties of spun cotton yarn. Four different 100% coarse count cotton yarns (Ne 10/1, Ne 12/1, Ne 14/1, and Ne 16/1) were spun with various front roller rubber cot hardness from 65°, 68°, 70°, 75 °, and 83° shore hardness with constant back cot roller shore hardness. Cotton yarn properties were investigated, including mass uniformity, imperfection level, unevenness, hairiness, tensile strength, and elongation. The results showed that as the shore hardness increased, the unevenness of mass and imperfection index of the yarns proportionally increased. It does not apparently influence the yarns' hairiness, tensile strength, or elongation properties.

1 Introduction

Yarn quality is essential for the economic success of spinning plants. International competition and market requirements dictate the necessity to produce quality yarns at an acceptable price. There have been numerous attempts to improve the yarn quality, including process modification, process optimization, and investigations [1]. The raw material, process control techniques, and spinning parameters are major factors that influence the final yarn properties [2-4]. Drafting is the most important and fundamental operation in ring spinning. It has substantial contribution to the yarn quality, particularly in spinning the top roller covers, i.e. cots and drafting aprons [5]. Cots and aprons in the drafting system contribute to the overall spinning performance and yarn quality, along with other components. Cots are used in draw frames, combers, speed frames, and ring frames. Aprons are used only in speed frames and ring frames. Cots work to provide uniform pressure on the fiber strand. This facilitates efficient drafting with the assistance of aprons, which results in a better grip and control of the fibers, particularly floating fibers. A front-line cot in the ring spinning also offers a sufficient pulling force. A minimum pulling force that is greater than or equal to the sum of frictional resistance between fibers and force exerted by the aprons on fibers is required at the front line cot to overcome drafting resistance. Special rubber compounds with shore hardness ranging from 63° to 90° are used as raw materials for the cots [6, 7]. The raw material composition determines the properties of the cot. This includes the shore hardness of the rubber cot, abrasion resistance, tensile strength, swelling resistance, resilience properties, surface characteristics (like grip offered on fiber strands), low compression set values, and elasticity of the cot and the color. The top roller cots should have good ageing stability, minimal film formation, no lap formation, good fiber guiding, and long working life for use as a top roller cover [8].

Synthetic top roller cots are typically available in cylindrical form. The length or width, the finished outer diameter, the bare roller diameter, and the construction like PVC (core and shore hardness) are technical requirements of a top roller cot. The degree of the shore hardness is a main property of the top roller cot. It varies within different types of fibers and applications [8]. Among the yarn quality parameters, hairiness is one of the most important factors in determining the quality of the spun yarn, or the fibers that protrude from the body of yarn as loops, wild fibers, and extra fiber ends [9]. The performance of the woven or the knitted fabric are greatly affected by these factors [10]. Hairiness can cause problems within machine stoppages and breakage during spinning, since it makes a weak point in the yarn. It can also affect the appearance, the pilling, and the handling of the fabric [10, 11]. It also reduced the dye uptake of the fabric during dyeing [12]. This makes yarn hairiness a point of interest for researchers who aim to produce better quality yarn and fabric [10]. The imperfection index (50% thick, 50% thin, and 200% nep) values, the uniformity of mass, the tensile strength, and the

[a] Corresponding author: linalin@wtu.edu.cn

DOI: 10.1051/matecconf/201713002003

elongation of the yarn are important factors determining its quality and usage in the subsequent process [13].

2 Experimental

2.1 Materials

Yarns were made of cotton fibers. The cotton fiber had a strength of 30 g/tex, a Micronaire value of 4.5-5.0, a fiber elongation of 6.0%, a tenacity of 3.0-4.9 g/Den, a moisture regain of 8.5%, and a specific gravity of 1.54.

2.2 Procedure

The production of the yarn samples was done with a Toyoda ring frame (Model: RX-240) auto-doffing linked with an auto-coner. Rubber cots with various hardness (range: 65°, 68°, 70°, 75° to 83°) shore were used on the front-line rollers in the RX-240 ring frame drafting system. The hardness of the back-line rollers in the drafting system were kept constant with a standard value of 83° shore across all experiments. The other process parameters during the yarn production were kept constant at the opening machine, the cleaning machine, the carding machine, the breaker drawing machine, and the finisher drawing machine. The hank roving count if the simplex machine was changed to spin a specific count (yarn fineness). The vertical pneumatic mounting machine was used to mount the cots. The semi-automatic grinding machine was used to grind after the cots were mounted. A grinding stone with a stone grid of 80 grit size and a porosity of 14 was used. An average roughness value of 0.8 ± 0.2 μm was applied to both the front and back line cots within ring spinning. The yarn properties were analyzed after completion.

2.3 Testing Instruments

The unevenness of mass, the yarn hairiness (hair length greater and equal to 3 mm), and the imperfection index of the yarn were determined with an USTER® evenness tester (UT-5, Uster Technologies AG, Switzerland). The yarn tensile strength properties were tested with a MESDAN lab strength tester (Italy).

3 Results and discussion

3.1 Preparatory Process

The cotton fibers need to be treated at different stages to make yarn. The fibers are treated on a card machine, a draw frame (breaker), a finisher draw frame, and a simplex before they are processed on the ring machine. These processes are called the preparatory processes. Each preparatory stage influences the quality of the final yarn. The fiber strand was tested after every stage to ensure the working conditions were suitable for producing good quality yarn. The unevenness percentage (U%) of the semi-product at each stage were investigated, as shown in Figure 1.

The variation in mass per unit length of resultant yarn was caused by the variation in fiber assembly from the preparatory processes. The U% of four different carded slivers were tested at UT-5. The mean value of the U% measured is 2.78. The factors affected the U% of the fiber strand are the wire conditions of the machine parts and the setting of the machine parts with respect to each other. It can be controlled by using wires that are in its best working condition and proper settings of the machine. The U% of four drawn slivers from different breaker draw frame machines is tested to check the quality of operation. The actual mean value of the unevenness comes out to be 2.88%, which is less than the acceptable limit of 3%. The unevenness depends upon drafting system; therefore, the optimum settings should be maintained in the drafting arrangement to get maximum uniformity in the product.

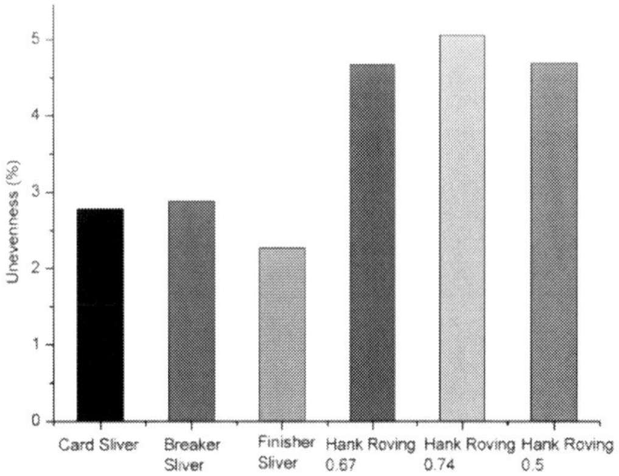

Figure 1. Unevenness after different processes during yarn spinning.

The finisher draw-frame is used to reduce the variations in the sliver. The finisher draw-frame has auto-leveling system that helps to control the variations in the sliver. The sensing device i.e. tong and groove roller checks the variations, and the signals are given back to control the variations. The mean value of U% for four samples is 2.26 that is good as compared to the values of breaker sliver.

In processing in the spinning mill, the unevenness of the product increases from stage to stage after draw frame [15]. Four different samples of roving with same hank roving from four different simplex machines have been tested. The actual mean value of the U% obtained is 4.67. The variation may arise due to the setting problems between the rollers, draft distribution, parts used within the drafting arrangement and some other factors. The U% is one of the important factor determining the quality of the product being produced on the machine. The lower it is, better the quality and vice versa. In order to get the best results out of the simplex machine all factors that affect quality should be taken into account.

Another parameter that determine the quality of the product being produced is coefficient of variation of mass (CVm%). Co-efficient of variation in the semi-product should be within acceptable range to get a product with desired properties and quality. Higher CVm in the yarn means lower product quality. Generally during spinning, the CVm% of the fiber strand decreases from carding to finisher but it drastically increases at simplex. The

CVm% of the fiber strands were noted after each step. The mean value for the four samples were taken at each stage. The results are shown in Figure 2.

The coefficient of the mass variation (3.6%) was measured for the slivers after the carding operation. The variation reduced further to 3.4% after the first draw frame passage, and then again to 2.9% after the material passed through the finisher draw frame. The step wise decrease in mass variation was due to doubling that occurred at the draw frame. Doubling refers to adding additional slivers (6 to 8 typically) to obtain an additional equalized sliver after drafting [14]. The CVm of the fiber strand was also checked after processing through a simplex machine. The obtained value was approximately 6%. Simplex attenuated the fiber strand obtained the desired roving counts (i.e. 0.67, 0.74, and 0.5 hank roving).

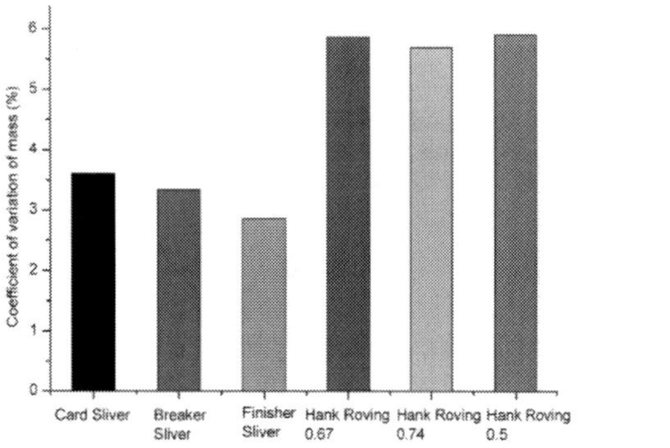

Figure 2. The coefficients of variation after various processes during yarn spinning.

3.2 Influence of rubber cots hardness on yarn unevenness of mass

Mass uniformity is required for good quality yarn [14]. The unevenness of the mass increase with the increase of rubber cots shore hardness are shown in Figure 3. This is particularly apparent for the 14/1 and 16/1 count yarn (fineness of single yarn). The mean values of the unevenness were 10.5%, 10.62%, 10.84%, 10.92%, and 11.32% for Ne 14/1 and Ne 16/1. The hardness shore was 65°, 68°, 70°, 75°, and 83°. The lower shore hardness gave more control of the fibers, due to an increased contact area between the front roller cot and the bottom fluted rollers. The increased contact area shortened the uncontrolled length of the fiber strand between the cot nipping point and the apron. Therefore, softer cots produce more even yarn [8]. Note, the highest uniformity of yarn mass occurred at 68° shore of the cot hardness.

3.3 Influence of rubber cots hardness on imperfection Index

The imperfection index values (provided in Figure 4) depicts the directly proportional relationship with the cots hardness. This indicated that softer cots produced a lower imperfection index of yarn. The mean values of the imperfection index for the four different yarn counts were 86.45%, 100.23%, 119.15%, 122.8%, and 149.77% with a shore hardness of 65°, 68°, 70°, 75°, and 83°. This was attributed to the decrease in the uncontrolled length between the aprons and the nipping point of the front roller cots while using softer cots. This led to a more controlled flow of the material. The uncontrolled length was the length where the material had no support to be carried. The fiber strands then moved and slid apart during drafting [15]. This uneven flow of material was reduced by the decreased uncontrolled length using relatively softer cots. The softer cots shift the nipping point of the front rubber rollers slightly forward, due to the increased cross-sectional contact. It reduced the length of the spinning triangle, which caused the twist to be inserted in a better way for the fiber strand to exit the rollers. The improved twist realization led to an improved imperfection index [8].

The yarn quality with lower imperfection index values were achieved by having less co-efficient of variation of mass and good mass uniformities, resilience property and low compression set values, and anti-lapping characteristics. These played a crucial role, since they affected the spinning performance. The lapping characteristics were important for the rubber cots because they influenced the quality of the product and the performance of the operation. A majority of the lapping originated after an end break. The finer fibers were more likely to follow the profile of the roller and had more of a lapping tendency. The finer the fibers, the more likely they were in the cross-section of the yarn. This led to an increased frequency of lapping. The applied pressure on the top rollers also influenced the lapping behavior of the fibers, resulting in a deterioration of the quality. Finer and longer fibers had more of a tendency to lap, since more pressure was needed to overcome the drafting resistance during spinning. Pneumatic suction is another parameter that was affected. The lower values of suction pressure had an increased tendency of lapping and vice versa. It could be adjusted by varying the different parameters, such as\ fan size, speed, and machine length. The mini pores, or pinholes in the rubber cots or impurities in the cots, also caused lapping. Softer rubber cots had more of a tendency for lapping, due to increased contact area. Therefore, frequent buffing of the cots (every 30 days) and after treatment could improve the quality of the process and the product.

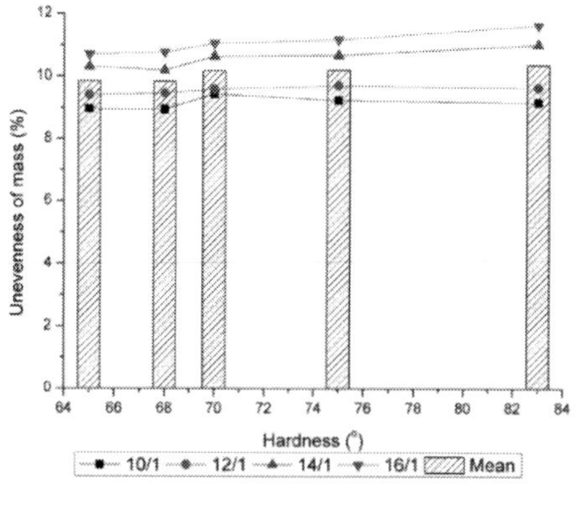

Figure 3. Unevenness of the mass of yarns spun with various rubber cots shore hardness at 65°, 68°, 70°, 75°, and 83° shore.

 DOI: 10.1051/matecconf/201713002003

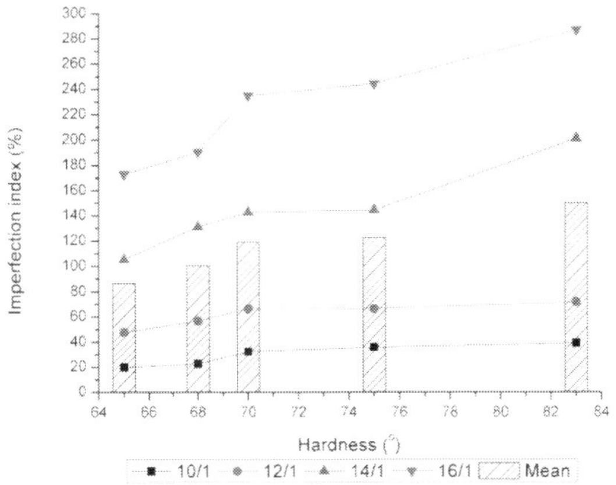

Figure 4. Imperfection index of yarns spun with various rubber cots shore hardness at 65°, 68°, 70°, 75°, and 83° shore.

3.4 Influence of rubber cots hardness on hairiness

Figure 5 summarizes the hardness levels of the rubber cots and the variation in the yarn hairiness. Rubber cots with a shore hardness of 68° was considered ideal for ring spinning. Our results confirmed this finding. The 70° shore hardness had the highest hairiness out of all the counts. There was no specific trend for the increase or the decrease in hairiness values with variation of the rubber cot hardness. There should be an optimum level of hardness to achieve lowest hairiness [16].

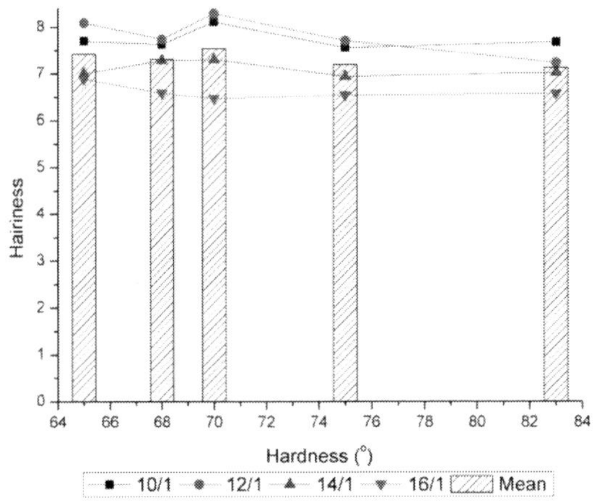

Figure 5. The hairiness of the yarns spun with various rubber cots shore hardness of 65°, 68°, 70°, 75°, and 83° shore.

The hairiness increased at two points during ring spinning, i.e. the front roller delivery point and the ring traveler junction. The twist did not flow directly to the nip point of the spinning triangle. The selvedge fibers in the strand did not fully integrate into the yarn. They appeared as hairs. An abrasion at the traveler/ring junction led to portions of the fibers to travel into hairs. The count and the twist had a considerable influence on the hairiness. The courser count yarns had increased amounts of hairiness than the finer count yarns because they had more cross-sectional area. The proportion of fiber ends that protruded from the yarn surface were about 31% of the actual number of ends present in the yarn. This was measured microscopically. The mean value of the hairiness increased as the cross-sectional area increased when the length of the protruding fiber ended and the loops were considered. An optimum value of hardness was necessary, as harder rubber cots damaged the fibers when the high compression increased the hairiness breakage occurred. The lower rubber hardness lapped the fibers around the cots, which led to breakage and low yarn quality [8].

3.5 Influence of rubber cots hardness on tensile strength and elongation

Tensile testing of the yarns was performed to determine the breaking force, elongation, and toughness properties. Tensile testing of yarn was performed by a tensile testing machine that used a unique grip to hold the yarn. Resistance against the pulling force was referred as the tensile strength [4, 17, 18]. This is an important characteristic of the yarn, which influenced its processing quality in the subsequent processes. The yarn should have an optimum level of tensile strength with optimum elongation. The influence of the changes in hardness values of the front roller rubber cots over the tensile strength and the elongation values are shown in Figure 6. There was no specific pattern found between the values of tensile strength and elongation with various hardness levels. The increased rubber cot hardness did not have an apparent direct influence on the yarn tensile strength and elongation properties. The mean values were approximately 211 lbs. (219 lbs. for 68° shore) and 13% under the various hardness conditions. The coarse yarn properties analysis showed that the optimum rubber cot hardness during ring spinning was 68° shore [8, 19].

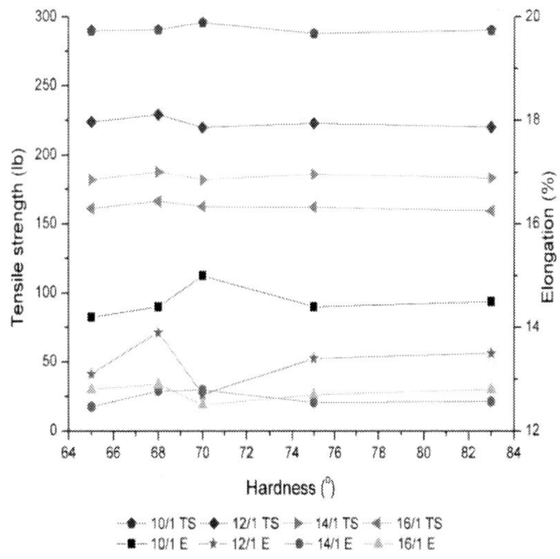

Figure 6. The tensile strength and the elongation of yarns spun with various rubber cots shore hardness at 65°, 68°, 70°, 75°, and 83° shore.

4 Conclusion

Four different 100% coarser count cotton yarns (Ne 10/1, Ne 12/1, Ne 14/1, and Ne 16/1) were spun with rubber cots of various hardness. These ranged from 65°, 68°, 70°, 75°, to 83° shore on the front-line rollers. As hardness increased, the unevenness of the mass and the imperfection index of yarns proportionally increased. For hardness ranging 65°, 68°, 70°, 75°, and 83° shore, the average values for four yarn counts were 9.8%, 9.8%, 10.2%, 10.2%, and 10.4%. The imperfection index mean values for four counts were 86.5%, 100.2%, 119.2%, 122.8%, and 149.8%. The increase of rubber cot hardness did not have an apparent, direct influence on the properties of yarn hairiness, tensile strength, or elongation. The mean values were approximately 7 lbs., 116 lbs. (220 lbs. for 68° shore) and 13% under the varied hardness conditions. According to the analysis of the coarse yarn properties, the optimum rubber cot hardness that should be applied in ring spinning was 68° shore.

5 Acknowledgements

This work was financially supported by the China National Textile & Apparel Council (2013 "Textile Vision" Applied Basic Research, 2013-153); Hubei Province Science and Technology Support Program (Grant No. 2013BAA043) and the Collaborative Innovation Plan of Hubei Province for Key Technology of Eco-Ramie Industry (2014-8).

6 References

1. N. Balasubramanian, Text. Res. J. **39**, 155 (1969)
2. P.K. Majumdar, A. Majumdar, Text. Res. J. **74,** 652 (2004)
3. P.R. Lord, *Handbook of Yarn Production: Technology, Science and Economics* (Elsevier, 2003)
4. C.A. Lawrence, *Fundamentals of Spun Yarn Technology* (CRC Press, London, 2003)
5. S. Sundaresan, A. Arunraj, K. Thangamani, International Journal of Innovative Research in Science, Engineering and Technology **5,** 2026 (2016)
6. G. Grover, P.R. Lord, J. Text. Inst. **83,** 560 (1992)
7. V. Subramaniam, A.P. Mohamed, Text. Res. J. **61,** 280 (1991)
8. A.S. Bagwan, R. Policepatil, S. Pawar, J. Text. Sci. Eng. **6,** 238 (2016)
9. X.X. Huang, X.M. Tao, B.G. Xu, Text. Res. J. **84,** 1121 (2014)
10. N. Haleem, X. Wang, Text. Res. J. **85,** 211 (2015)
11. Z. Xia, X. Wang, W. Ye, W. Xu, H.A. Eltahir, Fiber Polym. **12,** 534 (2011)
12. X. Wang, L. Chang, Text. Res. J. **73,** 327 (2003)
13. R. Furter, *The Standardization of the Quality Characteristics in the Textile Supply Chain*, Uster Technologies AG, (2009)
14. W. Klein, *The Rieter Manual of Spinning. Volume 1. Technology of Short-Staple Spinning*, Winterthur: Rieter Machine Works Ltd, (2008)
15. W. Klein, H. Stalder, *The Rieter Manual of Spinning.Volume 4. Ring Spinning*, (Rieter Machine Works, Winterthur, 2008)
16. A. Barella, Text. Prog. **13,** 1 (1983)
17. E. Oxtoby, *Spun Yarn Technology* (Butterworth-Heinemann, London, 2013)
18. C. Horst, S. Tetsuya, S. Leslie, *Handbook of Materials Measurement Methods* (Springer, Berlin, 2006)
19. V. Subramaniam, A.P. Mohamed, J. Text. Inst., **82,** 333 (1991)

Preparation and properties of silk sericin/cellulose cross-linking films

Kunyan Wang[a], and Fangyu Zhan

Department of Materials Chemistry, Huzhou University, zhuzhou, 313000, China

Abstract. Silk sericin/cellulose cross-linked films were successfully prepared using glutaraldehyde as cross-linkinger. FTIR was applied to characterize the chemical structure of films. Cross-linked silk sericin film was found the peak intensity of FTIR for cross-linked film decreased markedly compared to pure silk sericin, which indicating cross-linking reaction has been occurred. The increasing value of swelling ratio also indicated the cross-linking has been happened. The cross-linking reaction increased the thermal decomposition temperature.

1 Introduction

Biomaterials such as silk sericin and cellulose are considered as potential substitutes for synthetic polymers. Silk sericin delivered from silkworm silk, has been explored as a versatile biomaterial for various fields. It has excellent mechanical properties with high elastic modulus, elongation to break, and toughness [1-3]. In recent years biomaterials using flexible electronics based on organic field-effect transistors (OFETs) have attracted considerable attention. It have been introduced as gate dielectrics with low dielectric losses and leakage currents as well as sufficiently high dielectric breakdown strength. It has been introduced as gate dielectrics. B. Singh et. al. [4] fabricated an OFET device using a deoxyrionucleiv acid-based biopolymer for gate dielectric region. J.W. Chang et. al. [5] used albumen, as a gate dielectric in pentacene- and C_{60}-based organic field-effect transistors.

However, Silk sericin materials are either too weak in wet state or too brittle and stiff in the dry state to application. Blending is an effective possible way to combine advantanges of each componet[6]. Blending of silk sericin and cellulsoe have been prepare to acquire biomaterial that are able to obtain good mechanical properites[7]. Meanwhile, to maintain water-insolubility chemical crosslinking reagents were added to blending film. Glutaraldehyde is an effective synthetic crosslinking agents that can react with amino acids or proteins [8]. It has been reported biopolymers using glutaradehyde as crosslinking agents show good mechnical properites.

In this work, silk sericin and cellulose cross-linking films were prepared. The effects of diffenernt cellulose content on the chemical structure of the silk sericin and the thermal behavior were confirmed by FTIR and DSC.

2 Experimental Section

2.1 Materials

Silk sericin powder was purchased from Huzhou Xintiansi Biotec Co. (China). Glutaraldehyde was supplied by LingFeng Chemical Reagent Co. Ltd.. Cellulose was purchased from Aladdin Industrial Corporation (China). All chemical were of analytical grade, and used as received without further purification.

2.2 Specimens preparation

First 8% carbamide, 8% sodium hydrate and 6.5% thiocarbamide were mixed into conical flask with continuous magnetic stirring to make mixed solution. Silk sericin and cellouse with 2% contents of glutaraldehydewere dissolved into above mixed solution for 2h to generate a homogeneous solution. Finally the homogenous solution was poured into a smooth plate and placed in an oven at 60°C until dried silk sericin/cellouse samples were obtained.

2.3 Characterizations

Fourier-transform infra-red (FTIR) spectroscopy has been used to monitor the chemical reactions. FTIR were recorded by nicolet NEXUS-670 IR spectrometer in the wavelength range of 4000-400cm^{-1}.

X-ray diffraction patterns were recorded in XD-6 X-ray diffractometer (China). The length and width of all test samples are 20mm and 15mm, respectively. The wavelength of the monochromated X-ray from Cu Kα radiation is 0.154056nm. Scanning 2θ range was 10-60° with a scanning rate of 2°/min.

Differential scanning calorimetry was used to measure the melting temperature of samples. Sample weights ranged from 3 to 5 mg. All samples were heated

[a] Corresponding author: wkyan01@163.com

from room temperature to 400°C at a rate of 20°C/min. Heating curves was recorded.

The weight of 0.1g dry sample was immersed into adequate deionizer water at room temperature for 24h to reach the swelling equilibrium. Then, the swelling ratio (g/g) of sample was calculated.

3 Results and discussion

3.1 Chemical structure of silk sericin/cellouse cross-linked films

The FTIR spectra of pure silk sericin, pure cellouse and silk sericin/cellouse cross-linked films with 2% glutaraldehyde were shown in Fig.1. Silk sericin exhibited characterisitc amide absorption bands of protein, amide, and at 1635, 1415, and 1093 cm^{-1}, respectively, which indicating random coil structures of silk sericin [6, 7]. Cellouse exhibited characterisitc bands at 1635, 1419, 1310 and 1084 cm^{-1}, respectively. The peak at 1635 cm^{-1} was mainly corresponded with the stretching vibrations of C=O and –NH group. Compared to pure silk sericin and cellouse, cross-linked film with 2% glutaraldehyde was found the peak intensity decreased markedlly, which indicating crosslinking reaction has been occurred.

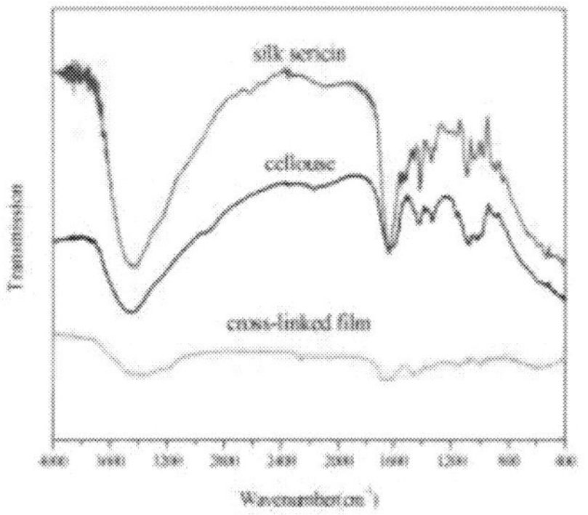

Figure 1. Infrared spectrums of pure silk sericin, pure cellouse and cross-linked film with 2% glutaraldehyde.

3.2 Crystal structure of silk sericin/cellouse cross-linked films

Fig. 2 shows the XRD patterns of pure silk sericin, pure cellouse and 2% glutaraldehyde cross-linked silk sericin films. Diffraction peaks of 2θ=21.9°, 28.0° for pure silk sericin are assigned to the silk II crystal planes, which is the spun silk state (β-sheet silk structure). Pure cellouse shows diffraction peaks at 2θ=22°, 34.6°. For 2% glutaraldehyde cross-linked film, it can be found that the diffraction peaks of 2θ=21.9° disappears. The results show that the silk sericin crystalline structure affected by glutaraldehyde and cellouse.

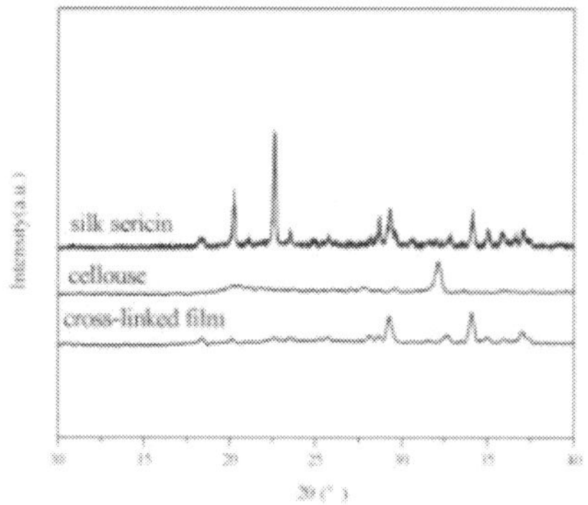

Figure 2. XRD patterns of of pure silk sericin, pure cellouse and cross-linked film with 2% glutaraldehyde.

3.3 Thermal behavior of silk sericin/cellouse cross-linked films

Fig. 3 shows the DSC thermograms of heating scan for pure silk sericin, pure cellouse and cross-linked film with 2% glutaraldehyde from room temperature to 400 °C. Both silk sericin and cross-linked film display a peak nearly 150 °C, which is caused by desorption of water and motivation of molecular chain in amorphous region [8, 9]. Meanwhile, an weak endothermic peak at 300 °C have appeared for pure silk sericin , representing the thermal decomposition of silk sericin. And then, an interse endothermic peak at 300 °C comes out in cellouse, representing the thermal decomposition of cellouse [10]. The peak at 300 °C disappeared for cross-linked film compared to pure silk sericin, which indicating the cross-linking reaction increasing the thermal decomposition temperature.

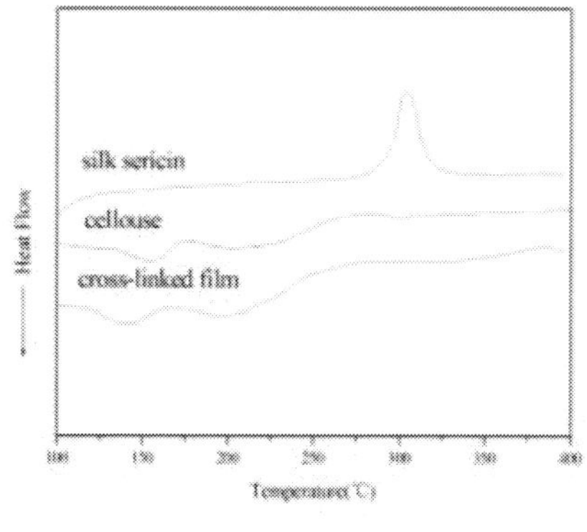

Figure 3. DSC thermograms of the second heating scan of pure silk sericin, pure cellouse and cross-linked film with 2% glutaraldehyde.

3.4 Swelling behaviors of silk sericin/cellouse cross-linked films

The swelling capacity in deionized water has been studied to evaluate the water-holding capacity. The

DOI: 10.1051/matecconf/201713002004

swelling capacities of pure silk sericin and cross-linked silk sericin film were shown in Fig. 4. The swelling ratio of cross-linked silk sericin film increased compared to pure silk sericin. Swelling behavior depends on hydrophilicity/hydrophobicity, crosslinking density and the characteristics of the external solution. The increasing value of swelling ratio indicated the crosslinking have been happened.

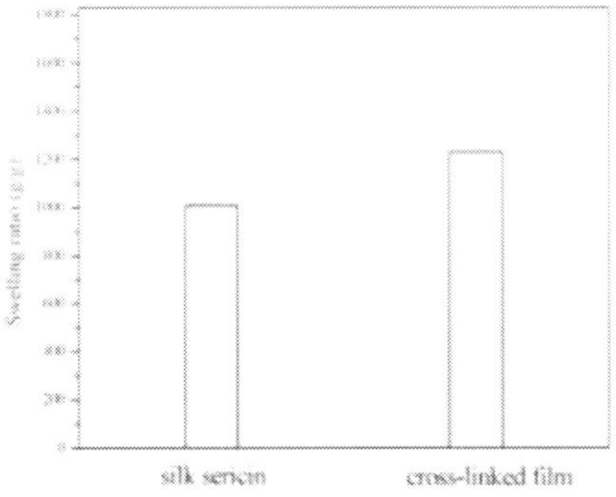

Figure 4. Swelling ratios of pure sericin and cross-linked film.

4 Conclusions

In this work, we report silk sericin/cellouse cross-linked films based on glutaraldehyde as crosslinking agent. FTIR was used to confirm the cross-linking structure of cross-linked films. The cross-linking reaction increased the thermal decomposition temperature. The increasing value of swelling ratio indicated the crosslinking have been happened.

Acknowledgements

The authors would like to acknowledge financial support by the Natural Science Foundation of Zhejing province of China (LQ14B040003).

References

1. Y.N. Lin, X.X. Xia, K. Shang, R. Elia, W.W. huang, P. Cebe, G. Leisk, F. Omenetto, D.L. Kaplan, Biomacromolecules. **14**, 2629 (2013)
2. Y.P. Huang, B.P. Zhang, G.W. Xu, W.T. Hao, Composite. Sci. and Tech. **84**, 15 (2013)
3. J. Luo, L. Zhao, Y.Y. Yang, G.L. Song, Y. Liu, L.J. Chen, G.Y. Tang, Solar Energy Mater. And Solar Cells. **147**, 144 (2016)
4. B. Singh, N.S. Sariciftci, J. Appl. Phys. **100**, 024514 (2006)
5. J.W. Chang, C.G. Wang, C.Y. Huang, T.D. Tsai, T.F. Guo, T.C. Wen, Adv. Mater. **23**, 4077 (2011)
6. T.D. Thanh, L.Kapralkova, J. Hromadkova, I. Kelnar, Eur. Polym. J. 50, 39 (2014)
7. L. Zhou, Q. Wang, J.C. Wen, X. Chen, Z.Z. Shao, Polymer. 53, 5035 (2013)
8. A. Vasconcelos, A.C. Gomes, A. Cacaco-paulo, Acta. Biomater. **8**, 3049 (2012)
9. E. S. Sashina, G. Janowska, M. Zaborski, A.V. Vnuchkin, J. Therm. Anl. Calorim. **89**, 887 (2007)
10. L. Zhou, Q. Wang, J. C. Wen, X. Chen, Z. Z. Shao, Polymer. **54**, 5035 (2013)

Accident tolerant composite nuclear fuels

Barbara Szpunar [1, a], Linu Malakkal[2], Seanne Chung[1], Momina Mateen Butt[1], Ericmoore Jossou[2] and Jerzy A. Szpunar[2]

[1]*Department of Physics and Engineering Physics, University of Saskatchewan, Canada*

[2]*Department of Mechanical Engineering, University of Saskatchewan, Canada*

Abstract. Investigated accident tolerant nuclear fuels are fuels with enhanced thermal conductivity, which can withstand the loss of coolant for a longer time by allowing faster dissipation of heat, thus lowering the centerline temperature and preventing the melting of the fuel. Traditional nuclear fuels have a very low thermal conductivity and can be significantly enhanced if transformed into a composite with a very high thermal conductivity components. In this study, we analyze the thermal properties of various composites of mixed oxides and thoria fuels to improve thermal conductivity for the next generation safer nuclear reactors.

1 Introduction

The novel, composite fuels with enhanced thermal conductivity can withstand the loss of coolant for an extended time by allowing faster dissipation of heat, thus lowering the centerline fuel temperature. This consequently prevents the melting of a reactor core, minimizes oxidation of cladding and related hydrogen production and therefore prevents hydrogen explosion that occurred during Fukushima accident. The composite fuels allow reduction of the thermal gradients and thermal stresses that contribute to cracking, thereby increasing the longevity of the fuel.

Although many factors need to be investigated before nuclear fuel composites are adapted to the service in the harsh environment in the nuclear reactor the suitable fuels must have a high thermal conductivity. Traditionally metal oxides were used as fuel since they have a high melting point [1]. Nitrides and Borides (with the appropriate isotopic composition that excludes strong neutron absorber ^{10}B) are also attracting attention due to their large thermal conductivity from the electronic contribution that increases with increasing temperature [2,3]. In this work, we evaluate the thermal conductivity of ceramic composites where the heat carriers are primarily phonons. In our previous work [4] we have found that the phonon contribution to the thermal conductivity of thoria, (an alternative fuel with a high melting point) can be well reproduced up to very high temperatures by a simplified Slack model [5] with the input parameters calculated using first principles molecular dynamics. Silicon Carbide (SiC) and Beryllium Oxide (BeO) are materials proposed to be used in accident tolerant fuel. Within the lower temperature range where quasi-harmonic approximation (QHA) applies we have shown [6] using the first principle code Quantum ESPRESSO (QE) [7] that for SiC and BeO both Slack model and the ShengBTE (BTE) a solver for phonon thermal conductivity [8] predict similar thermal conductivities in agreement with the experiment except for cubic SiC where Slack model over-predicts thermal conductivity by a factor of two. Here, we investigate it further for other candidate components and evaluate the effect of high thermal conductivity of the composite fuel on the thermal performance.

2 Methodology

A straightforward and efficient interface (QE_nipy_advanced), developed in ipython by our group [9], is used to operate Quantum ESPRESSO (QE), an open source code for materials simulation [7]. QE is a first principles code using density functional theory, plane waves and pseudopotentials; and it can predict ground state material properties in agreement with experiment provided the parameters like plane wave cut-off energies, k point optimal mesh and equilibrium structure are optimized accurately as described earlier [10]. We also found previously [9,10] that new generalized gradient approximation (GGA) functional for solids, PBEsol used with the pseudopotentials generated for PBE or PBEsol functional lead to equilibrium lattice constants that agree very well with experiment and therefore it is used here in our new calculations. Equally good results are obtained with local density approximation (LDA) and norm conserved pseudopotentials which we have been used predominantly in our previous calculations [10].

We calculate lattice governed thermal conductivity using QHA as implemented in QE code and two methods as discussed above: Slack model [5] and BTE code [8].

Once the thermal conductivity is calculated for the limited temperature range within QHA, we fit it into an

[a] Corresponding author: B.Szpunar@usask.ca

analytical form developed for phonons contribution as a function of temperature as:

$$k_L = \frac{1}{a+bT} \quad [W\ m^{-1}\ K^{-1}] \tag{1}$$

where "a" and "b" parameters are independent of the temperature (T), and are associated with the contribution of phonon defects (a) and phonon-phonon (b) scattering to the mean path. The calculations and the related inverse fits (Eq. 1) are for fully dense (100 % TD) nuclear materials and therefore a porosity factor must be added [4] when comparing with experimental results [1]. We demonstrated that this equation describes phonon assisted thermal conductivity for up to a very high temperature [4].

The phonon contribution to thermal conductivity as described by Eq. 1 decreases with temperature, and therefore it is of interest for reactor safety analysis to estimate the minimum thermal conductivity (k_{min}). We evaluate it here, using Young's modulus and Clark's model [11], from the equation:

$$kmin = 0.87 k_B M_{av}^{-2/3} Y^{1/2} \rho^{1/6} \tag{2}$$

where M_{av} is the average mass per atom ($M_{av} = M/(n\ N_A)$), where M refers to the molar mass, n is the total number of atoms per formula, N_A is the Avogadro's constant), ρ is the density, and k_B is the Boltzmann's constant.

The composite fuel can be fabricated as fibers or as particles of high-temperature conductivity component dispersed in fuel matrix. A simple analysis of thermal conductivity for these various arrangements for assumed here small volume fractions (5%-15%) of additivities shows that the thermal conductivities are comparable. We assumed therefore that the thermal conductivities of the composites could be approximated as an average over volume concentrations (C) of the thermal conductivities of their constituents (e.g. for ThO_2-BeO):

$$\kappa_L^{BeO/ThO_2} = (C_{vol\%}^{BeO} \kappa_L^{BeO} + C_{vol\%}^{ThO_2} \kappa_L^{ThO_2})/100 \tag{3}$$

3 Thermal conductivity

We carried out calculations of the lattice thermal conductivities as described in Section 2. In Fig. 1 we compare the theoretical results for the selected materials (SiC, BeO, and BAs) with experimentally observed one order of magnitude higher thermal conductivity than Urania and Thoria.

According to the previous studies [6], we observed in Fig. 1 that with the input parameters derived using QE espresso code Slack model predicts one order of magnitude higher lattice thermal conductivity than Urania [1] for the presented calculations for cubic SiC, BAs and wurtzite BeO.

The Slack model calculations for SiC and BeO were performed for the maximum strains: 0.99 to 1.04 and 25 number of deformations.

The pseudopotentials used for SiC were norm conserved, and the electronic exchange-correlation was described by LDA with Perdew-Zunger parametrization. The total energy convergence was obtained for a Monkhorst-Pack sampling of 8 × 8 × 8 and the plane wave's cut-off energy of 150 Ry. Kinetic energy cut-off of plane waves basis describing charge density was taken four times the energy cut-off of the used wave functions.

A mesh of 5 × 5 × 5 q-points was used to calculate the second order interatomic force constants (IFCs) needed to compute phonons within density functional perturbation theory (DFPT) as implemented in QE and the lattice thermal conductivity of SiC for both BTE and Slack model. A supercell of 4 × 4 × 4 was used to obtain the third order IFCs needed for BTE calculations. The force cut-off distance was set such that the interaction range is up to the third nearest neighbours.

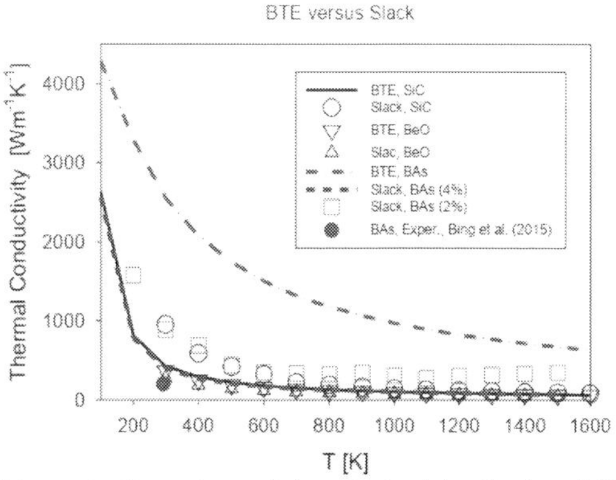

Figure 1. Comparison of the calculated by Slack and BTE methods lattice thermal conductivities of SiC, BeO and BAs. The shown experimental result is for BAs [12].

The total energy convergence of the BeO and structure in QE code was obtained using a kinetic energy cut-off of 75 Ry and the Brillouin zone integration over a Monkhorst–Pack of 8 × 8 × 5 mesh. Since we used the ultra-soft pseudopotentials, we need an augmented charge around the ion core. Hence the kinetic energy cut-off of plane waves basis describing charge density was taken twelve times the energy cut-off of the used wave functions. The dynamical matrix for phonon density of states of BeO was calculated on a mesh of 6 × 6 × 4 q-points in the irreducible Brillouin zone respectively. While the input to Slack model was obtained using DFPT QHA as implemented in QE for ShengBTE code we used a supercell of 4 × 4 × 4 to calculate the third order interatomic force constants (IFCs) for BeO by solving BTE. The force cut-off distance was set such that the interaction range is up to the five nearest neighbours for BeO. A mesh of 6 × 6 × 4 q-points was used to calculate the second order IFCs using QE and needed to compute the k_L of BeO.

For the simulation of the thermal properties of BAs and BP, we used available now PAW pseudopotentials developed with the new PBEsol functional. The optimal parameters for BAs were found to be 100 Ry a cut-off energy and mesh of 7 × 7 × 7 k-points, and we used cut-off of plane waves basis describing charge density four times the plane wave energy cut-off. However, in case of BP, the convergence of energy was achieved with a mesh of 8 × 8 × 8 k-points and a plane wave cut-off energy of 100 Ry. Also, we observed that setting the plane waves basis describing charge density at a high value (1000 Ry) has led to a non-physical linear thermal expansion and therefore we have maintained the value at four times larger than the energy cut-off. Furthermore, we used a 5×5×5 q-points mesh for the second order force constants required for the thermal expansion calculation of BAs and BP.

A mesh of 5 × 5 × 5 q-points was used for BAs and BP to calculate the second order interatomic force constants (IFCs) used in thermal expansion calculations for the Slack model. The same mesh was used for BTE calculation for BAs with 5 × 5 × 5 supercells for third order IFCs.

We note however that due to the usage by the Slack equation of the thermal expansion coefficient, which is derived from the second derivative over the temperature of volume, the derived thermal conductivity shows numerical accuracy dependence. Therefore, for example, the calculations of the thermal conductivity of BAs, using thermal expansion coefficients calculated considering the same 14 deformations, 100 Ry plane-wave cut-off energy and different strains from 0.99 to 1.04 or 1.02 (indicated as 4% and 2% in Fig. 1) differ by a factor of two.

Additionally, we performed a comprehensive testing on the dependence of the number of deformations, cut-off energy and k points mesh on the thermal conductivity of BP predicted by Slack model (Fig. 2.). We can see that the result (indicated by dotted line) for strain 0.99-1.02, 20 deformations, cut-off energy 80 Ry and 8 × 8 × 8 k point mesh significantly overpredicts the thermal conductivity, and this may be due to limited numerical accuracy at this strain. Therefore, we increased maximum strain range to 0.99-1.04.

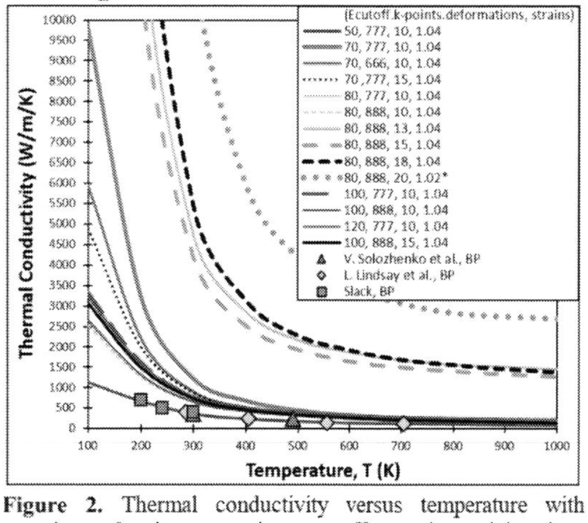

Figure 2. Thermal conductivity versus temperature with experimental values at various cut-off energies and k-points, and deformation number as indicated. The used strains were between 0.99 and 1.04 and 0.99 and 1.02, as indicated.The purple dash-dot curve agrees with experimental data, however, at a very low cut-off energy of 50 Ry. The best, desired curve is the black solid curve at 15 deformations, 100 Ry cut-off energy and mesh of 8 × 8 × 8 k-points.

In Figure 3, we present for BP the selected results as indicated for a larger range of maximum strains 0.99-1.04, which were consistent with each other. We note that the calculated by Slack thermal conductivities for maximum strains 0.99-1.04 agree with experiment [12] for our present calculations for BAs (Figure 1) and as shown in Figure 3 BP results also agree with experiments [13,14] at temperature 300 K and above.

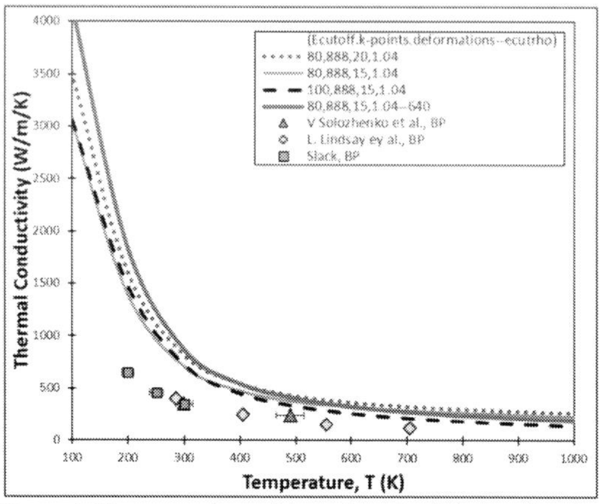

Figure 3. The comparison of thermal conductivity as a function of temperature with experimental values [13, 14] at various deformation numbers, cut-off energies, and k-points. The best, selected curve is the black-dashed plot at 15 deformations. The maximum strain has been fixed to 0.99-1.04.

The Slack model predicts the thermal conductivity of BAs in agreement to the experimental study by Bing et al. [12]. However, the most interesting fact is that the thermal conductivity presented by Lindsay et al. [14], and ours own simulations (indicated by dot dashed red line in Fig. 1) using BTE predicted the value an order of magnitude higher. Though our BTE results for BAs (Fig. 1) predict lower thermal conductivity than published later for a larger temperature range by the same authors (4288 $Wm^{-1}K^{-1}$ versus ~10000 $Wm^{-1}K^{-1}$ at 100 K) [15]. The calculated here thermal conductivity at 100 K for BP using Slack method is between 3000 $Wm^{-1}K^{-1}$ and 4000 $Wm^{-1}K^{-1}$ as shown in Fig. 3, while it was estimated previously [15] to be also much higher ~10000 $Wm^{-1}K^{-1}$. The BTE calculations for BAs were calculated using the same PAW-PBEsol pseudopotentials and optimal parameters as for Slack model as described above and the same PBEsol functional. In the previous BTE calculations, local density approximation (LDA) was used with norm conserved pseudopotentials and only 80 Ry cut-off energy and 6 × 6 × 6 k points mesh, which we found not sufficient for the used here PAW-PBEsol pseudopotentials. Our comprehensive investigation indicates that the predicted lattice thermal conductivity by both BTE and Slack model for some materials may be very sensitive to the tuning optimization of parameters used in DFT calculation, selected type of DFT functional and the respective pseudopotentials.

In Figure 4 we present calculated here phonon spectra for BP and BAs as indicated.

BTE results of a very high thermal conductivity of BAs, originated from the limited thermal resistance of three phonon scattering, umklapp (e.g. two acoustic phonon frequencies (a) must be equal to that of optical phonon: a+a=o) and which is largely restricted in BAs due to a large gap between acoustic (a) and optical (o) phonons remains to be confirmed experimentally. Acoustic phonons bunching also prevents umklapp involving three acoustic phonons (a+a =a) due to absence of high frequency acoustic phonons in BAs. As presented in Figure 4 the a-o gap is smaller in BP than in BAs, therefore its thermal conductivity is expected to go down with temperature according to Eq. 1 and as shown in Figure 3 and in agreement with experiment [13,14] and BTE calculation [15].

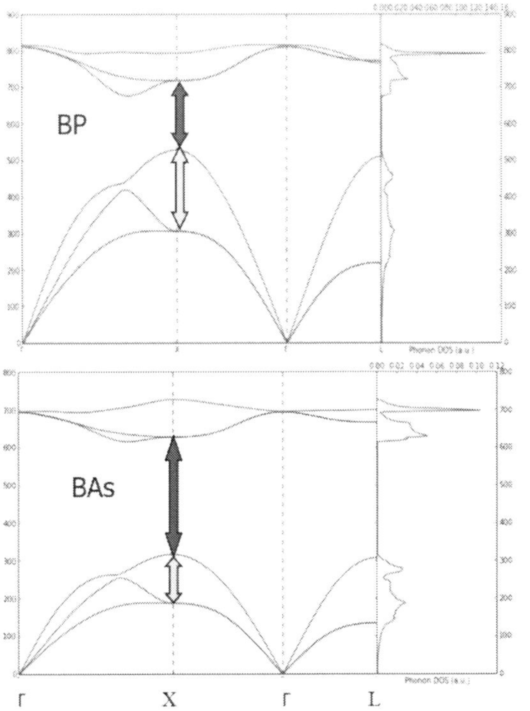

Figure 4. The phonon dispersion curves (frequency in cm^{-1}) of BP and BAs along high-symmetry direction path (Γ-X-Γ-L) and the corresponding phonons density of states in a.u. vs. the same frequency on the right panel.

3.1 Thermal conductivity correlation

Tables 1a,b show the results for SiC, both cubic and hexagonal, wurtzite BeO, cubic thoria, ceria, BP and BAs.

The presented in Table 1a parameters: a and b, are for the analytical expression (Eq. 1), and in the last column we present thermal conductivity calculated using Eq. 1 for 400 K. The presented properties in Table 1b were calculated at 400 K, except for the minimal thermal conductivity. The minimal thermal conductivity was evaluated using Eq. 2 and both Young's modulus and density were calculated at 400 K. Both SiC and BeO have about three times larger minimal thermal conductivity than ceria and toria and comparable to that of BP, while BAs has smaller minimal thermal conductivity as predicted by Eq. 2. Table 1b shows that in general the lattice origin thermal conductivity is larger for compounds with a larger Debye temperature (θ), larger Young's modulus (and related minimal thermal conductivity) and smaller thermal expansion coefficient (α). This observation may be used as a guideline in searching for higher thermal conductivity materials. In general the thermal conductivity is higher for compounds with lighter atoms (average atomic mass is listed in Table 1a, column two). However, the thermal conductivity of thoria and ceria, are almost the same, while the average mass of atom in thoria is significantly higher (88 u versus 57 u). Slack also noted the correlation between the lattice thermal conductivity value and the Debye temperature [5]. We used here generalised Slack model, where we include contribution to the thermal conductivity from both acoustic and optical phonons and all inputs to the model are calculated as described previously [4,10] and above. Slack model predicts large thermal conductivity for both BP and BAs and BP may be considered for a future application provided it has a proper isotopic contribution that is not absorbing neutrons.

Table 1a. The properties of selected compounds with the average atomic mass (in u) per unit cell listed in column two. The lattice thermal conductivity [$Wm^{-1}K^{-1}$], calculated at 400 K (column 5) using Eq. 1, and the respective a and b parameters (in $W^{-1}mK$ and $W^{-1}m$ units respectively) are shown in columns 3-4.

Compound	A_{avg}	$a\times 10^{12}$	$b\times 10^{6}$	k_L
SiC (cubic)	20	7.66	8.73	286
SiC (hex.)	20	9.55	9.69	258
BeO	25	6.84	31.9	221
ThO_2	88	0.93	142	18
CeO_2	57	2	211	12
BP cubic	21	10.4	5.59	447
BAs cubic	43	6.89	9.17	273

Table 1b. The properties of the same as in Table 1a selected compounds with Debye temperature (θ in K), thermal expansion coefficient (α, in K^{-1}), Young's modulus (Y in GPa) and the minimal thermal conductivity, calculated using Eq. 2 (k_{min} in $Wm^{-1}K^{-1}$), are listed in subsequent columns as indicated.

Compound	θ	$\alpha\times 10^{6}$	Y	k_{min}	k_L
SiC (cubic)	1179	3.24	456	3.01	286
SiC (hex.)	1155	3.33	440	2.96	258
BeO	1186	6.09	339	2.47	221
ThO_2	434	7.68	267	1.04	18
CeO_2	476	9.72	216	1.19	12
BP cubic	1086	4.22	379	2.64	447
BAs cubic	694	4.6	303	1.6	273

4 Temperature profiles, gradients and thermal stress

 DOI: 10.1051/matecconf/201713003001

The solution of the equation quantifying steady state heat conduction for cylindrical symmetry (radius of the fuel rod equal to 0.006 m is assumed) provides the temperature profiles (T) and temperature gradients (dT/dr) in an operating fuel rod as a function of radius (r). We investigate various nuclear fuels: thoria, MOX, and the composites with one order of magnitude higher thermal conductivity constituents like BeO and BP.

The temperature profiles for various composites in the shape of a cylindrical fuel rod were calculated using the thermal conductivity evaluated by Eq. 4 and by solving the steady state heat conduction equation for cylindrical symmetry:

$$\frac{1}{r}\frac{d}{dr}\left(r k_L^{BeO/ThO_2}\frac{dT}{dr}\right)+H=0 \qquad (4)$$

where "r" is the radial coordinate of a cylinder and "H" is the volumetric heat generation rate from fission as described previously [16]. Eq. 4 is subject to the boundary conditions: zero temperature gradient at the centerline of the pellet and the fixed temperature value at the surface of the pellet, assumed to be the same as used before [17]: 870 K.

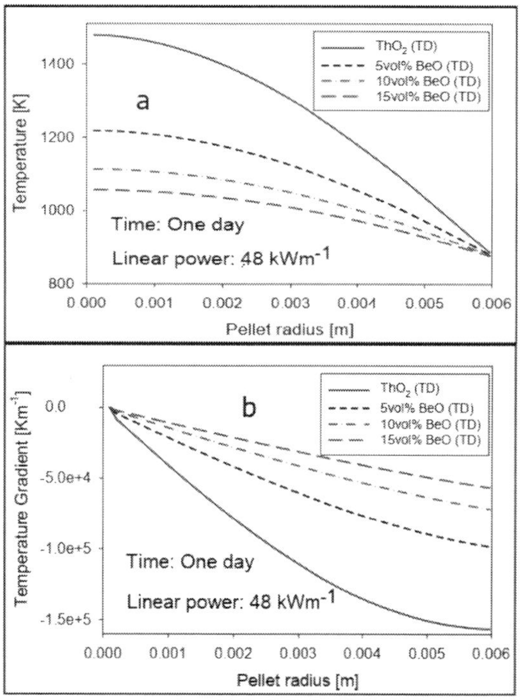

Figure 5 Predicted fuel temperature (a), temperature gradients (b) profiles for thoria and 5vol%, 10vol% and 15vol% BeO/ThO_2 composites, respectively. The a, b parameters used for inverse fit to thoria thermal conductivity were 9.25×10^{-13} and 1.42×10^{-4}, while for BeO: 6.84×10^{-12} and 3.19×10^{-5} (in $W^{-1}mK$ and $W^{-1}m$ units respectively) values were used (Table 1a).

We also assume that the used fuel enrichment, which leads to the same linear power (48 kWm^{-1}) does not affect thermal conductivity significantly and therefore can be neglected.

In Figs. 5a,b, we present the respective solutions by Maple code in a fuel pellet with thoria fuel and its composite with BeO operating for one day with linear power of 48 kWm^{-1}.

We assumed that the fuels were fully dense and the performance of pure thoria was compared with 5vol%, 10vol% and 15vol% BeO/ThO_2 composites. Replacement of 5 vol% BeO in thoria by SiC gives very similar, about 200 K centreline temperature reduction and similar reduction of temperature gradient [18,19].

The presented in Figure 6a centreline temperature reduction by replacing 5 vol% of MOX by BP predicts much higher reduction (1000 K) due to both lower thermal conductivity of MOX, which gives much higher centreline temperature for pure MOX (2200 K versus 1600 K) and predicted higher thermal conductivity of BP than that of BeO.

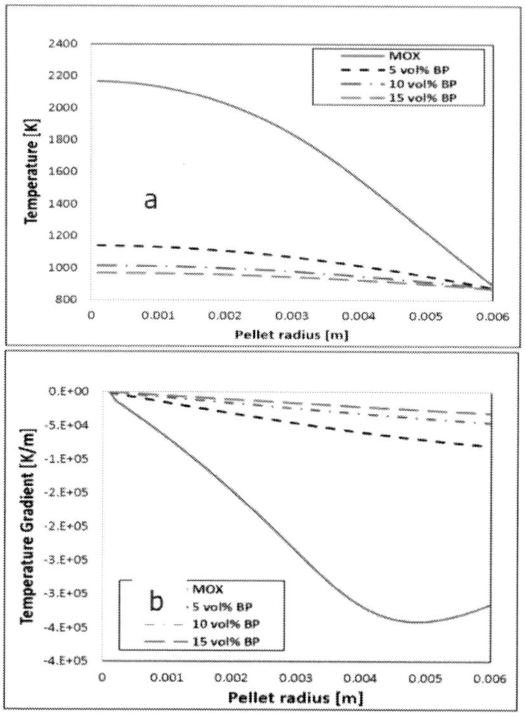

Figure 6 Predicted fuel temperature (a), temperature gradients (b) profiles for MOX [1] and 5vol%, 10vol% and 15vol% BP/MOX composites, respectively. The a and b parameters used for inverse fit to MOX thermal conductivity were 0.035 and 0.000286, while for BP: 10.4×10^{-12} and 5.59×10^{-6} (in $W^{-1}mK$ and $W^{-1}m$ units, respectively) values were used (Table 1a).

As shown in Figs. 5 and 6 the composite fuels using high thermal conductivity materials allow a significant reduction of the central fuel line temperature and the temperature gradient within the fuel. Therefore composite fuels would not only be safer, but would have higher longevity.

We have not included it here in our analysis but BN may be even better component of composite than BP since N is lighter than P and therefore has higher thermal conductivity [15].

6 Conclusion

The application of the discussed software enabled us to derive at the lattice thermal conductivity of various nuclear fuels using the thermo-mechanical properties obtained from first principles calculations. Slack model and the ShengBTE (BTE) a solver for phonon thermal conductivity predict similar thermal conductivities in agreement with the experiment except for cubic SiC where Slack model over-predicts thermal conductivity by a factor of two and BAs where BTE predicts order of magnitude higher thermal conductivity than Slack model and available currently experimental result. The studied here Borides (with the appropriate isotopic composition that excludes strong neutron absorber ^{10}B) like BAs and BP are predicted to have high thermal conductivity and BN may be even better component of composite since N is lighter than As and P.

Furthermore, we calculated temperature profiles and gradients using a simplified, exemplary scenario of an operating fuel rod.

Our studies also help to identify important factors that contribute to enhancing lattice thermal conductivity of nuclear fuel and its implications. The findings are helpful in developing accident tolerant nuclear fuels. The composite fuels using high thermal conductivity materials allow a significant reduction of the central fuel line temperature and the temperature gradient within the fuel. Therefore, composite fuels would not only be safer but also have higher longevity.

The presented here multidisciplinary simulation may be implemented for educational purposes to enhance reactor safety analysis.

5 Acknowledgements

The authors gratefully acknowledge the financial support of Canadian National Science and Engineering Research Council (NSERC) and the Canada Research Chair program. The authors acknowledge access to high performance supercomputers at Compute Canada: Calculquebec, Westgrid and Plato. The authors are also grateful to Ming-Hsien Lee for providing norm-conserving pseudopotentials for thorium and oxygen that were used in this work. We thank Dr. Jochym Pawel for his preliminary development of qe_nipy and initial calculations. We also thank Dr. Jesús Carrete Montaña for his help in doing the BTE calculations and the computing facilities provided by the ICT Department of the University of Saskatchewan.

References

1. J.J. Carbajo, G. Yoder, S.G. Popov, V.K. Ivanov, J. Nucl. Mater. **299**, 181 (2001)
2. B. Szpunar, L. Malakkal, E. Jossou, J.A. Szpunar, *Materials for Nuclear Energy, Proceedings of 2017 TMS Annual Meeting & Exhibition*, San Diego February 26 – March 2, 2017 (Springer, ISBN 978-3-319-51647-9), 367 (2017)
3. E. Jossou, L. Malakkal, B. Szpunar, D. Oladimeji, J.A. Szpunar, J. Nucl. Mater., **490**, 41 (2017).
4. B. Szpunar, J.A. Szpunar, Ki-Seob Sim, J. Phys. and Chem. Solids, **90**, 114 (2016)
5. G.A. Slack, Solid State Phys., **34**, 1 (1979)
6. L. Malakkal, B. Szpunar, J.A. Szpunar, *Materials for Nuclear Energy, Proceedings of 2017 TMS Annual Meeting & Exhibition*, San Diego February 26 – March 2, 2017 (Springer, ISBN 978-3-319-51647-9), 377 (2017)
7. P. Giannozzi, S. Baroni, N. Bonini, M. Calandra, R. Car, C. Cavazzoni, D. Ceresoli, G. L. Chiarotti, M. Cococcioni, I. Dabo, A. Dal Corso, S. de Gironcoli, S. Fabris, G. Fratesi, R. Gebauer, U. Gerstmann, C. Gougoussis, A. Kokalj, M. Lazzeri, L. Martin-Samos, N. Marzari, F. Mauri, R. Mazzarello, S. Paolini, A. Pasquarello, L. Paulatto, C. Sbraccia, S. Scandolo, G. Sclauzero, A. P. Seitsonen, A. Smogunov, P. Umari, and R. M. Wentzcovitch, J. Phys. Condens. Matter, **21**, 395502, (2009).
8. W. Li, J. Carrete, N.A. Katcho, and N. Mingo, Comput. Phys. Commun, **185**, 1747 (2014).
9. L. Malakkal, B. Szpunar, J. C. Zuniga, R. K. Siripurapu, and J. A. Szpunar, *Proceedings of the 3rd World Congress on Integrated Computational Materials Engineering (ICME 2015)*, (John Wiley & Sons, Inc., ISBN 978-1-119-13949-2), 155 (2015)
10. L. Malakkal, B. Szpunar, J. C. Zuniga, R. K. Siripurapu, and J. A. Szpunar, Int. J. Comput. Mater. Sci. Eng., **5**, 1650008, (2016)
11. D.R. Clarke, Surf. Coatings Technol. **163**, 67 (2003).
12. Bing L., Yucheng L., Xiqu W., Qian Z., Yongjie H., Allan J., David B., Gang C., Zhifeng R., and Ching-Wu C., Appl. Phys. Lett. **106**, 074105 (2015),
13. V.L. Solozhenko, O.O. Kurakevych,2 Y. Le Godec, A.V. Kurnosov, A.R. Oganov, J. App. Phys. **116**, 033501 (2014)
14. L. Lindsay, D.A. Broido, T.L. Reinecke, Phys. Rev. Let. **111**, 025901 (2013)
15. D.A. Broido, L. Lindsay, T.L. Reinecke, Phys. Rev. **B88**, 214303 (2013)
16. B. Szpunar, J.A. Szpunar, Phys. Int. **4**, 110, (2013
17. B.J. Lewis, B. Szpunar and F.C. Iglesias, *J. Nucl. Mater.*, **306**, 30 (2002)
18. Szpunar B., Malakkal L., Siripurapu R., J. Szpunar, J. Zuniga, *Proceedings of 3rd International Conference on Manufacturing Engineering and Technology for Manufacturing Growth,* August 1-2, 2015, Vancouver, (IERI & PRESS. USA, ISBN 978-1-61275-074-3), 165 (2015)
19. B. Szpunar, L. Malakkal, E. Jossou, D. Oladimeji, J. Ranasinghe, I. Rossland and J.A. Szpunar, *13th International Conference on CANDU Fuel, Conference Proceedings*, Kingston, August 15-18 (2016)

Fabrication and property evaluation of WO_3 particles dispersed Al-based composite material

Naoto Nakamura[1], Seungwon Lee [2], Tomohiro Takida[3] Susumu Ikeno[4] and Kenji Matsuda[2]

[1]Graduate School of Science and Engineering for Education, University of Toyama, Japan
[2]Graduate School of Science and Engineering for Research, University of Toyama, Japan
Allied materials Cooperation Limited
[3]Professor of Emeritus, University of Toyama, Japan

Abstract. Advantage of composite material is to take the outstanding mechanical and functional properties of fibers and powders. However, it is not easy to disperse fine particles homogeneously in metallic matrix. 3-dimentional penetration casting (3DPC) method enables to disperse particles in the matrix homogeneously without segregation. The present study puts its focus on WO_3 particle. Several reports are available about photocatalytic property of WO_3 in the visible ray region. Photocatalytic is one of the most promising matter that can be used for resolving environment pollution because photocatalytic cause the oxidation reaction that change carbon contained in organic compound (for example acetaldehyde, toluene) into CO_2 by light. The aim of this work is to establish the fabrication method using 3DPC, to evaluate the mechanical, photocatalytic properties, and to observe the microstructure using scanning electron microscopy (SEM), transmission electron microscopy (TEM).

1 Introduction

Metal matrix composites (MMCs) have wide range of applications in industrial fields such as automotive and aerospace industry because MMCs have the advantage to take the outstanding mechanical properties and functional properties of fibers and particles. There are some fabrication methods of particle reinforced MMCs, for example powder metallurgy [1], stir casting [2], squeeze casting and spray forming [3]. It is known that fabricating MMCs by casting method is difficult to obtain without particle aggregation, a high-volume fraction of particle, fine particles below 5μm [4]. However, three-dimensional penetration casting method (3DPC) enables to resolve the problems as mentioned above. In our previous work, we fabricated the MMC using photocatalysis particles (TiO_2) and pure Al. Photocatalytisis has been attracted and a lot of researches have done because photocatalysis is expected to resolve this environment pollution problems due to ability of decomposition for organic pollution by light irradiation. TiO_2 is the most widely used as photocatalysis because of its application for environmental purification [5,6]. However, the band gap of TiO_2 about 3.2eV limits its application only in ultraviolet region [7]. To response in visible ray region, we focused on WO_3. WO_3 has a wider response in the solar spectrum because of its lower band gap of 2.8eV [8] compared to TiO_2, and it is expected as a visible light active photocatalyst [9].

In this research, Al-based WO_3 particles dispersed composite material was fabricated using 3DPC process and micro-structure observation was performed using scanning electron microscopy (SEM) and transmission electron microscopy (TEM). Mechanical and functional properties were estimated by means of hardness measurement and photocatalytic property test.

[a] Corresponding author: matsuda@eng.u-toyama.ac.jp

2 Experimental procedure

WO_3 powders (A.L.M.T Corp) with average size of 5μm~50μm, and 99.99% purity aluminium were employed. High WO_3 volume fraction of Al-based composite material was fabricated by using 3-dimentional penetration casting (3DPC) process. WO_3 powders were packed and put on the steel mould. Molten Al heated up to 1053K was poured into steel mould, and pressed from a top. After cooling, the composite material was removed from the steel mould. Volume fraction of WO_3/Al composite material was measured by Archimedes method. The microstructure observation was conducted by SEM, (Hitachi S3500H) equipped with an energy dispersive spectroscope (EDS). To investigate the reaction products at interface between WO_3 and Al with high magnification, TEM (JEOL 4010T operated at 400KV) was used. TEM sample was prepared by focus ion beam (FIB). The mechanical property was evaluated by Vickers micro-hardness test. Vickers micro-hardness test was carried out on polished sample using Mitutoyo HM-101 with a load of 100g and holding time of 15s. Photocatalytic property was evaluated photocatalytic property evaluation checker PCC-1 (ULVAC PIKO Co. Ltd.) that measured the change amount of ultraviolet radiation absorbance of methylene blue aqueous solution on the surface of composite material with ultraviolet irradiation. 20×10^{-3} mol/L methylene blue aqueous solution was prepared by dissolving methylene blue trihydrate in distilled water. Samples were dipped in methylene blue aqueous solution for 1h and drying for 10h. To improve the photocatalytic property by increasing WO_3 ratio at surface area, the electro-polishing was performed for WO_3/Al composite material at 253K with a solution of 10% perchloric acid and 90% ethanol.

3.Result and discussion

3.1. Microstructure observation

Figure 1 (a) SEM image of powder, (b) cross-sectional image of WO_3/Al composite material and SEM image of (c) middle area and (d) bottom area of WO_3/Al composite material.

Figure 1 (b) shows the cross-sectional image of WO_3/Al composite material and SEM images of (c) middle area and (d) bottom area. The dark and white contrast regions correspond to Al matrix and WO_3 particles, respectively. It is noted that there are no clack, sink and unpenetrated part of Al. WO_3 particles were dispersed homogeneously in Al matrix after casting. It is measured that the volume fraction of WO_3 particles was 66% by Archimedes' principle. WO_3 particles size of composite material maintain the initial particle size, Fig.1(c), (d). This result shows that WO_3 particles were dispersed into Al matrix without aggregation by 3DPC process.

3.2 Mechanical property

The results of Vickers micro-hardness measure for pure Al and WO_3/Al composite material are 23HV and 101HV, respectively. Hardness level of WO_3/Al composite material is about 4 times higher than pure Al. Figure 2 shows the indentations after hardness test of WO_3 area and Al area. The hardness of WO_3 and Al were 163HV and 48HV in Fig.2, respectively. The hardness level of WO_3 is much lower than that of WO_3 sintered sheet [10] and it was caused by aggregation of WO_3 particles. Above 50 μm particles was observed in Fig.1 (a) because of aggregation, and it results in reduction the hardness of WO_3/Al composite material. It should be noted that the hardness of Al area was twice higher than pure Al. That was explained by results of an EDS analysis. Mainly WO_3 area and Al area were separated clearly, however fine WO_3 particles were detected in Al area, which are marked by white arrows in Fig.3 and it is possible to enhance hardness of Al matrix.

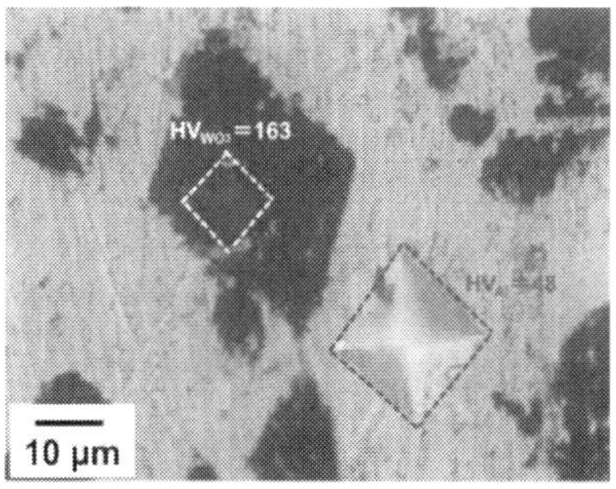

Figure 2. Indentations of Vickers micro-hardness test for WO_3 and Al area.

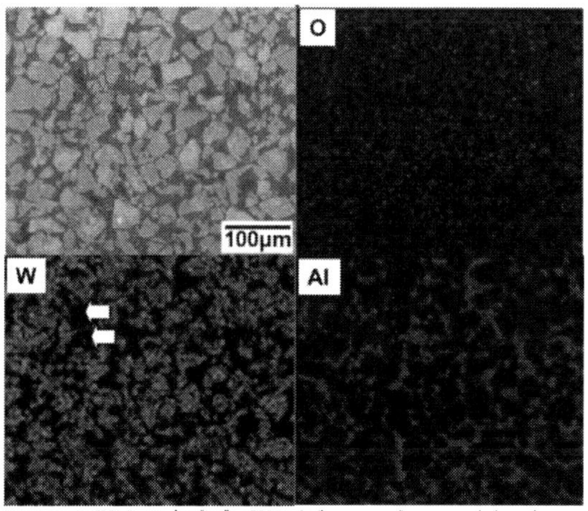

Figure 3. EDS analysis for WO_3/Al composite material and white arrows exhibit fine WO_3 particles detected in Al area.

3.3 Photocatalytic property

Figure 4 shows the result of photocatalytic property evaluation. The lower ΔABS (the varied amount of absorbance) is, the higher photocatalytic property exhibits. Lowering of ΔABS was also observed in WO_3/Al composite material. However, photocatalytic property level of WO_3/Al composite material was inferior WO_3. It can be considered that the effect of reaction products between WO_3 and Al that were caused during casting process. To clarify the existence of reaction products at interface between WO_3 and Al, TEM observation and an EDS analysis were conducted to investigate the reaction products at interface between WO_3 and Al. Figure 5 shows the TEM image of interface between WO_3 and Al and the results of EDS analysis that was carried out on WO_3 and Al area close to interface. The region dark and white contrasts correspond to WO_3

DOI: 10.1051/matecconf/201713003002

matrix and Al particles, respectively. As Figure 6 indicated, the interface between WO_3 and Al was relatively clear. In EDS analysis, W, O and Cu peaks were arisen in area 1 (WO_3 area) and Al, O and Cu peaks were detected area 2 (Al area) in Fig.6 both Cu peaks came from Cu mesh. Based on these results, it is considered that there were no reaction products between Al and WO_3.

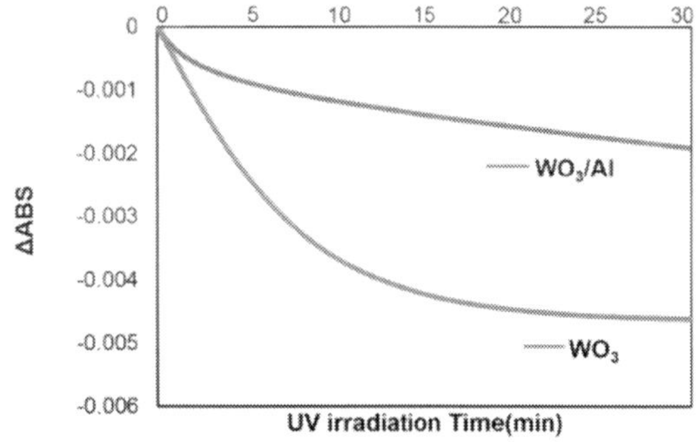

Figure 4. Photocatalytic property evaluation for WO_3 and WO_3/Al composite material.

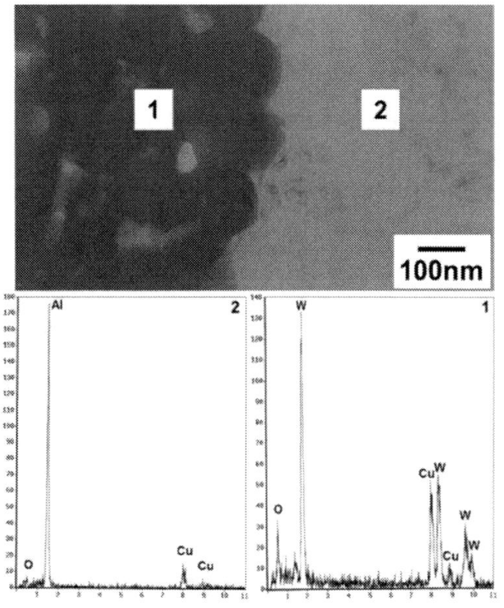

Figure 5. TEM observation at interface between WO_3 and Al and EDS analysis in WO_3 area and Al area close to interface.

To improve the photocatalytic property level, the electro-polishing was conducted. It enables to cause preference etching of Al and to expose more WO_3 particles on surface. Figure 7 shows microstructure changing by polishing method from (a) mechanical polishing to (b) electro-polishing. Green contrast and red contrast correspond to Al area and WO_3 area, respectively. WO_3 ratio on the surface in mechanical polishing and electro-polishing were 66% and 91%, respectively. In comparison with mechanical polishing, WO_3 ratio on the surface in electro-polishing indicated about 30% higher. Photocatalytic property evaluation of WO_3/Al composite material after electro-polishing appears in Figure 8. ΔABS of electro-polished sample became better than mechanical polishing. It is interpreted from this result that electro-polishing enhances photocatalytic property of WO_3/Al composite material by increasing WO_3 ratio at surface area.

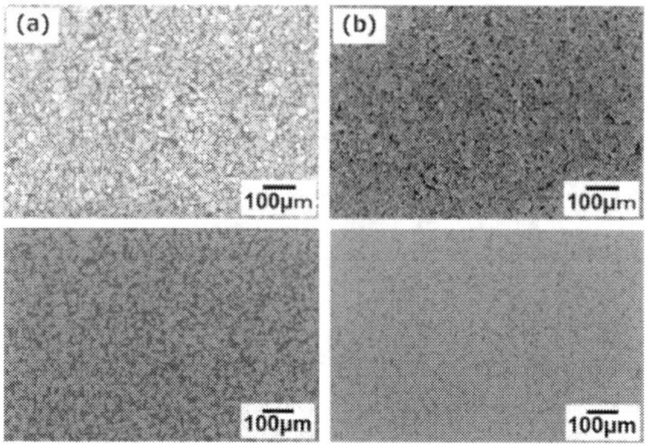

Figure 6. SEM images of WO_3/Al composite material changing polishing method from (a) mechanical polishing to (b) electro-polishing. Green and red contrasts correspond to WO_3 and Al area.

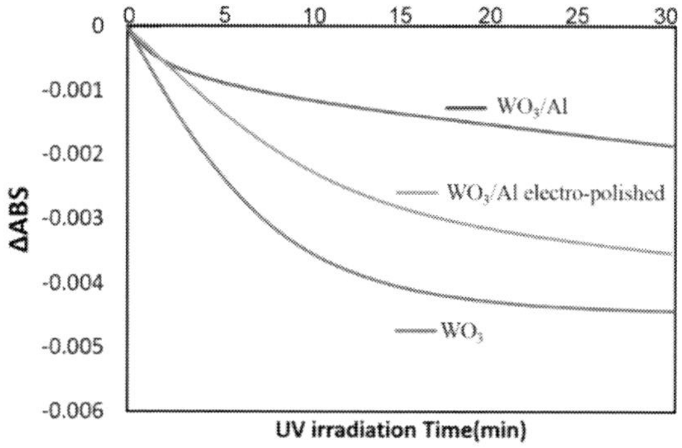

Figure 7. Photocatalytic property evaluation for WO_3, WO_3/Al composite material and after electro-polishing.

4 Conclusions

(1) WO_3 particles dispersed Al-based composite material was successfully fabricated by 3DPC process. In results of micro-structure observation, it was found that there is no unpenetrated part of Al, and WO_3 particles were dispersed into Al matrix homogeneously. The difference of distribution of WO_3 particles on places was not confirmed and it results in distribution of WO_3 particles without segregation.

(2) The hardness of composite material shows four times higher than pure Al. The hardness of Al area in composite material was twice higher than pure Al because fine WO_3 particles penetrated Al matrix. Those particles enhanced hardness level of Al matrix.

(3) The photocatalytic property was confirmed in composite material. By electro-polishing, WO_3 particles were exposed from surface because of preference etching of Al. electro-polished sample

showed better photocatalytic property than mechanical polishing.

References

1. D.M. Lee, B.K. Suh, B.G. Kim, J.S. Lee, C.H. Lee. Master Sci Technol, **13**, 590 (1997)
2. A. LuO, Metall Trans, **26,** 2445 (1995)
3. T. Ebert, F. Moll, K.U. Kainer, Powder Metal, **40**, 126 (1997)
4. Q.C. Jiang, X.L. Li, H.Y. Wang, Scr. Mater. **48**, 713 (2003)
5. X. Zhang, L. Zhang, J.S. Hua, C.L. Pan, C.M Hou, Appl Surf Sci **346**, 33 (2015)
6. H. Yin, X. Wang, L. Wang, Q. Nie, Y. Zhang, W. Wu, Master Res Bull, **72** 176 (2015)
7. S. Song, F.Y. Hong, Z.Q. He, Q.L. Cai, J.M. Chen, J Colloid Interf Sci **378** 159 (2012)
8. A. Srinivasan, M. Miyauchi, J Phys Chem C, **116** 15421 (2012)
9. S. Tao, X. Zhang, F. Qu, A. Umar, X. Wu, J Alloys Compd, **689** 570 (2016)
10. S. Dabbous, T.B. Nasrallah, A. Amlouk, K.Boubaker, M.Bouhafs, M.Amlouk, Sci Eng A, 527 862 (2010)

DOI: 10.1051/matecconf/201713003003

Production of Al-based composite materials including stress-luminescent particles using 3-dimensional penetration casting (3DPC)

Seungwon Lee[1], Yuta Kusunoki[2], Naoto Nakamura[2], Kenji Matsuda[1,a], Yuukou Horita[1], Takahisa Ohji[1], Kenji Amei[1], Keiji Shibata[1] and Koji Okino[1]

[1]Graduate School of Science and Engineering for Research, University of Toyama,3190, Gofuku, Toyama, Toyama, 930-8555, Japan
[2]Graduate School of Science and Engineering for Research, University of Toyama,3190, Gofuku, Toyama, Toyama, 930-8555, Japan

Abstract. A stress-luminescent powder can be applied as sensors because of luminous characteristic under stress. Key point in fabrication is how uniformly to disperse powders in metallic matrix. Three-dimensional penetration casting (3DPC) process is a good application to fabricate the composite materials using stress-luminescent powders and molten Al. A commercial stress-luminescent powder was prepared (a)stress- luminescent powders (Taikourozai Co.) and (b)EZ bright (EZ bright Cooperation). 3DPC process was conducted using a special mould has dimensions of 42mm height and 30mm diameter. Mould was heated up to 750℃ before 3DPC. After 3DPC, particles were well dispersed in Al matrix without crack generations with optical microscopy observation. Also, samples were emitted when they were pressed and scratched.

1 Introduction

A highly-developed material civilization needs more buildings and structures for comfortable living. However, it is getting difficult to maintain and protect buildings and structures from natural disasters, specially, earthquake due to their size and complexity of exterior. The real-time monitoring of stress using sensors can be solution. These composites are fabricated as a sensor for high buildings or structures in locations where safety inspectors hardly access. There are some applications of stress-luminescence powders by painting directly on concrete and pipe surface directly [1,2]. J.S. Kim *et al.* reported the use of stress-luminescence powders by spraying on alumina ceramics [3]. T. Nakanishi reported the fabrication of composite materials using both resin and stress-luminescence powder [4]. Stress-luminescent powers are difficult to use as itself because it is not easy to consolidate. So, we adopted the three-dimensional penetration casting (3DPC) process in Fig. 1 for fabricating composite. It can be expected that the Al composite of stress-luminescent can prevent failure of building and structure, or minimalize damages. Pure Al can be a good binder for composite material due to lightness, high ductility and low melting temperature compared to Fe, Cu, Ni.

2 Experimental procedures

Two kinds of commercial (a) stress-luminescence powders (Taikourozai Co.) and (b) EZ bright (EZ bright Cooperation) have respectively an average particle size of ~7 μm and 15μm as shown in Fig. 2(a), (b). Powders were prepared as preform and have shown emitting under black light in Fig. 2(c), (d). The 800°C molten Al prepared and poured into steel mould heated up to 750°C. Then, the mould covered graphite cover for pressing molten Al. Press machine (4 tons of load) introduced pressure through graphite cover, and molten Al penetrated the preform powders. Mould after 3DPC process cooled down into chilly water. To observe microstructure, SEM was used. To estimate the stress-luminescence, manual pressing machine was used.

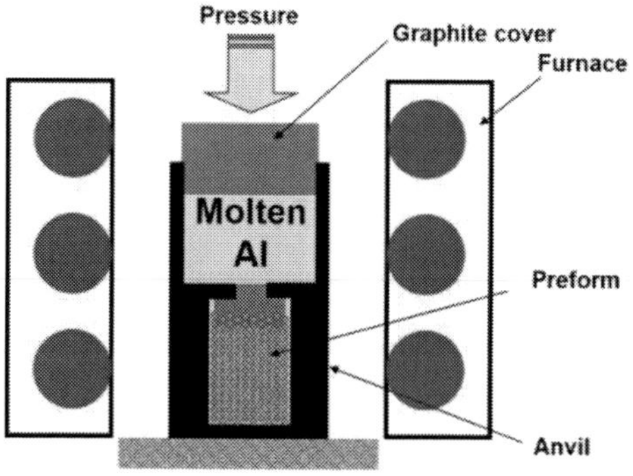

Figure 1. Schematic illustration of 3DPC facility.

[a] Corresponding author: matsuda@eng.u-toyama.ac.jp

Figure 2. SEM images of (a) stress-luminescence powders (Taikourozai Co.), (b) EZ bright (EZ bright Cooperation), and (c), (d) preforms of powders, respectively.

3 Results and discussion

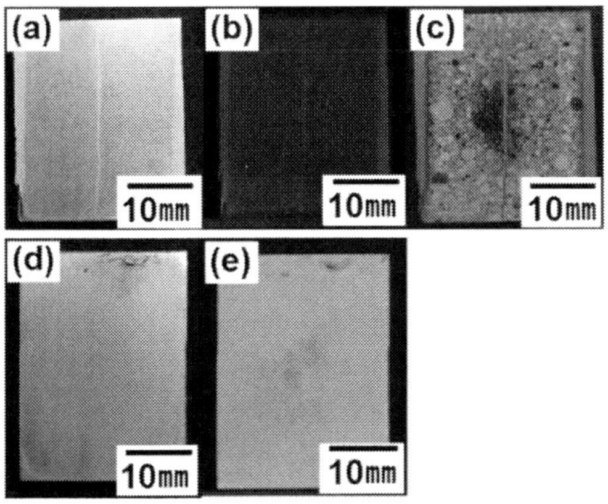

Figure 3. Cross sectional images of composite materials containing stress-luminescence powders (Taikourozai Co.) (a~c); (a) under visible ray, (b) mechanical polished sample under black light, (c) electro-polished sample under black light; composite materials containing EZ bright (EZ bright Cooperation) (d,e); (d) under visible ray and (e) mechanical polished sample under black light.

Figure 3(a) shows cross sectional image of composite material fabricated using 3DPC process. White contrasts, both side and centre are Al in Fig 3(a~c). With mechanical polishing, faint light emitted under black light. However, composite materials emitted strong light due to exposing more photo luminescence powders after electro polishing in Fig. 3(c). In the case of composite material with EZ bright powder, it emitted light under black light without electro polishing in Fig. 3(d, e). Figure 4 shows SEM-EDS analysis results. Powder size of both composite materisls after 3DPC process does not changed significantely. With EDS results, in Fig. 4 (a), smaller powder are well dispersed all over the composite materials, however, in the case of EZ bright, the boundaries between powder and Al matrx are clearly distinguished in Fig. 4(b).

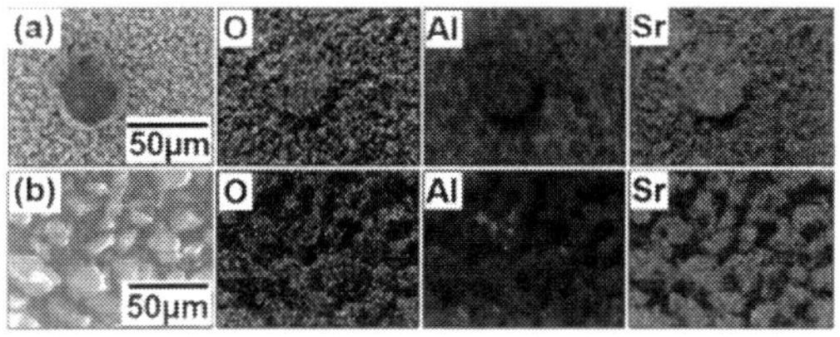

Figure 4. EDS analysis of both composite materials, (a) stress-luminescence powders (Taikourozai Co.), (b) EZ bright (EZ bright Cooperation).

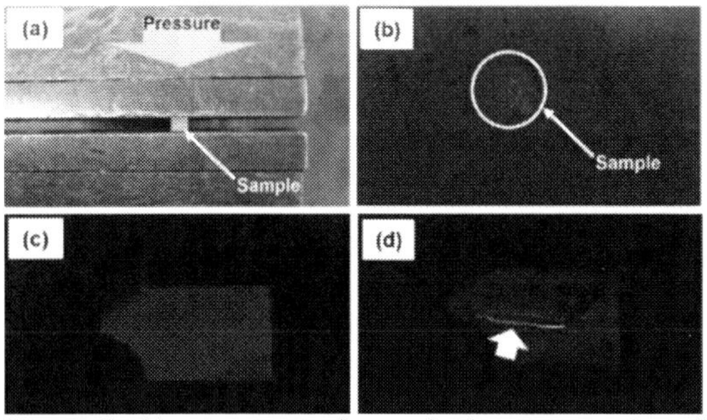

Figure 5. (a) appearance of stress-luminescence testing of stress-luminescence powders (Taikourozai Co.) using manual press machine, (b) emitting composite material under stress, (c) composite material with EZ bright (EZ bright Cooperation) under black light and (d) scratch test of (c).

Figure 5(a), (b) show stress-luminescence testing and result. 1 cubic centimetre (1×1×1cm) sample was set between vice in manual pressing machine in Fig. 5(a). Then, compression stress applied to the sample, and sample started to emit under compression stress. Also, scratch test was conducted in Fig. 5 (c), (d). Before scratch testing, sample was exposed under black light for 5 min in Fig. 5(c), then, scratch test conducted without black light condition. It was possible to observe the emitting along scratch path marked with white arrow in Fig. 5 (d).

4. Summary

Al composites with stress-photo luminescent powders was fabricated by using 3DPC. After casting processing, powders maintained its stress luminescence property. That was confirmed using compression and scratch tests.

References

1. D. Ono, C.N. Xu, C. Li, N. Bu, *J. JSEM* **10** (2010) (written in Japanese)
2. Y. Kawabata, D. Ono, C. Li, E. Kawasaki, M. Iwanami, E. Kato, N. Ueno, C.N. Xu, *J. JSEM* **67** (2011) (written in Japanese)
3. J.S. Kim, Y.N. Kwon, N. Shin, K.S. Shon, *Acta Mater* **53** (2005)

 DOI: 10.1051/matecconf/201713003003

4. T. Nakanishi, *J. Ceram. Soc. Japan* **123** (2015)

Synthesis and Evaluation of Zeolite Surface-Modified Perlite

Makoto Kasai[1, 2, a], Yosei Kobayashi[1], Masakazu Togo[2] and Atsushi Nakahira[2, 3]

[1] *Perlite Division, Mitsui Mining & Smelting Co., Ltd., Japan*

[2] *Department of Materials Science Biomaterial Group, Osaka Prefecture University, Japan*

[3] *Trans-Regional Corporation Center for Industrial Materials Research, IMR, Tohoku University, Japan*

Abstract. Perlite is volcanic glass mainly composed of amorphous aluminum silicate, mainly composed SiO_2 and Al_2O_3 with less impurities such as heavy metals. Amorphous (glassy) perlite is used in lightweight aggregate and insulation. In addition, it has also been used as a filter aid by grinding the expanded perlite. However, it has not been used as environmental cleanup materials, because the ion exchange capacity of the perlite is very low. In this study, we tried to synthesize the hybrid filter aid with chemical adsorption capacity by synthesizing the zeolite on the surface of the perlite. As a result, by using the hydrothermal synthesis method, zeolite surface modified perlite was synthesized in which the LTA type zeolites were generated on the surface of the perlite.

1 Background

Perlite is an aluminum silicate mainly composed of perlite ore. This perlite (expanded perlite ore) has porous structure. So it is a lightweight glass material with superior physical properties, such as an excellent thermal, flame, and chemical resistance, as well as thermal insulation. In addition, it is used as a heat-insulating material in building materials and large tanks, such as LNG tanks, as well as soil improvement material.

Perlite is produced by crushing the raw material and then subjecting it to heat processing. As volcanic glass, such as perlite ore, contains the water in the glass structure, the water steam generates on the heat-treatments [1], [2]. When the generation temperature of the gas is equal to or higher than the glass softening point, the glass is expanded with the generated gas. This means that by rapidly heating perlite ore using a rotary kiln, the volatile constituents (mainly water) of the raw material rapidly become volatile and expand when the glass softens.

Also, expanded perlite that has been crushed and classified can be used as a filter aid. Filter aids are additives used to increase the filtering performance for solid and liquid separation. Perlite is generally used in filter aid because this material is required to have a low bulk density, settleability, and must be incompressible with small cake specific resistance.

Separation of solids from liquids, on the other hand, can be accomplished using filtering equipment such as filter presses. At that time, a filter aid is added to improve the filtration efficiency. However, since the filter aid itself has no adsorption ability, it is necessary to add an adsorbent to remove contaminants such as heavy metals. Therefore, it was a problem to increase the amount of generated waste cake. In this regard, if adsorption capacity can be imparted to the perlite filter aid, it is possible to achieve excellent filtration efficiency without increasing the amount of waste cake produced. This has been considered a potential means to provide a hybrid purifying material.

Zeolite is a generic term of porous aluminosilicate, the basic unit is three-dimensional framework structures consisting of tetrahedrally coordinated SiO_4 and AlO_4 units. When a part of Si^{4+} is replaced with Al^{3+}, lack of positive charge occurs, cation exchange capability of the zeolite is obtained by supplementing the shortage cation such as an alkali metal[3].

Taking into consideration that perlite is a powder of aluminum silicate, it can be considered a raw material that can be used as a source of silicon ion and aluminium ion in zeolite synthesis. The focus of this research was to synthesize zeolite from perlite, especially to synthesize zeolite on the surface of perlite for filter aid. This latter material can then be used to synthesize a perlite filter aid with ion exchange and adsorption capacities to increase its value as an environmental purification material. In past studies, perlite ore and expanded perlite have been used as starting materials in an attempt to synthesize a variety of zeolites[4]-[8] However, there has not been any report made on using perlite as a raw material to produce compounds with zeolites, while sustaining its characteristics. In this work, a perlite filter aid was fabricated with added chemical characteristics, such as adsorption and ion exchange capacities, both of which are not features of conventional filter aids.

[a] Corresponding author: m_kasai@mitsui-kinzoku.co.jp

2 Methods

2.1 Specimen

The perlite filter aid samples used for the experiment were manufactured using perlite ore produced in Kitakata City of the Fukushima Prefecture. The perlite used was prepared by grinding and classifying the perlite ore, which was then heated with a rotary kiln, and lastly expanded and crushed to prepare the perlite filter aid. The derived perlite was analyzed for its chemical composition by using a fluorescent X-ray analysis system (XRF; model RIX-3000 manufactured by Rigaku Corporation). The particle diameter of the perlite used in this synthesis experiment was measured using laser diffraction and scattering particle diameter distribution measurement equipment (model MT3000 II manufactured by MicrotracBEL Corp.). Furthermore, the specific surface area of the perlite was measured using an automatic specific surface area measurement apparatus (model BELSORP-mini II manufactured by MicrotracBEL Corp.) and the BET method.

2.2 Synthesis of zeolite

For the filter aid, 0.70 g of perlite was collected in a 22 ml Teflon container. To adjust the Si / Al ratio to 1.0 and 2.0, a prescribed quantity of sodium aluminate (Wako Pure Chemical Industries, Ltd.) was added there. Afterwards, 15 ml of a 1.0 M NaOH solution was added to the container. The solution was mixed and aged at 25°C for 24 hours. An autoclave system was then used to perform the hydrothermal synthesis at 55°C for 72 h.

The hydrothermal synthesis was also performed using a fixed silicon to aluminum ratio of 1.0, a synthesis temperature of 55 °C, and different reaction times of 72 and 96 h. The derived product was filtered, then cleansed with distilled water, and lastly dried in an oven at 50 °C for 24 h. The dried specimen was then crushed in an agate mortar to prepare the powder specimens.

2.3 The heavy metal removal capability

The heavy metal removal capability tests were done for the zeolite surface modified perlite and zeolite reagent. Each sample of 0.2g was immersed for 1 to 12 hours to cadmium nitrate solution in each 20ml of 10 ppm. Then, the solutions were filtered and then the residual metal ion concentration were measured using the ICP emission spectral analysis (ICP; model PS7800 manufactured by Hitachi High-Tech Corporation) to evaluate the ion removing capacity.

2.4 Characterization of the synthesized sample

The structural analysis of the samples obtained was carried out using the powder X-ray diffraction method (XRD; model Mini Flex II manufactured by Rigaku Corporation). The measurement range was set to fall between 2° to 60°, at a scanning speed of 4°/min, a voltage of 30 kV, a current of 15 mA, and a sampling width of 0.02°. Their structure and microstructures of obtained synthesis products were observed with a scanning electron microscope (SEM; model VE-8800 manufactured by Keyence).

Furthermore, the vapor adsorption isotherm of the derived specimens were measured using a high precision, fully automated, gas adsorption apparatus (model BELSORP-18 manufactured by MicrotracBEL Corp.); and the BET specific surface area was measured using the BET multi-point method.

The structural analysis of the detailed fine structures of the silicon and aluminum in the derived specimens was performed using X-ray adsorption fine structure analysis (XAFS), using the beam lines from two different radiation facilities (BL6N1 at Aichi Synchrotron Radiation Center and BL-10 at the SR Center of Ritsumeikan University). In both instances, the measurements were performed using the fluorometric method, with the measurements for the K-edge adsorption of silicon performed using BL6N1 at Aichi Synchrotron Radiation Center; and the measurements for the K-edge adsorption of aluminum performed with BL-10 at the SR Center of Ritsumeikan University.

The structural analysis of the detailed fine structures of the cadmium in the heavy metal removal test samples were performed using X-ray adsorption fine structure analysis (XAFS), using BL14B2 at Spring-8. The measurements were performed using the fluorometric method, with the measurements for the K-edge adsorption of cadmium performed using BL14B2 at Spring-8.

3 Results

3.1. Physical properties of perlite

Table 1 shows the chemical composition of perlite. The perlite consisted of 75 wt% SiO_2, 14 wt% Al_2O_3 and, from the table, it is evident that SiO_2 and Al_2O_3 comprised approximately 90% of the perlite. Furthermore, while Na_2O, K_2O, and CaO were also present, their contents were low. Similarly, the contents of other metal elements were also very small.

Table 1. Composition of perlite

Chemical composition [wt%]					
SiO_2	Al_2O_3	Fe_2O_3	Na_2O	K_2O	CaO
75.1	14.3	0.9	3.5	5.2	0.3

DOI: 10.1051/matecconf/201713003004

Fig.1 shows the particle diameter distribution of the perlite used in the test. The median diameter (D50) of the perlite used in the synthesis was 30.1 μ m, while its specific surface area (by BET method) was 3.2 $m^2 \cdot g^{-1}$.

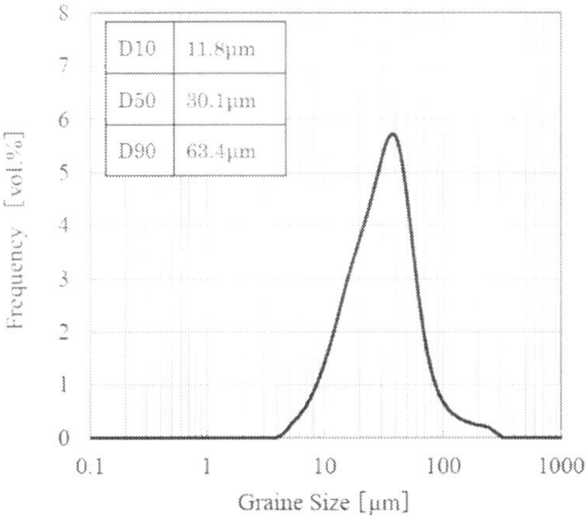

Fig. 1 Grain size distribution of perlite.

Furthermore, the XRD measurement results for perlite are shown in Fig. 2. The derived spectrum contained a halo peak, which confirmed that the perlite had an amorphous structure.

3.2. Synthesis of zeolite using perlite

Table 2 shows the different amounts of compounds used to prepare the various systems.

Table 2. Quantity of starting materials

Perlite [g]	NaOH		Chemical Composition ratio				Si/Al	Na/Si
	[mol/L]	[ml]	SiO_2	Al_2O_3	Na_2O	H_2O		
0.7	1.0	15	1.0	0.26	1.12	95.32	2.0	2.2
0.7	1.0	15	1.0	0.53	1.51	95.32	1.0	3.0

The silicon to aluminum ratio was controlled to be 1.0 to 2.0 using sodium aluminate.

Fig.2 shows the XRD results of the specimen obtained from the hydrothermal synthesis performed at 55 °C. Fig.3 shows the SEM images. The XRD results show that peaks attributed to LTA type zeolite. Furthermore, as the silicon to aluminum ratio became lower, the peak intensity attributed to the LTA zeolite became stronger. The broad peak derived from perlite remained between 15° to 40° degrees, indicating that the perlite had not been entirely converted into zeolite.

Results of the SEM observations confirmed that for samples between silicon to aluminum ratios ranging from 1.0 to 2.0, the squamation particles of the perlite filter aid remained. Furthermore, precipitation of the three-dimensional LTA type zeolite was confirmed on the surface and edges of the perlite filter aid. The amount of LTA type zeolite that precipitates increased as the silicon to aluminum ratio became lower, while the particle diameters of the precipitated LTA type zeolite were observed to become microscopic, decreasing from about 2 μ m to 0.5 μ m. Furthermore, as the silicon to aluminum ratio became lower, the LTA type zeolite particles were observed to assume organized three-dimensional shapes.

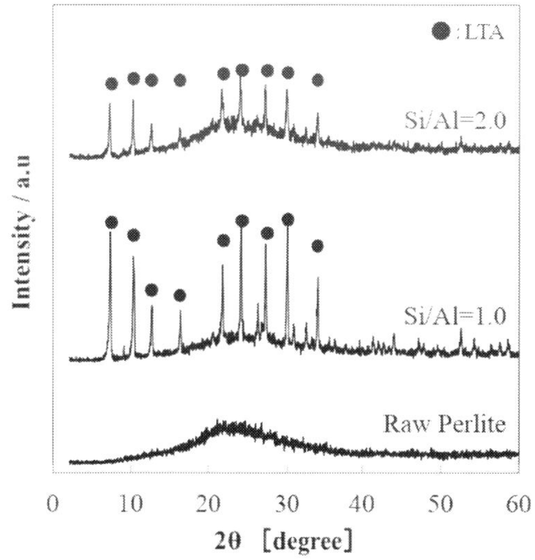

Fig.2 XRD patterns of the products obtained from the hydrothermal synthesis carried out at 55℃.

Fig.3 SEM images of the products obtained from the hydrothermal synthesis carried out 55℃, and at different Si/Al ratios (a: 1.0; b: 2.0; c: raw perlite).

Fig.4 shows the XRD results of specimens prepared at reactions times of 72 and 96 h, conducted with a silicon to aluminum ratio of 1.0 and at 55 °C. Fig.5 shows their SEM images for the products. The XRD measurement results revealed that the intensity of peaks attributed to LTA type zeolite became stronger as the reaction time increased, and reached about one thrid peak intensity of the LTA type zeolite reagent (zeolite A-4 manufactured by Tosoh Corporation) when the reaction time was 96 h.

DOI: 10.1051/matecconf/201713003004

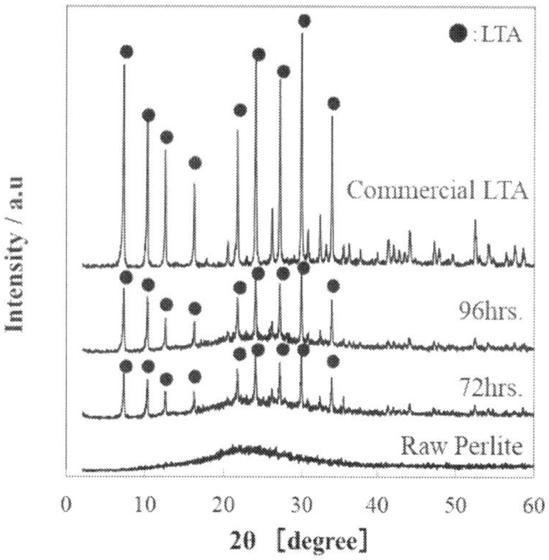

Fig.4 XRD patterns of the products from the hydrothermal synthesis carried out at 75 °C for 72 and 96 h.

A verification of the SEM images confirms that the number of LTA type zeolites generated on the surface of perlite increased as the synthesis time increased. The diameters of the generated LTA type zeolite particles were about 1 to 2 μ m, even when the synthesis time was increased. For these samples, the generation of LTA type zeolite was minute, even when compared with the diameters of LTA type zeolite reagent particles (about 5 μ m).

Fig.5 SEM images of the products obtained by carrying out the hydrothermal synthesis at different reaction times of (a) 72 h, (b) 96 h, (c) LTA zeolite reagent and (d) raw perlite.

Table 3 shows the BET specific surface area. It was calculated using the BET multi-point method on the vapor adsorption isothermal line. The BET specific surface area of the perlite filter aid, which was used as the raw material, was extremely small with a value of 3.2 $m^2 \cdot g^{-1}$. However, this value increased to 194 $m^2 \cdot g^{-1}$ after performing the hydrothermal synthesis for 72 h. When the hydrothermal synthesis time increased to 96 h, the specific surface area increased to 246 m2 · g-1, with the latter specific surface area reaching about one third the value of the specific surface area of the LTA type zeolite reagent. A correlation was observed between this increased specific surface area and the amount of precipitates on the LTA type zeolite generated on the surface of the perlite filter aid. In addition, the ratio of the BET specific surface area and the peak intensity ratio from the XRD results were in accordance with each other.

Table 3. Specific surface area of commercial LTA, perlite, and the products of the hydrothermal synthesis.

	Specific surface arc / $m^2 \cdot g^{-1}$
Commercial LTA	624
Perlite	3.2
Perlite – Zeolite LTA (72hrs.)	194
Perlite – Zeolite LTA (96hrs.)	246

3.3 Cd ion removal capability

Fig.6 shows the cadmium ion removal capability. It was confirmed that perlite has no cadmium ion removing ability. On the other hand, it was confirmed that zeolite surface modified perlite (perlite-zeolite) obtained by synthesizing on the surface of perlite has cadmium ion removing capability.

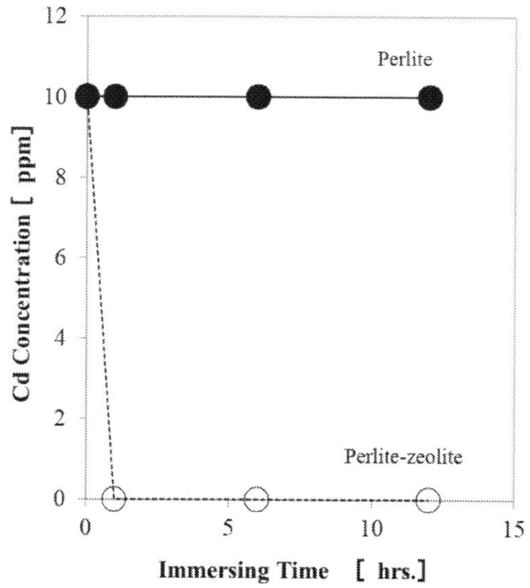

Fig.6 Relationship between Cd ion removal performance and immersing time of perlite and perlite-zeolite.

3.4 Characterization of the synthesized sample

Fig.7 shows the K-edge XANES adsorption spectra of silicon. They were obtained in order to further examine the conditions of silicon in the derived specimen. Peaks of perlite, zeolite surface modified perlite (perlite-zeolite), as well as the four coordinate structure of silicon dioxide in the vicinity of 1847 eV, along with the LTA type zeolite reagent, were derived from Fig.7. A peak shift to lower energies was observed as the zeolite was synthesized from perlite, and the peak position was located between the perlite and LTA type zeolite reagent. A shoulder of the LTA type zeolite reagent was observed between 1850 eV to 1853 eV, and a spectrum of perlite with LTA type zeolite on its surface was also observed. This spectrum had characteristics of perlite and the LTA type zeolite reagent, which suggest that the conditions of the electrons in the vicinity of the silicon atoms in perlite differed from the conditions of electrons in the LTA type zeolite. In addition, zeolite surface modified perlite had the electron conditions of the perlite and LTA type zeolite. Therefore, the spectrum had features of both the perlite and LTA type zeolite.

Fig.8 shows the K-edge XANES spectra of aluminum. These spectra were used to further examine details of the conditions of aluminum in the derived specimen. Peaks of perlite, zeolite surface modified perlite (perlite-zeolite), as well as the four-coordinated structure of aluminium oxide in the vicinity of 1565 eV. A peak shift towards higher energies was observed when the zeolite was synthesized from perlite, and the peak position at 1565 eV matched with the position of LTA type zeolite. Furthermore, the results showed a significant difference between perlite and zeolite with regards to the spectra between 1568 eV and 1580 eV. On the other hand, zeolite surface modified perlite showed a spectrum similar to that of the zeolite reagent. This indicates that the conditions of the electrons in the vicinity of the perlite aluminum atoms differ from the conditions of those in zeolite. Furthermore, since the zeolite surface modified perlite had electron conditions that were similar to those of LTA type zeolite, a prioritized structural change of aluminum in perlite is believed to have occurred when LTA type zeolite was generated from perlite to generate the LTA type zeolite.

Fig.9 shows the K-edge XANES spectra of cadmium. These spectra are the K-edge XANES spectra of cadmium ions adsorbed on zeolite surface modified perlite (Perlite-zeolite) and LTA type zeolite reagent. It was revealed that cadmium ions were adsorbed in divalent state to zeolite surface modified perlite (perlite-zeolite) and LTA type zeolite reagent. Also, in the spectrum between 26730 eV and 26750 eV, a difference was confirmed between the zeolite surface modified perlite and the LTA type zeolite reagent. It indicates the possibility that cadmium ions are adsorbed to the adsorption sites obtained on the surface of the perlite, which were treated during zeolite synthesis.

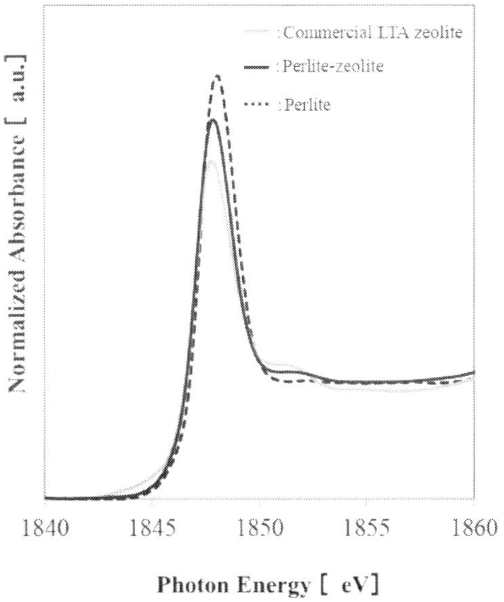

Fig.7 Si K-edge XANES spectra of perlite, perlite-zeolite, and LTA type zeolite reagent.

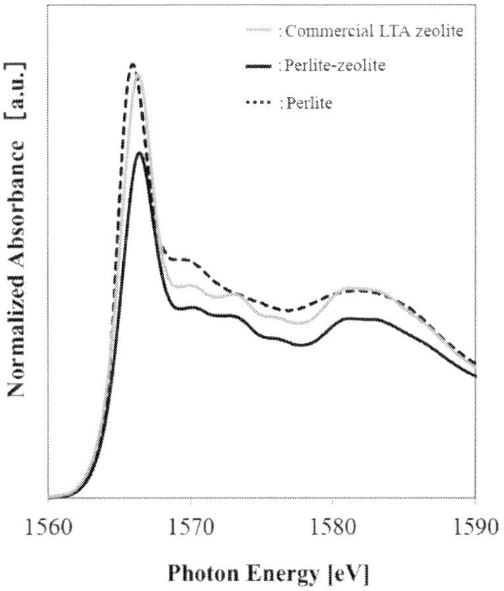

Fig.8 Al K-edge XANES spectra of perlite, perlite-zeolite, and LTA type zeolite reagent.

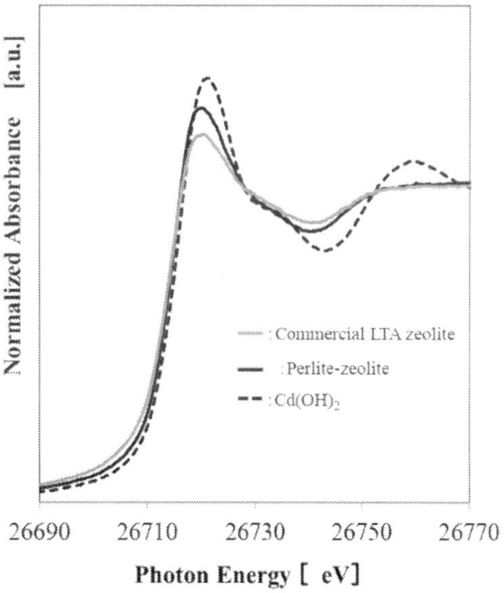

Fig.9 Cd K-edge XANES spectra of Cd absorbed perlite-zeolite, and Cd absorbed LTA type zeolite reagent.

4 Conclusions

The aim of this work was to synthesize zeolite on the surface of perlite filter to improve its functionality, and the following results were obtained.

I. LTA type zeolite, with high crystallinity precipitates on the surface of perlite filter aid, was obtained through a hydrothermal synthesis using silicon to aluminum ratios of 2.0, or lower, and a reaction temperature of 55 °C.

II. A correlation was observed between the amount of precipitated zeolite and the reaction time. The amount of precipitated LTA type zeolite increased as the synthesis time increased.

III. Based on the SEM observations, it became evident that LTA type zeolite was precipitating from the surface of perlite.

IV. The results from the XAFS measurements also confirmed a structural change from perlite to LTA type zeolite based on the differences in the conditions of the electrons in the vicinity of silicon and aluminum, indicating that a portion of the perlite was converted into zeolite.

V. The results from the XAFS measurements of cadmium indicated the possibility that cadmium ions are adsorbed to the adsorption sites obtained on the surface of the perlite.

We succeeded in synthesizing LTA type zeolite on the surface of perlite filter aid, and derived perlite filter aid with added chemical characteristics that were not possible in the past. We intend to continue this work for the future employment of this material in environmental purification applications.

5 Acknowledgments

This research was conducted with the assistance provided by the Adaptable and Seamless Technology Transfer Program Through Target-driven R&D (A-STEP) of the Japan Science and Technology Agency.

The light emission experiments of this research were conducted with the assistance of the "Photon Beam Platform Formation Project" of the Ministry of Education, Culture, Sports, Science and Technology, and were performed using the BL6N1 at the Aichi Synchrotron Radiation Center of the Aichi Science & Technology Foundation (Experiment No: 2015PA003).

Furthermore, assistance was also provided by the "Advanced Research and Foundation Joint Platform Formation Project" of the Ministry of Education, Culture, Sports, Science and Technology, and the BL10 at the SR Center of Ritsumeikan University (Experiment No: R1517).

The synchrotron radiation experiments were performed at the BL14B2 of SPring-8 with the approval of the Japan Synchrotron Radiation Research Institute (JASRI) (Proposal No. 2015A1992, 2015B1595, and 2016A1540).

6 References

1. N. Koie, Introduction to ceramic materials and technologies, Industrial technology institute, 815-817 (1979).
2. K. Kimura, K. Hamano, Yogyo-kyokai-Shi, **84**, 70-75 (1976)
3. N. Hara, H. Takahashi, Fundamentals and Applications of zeolite, Kodansha (1975)
4. S. Tangkawanit, K. Rangsriwatananon, Suranaree J. Sci. Technol, **12(1)**, 61-68 (2004)
5. P. Kongkachuichay and P. Lohsoontorn, Science Asia, **32**, 13-16 (2006)
6. A. Dyer, S. Tangkawanit, K. Rangsriwatananon, Microporous and Mesoporous Materials, **75**, 273-279 (2004),
7. A. Rujiwatra, Material Letters, **58**, 2012-2015 (2004),
8. G.E. Christidis, I. Paspaliaris, A. Kontopoulos, Applied Clay Science, **15**, 305-324 (1999)

 DOI: 10.1051/matecconf/201713003005

Electrochemical performance and stability of $Ni_{1-x}Co_x$-based cermet anode for direct methane-fuelled solid oxide fuel cells

Wongsawatgul Nicharee[1], Soamwadee Chaianansutcharit[2] and Kazunori Sato[3,a]

[1] Program of Petrochemistry, Chulalongkorn University, Thailand & Program of Materials Science, Nagaoka University of Technology, Japan
[2] Department of Chemistry, Faculty of Science, Chulalongkorn University, Bangkok, Thailand
[3] Department of Materials Science and Technology, Nagaoka University of Technology, Nagaoka, Niigata, Japan

Abstract. Carbon deposition on Ni-based anode is well-known as a major barrier for the practical use and commercialization of hydrocarbon-fuelled solid oxide fuel cells (SOFCs). In this work, Co alloying in Ni-YSZ was studied as an alternative anode material for using CH_4 as a fuel. The Ni-YSZ and Ni-Co alloyed-YSZ were prepared by the traditional impregnation method without further mixing processes. After sintering and reduction in H_2 atmosphere, the introduced Co can completely dissolved into the Ni lattice and changed the morphology with an increase in the Ni-YSZ grain size and showed a better uniform microstructure. The Co alloying also enhanced the electrochemical performance under CH_4 fuel by reducing the resistance and anodic overvoltage. Moreover, the Co addition enhanced the stability of the cell with CH_4 a constant load current of 80 mA for 60 h. This performance related to the carbon deposition on the anode surface. The Co alloying showed a high efficiency to suppress the carbon deposition and improved the electrochemical performance of an SOFC cell operating under CH_4 fuel.

1 Introduction

Solid oxide fuel cells (SOFCs) are one of the most effective devices to convert chemical energy to electricity. With its high operating temperature (normally >600 °C), SOFCs can run with both hydrogen and hydrocarbon fuels, such as natural gas and ethanol. However, unsolved problems of hydrogen storage have driven a lot of researches on alternative fuels for SOFCs [1-6]. Methane (CH_4) is the simplest hydrocarbon, which is one of the most interested alternative fuels for SOFCs due to its high hydrogen-to-carbon molar ratio and it has a small amount of N and S impurities. Moreover, it is not only the main component of gas and coal-bed gas but also the biogas [7-10]. CH_4 can be therefore considered as a renewable energy.

Ni-YSZ cermet has been known as an effective anode for SOFCs due to its high electrochemical catalytic activity for hydrogen. However, Ni-YSZ cermet showed a coarsening of Ni grains at high temperatures and carbon deposition when hydrocarbons are directly used as fuels. Carbon is easy to form and deposit on a Ni-YSZ anode due to a high catalytic activity of C-H bond braking in hydrocarbons, resulting in reduce fuel utilization but also deactivation of the anode [11]. An addition of water vapor to hydrocarbon fuels is an effective way to moderate the carbon deposition problem [12]. Unfortunately, high H_2O content will deactivate the cell and the system will be more complexed [13]. The replacement of Ni with electronic conductors is another way to avoid the carbon deposition. Cu was reported to avoid the coke formation in methane [14, 15]. However, the diffusion of Cu into YSZ electrolyte results in a YSZ degradation, and the low sintered temperature of Cu limits the operation temperature and decrease catalytic activity for methane conversion [16-18]. With a similar character to Ni, Co has been interested to dope in Ni anode [19]. Co alloying showed a retardation of Ni particle coarsening and enhanced the cell performance in H_2 [20]. Catalytic activity and stability of Ni-Co alloying was enhanced as compared to that of Ni in H_2 containing CO or H_2S as a fuel [21]. Co alloying also reduced the anodic polarization resistance and the amount of carbon deposition on Ni-GDC under a CH_4 fuel [22].

However, the precise mechanism of preventing carbon deposition by Co-alloying to Ni in the cermet anode is still unclear and the electrochemical performance stability has not been carefully studied. In this research, we have investigated the influence of Co alloying in conventional Ni/YSZ cermet in term of morphology and the electrochemical performance, especially the stability of the cell in dry methane.

2 Experimental

The $Ni_{1-x}Co_xO$-YSZ cermet (x=0 and 0.15) was prepared by the traditional impregnation method. This process was started with on appropriate amount of $Ni(NO_3)_2{\bullet}6H_2O$ (98% purity, Nacalai Tesque) and $Co(NO_3)_3{\bullet}9H_2O$ (98%

[a] Corresponding author: sato@analysis.nagaokaut.ac.jp

purity, Nacalai Tesque) were dissolved in distillated water, and then YSZ (8mol% Y_2O_3-ZrO_2, 99.9% purity, Tosoh Corp.) powder was proportionately added into the solution with the composition of 60%$Ni_{1-x}Co_xO$ + 40%YSZ (wt.%). After dehydration, the obtained powder was calcined at 800 °C for 5 h in air to decompose the nitrate acid. To obtain fine particle, the calcined powder was manually ground for 1 h in ethanol and then dried. The ground powders of $Ni_{1-x}Co_xO$-YSZ were mixed with glycerol to form a slurry and then painted as the anode on one side of the YSZ electrolyte disk (15 mm diameter and 0.3 mm thick, Nikkato) with an area of 0.28 cm^2. After painting, the electrode was fired at 1300 °C for 3 h. The other side of the YSZ disk was paint with $La_{0.8}Sr_{0.2}MnO_3$-YSZ composite cathode slurry. The cathode slurry was prepared from the powder of 70%$La_{0.8}Sr_{0.2}MnO_3$ (LSM, Seimi Chemical) + 30%YSZ (wt.%) mixed with glycerol. The cathode was fired at 1200 °C for 3 h in air. The thickness of each electrode was approximately 70 μm. The platinum paste, which used as reference electrode, was pasted as a small circle closed to working anode with a small gap (2-3 mm) on the anode side. The 0.3 mm diameter of platinum mesh (#100 mesh) was used as a current lead. A Pyrex® glass ring was used to seal between the single cells and the alumina tube. The cells were fired at 850 °C for 30 min and the anode was then reduced at 800 °C for 2 h in a H_2 atmosphere prior to the measurement of the single cell performance at 750 °C under 20 v/v%CH_4 in He as fuel at a total flow rate of 20 cm^3/min. The cathode was fed with oxygen to use as an oxidant with the same rate as CH_4. The single cell performance was evaluated by the current-voltage (*I-V*) characterization. Impedance spectra were measured with a frequency response analyzer (FRA5097). After the single cell performance test, the operating temperature was maintained at 750 °C at a constant current, at 80 mA, to investigate the stability for 60 h by evaluating of impedance at 3, 6, 12, 24, 36, 48 and 60 h.

The $Ni_{1-x}Co_xO$-YSZ powders were also fired and reduced under the same condition as the anode to analyze the phase by x-ray diffraction (XRD). Morphology and carbon deposition were analysed by scanning electron microscopy (SEM) and energy-dispersive x-ray (EDX) analysis, respectively.

3 Results and discussion

3.1. Phase identification

Fig. 1 shows the XRD pattern of NiO-YSZ and $Ni_{0.85}Co_{0.15}O$-YSZ after reduction. The reduced NiO-YSZ composite (Fig. 1(a)) exhibited the diffraction pattern of face-centered cubic (fcc) Ni metal, according to ICDD 01-070-1849, and the diffraction pattern of YSZ, according to ICDD 01-070-4431, without impurities. The 111 and 200 diffraction peaks from the fcc Ni metal in the reduced $Ni_{0.85}Co_{0.15}O$-YSZ sample (Fig. 1(b)) ~~was~~ slightly shifted towards the direction of low 2*θ* values. This result indicated that NiO and $Ni_{0.85}Co_{0.15}O$ were completely reduced, and the Co was dissolved into the Ni phase and form a Ni–Co alloy. This result agreed to the binary Ni-Co phase diagram [23]. The shifts observed in the 111 and 200 peaks showed that the lattice constant of Ni was increased by Co alloying. The increase in the lattice parameters (*a*) was due to the ionic radius of Ni (0.69 Å), which is larger than that of Co (0.74 Å) [24].

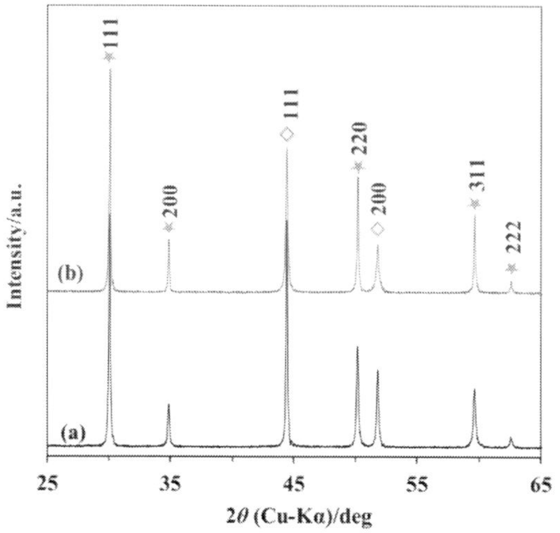

Figure 1. The XRD patterns of (a) NiO-YSZ and (b) $Ni_{0.85}Co_{0.15}O$-YSZ after reduction under an H_2 atmosphere at 800 °C for 2 h (★: $Ni_{1-x}Co_x$, ◇: YSZ)

3.2 Microstructure of anode

The cross-sectional microstructures of reduced $Ni_{1-x}Co_x$-YSZ cermet samples after sintering at 1300 °C for 3 h in air is shown in Fig. 2. Both anodes had a porous structure with well dispersed small metal particles. The $Ni_{0.85}Co_{0.15}$-YSZ showed a large grain size both in the metal alloy and YSZ compared to the Ni-YSZ. A coarsening of the YSZ particles was observed by Co alloying since Co_3O_4 can act as sintering aids [25].

3.3 Performance of SOFC

The *I-V* curves and power densities for Ni-YSZ and $Ni_{0.85}Co_{0.15}$-YSZ anodes at 750 °C in 20%v/vCH_4 in He are shown in Fig. 3. The $Ni_{0.85}Co_{0.15}$-YSZ anode showed a significantly higher performance than the Ni-YSZ. The maximum power densities of the single cells with the $Ni_{0.85}Co_{0.15}$-YSZ anode was 136 mW/cm^2, which was increased from 94 mW/cm^2 for the Ni-YSZ, approximately 44%.

The AC impedance spectra for the Ni-YSZ and $Ni_{0.85}Co_{0.15}$-YSZ anodes at 750 °C in CH_4 are compared in Fig. 4. The left intercept of the impedance spectra at high frequencies on the real axis (Re(*Z*)) indicated the ohmic resistance (R_o) which was mainly contributed by the ionic resistance of the electrolyte and electrodes-electrolyte connectivity, while the right intercept on the Re(*Z*) corresponded to the total resistance (R_{tot}). The subtraction of the ohmic resistance from the total resistance (R_{tot}-R_o) was the polarization resistance (R_p) which is mainly generated from a charge-transfer process (corresponding to the electrode reaction) and a mass-

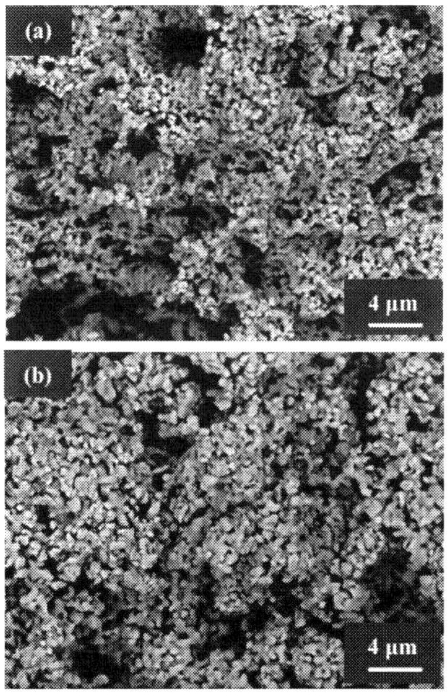

Figure 2. SEM images of (a) Ni-YSZ and (b) $Ni_{0.85}Co_{0.15}$-YSZ after reduction under an H_2 atmosphere at 800 °C.

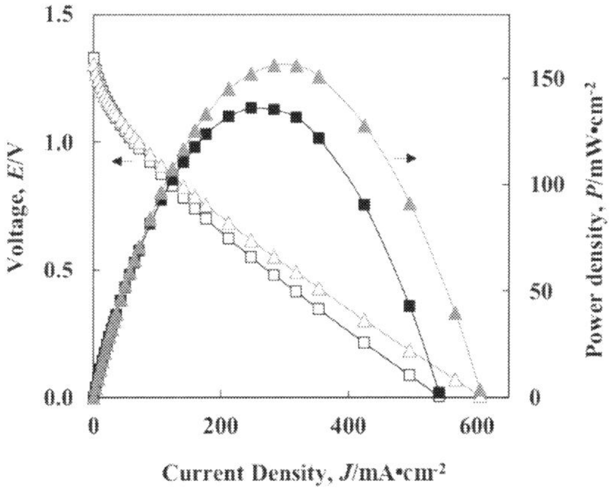

Figure 3. The current-voltage (*I-V*) curves (open symbols) and power densities (solid symbols) for (□,■) Ni-YSZ and (△,▲) $Ni_{0.85}Co_{0.15}$-YSZ anodes under an CH_4 atmosphere at 750 °C.

transfer process (corresponding to the gas diffusion, adsorption and dissociation on electrodes surface process) [26,27]. Since we used the same cathode and electrolyte, the difference in impedance arc was influenced by the anode resistance and its interface. The R_o and R_p decreased with the Co-alloying, as shown in Fig. 4, indicating that the alloying of Co in the Ni-YSZ can enhance the cell performance. The decrease in R_o by Co alloying is related to the microstructural change of the anode, as shown in Fig. 2. An improvement in the uniformity of grain connectivity most likely contributed to the decrease in the anode-electrolyte material resistance. R_p was also decreased by Co alloying, which was probably caused the formation of porous structure favorable for the diffusion of CH_4 molecules to the reaction site with the O^{2-} ions transported through the electrolyte. On the contrary, the Ni-YSZ showed the small grain size, which packed densely and provided a small porous structure resulting in a limited CH_4 diffusion.

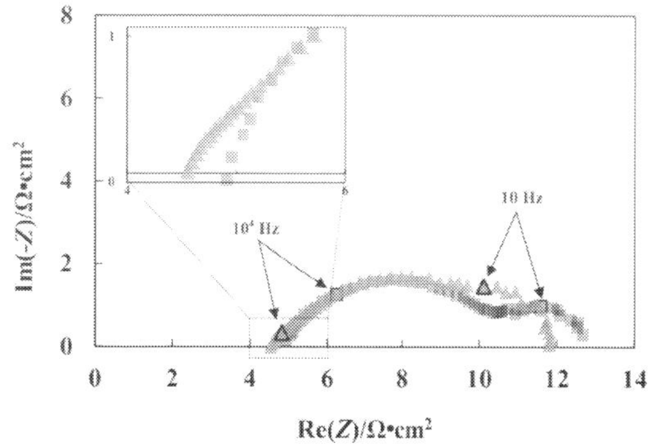

Figure 4. The impedance spectra for (■) Ni-YSZ and (▲) $Ni_{0.85}Co_{0.15}$-YSZ anodes under an CH_4 atmosphere at 750 °C.

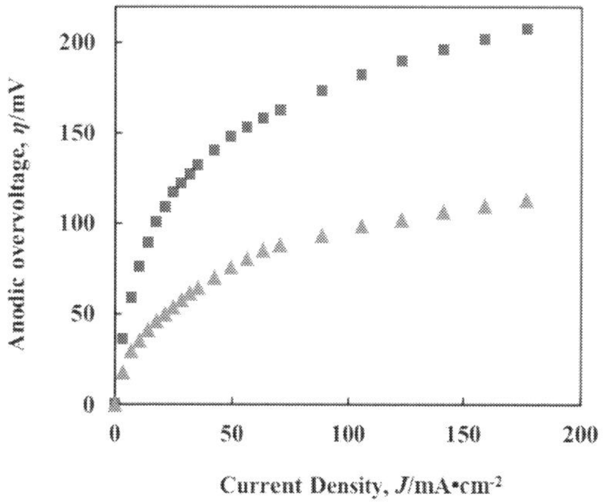

Figure 5. The anodic overvoltage for (■) Ni-YSZ and (▲) $Ni_{0.85}Co_{0.15}$-YSZ anodes under an CH_4 atmosphere at 750 °C.

Fig. 5 shows a comparison of the anodic overvoltage between the Ni-YSZ and $Ni_{0.85}Co_{0.15}$-YSZ. The anodic overvoltage for the $Ni_{0.85}Co_{0.15}$-YSZ was significantly lower than that for the Ni-YSZ, which corresponds to the behavior in the cell performance and R_p.

3.4 Stability of SOFC

To evaluate the long-term stability of the electrochemical performance, the impedance of the single cell for the Ni-YSZ and $Ni_{0.85}Co_{0.15}$-YSZ cells in the same CH_4 atmosphere were investigated at 750 °C for 60 h at a constant loaded current, 80 mA, for each cell. The ohmic resistance (R_o) and the polarization resistance (R_p) were

 DOI: 10.1051/matecconf/201713003005

plotted as a function of time, as exhibited in Fig. 6. $Ni_{0.85}Co_{0.15}$-YSZ showed an almost constant R_o and a slight increase in R_p, while the Ni-YSZ showed increases in both R_o and R_p. This result was caused by the carbon deposition onto the anode surface of Ni-YSZ, which limited the gas diffusion to the anode surface. Moreover, the carbon deposition at the triple-phase boundary (TPB) of the anode, where electrochemical reaction takes place, resulted in a decrease in the length of TPB. The surface morphologies and carbon deposition on the Ni-YSZ and $Ni_{0.85}Co_{0.15}$-YSZ anode were verified by SEM and a qualitative EDX analysis, respectively. The surface morphologies of the Ni-YSZ and the $Ni_{0.85}Co_{0.15}$-YSZ are illustrated in Figs. 7(a) and (b), while the EDX spectra of each anode were showed in the inlet of each micrograph. A significant amount of carbon was detected on the Ni-YSZ anode, whereas a very small amount of carbon was observed on the $Ni_{0.85}Co_{0.15}$-YSZ. This result corresponds to the previous report [22] that the rate of carbon deposition was decreased dramatically by increasing Co content in the Ni-GDC. This process was the most similar as Sn surface alloying in Ni [28] which can suppress the C-C bond formation and/or reduce the thermodynamic driving force for carbon nucleation on low-coordinated sites [29].

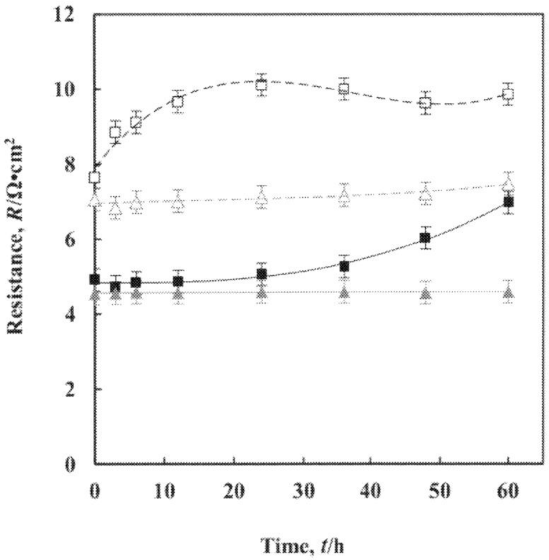

Figure 6. The ohmic resistance (solid symbols) and the polarization resistance (open symbols) as a function of time for (■,□) Ni-YSZ and (▲,△) $Ni_{0.85}Co_{0.15}$-YSZ anodes under a CH_4 atmosphere at 750 °C at a constant loaded current 80 mA.

4 Conclusion

The 15 mol% Co alloying in the Ni-YSZ anode has been investigated as an alternative SOFC anode material, which can be used for an operation by methane fuels. The introduced Co was completely alloyed in the Ni metal matrix phase. The alloying of Co in Ni affected the growth of Ni grains and reduced the electrical resistance and anodic overvoltage, which consequently improved the fuel cell performance of the Ni-YSZ cermet anode in CH_4 atmosphere. The doping of Co improved the stability of the electrochemical performance of the Ni-YSZ cermet anode in CH_4 fuel at 750 °C for 60 h, which was mainly caused by reducing the carbon deposition due to a suppression of the C-C bond formation and/or reducing the required thermodynamic driving force for the nucleation and growth of graphite.

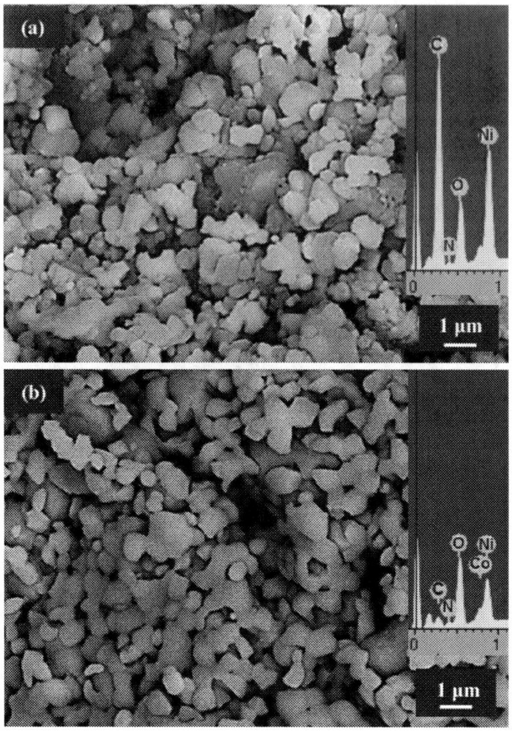

Figure 7. SEM surface micrographs and EDX spectra of Ni-YSZ and $Ni_{0.85}Co_{0.15}$-YSZ anodes after a stability test under an CH_4 atmosphere at 750 °C for 60 h.

References

1. A. Atkinson, S. Barnett, R. J. Gorte, J. T. S. Irvine, A. J. McEvoy, M. Mogensen, S. C. Singhal, J. Vohs, Nat. Mater., **3**, 17 (2004)
2. Z. Zhan, S. A. Barnett, Science, **308**, 844 (2005)
3. S. Park, J. M. Vohs, R. J. Gorte, Nature, **404**, 265 (2000)
4. J. Liu, S. A. Barnett, Solid State Ionics, **158**, 11 (2003).
5. E. N. Armstrong, J. Park, N. Q. Minh, Electrochem. Solid-State Lett., **15**, B75 (2012)
6. R. Muccillo, E. N. S. Muccillo, F. C. Fonseca, D. Z. de Florio, J. Electrochem. Soc., **155**, B232 (2008)
7. P. Kaparaju, I. Buendia, L. Ellegaard, I. Angelidakia, Bioresour. Technol., **99**, 4919 (2008)
8. C. H. Ting, D. J. Lee, Int. J. Hydrogen Energy, **32**, 677 (2007)
9. M. S. Fountoulakis, T. Manios, Bioresour. Technol., **100**, 3043 (2009)
10. N. Sanphoti, S. Towprayoon, P. Chaiprasert, A. Nopharatana, J. Environ. Manage., **81**, 27 (2006)
11. R. J. Gorte, J. M. Vohs, J. Catal., **216**, 477 (2003)

12. K. Sasaki, Y. Teraoka, J. Electrochem. Soc., **150**, A878 (2003)
13. D. Mogensen, J. D. Grunwaldt, P. V. Hendriksen, K. Dam-Johansen, J.U. Nielsen, J. Power Sources, **196(1)**, 25 (2010)
14. H. Kim, C. Lu, W. L. Worrell, J. M. Vohs, R. J. Gorte, J. Electrochem. Soc., **149**, A247 (2002)
15. A. Sin, E. Kopnin, Y. Dubitsky, A. Zaopo, A. S. Arico, D. L. Rosa, L. R. Gullo, V. Antonucci, J. Power Sources, **164**, 300 (2007)
16. A. Ringuedé, J.A. Labrincha, J.R. Frade, Solid State Ion., **141–142**, 549 (2001)
17. S. McIntosh, J. M. Vohs, R. J. Gorte, J. Electrochem. Soc., **150**, A470 (2003)
18. S. Jung, C. Lu, H. P. He, K. Ahn, R. J. Gorte, J. M. Vohs, J. Power Sources, **154**, 42 (2006)
19. N. Q. Minh, J. Am. Ceram. Soc., **76**, 563 (1993)
20. A. Ringuedé, D. Bronine, J.R. Frade, Electrochem. Acta, **48**, 437 (2002)
21. C.M. Grgicak, M.M. Pakulska, J. S. O'Brien, J. B. Giorgi, J. Power Sources, **183**, 26 (2008)
22. C.K. Cho, B.H. Choi, K.T. Lee, Journal of Alloys and Compounds, **541**, 433 (2012)
23. T. Nishizawa, T.B. Massalski (Ed.), *Binary Alloy Phase Diagrams,* **2**, 1214 (1990)
24. R. Shannon, Acta Crystallogr., 751 (1976)
25. G.C. Silva, E.N.S. Muccillo, Solid State Ion., **180**, 835 (2009)
26. M. Brown, S. Primdahl, M. Mogensen, J. Electrochem. Soc., **147**, 475 (2000).
27. H. Nakajima, T. Kitahara, T. Konomi, J. Electrochem. Soc., **157**, B1686 (2010)
28. E. Nikolla, J.W. Schwank, S. Linic, Catal. Today, **136**, 243 (2008)
29. E. Nikolla, J. W. Schwank, S. Linic, J. Catal., **250**, 85 (2007)

 DOI: 10.1051/matecconf/201713003006

EFFECT OF SINTERING ATMOSPHERE ON THE MECHANICAL PROPERTIES OF SINTERED TUNGSTEN CARBIDE

Salina Budin[1,a], Talib Ria Jaafar[1] and Mohd Asri Selamat[2]

[1]*Faculty of Mechanical Engineering, Universiti Teknologi MARA (Pulau Pinang), 13500 Permatang Pauh, Pulau Pinang, Malaysia*
[2]*Advanced Material Research Centre, SIRIM BERHAD,, Lot 34, Jalan Hi-Tech 2/3, Kulim Hi-Tech Park, 09000 Kulim, Kedah*

Abstract. One of the crucial stages in fabrication of WC-Co composite is sintering process. Inappropriate sintering state, especially sintering temperature will lead to an abnormal grain growth and thus reduce the mechanical properties of the sintered parts. Sintering atmosphere also plays a big role in oxidation, carburizing, decarburizing and nitriding on the surface of sintered parts. Prevention of any unwanted process during sintering is demanded to maintain the performance of WC-CO composites without any by-products. In this work, the effect of different sintering atmosphere and sintering temperatures was studied on the mechanical properties of WC-Co composites. 95% nitrogen (N_2) mixed with 5% hydrogen (H_2) gases was used as sintering atmosphere in a tube furnace whereas vacuum atmosphere was carried out in vacuum furnace. The sintering process was conducted at 1300°C and 1400°C, respectively. Higher relative density, hardness and transverse rupture strength (TRS) were observed on the sintered parts. Sintering under N_2-H_2 atmosphere exhibited better mechanical properties. Low mechanical properties observed in vacuum atmosphere was probably due to over-sintering which led to grain growth activities and formation of η-phase.

1 Introduction

Tungsten carbide-cobalt (WC-Co) is a well-known of hard and relatively composite materials which are expected to exhibit excellent hardness, wear resistance, thermal stability and corrosion resistance [1]. WC is bonded together with Co to form cemented carbides which have been widely used as cutting tools and wear resistant parts [2]. The cooperation of hard WC phase and the soft Co binder phase has contributed to unique good combination of high hardness and transverse rupture strength (TRS) [3].

WC-Co composite is typically fabricated using powder metallurgy route where WC and Co powders were mixed, compacted and sintered. During sintering, densification of the powders takes place where WC grains with higher solubility would dissolve in the Co-rich liquid. Many sintering methods are available such as liquid stage sintering, vacuum sintering, hot isostatic pressing, microwave sintering, spark plasma sintering and many more. One of the crucial aspects during sintering is high possibility of grain growth. The inappropriate sintering state will lead to abnormal grain growth and consequently reduce the mechanical properties of WC-Co composites.

The final properties of sintered WC-Co is determined by various processing parameters such as powder size, powder composition, compaction densities and furnace parameters including sintering temperature, sintering time and cooling rate. However, there is lack of attention on effect of sintering atmosphere. Shvab et al. [4] stated that furnace atmosphere is one of the important factors during sintering of stainless steel. During sintering process, the reaction between the furnace atmosphere and sintered parts took place. They are strongly associated with the chemical composition of the atmosphere.

In the liquid phase sintering, the sintering atmosphere consumed with nitrogen, hydrogen, argon or helium gas. The main goals were to prevent oxidation on the surface of sintered parts. Besides, on certain metal, it can avoid carburizing, decarburizing and nitriding. Another favorable sintering atmosphere is vacuum sintering. In the vacuum sintering, the chamber in which the compact parts are heated and cooled is connected to vacuum pump by means of which the chamber is usually pumped down to the desired pressure level prior to and during heating. These will equip sintering atmosphere with vacuum environment. The vacuum condition is nearly free of oxygen, nitrogen, and hydrogen. Hence, it is expected to produce better sintered parts. However, vacuum sintering is costly and only can be employed on a small scale in very special cases where it is essential for research work.

Many works related to sintering environment have been reported. Gao et al. [5] showed that synthesis of $LiFePO_4$ under vacuum condition exhibited higher charge-discharge capacity as compared to traditional Ar/N_2 atmosphere at constant pressure. Similarly, Zhang

[a] Corresponding author: salinabudin@ppinang.uitm.edu.my

et al. [6] concluded that the effect of carbon coating on $LiFePO_4/C$ prepared by vacuum thermal decomposition delivered better cycle and rate performances than $LiFePO_4/C$ by conventional sintering. Recently, Yao et al. [7] have also worked on $LiFePO_4/C$ (LPC/C) composites. They found that LPC/C composite being produced under vacuum sintering environment resulted in the formation of porous structure. The structure had excellent cycle performance as compared to argon sintering environment. Besides, the vacuum sintering reduced the optimum sintering temperature by nearly 100 °C. In another report, Kurlov et al. [8] stated that sintering of WC-Co composite in vacuum atmosphere is accompanied by the loss of a small amount of carbon. The composite composition was slight shifted to the brittle ternary phase (η-phase) of Co_6W_6C and Co_3W_3C phase transformation.

Shvab et al. [4] have investigated the effect of sintering atmosphere on microstructure of high Cr-alloyed sintered stainless steel. They observed that sintering in the nitrogen gas atmosphere produced coarse grain microstructures as compared to nitrogen-hydrogen atmosphere. On the other hand, Bao and Yi [9] reported that decarburizing on the sample surface occurred under both nitrogen atmosphere and mixed gases of nitrogen and hydrogen atmosphere. However, no effect on the sample surface was noticeable in argon atmosphere. Rui et al. [10] have studied the effect of sintering atmospheres on properties of stainless steel porous hollow fiber membranes. They observed that the air sintering environment and the CO_2 sintering environment could lead to metal oxidation while inert atmospheres (He and N_2) caused carbon remaining in SS hollow fibers. It is also observed that H_2 atmosphere was effective in removing organic additives without metal oxidation. In other work, Tian et al. [11] investigated the effect of sintering atmosphere on corrosion resistance of nickel-ferrite oxide based inert anode. It is reported that the corrosion rate of the anode prepared in the atmosphere containing the oxygen content is lower than the anode prepared in the vacuum environment.

The above reviews emphasized the important and significant effect of sintering atmosphere on the final properties of sintered parts. Hence, this study highlighted the effect of different sintering atmosphere on the mechanical properties of WC-Co composite. Two different sintering atmospheres which were 95% nitrogen (N_2) - 5% hydrogen (H_2) atmosphere and vacuum atmosphere were used at different sintering temperatures of 1300°C and 1400°C. The relative density, hardness and TRS of the sintered parts were also investigated.

2 Experimental Works

Starting materials containing 94wt% WC, 6wt% Co and heptanes were mixed in a tabular shaker mixer using wet mixing process. The duration of the mixing was three hours at a constant speed of 56 rpm. The average particle sizes of WC and Co powder were 0.8 μm and 3.51 μm, respectively. Upon completion, the mixed powders were subjected to drying in oven at a temperature slightly above the boiling point of heptanes for 2 hours to remove the heptanes from the mixture.

The dried mixture powders were sieved to obtain fine and uniform particle size before compacted into final shape at dimension of 16mm X 16 mm X 3 mm. The powder was compacted using Automatic Hydraulic Press Machine at a pressure of 20 tons.

The green samples were then subjected to cold isostatic pressure (CIP) to obtain denser and uniform samples. Finally, the samples were proceeded to sintering process. Sintering was conducted at two different atmospheres; 95% nitrogen (N_2) - 5% hydrogen (H_2) atmosphere in tube furnace and vacuum atmosphere in a vacuum furnace at the sintering temperature of 1300°C and 1400°C, respectively.

The dimension of sintered samples was 12.5 mm X 12.5 mm X 4.2 mm. The relative density of sintered samples was measured using an electronic densimeter. The hardness of the samples was measured by Vickers Hardness tester at a load of 500 g using a diamond pyramid indentation. The transverse rupture strength (TRS) of the sintered samples was obtained using the ultimate testing machine.

3 Results and Discussion

3.1. Relative Density

Figure 1 shows the comparison of relative density sintered at 1300°C and 1400°C in different atmosphere. It is noted that the relative density of the sintered parts at the sintering temperature of 1400°C is higher as compared to sintering temperature of 1300°C. During sintering process, elimination of the pore region occurs which lead to the improvement of parts density. As the temperature was increased to 1400°C, higher thermal energy was produced and gave better driving forces for the binder to completely fill out the pore region, thus increasing the relative density. It is also perceived that the relative density of the parts sintered in N_2-H_2 atmosphere was slightly higher as compared to parts sintered in vacuum atmosphere. The relative density of sintered parts sintered in N_2-H_2 atmosphere at the sintering temperature of 1300°C is 13.6 g/cm^3. While the relative density of sintered part sintered in vacuum atmosphere at similar sintering temperature is 12.3 g/cm^3. Respectively, the relative density of sintered parts sintered in N_2-H_2 atmosphere at 1400°C is 14.8 g/cm^3. While the relative density of sintered part sintered in vacuum atmosphere at similar sintering temperature is 13.2 g/cm^3. These could be associated with the window sintering band [7] sintering at vacuum atmosphere reduced the optimum sintering temperature as compared to inert gas atmosphere. Although the parts were sintered at similar sintering temperatures, the parts which were sintered at vacuum atmosphere could exceed the window sintering range and became over-sintered band. Consequently, it led to grains coarsening phenomena.

 DOI: 10.1051/matecconf/201713003006

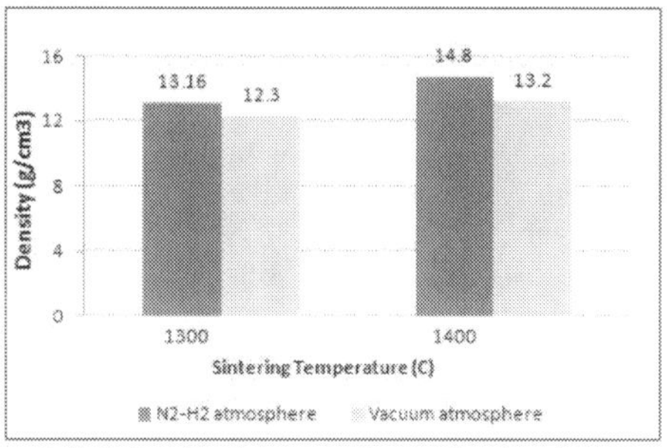

Figure 1. Relative density comparison at the sintering temperature of 1300°C and 1400°C

3.2 Hardness

Figure 2 shows the comparison of the hardness of sintered sample at different sintering temperatures and atmosphere. Generally, the hardness of the sintered samples increases with the increasing of sintering temperature. As the sintering temperature increases from 1300°C to 1400°C, the hardness was reported increases drastically from 743 HV to 1742 HV. The drastic increase on the hardness at sintering temperature of 1400 °C for both sintering atmospheres is probably resulted from full densification on the sintered parts. High temperature generates higher thermal energy which provides a better driving force for the binder phase, thus increases the relative density. The hardness trend was consistent with the relative density. The hardness results are seen to be correlated with the relative density. Low relative density on the sintered parts sintered at vacuum atmosphere was consequently contributed to low hardness.

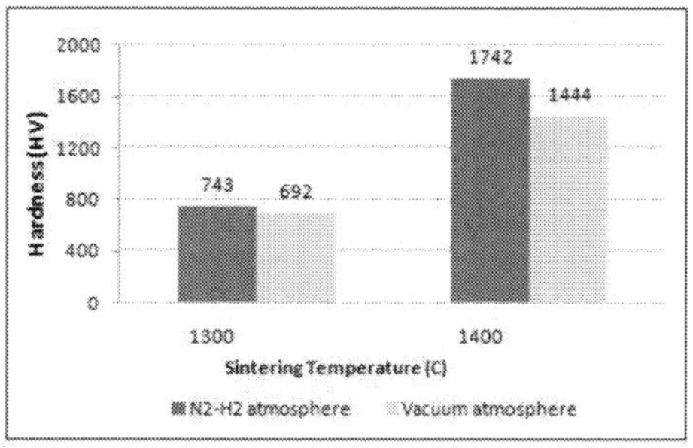

Figure 2. Hardness comparison at the sintering temperature of 1300°C and 1400°C

3.3 Transverse Rupture Strength (TRS)

The variation of TRS as a function of sintering temperature at different atmosphere is shown in Figure 3. The parts sintered at 1300°C in N_2-H_2 atmosphere shows a TRS value of approximately 1320 MPa. The TRS was increased to 1680 MPa as the sintering temperature was raised to 1400°C. As for vacuum sintering atmosphere, the TRS was observed at 617.6 MPa at sintering temperature of 1300°C and increased to 1383 MPa at the sintering temperature of 1400°C. It is believed that the gradually increasing of the sintering temperature helps in promoting sintering process and eliminating pore thus improves the TRS value. The full densification at 1400°C is entirely reducing the strain points along the rupture mechanism and increasing TRS value.

It is also clearly seen that the TRS values obtained in a N_2-H_2 sintering atmosphere exhibited higher values as compared to vacuum sintering atmosphere. This could be probably due to grains coarsening phenomena in parts which were sintered in vacuum atmosphere. As claim by Yao et al. [7] sintering under vacuum atmosphere required lower sintering. Thus, the grain growth occurred at the sintering temperature of 1400°C in vacuum atmosphere, was possibly due to over-sintered. In addition, the low TRS value could be also due to the formation of η-phase that made the WC-Co composite extremely brittle.

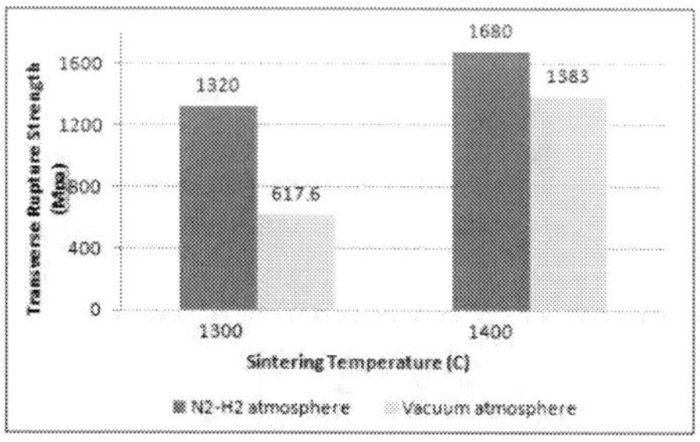

Figure 3. TRS comparison at sintering temperature of 1300°C and 1400°C

4 Conclusion

In this work, WC-Co composite was prepared by the powder metallurgy method and sintered at different sintering atmosphere; N_2-H_2 atmosphere and vacuum atmosphere at two different sintering temperatures; 1300°C and 1400°C. The results revealed that the relative density, hardness and TRS were increased as the sintering temperature was higher. High temperature promotes the densification process and effectively eliminates the pore. Simultaneously, higher relative density, hardness and TRS were observed on parts sintered in the N_2-H_2 atmosphere. Low mechanical properties exhibited on the parts sintered in vacuum atmosphere probably was due to over-sintering. Over-sintering led to grain growth activities and formation of η-phase which produced the brittle WC-Co composite. Further studies should be conducted to reveal a clear explanation on the relation between hardness properties and relative density, over sintering stage and mechanism of η-phase formation.

Acknowledgement

This work was financed by Government of Malaysia under the e-Science fund project of the Ministry of Science, Technology and Innovation (MOSTI) (Project No: 06-01-01-SF0702). The authors gratefully recognize all the support, scientific contributions and stimulating collaboration from the partners from Advanced Material Research Centre (AMREC).

References

1. Q. Yang, J. Yang, H. Yang, J. Ruan, The effects of fine WC contents and temperature on the microstructure and mechanical properties of inhomogeneous WC-(fine WC-Co) cemented carbides, *Ceramics International*, **Vol 42**, pp 18100–18107(2016)
2. X. Ren, Z. Peng, C. Wang, H. Miao, Influence of nano-sized La_2O_3 addition on the sintering behavior and mechanical properties of WC–La2O3 composites *Ceramics International*, **Vol 41**, pp 14811–14818 (2015)
3. C. B. Wei, X.Y. Song, J. Fu, X.M. Liu, Y. Gao, H.B. Wang, S.X. Zhao, Microstructure and properties of ultrafine cemented carbides – Differences in spark plasma sintering and sinter-HIP. *Material Science and Engineering*. **Vol A** 552, pp 427 – 433 (2012)
4. R. Shvab, E. Dudrova, O. Bergman, S. Bengtsson, Effect of sintering atmosphere on the microstructure of high Cr-Alloyed sintered stainless steel. *Powder Metallurgy Progress*, **Vol 13**, No 3 – 4, pp 103 – 108 (2013)
5. L. J. Gao, Y.L. Zhang, Preparation of LiFePO4 positive electrode material by vacuum calcination and its electrochemical performance in Lithium-ion batteries, *J. Nanchang Univ. (Nat. Sci.)*, **Vol 32**, pp 239–242 (2008)
6. L. L. Zhang, G. Peng, X.L. Yang, P.C. Zhang, High performance LiFePO4/C cathode for lithium ion battery prepared under vacuum conditions. *Vacuum*, **Vol 84**, pp 1319–1322 (2010)
7. Y. Yao, P. Qu, X. Gan, X. Huang, Q. Zhao, F. Liang, Preparation of porous-structured LiFePO4/C composite by vacuum sintering for lithium-ion battery. *Ceramics International*, **Vol 42**, pp 18303–18311 (2016)
8. A. S. Kurlov, A.I. Gusev, A.A. Rempel, Vacuum sintering of WC-8 wt.% Co hardmetals from WC powders with different dispersity. *International Journal of Refractory Metals and Hard Materials*, **Vol 29**, pp 221 – 231 (2011)
9. R. J. Bao, J. Yi, Effect of sintering atmosphere on microwave prepared WC–8wt.%Co cemented carbide. Int. *Journal of Refractory Metals and Hard Materials*, Vol 41, pp 315–321 (2013)
10. W. Rui, C. Zhang, C. Cai, X. Gu, Effects of sintering atmospheres on properties of stainless steel porous hollow fiber membranes. *Journal of Membrane Science*, **Vol 489**, pp 90–97 (2015)
11. Z. L. Tian, W. C. Guo, Y. Q. Lai, K. Zhang, J. Li, Effect of sintering atmosphere on corrosion resistance of cermet inert anode for aluminum electrolysis. *Trans. Nonferrous Met. Soc. China*, **Vol 26**, pp 2925–2929 (2016)

Effect of ultrasonic vibration and solution-aging treatment on microstructure and properties of in-situ $TiAl_3$/7050Al composites

Lili Chen[1], Zhiming Du[1, a], DanZhao[1], Lihua Chen[2] and Changshun Wang[2]

1School of Material Science and Engineering, Harbin Institute of Technology, Harbin 150001, China
2Technology Center, Beijing North Vehicle Group Corporation, Beijing 100072,China

Abstract. Ultrasonic vibration have positive effect on formation of in-situ $TiAl_3$/7050Al composites, the role of ultrasound in the reaction process was described systematically. The best solution-aging treatment parameters were discussed and effects of the selected treatment on the resultant microstructure and mechanical properties were investigated. The results show that ultrasonic vibration not only promote the reaction process, but also result in significant spreading particles fallen off from the reaction interface, therefore creating a finer and homogeneous distribution of $TiAl_3$ particles in the composite. In solution process, the skeleton second phase dissolve into matrix with increasing temperature and holding time, the transformation is more sensitive to solution temperature. Residual second phase transform into finer-size rod or discoid shape and a large amount of finer phase ($MgZn_2$ and $CuAl_2$)precipitates after aging-treated. As a result, the strength, hardness and elongation of the composite are slightly increased after using ultrasonic vibration and significantly increased after solution-aging treatment.

1 INTRODUCTION

Aluminium matrix composites (AMCs) have been recognized as the most popular materials, used for aerospace, automobile, marine and mining industries, and the most promising light alloy among types of metal matrix composites, due to their greater strength, reduced weight, improved high temperature properties, improved abrasion and wear resistance[1-5], etc. The form of reinforcement include particle, whisker or short fiber, continuous fiber and mono filament. Particle-reinforced AMCs are less expensive, superior machining properties, easily to form, more reinforcement to choose and isotropic in nature, compared with other AMCs. Many kinds of particle reinforcements have been studied widely, such as Al_2O_3[6], SiC[7], TiB_2[8],$MoSi_2$[9], of which in-situ $TiAl_3$ particles has received extensive attentions.

For the interface between reinforcement and Al matrix is good wettability and clean in in-situ $TiAl_3$/Al alloy composite, as well intermetallic compound TiAl3 is very easy to control in Ti-rich condition and suitable to be reinforcement, several methods to fabricate the in-situ $TiAl_3$/Al composites have been reported, such as powder metallurgy[10-12], mechanical alloying[13] and casting. Casting has been regarded as a lower cost and higher efficiency method for commercial production. However, composites fabricated by traditional casting usually have uneven distribution of reinforcement, so external energy is introduced. Jiao et al.[14] exerted magnetic field to make the melt a forced movement during the process of in situ generated Al_3Ti particles reinforced aluminum matrix composites, promoting changes in the conditions of melt dynamics, changing the morphology and distribution. Dinaharan et al. [15] applied friction stir processing to enhance the distribution and morphology of Al_3Ti and Al_3Zr particles.

In recent years, ultrasonic vibration has been more effective method to form homogeneous $TiAl_3$ in Al composite.Liu et al.[16] fabricated in situ Al_3Ti/Al composites via ultrasound assisted direct reaction between solid Ti powers and liquid Al at 780°C, the in situ formed Al_3Ti particles were blocky in morphology, and the size of most Al_3Ti particles was in the range of 2-7μm. Qin et al. [17] studied the formation process of $TiAl_3$ based on the in-situ reaction between globular solid Ti powders with almost uniform size and 2024 Al melt under ultrasonic vibration. Although these studies have been reported, most experiments have been focused on illustrating the formation mechanism of $TiAl_3$ particles during the solidification process, the effects of ultrasonic vibration on fabricating process of in situ $TiAl_3$/Al composites were still not discussed sufficiently.

Furthermore, in the case of cast Al alloy, solution heat treatment is frequently recommended to reduce porosity and enhance the properties[18]. However, law of solution-aging treatment on microstructure and mechanical properties of $TiAl_3$/Al alloy under selected conditions has not been reported. The experiments described in this work were designed to examine the microstructures and properties of $TiAl_3$/Al composites by introducing ultrasonic vibration in molten condition, and to describe the role of ultrasound in the reaction process systematically. Solution-aging treatment was applied to

[a] Corresponding author: duzm@263.net

explore the best treatment parameters by observing variation of microstructure and mechanical properties.

2 EXPERIMENTAL

2.1. Experimental materials and fabrication of starting composite

The $TiAl_3$/7050Al composites were fabricated by in-situ synthesis between globular solid Ti powders with almost uniform size and as-cast 7050Al molten alloy material under ultrasonic vibration. For elemental Ti powder, is sponge titanium powder, the mean particle size was 20 μm, and the purity is higher than 99.5%. For 7050Al alloy, the nominal chemical composition includes 6.624%Zn, 1.960%Mg, 2.280%Cu, 0.0975%Zr, 0.093%Si, 0.132%Fe, 0.074%Cr and balance of Al(in wt.%). The as-cast block 7050Al alloy was melt in crucible resistance furnace at 750°C and refined 1 minute by refining agent(55%KCl, 5%CaF_2, 40%NaCl, about 2% of melt, in wt.%). Subsequently, the Ti powders wrapped with Al foil were added into the melt, meanwhile ultrasonic equipment was opened. The amount of additive Ti corresponded to the composition of 6 wt% $TiAl_3$/2024 Al composite. Ultrasonic vibration with power of 1.5 kW and frequency of 20 kHz was introduced into the molten alloy, the reaction can be accelerated and adequately proceeded. After 5–20 min, the molten alloy was rapidly poured into a steel mold under gravity to cast an ingot with dimension of Φ50mm×h60mm. Samples for microstructure examination and following treatment and property-test were cut from the casting ingots, and prepared by standard metallographic technique of grinding with SiC abrasive and polishing with a diamond spray (0.5 μm).

2.2. Solution-aging processing

Solution treatment was carried out at 450-500°C in 30-120min for the $TiAl_3$/7050Al composites, quenching in water at 20°C and then artificial aging at 120−195°C for different time intervals from 3h up to 24 h. After above treatment, the samples were slightly chemically etched to remove the surface oxide layer. The etching agent was an aqueous solution of 2.5vol.% HNO_3+0.5vol.% HF.

2.3. Mechanical testing and microstructure characterization

Tensile tests were carried out (by universal material testing machine, Instron-5569R) at room temperature using round specimens machined from the ingot with a gage section of Φ6 mm×24 mm. The initial strain rate at room temperature was $1\times10^{-4}s^{-1}$. Hardness tests were performed on a HB-3000B (load of 62.5kg and 500+62.5kg, before and after treatment, respectively. 30s dwell time) on cross sections. Both above, at least five tests were performed and average values were reported.

The microstructures of in-situ and solution-aging treatment samples were carried out by scanning electron microscopy(SEM) (Quanta2000) equipped with energy dispersive spectroscopy(EDS), transmission electron microscopy (TEM) (Talos F200x, USA). Specimens for TEM observation were prepared using a standard procedure and an jet polisher.

3. RESULTS AND DISCUSSION

3.1. Microstructure of in-situ $TiAl_3$/7050Al composites with ultrasonic vibration

Microstructure of in-situ $TiAl_3$/7050Al composites with and without ultrasonic vibration are shown in Fig.1a and b, respectively. It can been clearly seen that the distribution of block-shaped $TiAl_3$ particles with regular appearance is more homogeneous and the particles are smaller in composite with ultrasonic vibration(Fig.1(a)), while network linked by skeleton-shape second phase (TEM analysis proved Al_2CuMg and $CuAl_2$) distribute in the composite evenly. In contrast, $TiAl_3$ particles present large-sized and gather with each other in the composite without ultrasonic vibration(Fig.1(b)), indicating ultrasonic vibration promote the reaction process and dispersion of Ti, therefore creating a homogeneous distribution of $TiAl_3$ particles in the Al matrix.

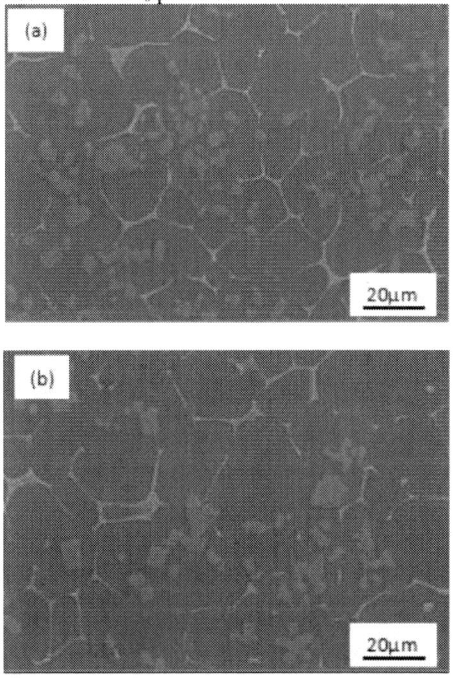

Fig.1 microstruture of in-situ $TiAl_3$/7050Al composites (a)with ultrasonic vibration; (b)without ultrasonic vibration

In Qin Y's work[11], the formation of $TiAl_3$ particles at the Al melt/Ti interface has been studied. For the reason of inter-diffusion and reaction of Ti and Al atoms, $TiAl_3$ layer is formed on the surface of Ti powder. Cracks can make the layer peel off when the layer thickness increases to a certain threshold. Consequently, the brittle $TiAl_3$ layer is ruptured into small particles more easily and quickly when we add ultrasonic vibration in the process of composite formation. Fig.2 shows high magnification SEM micrograph of $TiAl_3$ particle, the circle region is Ti-rich region, while the peripheral area Ti:Al approximately equal 3:1(at%) by EDS element scanning, indicating peripheral area is $TiAl_3$ layer which has not fall off, the thickness is about 2μm.

When ultrasonic act on the melt, due to cavitation effect, the reaction system will form a large number of hot spots which produce high temperature and high pressure. The temperature and pressure ensure reaction substances high activity by ionization and free radical reaction, which greatly accelerates the reaction process of Ti-Al system. The local high temperature regions, favourable wetting effect between Ti powders and liquid, decrease the surface tension of the liquid and thus promote the solid-liquid reaction. In addition, strong turbulence produced by high pressure impact on Ti/Al interface to make the reaction activity of Ti powder increased greatly. Moreover, the acoustic streaming and cavitation generated $TiAl_3$ falling off from the reaction interface quickly and more uniform particle distribution in the matrix, instead of converging together. With the increasing of ultrasonic power(CS0,0.6,0.9,1.2,1.5KW), the particle size distribution is more finer, larger size (>10μm) is less and less, intermediate size(5-10μm) and smaller size(<5μm)are more and more, as shown in Fig.3.

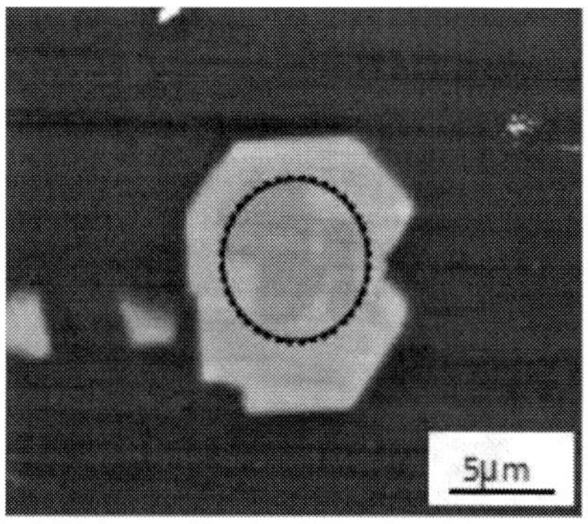

Fig.2 High magnification SEM of particle

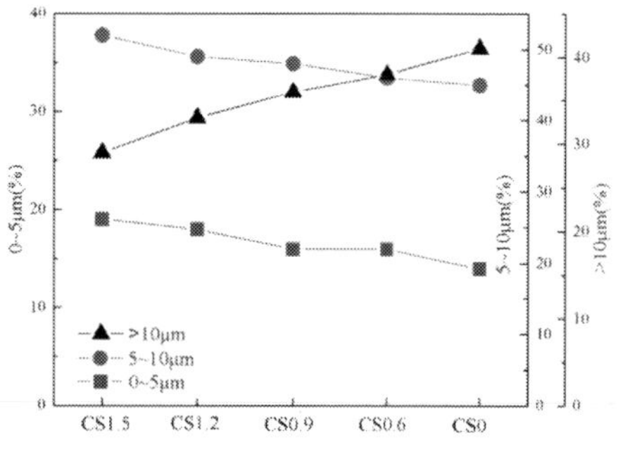

Fig.3 Particle size distribution with ultrasonic power

3.2. Composite microstructure evolution during solution treatment

Fig.4 shows microstructure images of solution-treated composite at 450°C in different time(30-60min). The second phase (TEM analysis proved Al_2CuMg and $CuAl_2$) network linked by skeleton-shape which distribute in the as-cast composite evenly break and decrease with time during heating at 450°C. The volume fraction of second phase is less than 0.5% after solution-treated in 120min, compared with 4.1% in the as-cast composite by Image-Pro Plus 6.0 software. As the heating time increase, the second phase dissolve into the matrix. Fig.5 represents microstructure images of solution-treated composite at different temperatures for 90min(470-500°C). We can conclude variation law of second phase with increasing temperature keep in touch with increasing holding time, the transformation is more sensitive to solution temperature. When the temperature reaches 480°C, only few skeleton-shape phase left. Table 1 lists the Matrix composition of composites heated at 450°C for different time, element Mg, Cu, Zn etc, resolve into the matrix. The weight percent of element in matrix augments with increasing the solution time, while the network second phase decreases.

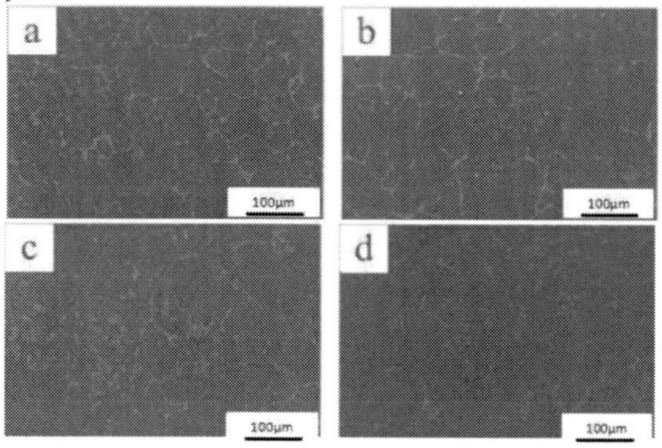

Fig.4 Microstructure image of solution-treated composite at 450°C in different time(a)30min;(b)60min;(c) 90min;(d) 120min

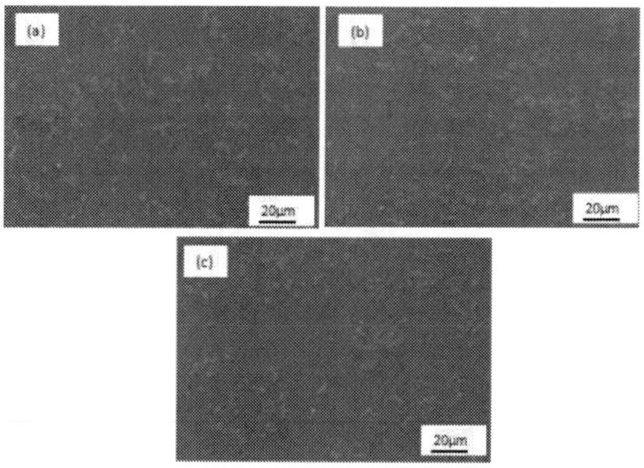

Fig.5 Microstructure image of solution-treated composite at different temperatures for 90min(a)470°C;(b)480°C;(c)500°C

Table 1 Matrix composition of composites heated at 450°C for different time (Wt%)

Heating time/min	Mg	Si	Ti	Cu	Zn	Zr	Fe	Cr	Al
30	1.41	0.04	9.87	1.49	5.08	0.35	-	-	81.76
60	1.50	0.05	9.65	1.62	5.38	0.038	0.08	-	81.68
90	1.62	0.07	9.02	1.80	5.62	0.042	0.07	0.02	81.74
120	1.78	0.083	9.80	1.83	5.64	0.05	0.10	0.02	80.7

3.3. Effect of aging treatment on microstructure and mechanical properties

Microstructure image of aging-treated composite are shown in Fig.6, residual skeleton-shape second phase generated in as-cast composite transform into finer-size rod or discoid shape after aging-treated (white-bright ones in Fig.6a), instead of dissolving into matrix absolutely. These residual second phase are usually distributed around grain boundary, while intragranular ones are redissolved into matrix completely during solution treatment process. Besides, a large amount of finer phase precipitates(with an average size of 20nm) along the grain boundary during the aging process, morphology of precipitates is shown in Fig.6b. Consequently, TEM analysis proves that precipitates are $MgZn_2$ and $CuAl_2$.

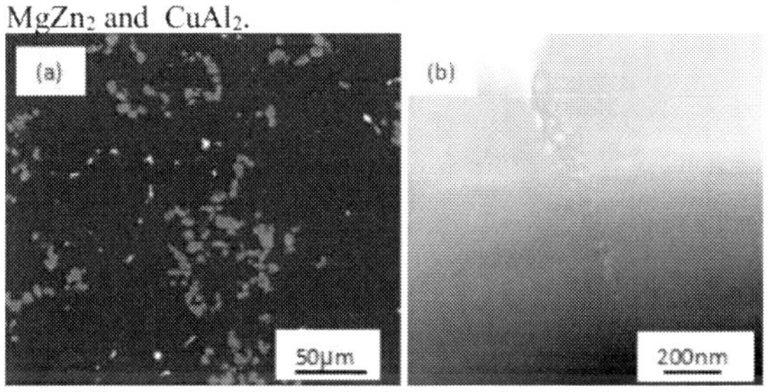

Fig.6 Microstructure image of aging-treated composite low magnification;(b)morphology of precipitates

Consult solution-aging treatment technology of 7050Al alloy substrate, the strength and hardness variation pattern are investigated at aging temperature 120℃、145°C、160°C、195°C for 12h followed by 480°C/60min solution-treated, as shown in Fig.7. Both variation present descend with improved temperature, therefore 120°C is concerned as the best aging temperature. Since grain size is coarsen under high temperature, performance degrade distinctly. In addition, the effect of aging time(from 3h to 24h) on properties is studied, as shown in Fig.8. When aging time is short, the curve of ultimate strength(UTS), yield strength(YS) and hardness ascend with increasing aging time. In fact, the composite is strengthen by precipitates continuously increasing. Until aging time reaches 18h, precipitate is no longer increased, the curves start to appear decline tendency. So 18h is concerned as peak aging time. Most important, elongation has relatively high valve(8.03%). As a result, the best treatment of in-situ wt.6%$TiAl_3$/7050Al composites with ultrasonic vibration is:500°C/60min+120°C/18h, as Table 2 shown, mechanical properties of composites in different treatment. For comparing, values of as-cast without ultrasonic vibration and un-treated samples are also listed. Evidently, the ultrasonic power influences slightly the strength and elongation of composite. Moreover, comparing with the properties of the composite without subject to solution-aging treatment, the UTS,YS, elongation and hardness of the composite processed solution-aging are improved greatly, by 174%, 165%, 143% and 129%, respectively.

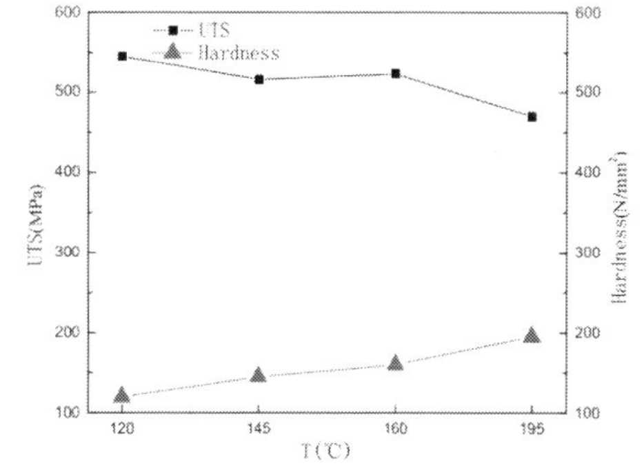

Fig.7 The strength and hardness variation pattern

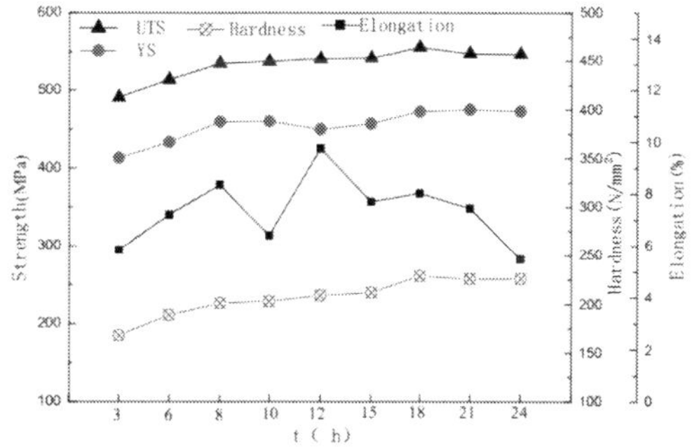

Fig.8 Effect of aging time on properties

 DOI: 10.1051/matecconf/201713003007

Table 2 Mechanical properties of composites in different treatment

sample	UTS(MPa)	YS(MPa)	Elongation(%)	Hardness(N/mm2)
480°C/60min+120°C/18h	555.5	472.7	8.03	229
480°C/90min+120°C/18h	541.0	459.1	7.20	212
480°C/120min+120°C/18h	572.2	467.0	10.42	232
500°C/30min+120°C/18h	558.1	484.2	6.31	229
500°C/60min+120°C/18h	582.0	496.1	11.20	236
As-cast with ultrasonic vibration	320	286	5.6	178
As-cast without ultrasonic vibration	237	224	4.28	113

3.4. Discussion

The properties of $TiAl_3$/7050Al composite are determined by $TiAl_3$ particles, Al matrix and their combination. The composite applied ultrasonic vibration possess much finer $TiAl_3$ particles and more homogeneous particle distribution, consequently their properties are slightly increased compared with those without ultrasonic vibration. After solution-aging treatment, morphology of $TiAl_3$ particles is stability without any change, on the contrary variation have taken place in matrix and second phase. These variation cause properties of $TiAl_3$/7050Al composite after solution-aging treatment enhanced greatly compared with the as-cast ones(shown in table 2).

The main strengthen reason obtained direct evidence from our work is as followed. Firstly, when the in-situ $TiAl_3$ particle reinforced composite fabricated, the interface bonding between $TiAl_3$ particle and Al matrix is investigated, as shown in Fig.9. Morphology of $TiAl_3$ particle in Al matrix(Fig.9a), combined with Fig.1 a, $TiAl_3$ reinforcement with high strength and hardness is distributed in matrix evenly, result in dispersion strengthening effect. Furthermore, HRTEM of interface bonding depicts coherent relationship between $TiAl_3$ particle and Al matrix(Fig.9b), owing to the in-situ synthesis between Ti powders and 7050Al molten alloy. In fact, the in situ reinforcements are ultra fine and more compatible with the matrix, the interface is cleaner than that of composites conventionally. Because the in-situ formed particulates are thermally stable, this will ensure that the composite matrix has sufficient strength to transfer stress.

Secondly, solid solution strengthening and supersaturation degree of the matrix is enhanced via solution-treated. Element Mg, Cu, Zn etc, resolve into the matrix, result in lattice deformation and energy raising. As a result, dislocation motion become difficult in plastic deformation. Thirdly, a large amount of finer phase precipitates along the grain boundary during the aging process, result in dispersion strengthening effect similarly.

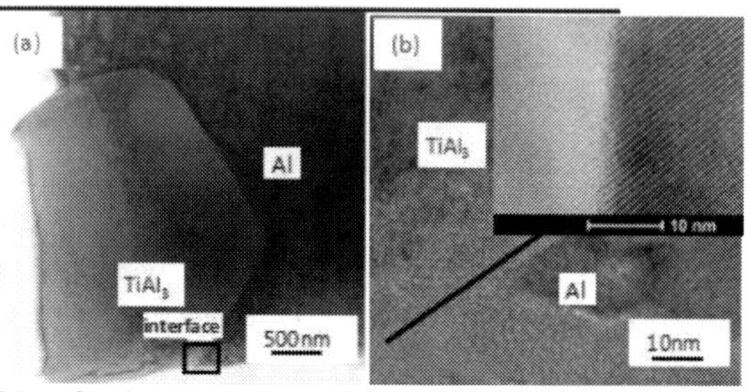

Fig.9 Interface bonding between $TiAl_3$ particle and Al matrix (a)morphology of $TiAl_3$ particle in Al matrix; (b)HRTEM of interface

4 CONCLUSIONS

$TiAl_3$/7050Al composites can be certainly obtained by using the selected conditions, the microstructure of as-cast composite consists of $TiAl_3$ particles, skeleton-shape second phase (Al_2CuMg and $CuAl_2$) and 7050Al matrix.

(1)In the fabrication of $TiAl_3$/7050Al composites, ultrasonic vibration have positive effect on formation of $TiAl_3$ particles, it can not only promote the reaction process, but also create finer particles and distribution of $TiAl_3$ particles more evenly.

(2)After solution treatment, skeleton-shape second phase break and disappear with increasing heating temperature and holding time, accordingly more and more element Mg, Cu, Zn etc. resolve into matrix. Solid solution strengthening and supersaturation strengthening enhance the composite. after aging, residual second phase transform into finer-size and a large amount of finer precipitates ($MgZn_2$ and $CuAl_2$) make the composite strengthen further. The best solution-aging treatment parameters for the designated composite were:500°C/60min+120°C/18h.

(3)The properties of the composite are slightly increased after applying ultrasonic vibration and significantly increased after solution-aging treatment. The best UTS,YS, elongation and hardness of the composite are improved by 174%, 165%, 143% and 129%, respectively.

Acknowledgements

The authors express their appreciation for financial support of Ministry of science and technology project of international cooperation under Grant(No.2014***50320).

References

1. Surappa M K . Sādhanā. 28,1(2003):319-334
2. Ervina Efzan M N, Siti Syazwani N, Mustafa A B A M. Key Eng. Mater.700(2016):102-110
3. Deaquino-Lara R, Soltani N, Bahrami A, et al. Mater. Des. 67,1(2015):224–231
4. Adamiak M, Tomiczek B, Górka J, et al.Arch. Metall. Mater. 61,2(2016):847-852
5. Kala H, Mer K K S, Kumar S. Procedia Mater. Sci. 6(2014):1951-1960
6. Su H, Gao W, Feng Z, et al. Mater. Des. 36(2012):590-596
7. Xu H, Palmiere E J. Composites Part A Applied Science & Manufacturing. 30,3(1999):203-211
8. Kumar S, Chakraborty M, Sarma V S, et al. Wear, 265,1–2(2008):134-142
9. Corrochano J, Lieblich M, Ibáñez J. Composites Part A Applied Science & Manufacturing. 42,9(2011):1093-1099
10. Nofar M, Hosseini H R M, Kolagar-Daroonkolaie N. Mater. Des. 30,2(2009):280-286
11. Qin Y, Chen T, Wang Y, et al.Materials. 9,3 (2016):199
12. Chianeh V A, Hosseini H R M, Nofar M. J. Alloys Compd. 473,1–2(2009):127-132
13. Al-Dabbagh J B, Tahar R M, Ishak M, et al. International Journal of Nanoelectronics & Materials.8,1(2015):23-32
14. Jiao L, Zhao Y T, Wu Y, et al. Advanced Materials Research.600(2012):199-203
15. [15]Dinaharan I, Kumar G A, Vijay S J, et al. Mater. Des. 63,21(2014):213-222
16. Liu Z, Han Q, Li J. Powder Technol. 247(2013):55-59
17. Qin J, Chen G, Wang B, et al. J. Alloys Compd.653(2015):32-38
18. Liu D, Atkinson H V, Kapranos P, et al. J. Mater. Sci. 39,1(2004):99-105

Oxide reinforced Ni base composite prepared by spark plasma sintering

Guoliang Zhu[1,2,a], Shiqi Zhao[3], Rui Wang[1,2], Anping Dong[1,2], Liang Zhang[4,5], Wenheng Wu[4,5], Wei Wang[1,2], Sheng Jiang[6] and Yilong Pu[6]

[1]School of Materials Science and Engineering, Shanghai Jiao Tong University, Shanghai 200240, China
[2]Shanghai Key Laboratory of Advanced High-temperature Materials and Precision Forming, Shanghai 200240, China
[3]School of Mechanical Engineering, Shanghai Jiao Tong University, Shanghai 200240, China
[4]Shanghai Engineering Research Center of 3D Printing Materials, Shanghai 200437, China
[5]Shanghai Research Institute of Materials, Shanghai 200437, China
[6]Jiangsu Longda Superalloy Material Co., Ltd, Wuxi 214105, China

Abstract. Y_2O_3 is a favourable reinforcement for the oxide dispersion strengthened superalloy due to its good high temperature stability, excellent strength and large elastic modulus. In this work, Y_2O_3 is introduced into the K4169 superalloy aiming to improve its high temperature properties. The high quality Y_2O_3 and K4169 powder mixture can be obtained with the addition of the process control agent of stearic acid after high energy ball milling, and the Y_2O_3/K4169 composite with uniformly dispersed reinforcement of nanoscale Y_2O_3 particles are finally synthesized with the ball-milled powders by using spark plasma sintering.

Introduction

Different from the technical route of the optimization of composition and microstructure, the oxide dispersion strengthening (ODS) is an effective method to improve the high temperature performances of materials [1]. Thus, the synthesis of ODS superalloys has attracted wide attention. Yttria (Y_2O_3) is a common reinforcement for ODS superalloy due to its high thermal stability, favourable strength and elastic moudulus [2]. Benjamin [3] has prepared the Y_2O_3 reinforced superalloy by using mechanical alloying in 1970s. Later, a few ODS superalloys have been invented, including MA754 and PM3000, in which the concentration of Y_2O_3 is 0.6wt.% and 0.9wt.% Y_2O_3 respectively. The superalloys strengthened by Y_2O_3 have been widely applied in aero engines. Moreover, Y_2O_3 also can react with the elements in matrix or other added elements to form new reinforcing phases, resulting in better strengthening effect. The $Y_3Al_5O_{12}$, $YAlO_3$ and $Y_4Al_2O_9$ phases are formed in the MA6000 alloy due to the reactions between Y_2O_3 and Al during the mechanical alloying [4]. the addition of Hf in PM1000 ODS superalloy leads to the formation of $Y_2Hf_2O_7$ which has better interface characteristics with the matrix compared to Y_2O_3 [5].

K4169 alloy is a commercial superalloy which has similar composition and microstructure with IN718C alloy. The high temperature strength of K4169 is not satisfactory though it has excellent comprehensive properties when the service temperature is lower than 650℃ [6-8].

In this work, Y_2O_3/K4169 composite is prepared by using high energy ball milling (HEBM) and spark plasma sintering (SPS) to improve the high temperature strength of matrix K4169 superalloy. The microstructures of the ball-milled Y_2O_3 and K4169 powders as well as the microstructure of Y_2O_3/K4169 composite after SPS are studied.

[a] Corresponding author: glzhu@sjtu.edu.cn

1 Materials and method

High-purity Y_2O_3 powders and K4169 powders are used as raw materials. Spherical K4169 powders (45 μm) and nanoscale Y_2O_3 powders (50 nm) are added to a ball milling jar. The weight ratio of Y_2O_3 in the powder mixture is 10wt.%, and 0.5wt.% of stearic acid is used as process control agent for HEBM. The weight ratio of steel balls to the powder mixture is 10:1. The HEBM is conducted with a rotating speed of 300 rmp, and the ball milling time is 12h. After HEBM, the passivation of powders is conducted under 0.2 atm for 30 min in an argon glove box to avoid self-burning due to the high activity of ball-milled powders. Then the ball-milled powder mixture is sintered in SPS apparatus at 1050 ℃ under the pressure of 50 MPa for 5 min.

The microstructure of ball-milled powders before SPS and the Y_2O_3/K4169 composite after SPS are investigated by using scanning electron microscope (JSM7600F). The phases of the ball-milled powders before SPS and the Y_2O_3/K4169 composite after SPS are analysed by X-ray diffractometer (Bruker AXS D8), Cu target and a scanning speed of 0.2 °/s are employed for XRD tests.

2 Results and discussion

Figure 1 shows the SEM images of the ball-milled powders. The size of the K4169 powders is greatly decreased from 45μm to 20μm. No obviously scattered Y_2O_3 particles are found in the ball-milled powders, indicating that the nanoscale Y_2O_3 powder and microscale K4169 powders are homogeneously mixed after HEBM.

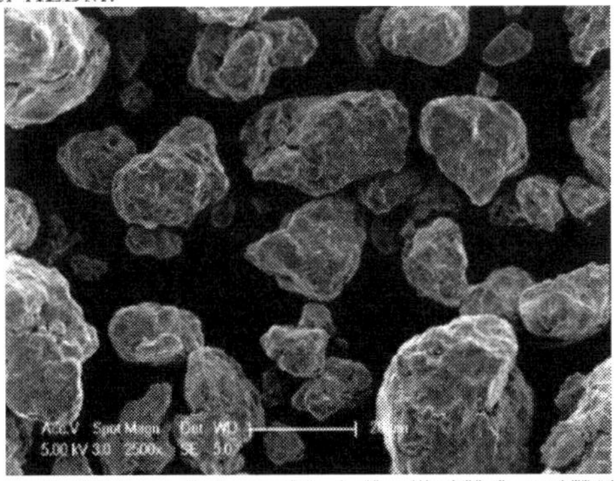

Figure 1. SEM morphology of the ball-milled Y_2O_3 and K4169 powders after HEBM.

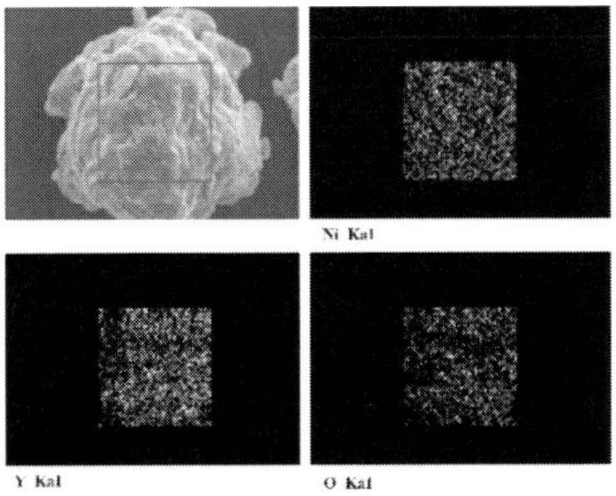

Figure 2. The distribution of elements in the ball-milled Y_2O_3 and K4169 powders after HEBM.

The energy dispersive spectrometry results of a ball-milled powder, as shown in Figure 2. Both oxygen and yttrium are uniformly distributed in the ball-milled powder, without obvious aggregation. The result confirms that high quality Y_2O_3 and K4169 powder mixture can be prepared after HEBM for 12h with the help of process control agent of stearic acid.

Figure 3 shows the XRD pattern of the ball-milled Y_2O_3 and K4169 powders after HEBM. The characteristic diffraction peaks of Ni and Y_2O_3 are clear and sharp. Moreover, no other characteristic diffraction peak is found in Figure 3, indicating that no impurity is formed during the HEBM.

The XRD pattern of Y_2O_3/K4169 composite after SPS is shown in Figure 4. The characteristic diffraction peaks of γ phase and Y_2O_3 are observed, indicating that the Y_2O_3 was thermal stable during the SPS. Considering that the standard heat treatment is not carried out for Y_2O_3/K4169 composite after SPS, it is reasonable that the characteristic diffraction peaks of γ′ and γ″ phases are not found in the XRD pattern.

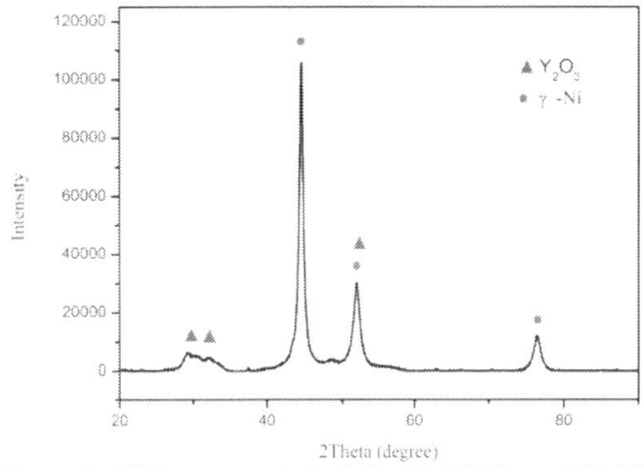

Figure 3. XRD pattern of the ball-milled Y_2O_3 and K4169 powders after HEBM.

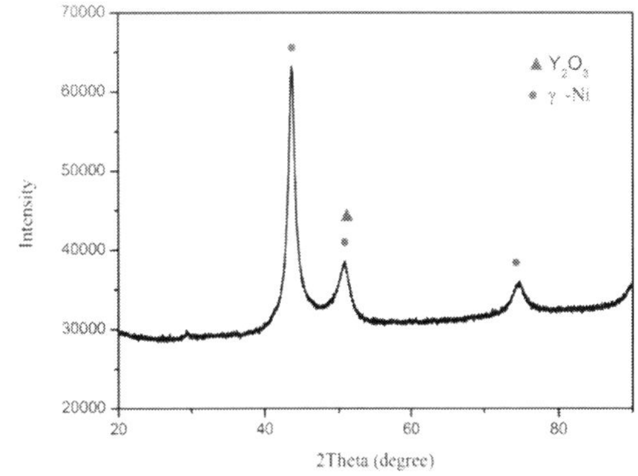

Figure 4. XRD pattern of Y_2O_3/K4169 composite after SPS.

It is believed that the nanoscale Y_2O_3 could enhance the material strength through the dislocation pinning mechanism. With the coarsening of γ′ and γ″ phases during long time service of Y_2O_3/K4169 composite, the strengthening effect from γ′ and γ″ is greatly decreased, but the Y_2O_3 could still work as reinforcement to strengthen the matrix, due to its high thermal stability. In the future, the deformation behavior and the microstructure evolution after heat treatment will be investigated to further optimize the microstructure of Y_2O_3/K4169 composite, aim to improve the high-temperature performance of Y_2O_3/K4169 composite.

3 Conclusion

The high quality Y_2O_3 and K4169 powder mixture can be obtained by HEBM which is performed with a rotational speed of 300 rmp for 12 h under argon protection, and the Y_2O_3/K4169 composite with uniformly distributed nanoscale Y_2O_3 particles is prepared by SPS of the ball-milled powders. It is believed that the nanoscale Y_2O_3 in Y_2O_3/K4169 composite could greatly enhance the material strength through the dislocation pinning mechanism during long time service due to its high thermal stability. The optimization of the microstructure of Y_2O_3/K4169 composite will be further investigated in the next research work.

4 Acknowledgement

This work was financially supported by the Joint fund of the Ministry of Education of China under Project No. 6141A02022210, the National Natural Science Foundation of China under Project No. 51704195, the Industry-University-Research Cooperation Annual Plan of Shanghai under Project No. CXY-2016-004, the National Industrial Basis Improvement Project under Project No. TC160A310-12-1 and Special fund for research and development platform of Shanghai Science and Technology Commission under Project No. 16DZ2260605.

References

1. I. Hilger, M. Tegel, M.J. Gorley, et al. The structural changes of Y_2O_3 in ferritic ODS alloys during milling. J. Nucl. Mater., **447**, 242-247 (2014)
2. T. Liu, H. Shen, C. Wang, et al. Structure evolution of Y_2O_3 nanoparticle/Fe composite during mechanical milling and annealing. Prog. Nat. Sci.: Mater. Int., **23**, 434-439 (2013)
3. J.S. Benjamin. Dispersion strengthened superalloys by mechanical alloying. Metall. Tran., **1**, 2943-2951 (1970)
4. G.B. Schaffer, M.H. Loretto, R.E. Smallman, et al. The nature of the dispersoids in INCONEL alloy M A6000. J. Mat. Sci., **24**, 3261-3266 (1989)
5. N. Oono, Q.X. Tang, S. Ukai. Oxide particle refinement in Ni-based ODS alloy. Mat. Sci. Eng. A, **649**, 250-253 (2016)
6. R. Cozar, A. Pineau. Morphology of γ' and γ'' precipitates and thermal stability of inconel 718 type alloys. Metall. Trans., **4**, 47-59 (1973)
7. E. Guo, F. Xu, E.A. Loria. Improving thermal stability of alloy 718 via small modifications in composition. Superalloy 718– Metallurgy and Applications, 567-576 (1989)
8. X. Liu, J. Dong, B. Tang, et al. Investigation of the abnormal effects of phosphorus on mechanical properties of INCONEL718 superalloy. Mat. Sci. Eng. A, **270**, 190-196 (1999)

MATEC Web of Conferences 130, 03009 (2017) DOI: 10.1051/matecconf/201713003009
CMPSE2017

Removing metal debris from thermosetting EMC powders by Nd-Fe-B permanent magnets

Yowching Liaw[1,a], Jung-Hua Chou[1] and Shouyen Chao[2]

[1]*Department of Engineering Science, National Cheng Kung University, Tainan, 70101, Taiwan*
[2]*Department of Electronic Engineering, Minghsin University of Science and Technology, Hsinchu, 30401, Taiwan*

Abstract. During the preparation of thermosetting encapsulation molding compounds (EMCs) for semiconductor packaging, metal debris are always present in the EMC powders due to the hard silica fillers in the compound. These metal debris in the EMC powders will cause circuit shortage and therefore have to be removed before molding. In this study, Nd-Fe-B permanent magnets are used to remove these debris. The results show that the metal debris can be removed effectively as the rate of accumulation of the metal debris increases as time proceeds in the removing operation. The removal effectiveness of the debris is affected by both the magnetic flux density and the flow around the magnet. The wake flow behind the magnet is a relatively low speed recirculation region which facilities the attraction of metal debris in the powders. Thus, the largest amount of the accumulated EMC powders occurs downstream of the magnet. Hence, this low speed recirculation region should be better utilized to enhance the removal efficiency of the metal debris.

1 Introduction

The electronic packaging industry plays the cap stone role in the semiconductor industry because it is the last step in the whole manufacturing process of making electronic components. Currently, these components are forced into the realm of higher input/output (I/O) counts due to the ever quest for electronic products with high performance and lower cost of consumers. Higher I/O counts imply the reduction of the I/O pitch of IC chips. The immediate consequence of this pitch reduction is that the tolerance of metal debris in the encapsulation molding compounds (EMCs) is extremely restricted in order to avoid short circuits. This type of short circuit is very different from that due to wire sweeping [1, 2] which can be resolved by proper wire bonding and molding. The short circuit caused by metal debris can only be resolved by removing them from the EMCs.

The metal debris in the molding compounds can originate from several sources such as metals in the raw materials or those generated during the manufacturing process [3]. The latter situation typically occurs because EMCs are thermosetting materials and presently silica is the most common EMC filler with a large weight percentage of approximately 75%. The hardness of silica of Mohs scale 7 is very large for the facilities used in making the compounds during blending, kneading, and transporting through the metal piping system. Thus, metal debris can be produced due to abrasion or erosion of the facilities and have to be removed for reliability requirement as described above.

Even though the removal of metal debris in the EMC powders is essential in the electronic packaging industry, the related issue was rarely reported in the literature. This situation is very unique comparing to other industries facing similar issues, such as those in making foods, drugs, general powders [4-8], and metal machining [9, 10]. For the food, drug, and general powder industries, the causes of metal debris are similar to those of EMCs. Whereas, for the metal machining industry, metal debris are a by-product. Hence, for these industries, various methods have been applied to remove metal debris. These methods of removal are generally based on permanent magnets as the debris are usually ferromagnetic materials [4-10]. However, comparing to these industries, the uniqueness of lacking in the literature of metal debris removal in the EMC industry deserves special attention. Thus, this study intends to fill in this gap by presenting a technique which can remove metal debris in the EMC powders effectively for industrial applications.

Typical metal debris generated in the production of EMCs as shown in Fig. 1 are of slender types, including ribbon, sheet, and bulk shapes. Because of different shapes, their weights are also very different. In reality, these debris are generally enclosed by the EMC powders and therefore need to be removed by proper solvant to obtain the metal

[a] Corresponding author: n9897121@mail.ncku.edu.tw

debris alone. In this study, the metal debris with the EMC powders are removed by magnets. Hence, in the following sections, design and implementation of the removing setup is presented first, followed by the removal experimental results and conclusions.

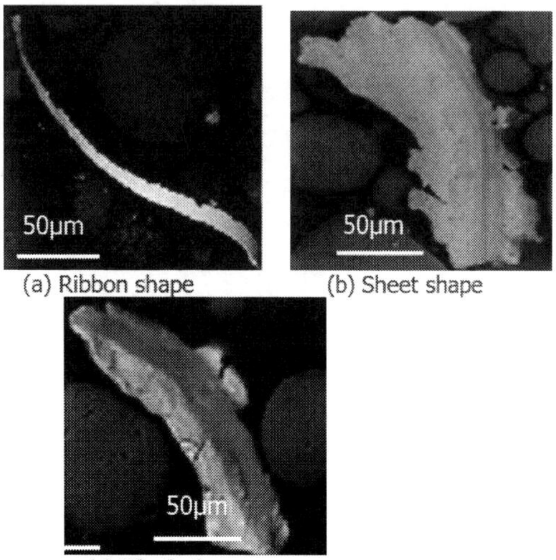

(c) Block shape

Figure 1. Typical shapes of metal debris in the EMC powders

2 Design and implementation of the Nd-Fe-B magnets

Due to energy considerations, the magnets adopted in removing metal debris are always of high strength permanent types, notably Nd-Fe-B magnets. In general, the geometry of the magnets is either flat or cylindrical. It is well known from fluid flow principles [11] that the form drag of a flat magnet is much larger than that of a cylindrical magnet. Moreover, the mesh type arrangement of the magnets tends to cause blockage [4, 7] because of aggregation due to the van der Waal force and accumulation of the powder particles to the magnets due to the magnetic force. Hence, the configuration of the parallel arrangement of magnets is selected in this study.

As the efficiency of separation of metal debris from EMC powders depends on the size, density and speed of the metal debris, and magnetic strength [8], a proper selection of the magnets and their arrangement are essential. Thus, the field strength of the proposed parallel configuration is examined first as follows.

2.1 Field strength of magnets versus distance

In order to determine the proper configuration of the magnets, the magnetic field strengths were measured first. Two types of magnets of different diameters of 134mm and 25mm, respectively were evaluated for their suitability. The magnets were formed by aligning the magnetic blocks with the same polarity inside a stainless shell of 1mm in thickness. The gap between the magnetic blocks chosen for this study was 5.5mm. The field strength in terms of magnetic flux density in Gauss was measured by a Gauss meter which was mounted on a height gauge. The magnet flux density was measured by varying the distance between the probe and the magnet surface. The measuring facility was setup on a granite platform as shown in Fig. 2.

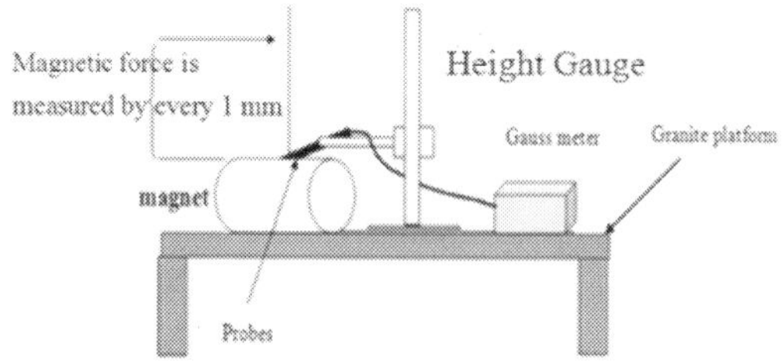

Figure 2. A schematic diagram of the magnetic field strength measurement facility

Fig. 3 depicts the magnetic field strengths of the two types of Nd-Fe-B permanents normalized by the field strength of the smaller magnet. In addition to the fact that the magnet flux densities of both magnets decrease as the distance between the Gauss meter and the surface of the magnet increases, the flux density of the magnet of 134mm in diameter is always larger than that of the 25mm magnet, irrespective of the distance from the surface of the individual magnet.

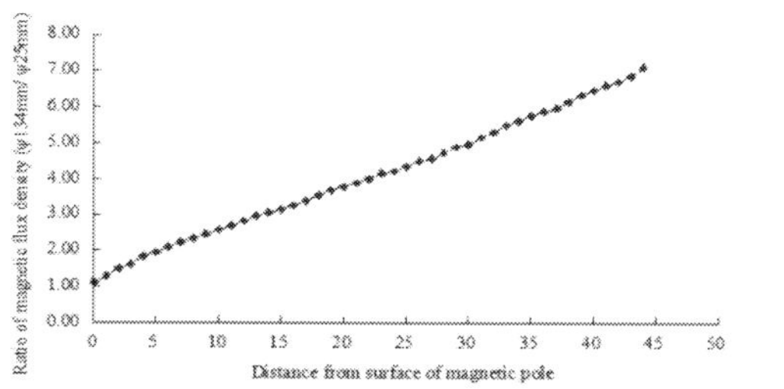

Figure 3. Normalized field strength versus distance from the surface of the magnet

Despite the fact that a higher magnetic flux density can be achieved by the larger magnet, the smaller magnet of 25mm in diameter was adopted in this study so that more magnets could be installed in a given space to generate a more uniform distribution of the magnetic flux density. Hence, the smaller magnet was selected in this study.

2.2 Field strength of magnets versus gap

For debris attraction, Nd-Fe-B magnetic blocks were arranged with the same polarity facing each other. Hence, the magnetic force is larger when the gap between the poles of the same polarity is smaller. To quantify this gap effect, a magnet with two magnetic blocks inside a SUS304 stainless tube with an inner diameter of 25mm was designed for this purpose. The gap between the poles was adjusted continuously so that the magnetic field intensity on the tube surface could be measured.

The results are portrayed in Fig. 4. For the gap between 0 and 40mm, two trends can be observed. One is that the smaller the gap, the larger the flux density, consistent with the results of the literature [12-14]. The other is that the rate of decreasing of the magnetic flux density reduces as the gap increases. That is, the flux density reduces more rapidly for smaller gaps and then gradually levels off to a constant value as the gap reaches 20mm and beyond. Thus, for the present study, the gap of 5.5mm was selected.

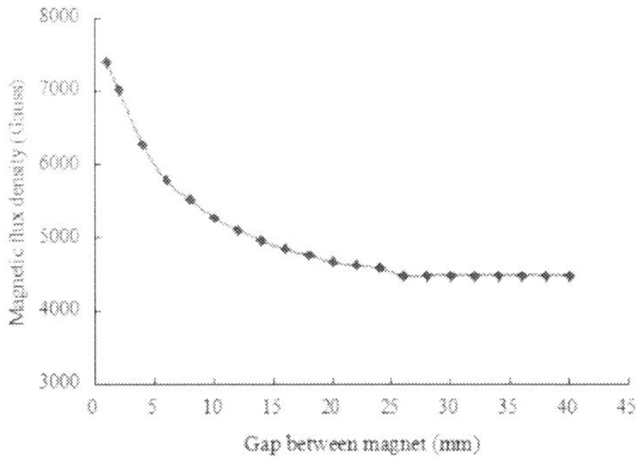

Figure 4. Surface Magnetic flux density versus pole gap

2.3 Facility for debris removal

The facility for removing the metal debris is drawn in Fig. 5. The facility consisted of a T-shaped upper hopper with a square cross-sectional area and a lower chamber with the magnets. The EMC powders were released from top of the hopper and moved through the hopper and the magnet chamber by free falling. The hopper was about 177mm in height with the entrance area of 350mm by 350mm and exit area of 255mm by 255mm. The magnet chamber housed five Nd-Fe-B magnets of 25mm in diameter and 208mm in length; the gap between the neighboring magnets was 32mm. Each magnet had seven Nd-Fe-B blocks with six gaps; the size of each gap was 5.5mm.

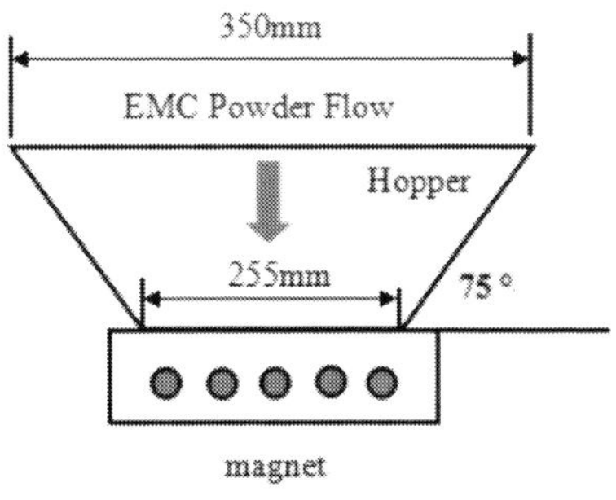

Figure 5. The facility for collecting EMC powders with metal debris

The EMC powder entered the hopper from the top at a mass flow rate of 1200kg/hr continuously for four hours. Afterwards, the magnets were removed from the magnet chamber to measure the thicknesses of the EMC powders attracted by the magnets. Since the shape of the EMC powders attached to the magnets was not circular due to the flow around the magnet, four dimensions, X1, X2, Y1 and Y2, as shown in Fig. 6 were used to quantify the attraction of the EMC powders. The symbol Y and X denote the gravitational direction and the direction perpendicular to gravity, respectively.

As the metal debris were covered by the EMC powders, the powders were removed from the magnets after thickness measurements. Then solvents were used to dissolve the EMC powders to separate the metal debris from the EMC powders. Then, the weight of the metal debris was measured afterwards.

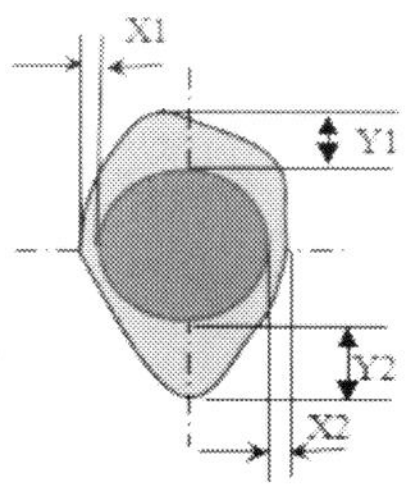

Figure 6. Measurement dimensions of EMC powders versus the magnet

3 Results and discussion

As described above, the magnets were removed from the magnet chamber to measure the thickness of the EMC powders attracted by the magnets after continuous operation for four hours. Typical shapes of the attached EMC powders are portrayed in Fig. 7 from the top view for visual observation. The numerals in the figures denote the numbers of the five magnets installed in the magnet chamber.

Two trends can be observed from the shape of these EMC powders on the magnets. One is that more powders were attracted by the two magnets at the two sides of the magnet chamber. This is due to the convergent geometry of the hoper to direct more EMC powders passing over the two magnets at the sides than those three in the middle. The other is that the powder shape varies fairly periodically along the longitudinal axis of the magnet. This phenomenon is caused by the periodic arrangement of magnet blocks inside the magnets. It is well known that the magnetic flux density is the largest at the pole locations [15]. Hence, the periodic peak corresponds to the pole positions, whereas the periodic valley occurs at the middle of each magnetic block.

Further quantification is made from the measured thickness tabulated in Table 1 for the five magnets represented by their corresponding numerals. Magnet numbers 1 and 2 denote the magnets at the two sides whereas 2, 3, and 4 are the middle ones. It is clear that the magnets at the two sides have a larger thickness, irrespective of the X and Y directions for the reasons explained above. In addition to the shape features described above, three more features can be noticed. First,

 DOI: 10.1051/matecconf/201713003009

the thicknesses in the X direction are always smaller than those in the Y direction, irrespective of the position of the magnet. Second, the thicknesses of X1 and X2 are about the same. Last, The Y2 thickness is the largest for all cases. It should be noted that a larger thickness represents a large amount of metal debris being attracted by the magnets.

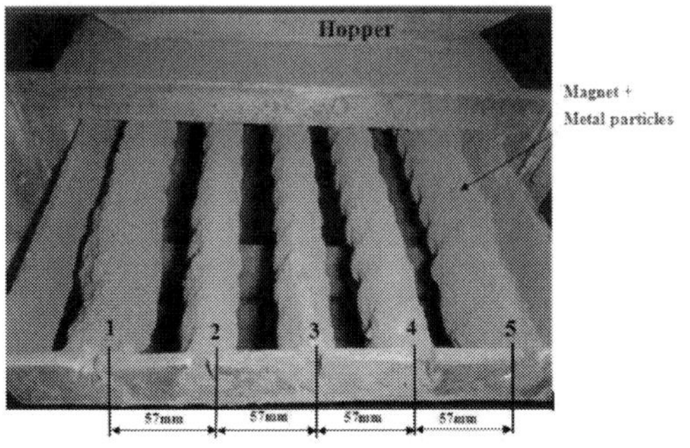

Figure 7. The shapes of the EMC powders attached to the magnets

Table 1. The thickness of EMC powders on each magnet

Magnet No.	1	2	3	4	5
X1(mm)	3.2	1.1	1.1	1.1	3.0
X2(mm)	3.2	1.1	1.1	1.1	3.0
Y1(mm)	4.3	1.5	1.5	1.5	4.3
Y2(mm)	8.0	6.0	6.0	6.0	8.0

As the magnetic flux density is symmetric with respect to the longitudinal axis of the magnet, the thickness distribution of the EMC powders around the magnet shows an interesting phenomenon. Apparently, debris attraction is not only influenced by the magnetic flux density but also by the flow around the magnet. For the circular magnet, the wake flow forms behind the magnet [11]. The powder velocity in the wake is smaller due to flow recirculation. In contrast, the velocity at the side of the magnet is the largest due to acceleration while that in front of the magnet is second due to flow retardation. Thus, the corresponding thickness is the smallest at the side and that in the front is slightly larger. That is, the debris removal efficiency can be enhanced by adjusting these two factors of field strength and wake recirculation.

To further understanding the accumulation rate of debris attracted by the magnets, the magnets were removed from the magnet chamber at each hour during the four hour operation period to measure the weight of the attracted metal debris. The results are shown in Fig. 8. Please note that the results shown in Fig. 5 are for the metal debris only without the covered EMC powders because the powders were dissolved by solvent already before measuring the weight of the metal debris. It can be seen that the rate of accumulation increases as the operation time proceeds. Specifically, at the first hour, the accumulation rate is 0.25g/hr; whereas at the last hour, the rate is 0.48g/hr. This situation is partly due to the van der Waals force [4] of the attracted metal debris and also due to the increased equivalent surface area of the magnet by the presence of the irregular shaped metal debris. The increased rate of accumulation also indicates that the metal debris are removed from the EMC powders effectively.

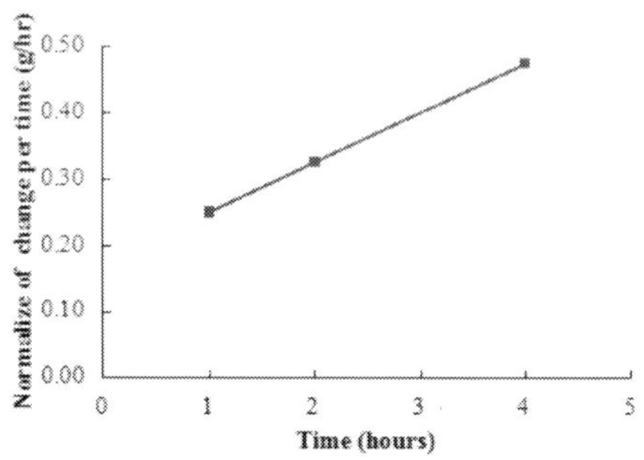

Figure 8. Accumulation rate of metal debris versus operation time

4 Conclusions

The thermosetting characteristic of current encapsulation molding compound generally leads to the generation of metal debris during its preparation. The situation is caused by the very hard silica fillers in the compound. The presence of these metal debris will cause circuit shortages and have to be removed.

In this study, Nd-Fe-B permanent magnets were used to remove these metal debris in the encapsulation molding compounds. The results show that the metal debris can be removed effectively as evidenced by the increased accumulation rate. More specifically, the metal debris removal rates are 0.25g/hr and 0.48g/hr at the first and fourth hour of the removal operation, respectively. Moreover, the backside of the magnet is the most effective region of debris removal due its recirculation flow feature.

References

1. L. T. Nguyen, 43rd ECTC, Orlando, FL, USA, 1993.
2. S.Han and K. K. Wang, IEEE Trans. Comp., Packag., Manufact. Technol. B, **18**, 4 (1995).
3. Y. Liaw and J. H. Chou, AMR, 126-128, 2010.
4. Y. Nakai, K. Senkawa, F. Mishima, Y. Akiyama, and S. Nishijima, {IEEE} Trans. Appl. Superconduct, **21**, 3 (2011).
5. F. Mishima, T. Terada, Y. Akiyama, Y. Izumi, H. Okazaki, and S. Nishijima, {IEEE} Trans. Appl. Superconduct, **18**, 2 (2008).
6. S. Hayashi, F. Mishima, Y. Akiyama, and S. Nishijima,

IEEE Trans. Appl. Superconduct, **20**, 3 (2010).
7. Y. Nakai, F. Mishima, Y. Akiyama, and S. Nishijima, IEEE Trans. Appl. Superconduct, **20**, 3 (2010).
8. F. Mishima, S. Yamazaki, K. Yoshida, H. Nakane, S. Yoshizawa, S. Takeda, Y. Izumi, and S. Nishijima, {IEEE} Trans. Appl. Superconduct, **14**, 2 (2004).
9. D. Ruttley (Rattler Tools, Inc.) Metal debris cleanout system and method US 20060049111 A1, 2006
10. D. Ruttley (Rattler Tools, Inc.) Metal debris cleanout system and method US 7410014 B2, 2008
11. M. Van Dyke, An Album of Fluid Motion, The Parabolic Press, California, USA, 1982.
12. N. I. Kasim, M. A. Musa, H. Ngah, A. R. Razali and M. Ishak, J. EAS, **10**, 17 (2015).
13. K. Ozturk , E. Sahin , M. Abdioglu , M. Kabaer , S. Celik , E. Yanmaz , T. Kucukomeroglu, J. Alloy Compd., **643**, 201–206, 2015.
14. S. Celik, J. Alloy Compd., **662**, pp.546-556, 2016.
15. T. Dimova, M. Marinova, B. Aprahamian, 19th SIELA, Bourgas, Bulgaria, 2016.

DOI: 10.1051/matecconf/201713004001

Engineering Properties and Microstructural Performance of Low Energy Super-Sulfated Cement Using Industrial Waste Anhydrite

Tsung-Yuan Huang[1], Hoang-Anh Nguyen[2], Ta-Peng Chang[3,*], Chun-Tao Chen[4] and Yu-Wen Liu[5]

[1,2,3,4]*Department of Civil and Construction Eng., National Taiwan University of Science and Technology (NTUST) (Taiwan Tech), Taipei, Taiwan 106*
[5]*Department of Civil and Water Resources Engineering, National Chiayi University, Chiayi, Taiwan*
[*]*Corresponding author, Department of Civil and Construction Engineering, Taiwan Tech, tpchang@mail.ntust.edu.tw*

Abstract. This study aims at proposing the mix proportions of low energy super-sulfated cement (SSC) concrete from industrial waste anhydrite from circulating fluidized bed combustion (CFBC) fly ash (CFA) as an alternative sulfate activator of ground granulated blast furnace slag (GGBFS/slag). The optimized mix proportion of the SSC was carried out by using mixture of different amounts of CFA in range of 25 − 45 wt.% and various quantities of ordinary Portland cement (OPC) in range of 0 − 10 wt.% to trigger the hydration of slag. Experimental results showed that with the expected slump at values of 190 − 220 mm, the 28-day compressive strengths of the concrete with low energy SSC reached 43.69 MPa which can be feasibly applied for widely advanced construction materials. The OPC in range of 3 − 5 wt.% and 25 wt.% of CFA were considered as the optimum ingredients of the activator and was suggested to be used for fabricating the low energy SSC concrete with the good performance on compressive strength, dynamic Young's modulus, UPV measurement, and stabilized change of length. The OPC additive up to 10 wt.% was encouraged to be used for producing the SSC concrete with significant reduction on creep.

1 Introduction

The ordinary Portland cement (OPC) has become one of the primary construction materials because of its lower energy consumption than those of others such as aluminum and steel. However, the cement industry has been remarked as an intensive consumer of natural raw materials, fossil fuels, energy, and a major source of multiple pollutants [1]. Indeed, during the manufacturing process of OPC cement, a great deal of amounts of lime stone, quartz, and clay are fed as the raw materials. As such, the OPC manufacture significantly causes the serious damage of natural resources, particularly the surface of the earth. During the past decades, super-sulfated cement (SSC), which is fabricated mostly without OPC, has become one of the promising candidates of alternative ordinary Portland cement and has been the research interest because of the lowered energy consumption and superior durability in terms of resistances to chemical attacks [2]. Indeed, the SSC has been commercially manufactured in Europe and India to reduce the impacts from OPC manufacture on high consumption of natural material and energy and air pollution caused by carbon dioxide emissions. In general, the SSC is comprised by mixture of 80-85% ground granulated blast furnace slag, 10-15% sulfate activator (anhydrite/gypsum) and low amount of OPC (normally 5%) [3-5]. Recently, to produce the SSC with lowered cost, the utilization of industrial waste gypsum such as flue gas desulfurization (FGD) gypsum was proposed to be used as sulfate activator instead of using commercial gypsum [6, 7]. In addition, the industrial waste anhydrite included in circulating fluidized bed combustion (CFBC) fly ash was also used as alternative sulfate activator in low energy SSC [8, 9]. However, the studies focusing on using the combination of CFBC fly ash and OPC as the activator for low energy SSC has not been found. Therefore, this study proposes a new way for manufacturing low energy SSC by using the CFBC fly ash totally replacing for commercial gypsum and small amount of OPC as the activator of slag. The proposed concrete with low energy SSC, in this study, plays an important role in enriching the categories of green cement mostly qualifying for the requirements from the development of sustainability.

2 Experimental program

2.1 Materials and mix proportions low energy SSC

The commercial Type I Portland cement in accordance to ASTM C150 and CFBC fly ash (CFA) were used as the alkali and sulfate activators, respectively. Ground granulated blast furnace slag (GGBFS) and CFBC fly ash

[a] Corresponding author: author@e-mail.org

(CFA) were used as the primary powder. The chemical and mineral compositions of raw materials were shown in Table 1 and Fig. 1, respectively. Crushed natural stone with maximum size of 20 mm and river sand with fineness modulus (FM) of 2.9 were used as coarse and fine aggregates, respectively. The specific gravities of sand and crushed stone are 2.65 and 2.67, and water absorptions are 0.8% and 1.0%, respectively. Both sand and crushed stone were carefully washed to ensure good binding in concrete mixtures.

Table 1. Physical properties and chemical compositions of three industrial solid by-products.

	Slag (GGBFS)	CFBC fly ash (CFA)
Specific gravity	2.90	2.70
Blaine fineness, cm^2/g	6000.00	3000.00
SiO_2, wt.%	34.90	5.22
Al_2O_3, wt.%	13.53	2.21
Fe_2O_3, wt.%	0.52	0.58
CaO, wt.%	41.47	56.80
MgO, wt.%	7.18	2.06
SO_3, wt.%	1.74	32.40

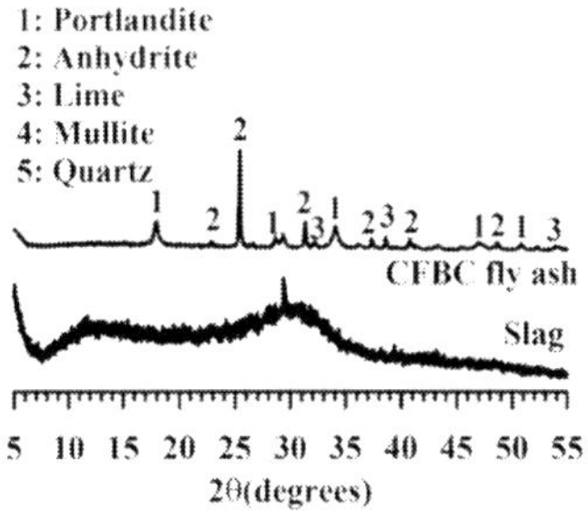

Figure 1. XRD patterns of two solid wastes.

In this study, the effects of ingredients on performance of low energy SSC concretes were estimated based on the various percentage of alkali (OPC) and sulfate (CFA) activators as replacements for the main powder (i.e., slag). Accordingly, three values of 25, 35, and 45 wt.% of CFA and four values of 0, 3, 5, and 10 wt.% of OPC as partial replacement for GGBFS were used to produce SFC-SSC. The water to binder ratio (W/B) of the low energy SSC concrete was fixed at 0.4. The volume of paste and weight ratio of sand to mixture of sand and stone were fixed at 37% and 0.41, respectively. To control the workability of fresh concretes with the slump value in range of 190−225 mm, Type G superplasticizer (SP) was used. The mix proportions of SFC-SSC concretes are shown in Table 2.

2.2 Specimen preparation and test methods

The property of fresh low energy SSC concrete was adjusted by using slump test by adapting ASTM C143. The cylindrical specimens of the concretes with diameter of 100 mm and length of 200 mm were cast for the tests of compressive strength in accordance to ASTM C39, dynamic modulus standardized by ASTM C215, ultrasonic pulse velocity (UPV) suggested by ASTM C597, and creep behavior in accordance with ASTM C512. After 24 hours of curing in the molds at room temperature of 25±2 °C, all specimens were removed and cured in the room temperature of 25±2 oC and 50 % RH. For the shrinkage/expansion test, the SFC-SCC concrete prisms with dimensions of 75×75×285 mm were cast. The specimens were demolded after 24 h of casting and followed the same curing regimes to the cylindrical specimens. The test would be stopped as the changes of shrinkage/expansion of the bars were no longer significant at all the curing regimes. In this work, the shrinkage/expansion of specimens was monitored until 60 days. The procedure of the length change of concretes was followed ASTM C596. For microstructural examination, the mortar collected from fractured concrete was soaked in to the alcohol for at least 7 days in advance and tested by using scanning electron microscopy (SEM) and X-Ray diffraction (XRD).

Table 2. Mix proportions of the low energy SSC (kg/m^3).

Mixture	Slag	OPC	CFA	FA	CA	W
S75-C0-CF25	350	0	117	688	990	184
S72-C3-CF25	336	14	117	688	990	185
S70-C5-CF25	327	23	117	688	990	185
S65-C10-CF25	304	47	117	688	990	185
S62-C3-CF35	289	14	163	688	990	184
S52-C3-CF45	241	14	209	688	990	183

Note: OCP: Ordinary Portland cement; CFA: CFBC fly ash; FA: Fine aggregate; CA: Coarse aggregate; W: Tap water; Superplasticizer (SP) was fixed at 2.1 kg/m^3.

3 Results and discussions

3.1. Fresh properties

The fresh properties of the resulting low energy SSC concretes were shown in Table 3. According to the table, the effects of both CFA and OPC as partial replacements of slag on the workability of the fresh low energy SSC concretes were insensitive because Table 2 shows that the expected slump values of the fresh SSC concretes were approached by using the equivalent amount of SP value. On the other hand, the increases of either OPC or CFA replacing for slag led to the decrease in both the initial and final setting time of the resulting low energy SSC concretes, which obviously implies that the hydration rate of the SSC binder was accelerated with the increase of the activators.

Table 3. Fresh properties of the low energy SSC

Mixture	Setting time (min) Initial :Final	Slump (mm)
S75-C0-CF25	825 : 990	190
S72-C3-CF25	860 : 1006	225

 DOI: 10.1051/matecconf/201713004001

S70-C5-CF25	837 : 987	225
S65-C10-CF25	797 : 944	220
S62-C3-CF35	799 : 963	220
S52-C3-CF45	746 : 887	190

3.2 Compressive strength

The compressive strengths of SSC concretes are shown in Fig. 2. Accordingly, the compressive strengths of the concretes specimens increased with the increase of curing ages. To produce the low energy SSC concretes with the expected compressive strength, a range of 3 − 5 wt.% of OPC accompanying with 25 wt.% of CFA was the optimum value of the activator. According to Fig. 2(a), the increase of OPC from 0 to 5 wt.% significantly increase the compressive strength of the hardened SSC concretes at all ages of curing. Such result could be due to the accelerated hydration of slag with the increase of OPC addition. On the other hand, Fig. 2(b) showed that the increase of CFA beyond 25 wt.% led to the significant decrease in compressive strength of the SSC concrete irrespective of ages of curing. The reason could be resulted from the destroyed structure of the SSC concrete with the excessive addition of anhydrite (i.e. from CFA) which causes the expansion of the hardened concrete specimens induced by the generation of gypsum and ettringite (AFt).

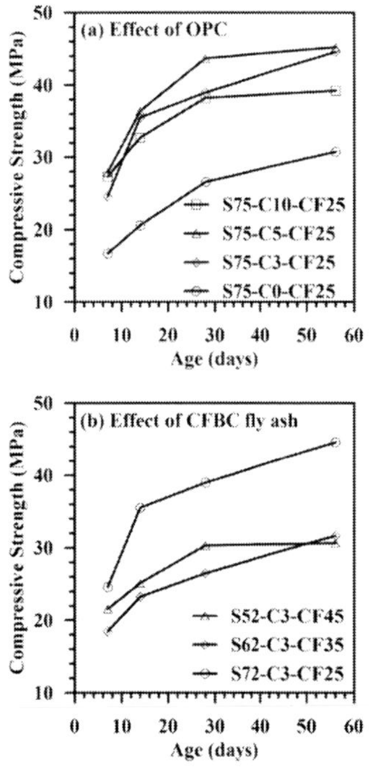

Figure 2. Compressive strength of low energy SSC concretes.

3.3 Dynamic Young's modulus

The dynamic Young's moduli of SSC concretes were conducted and the results were shown in Fig. 3. According to the figure, the dynamic Young's moduli of the concretes specimens increased with the increased ages of curing due to the increase of compressive strength with the days. In this study, the result of dynamic Young's moduli of the SSC concretes matched with the observation from the compressive strength results. As such, OPC amount in range of 3 − 5 wt.% and CFA with 25 wt.% was the optimum quantities of the activators because their additions led to the hardened concrete samples with the improved dynamic Young's moduli accompanying with the increased compressive strengths as aforementioned.

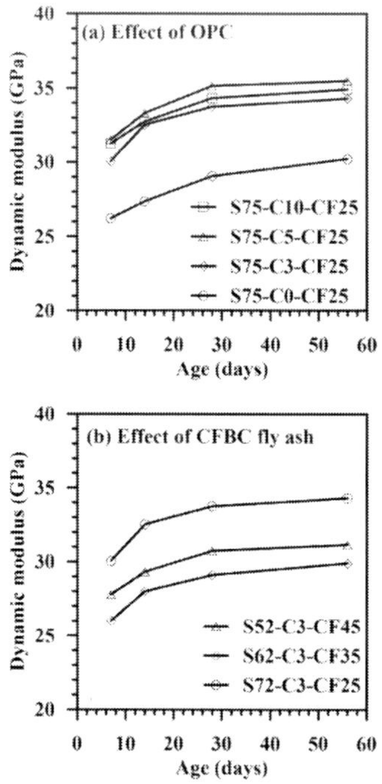

Figure 3. Dynamic modulus of low energy SSC concretes.

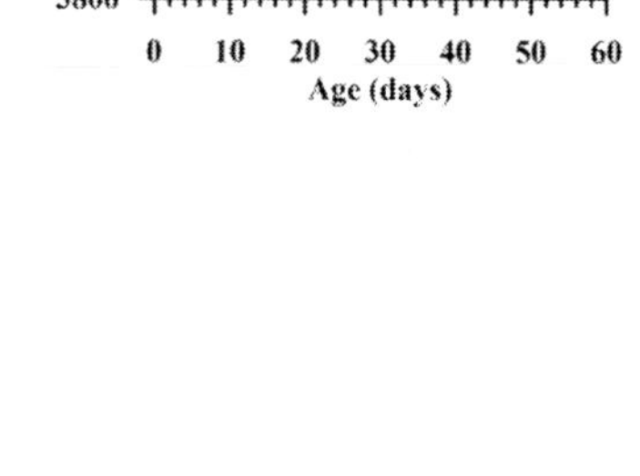

 DOI: 10.1051/matecconf/201713004001

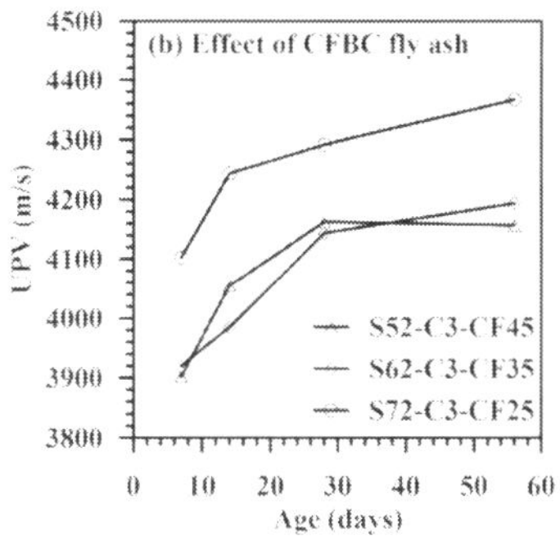

Figure 4. UPV measurements of low energy SSC concretes.

3.4 Ultrasonic pulse velocity (UPV)

The UPV measurements of the hardened SSC concretes as shown in Fig. 4 apparently proved for the observed results from the strength and modulus of the hardened concrete specimens. Generally, the detected UPV values of the hardened SSC concretes increased proportionally with the increase of ages of curing because of the increased compressive strengths of the samples. As can be seen in Fig. 4(a), the increase of OPC in range of 0−5 wt.% significantly increased the UPV value of the hardened concrete samples due to the more condensed structure of the binder system resulted from the accelerated hydration process of slag with the increase of OPC addition. On contrary, the damaged interior microstructures of the hardened concrete samples were obviously observed with excessive addition of CFA because of the remarkable reduction of UPV measurements (Fig. 4(b)).

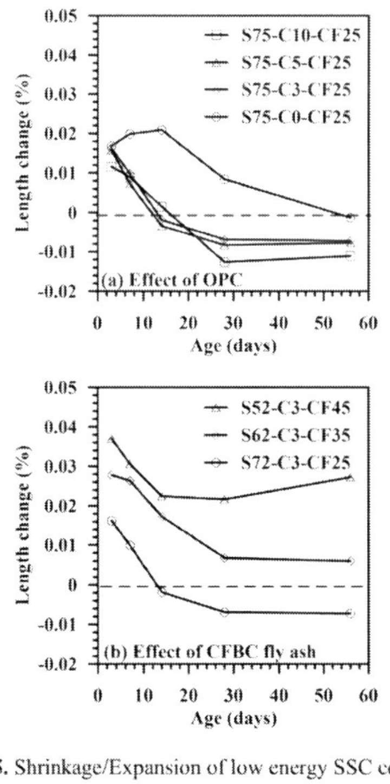

Figure 5. Shrinkage/Expansion of low energy SSC concretes.

3.5 Shrinkage/Expansion

The shrinkage/expansion values of the hardened SSC concretes as shown in Fig. 5 was also the valuable evidence supporting for the engineering performance of the concrete specimens. It was obvious to point out from the figure that both OPC and CFA additions had the distinguishing influence on shrinkage/expansion of the hardened SSC concretes. From Fig. 5(a), the addition of OPC increased the shrinkage of the hardened SSC concretes due to the reduction of AFt crystals induced by the reduction of alumina supplied from slag. However, the increase of the CFA led to the hardened SSC concretes with the significant increase in expansion possibly caused by the increases of AFt precipitation and delayed generated gypsum. As such, the addition of CFA further from 25 wt.% caused the impact of structure of the hardened SSC concrete and led to the concrete specimens with the reduction on both engineering indicators such as compressive strength, dynamic Young's modulus and UPV measurement.

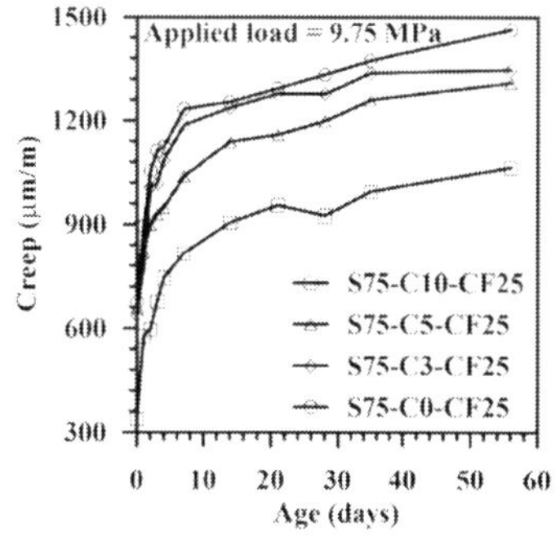

Figure 6. Creep behavior of low energy SSC concretes.

3.6 Creep behavior

The creep values of SSC concretes are shown in Fig. 6. Accordingly, the creep values of the concretes specimens increased with the increase of curing ages. For fabricating the low energy SSC concretes with the reduced creep, the addition of OPC up to 10 wt.% as replacement for slag was encouraged. Such result apparently implies that the cementing system of the SSC binder was strengthened with the addition of OPC. In this study, however, the influence of the CFA addition on creep behavior of the hardened SSC concretes was not conducted.

Figure 7. SEM image of low energy SSC concrete with S70-C5-CF25 mix at 28 days (top) and 56 days (bottom) of curing.

3.7 Microstructural examination

The hydration products of the hardened mortar collected from the fractured pieces of the low energy SSC with S70-C5-CF25 mix were conducted as shown in SEM image (Fig. 7). Accordingly, the hydration products of the hardened low energy SSC mainly included rod-like crystal of ettringite (AFt) blending with amorphous gels of C-S-H/C-A-S-H. From Fig. 7, presence of gypsum was also obviously detected. Such result could be resulted from the hydration of remaining anhydrite from CFBC fly ash with free water. The XRD patterns of the powder hydrated mortar with S70-C5-CF25 mix significantly clarified the observation from SEM images. Indeed, the peaks indicating the precipitations of AFt and gypsum were apparently detected from the figure. Because the C-S-H/C-A-S-H gels are amorphous, their presence was not obviously observed from XRD pattern. In this study, the peaks revealing for the presences of quartz and calcite was due to the presence of fine aggregate and unavoidable carbonation of the sample, respectively.

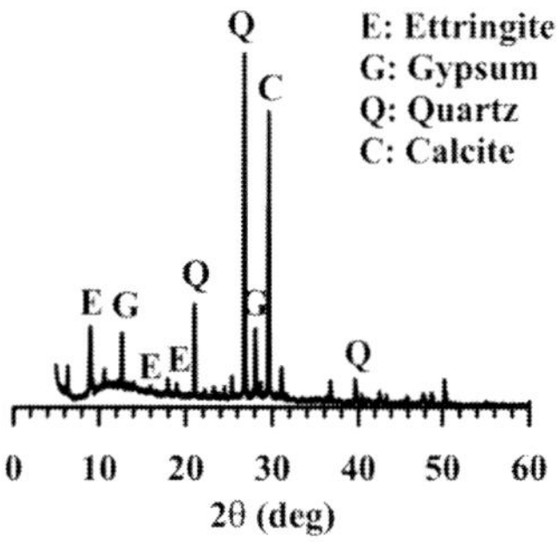

Figure 8. XRD patterns of low energy SSC concrete with S70-C5-CF25 mix at ages of 28 days.

4 Conclusions

The new manufacture of low energy super-sulfated cement (SSC) by using industrial waste anhydrite from circulating fluidized bed combustion (CFBC) fly ash (CFA) has been proposed for the future cement/concrete industries reaching the requirements for sustainable construction materials. The activator combining 3 – 5 wt.% of OPC with 25 wt.% of CFA was suggested to be used for manufacturing the SSC concrete with the expected compressive strength, dynamic Young's modulus, UPV measurement, and stabilized change of length. The 28-day compressive strengths of the low energy SSC concrete reached value of 43.69 MPa which possible applied for advanced construction materials. The SSC concrete with optimized ingredients illustrated the dynamic Young's modulus higher than 35 GPa at ages of 28 days. The observed UPV measurement of the SSC concrete was much higher than 4000 m/s which apparently implying the concrete with high quality. The addition of OPC was encouraged to be used for producing the SSC concrete with stabilized length change and significant reduction on creep. The excessive addition of CFA was avoided to assure the resulting concrete with eliminated impact on engineering performance induced by expansion phenomenon. In this study, the OPC addition of up to 10 wt.% replacing for slag led to the hardened SSC concrete with creep value 27.5% less than that of the SSC concrete samples without addition of OPC. Microstructure examination showed that the hydration products mainly included ettringite (Aft), C-S-H/C-A-S-H gels, and gypsum.

References

1. Y. Elakneswaran, E. Owaki, S. Miyahara, M. Ogino, T. Maruya, T. Nawa, Hydration study of slag-blended cement based on thermodynamic considerations, Construction and Building Materials **124** (2016) 615-625.
2. M.C.G. Juenger, F. Winnefeld, J.L. Provis, J.H. Ideker, Advances in alternative cementitious binders, Cement and Concrete Research **41**(12) (2011) 1232-1243.
3. A. Gruskovnjak, B. Lothenbach, F. Winnefeld, R. Figi, S.C. Ko, M. Adler, U. Mäder, Hydration mechanisms of super sulphated slag cement, Cement and Concrete Research **38**(7) (2008) 983-992.
4. M.S. Imbabi, C. Carrigan, S. McKenna, Trends and developments in green cement and concrete technology, International Journal of Sustainable Built Environment **1**(2) (2012) 194-216.
5. S. Ding, Z. Shui, W. Chen, J. Lu, S. Tian, Properties of supersulphated phosphogysumslag cement (SSC) concrete, Journal of Wuhan University of Technology-Mater. Sci. Ed. **29**(1) (2014) 109-113.
6. F.-Q. Zhao, W. Ni, H.-J. Wang, H.-J. Liu, Activated fly ash/slag blended cement, Resources, Conservation and Recycling **52**(2) (2007) 303-313.
7. S. Zhong, K. Ni, J. Li, Properties of mortars made by uncalcined FGD gypsum-fly ash-ground granulated blast furnace slag composite binder, Waste management **32**(7) (2012) 1468-72.
8. N.T. Dung, T.-P. Chang, C.-T. Chen, Engineering and sulfate resistance properties of slag-CFBC fly ash paste and mortar, Construction and Building Materials **63** (2014) 40-48.
9. I.M.A.K. Salain, P. Clastres, J.M. Bursi, C. Pellissier, Circulating Fluidized Bed Combustion Ashes as an Activator of Ground Vitrified Blast Furnace Slag, Special Publication **202** (2001) 225-244.

Investigation of the Mechanical Property of Low-Activation Concrete

Yu-Cheng Kan [1, a], Kuang-Chih Pei [2] and Meng-Hwa Cheng [1]

[1] Department of Construction Engineering, Chaoyang University of Technology, Taichung, Taiwan

[2] Institute of Nuclear Energy Research, Taiwan

Abstract. This paper presents the strengths and fracture toughness of low-activation concrete (LAC) which can be used in the nuclear shielding structures and possibly in the nuclear decommissioning project. The research used four mixtures which include high alumina cement with 20%, 40% and 60% aluminum oxide as fine aggregate replacement and compared with those of normal concrete. All of the concrete strengths were evaluated and compared. The test results showed that LAC has high early strength and decays later due to conversion effect. The other fundamental mechanical properties will also be presented in this paper.

1 Introduction

Although the Fukushima nuclear power plant accident in 2011 has attracted plenty attentions in nuclear safety issue, nuclear energy still plays a significant role in the world due to its high cost-efficiency and little carbon emission. Therefore, the structural technology dealing with the nuclear facilities still needs to be innovated. Conventionally, heavy concrete is mostly used in biological shields or shielding walls in a reactor vessel[1]. Kan and Pei[2] have found that the compressive and tensile strengths of heavy concrete did not differ from those of regular mortar at the same water to cement ratio; however, the modulus of elasticity increases with the heavy aggregate aggregates. Kan and Yang [3] later also investigated the toughness of heavy concrete strengthened by the inclusion of steel fibers. Recently, M. Kinno et al. [4-5] has attempted to develop many mixtures of low-activation concrete and reported that the concrete including fused alumina aggregate and high alumina cement contribute led to the least activation among all the designated mixtures. Thus, it seems reasonable and necessary to access the feasibility to use this type of material in shielding structure design. Thus, the research investigated many engineering properties and strengths of low-activated concrete influenced by the various aggregates of high aluminum inclusion

2 Experimental Program

2.1 Materials and Mixtures

A Type I Portland cement and an alumina cement produced in France were used in this study. The typical constituent of both cements are shown in Table 1. The mechanical properties of concretes made were tested and compared. Mixtures of both concrete have a same water/cement ratio of 0.5. Table 2 shows the all mixtures. The amount of boron carbide B_4C is fixed at 10 % by weight for all LAC mixtures; while the amount of Al_2O_3 aggregates is varied. The low-activated concretes were designated with 20 %, 40 % and 60 % fine aggregate replacement by fused aluminum oxide (Al_2O_3 aggregates) by weight, to examine the effect of the high alumina cement and various amounts of Al_2O_3 aggregates on the mechanical properties of LAC. The LAC mix with 40% sand replacement by Al_2O_3 aggregates, for example, is specified as LAC40. It is noted that the unit weight of concrete increases with the Al_2O_3 aggregates, as shown in Fig. 3. And, it is found that the placement in concrete with 60% of Al_2O_3 aggregates by weight produced a harsh mix in workability.

Table 1. Constituents of cements.

Composition	Type I	HAC
Al_2O_3	4.95	50.8-54.2
CaO	61.96	35.9-38.9
SiO_2	20.42	4.5-5.6
Fe_2O_3	3.09	1.0-2.2
MgO	3.29	<1.0
TiO_2	-	<4.0
K_2O+Na_2O	-	<0.5
SO_3	-	<0.1

[a] Corresponding author: yckan@gm.cyut.edu.tw

2.1.1 High Alumina Cement

In high alumina cement, the predominant compounds are calcium aluminates; calcium silicates account for no more than a few percent. The calcium aluminates react with water and the primary product is calcium aluminate decahydrate (CAH_{10}) which has essentially a hexagonal crystal form. This material is unstable and changes to tricalcium aluminate hexahydrate (C_3AH_6) spontaneously. The process occurs at room temperature and is accelerated by an increase in temperature. The effect of conversion, as it is known, is accompanied by an increase in porosity and a decrease in strength.

2.1.2 Fused Alumina Aggregates

Unlike normal concrete, the main feature of heavy concrete is the inclusion of metallic fillers, which are usually ilmenite, limonite-goethite, serpentine, magnetite, barite, ferrophosphorous, steel aggregate and iron shot. In this study, fused aluminate aggregates were adopted as part of fine aggregates in concrete. It has an average grain size of 1 mm (ANSI No. 20) with a specific gravity is 3.90 and a hardness of 9 in Moh's scale.

2.1.3 Boron carbide

The ability of boron carbide to absorb neutrons without forming long-lived radionuclides makes it attractive as an absorbent for neutron radiation arising in nuclear power plants. Nuclear applications of boron carbide include shielding, control rod and shut down pellets. Within control rods, boron carbide is often powdered, to increase its surface area. In this study, boron carbide powder was used as part of concrete inclusion. It has an average grain size of 0.045 mm (ANSI No. 240) with a specific gravity is 2.52 and a hardness of 9.6 in Moh's scale.

Table 2. Concrete mixtures.

	NC	HAC		
Component		20%	40%	60%
Water	193§	196	200	204
Cement	386	353	361	368
Aggregate	995	1012	1032	1054
Sand	762	619	474	322
B4C	-	39	40	41
Al2O3	-	155	316	484
Water	193§	196	200	204

§All units are in kg/m³. *Type I Portland cement concrete; **high aluminum concrete.

2.2 Testing Specimens

Four types of specimens made from the above four mixtures were fabricated in this testing program. The ϕ 10 x 20 cm cylinders were used to determine the mechanical properties of concrete, including the compressive strength and tensile strength, modulus of elasticity and wave velocity.

3 Results and Discussion

The fundamental mechanical properties of concrete including compressive strength fc', splitting tensile strength ft'and elastic modulus Ec are presented in Table 3 and discussed as following.

Table 3 Mechanical Properties of concrete

Mixture	day	f c' MPa	f t' MPa	Ec GPa
NC	28	30.1	1.98	20.7
	91	38.0	3.6	18.0
LAC20	28	43.7	2.97	24.4
	91	35.9	3.33	17.9
LAC40	28	50.2	3.42	26.9
	91	51.9	2.61	24.3
LAC60	28	44.3	2.16	27.3
	91	31.1	3.33	24.5

3.1 Compressive Strength

. The compressive strength of LAC exceeds that of normal concrete and increases with Al_2O_3 aggregates for a given water/cemnt ratio. The compressive strengths for all mixes are shown in Fig. 2. It reveals that an increase in Al_2O_3 aggregates causes the increase in the

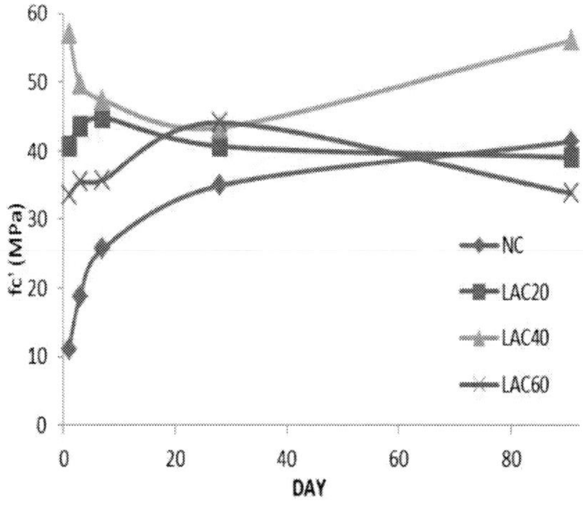

Fig. 2 Compressive strength of concretes

compressive strength. However, the mixture with an inclusion of 60% of Al_2O_3 aggregates by weight turns out the lower compressive strength and can be attributed to the bad workability for placing the concrete. The mixture with 40% of Al_2O_3 aggregates appears the highest strengths in all the curing duration. The compressive strength of LAC turns out higher in its earlier age (before 7 days of curing), but decays later due to the effect of conversion.

3.2 Wave Velocity

The mechanical properties of concrete can be estimated from its wave velocity by means of an ultrasonic wave tester. The test results revealed that the longitudinal wave velocity of LAC increases as the Al_2O_3 aggregates increases, as shown in Table 4 and Figure 3, which show a consistence with that of compressive strength of LAC. The mixture with 60% of Al_2O_3 aggregates by weight turns out the lower wave velocity and can be still attributed to the bad workability in placing the concrete. The mixture with 40% of Al_2O_3 aggregates appears the highest wave velocity before 28 days of curing duration. It is noticed that the wave velocity of all mixtures of LAC turn out decay in 91 day, which can be explained by the conversion effect.

Table 4 Wave velocity of concrete

Mixture	1 day	3 days	7 days	28 days	91 days
NC	3501*	4206	4344	4455	4481
LAC20	4561	4578	4604	4787	4021
LAC40	4760	4784	4823	4858	4397
LAC60	4633	4734	4792	4591	4330

*Unit: m/sec

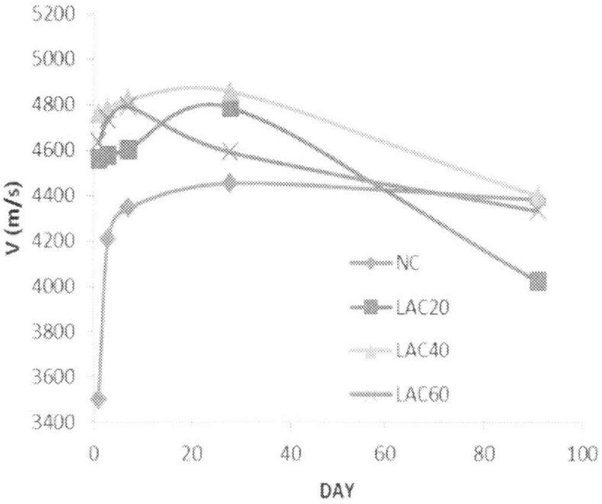

Fig. 3 Wave velocity of concrete

3.3 Splitting Tensile Strength

The splitting test was used to determine the tensile strengths of concrete with various amounts of Al_2O_3 aggregates. Figure 4 shows that the tensile strength of NC is lower than LAC in 28 days, but becomes higher in 91 days of age. Furthermore, the tensile strength of LAC decreases as the Al_2O_3 aggregates increase. Obviously, the voids resulted from conversion dominate the tensile strength of LAC. The tensile strength reduces in 91 days of curing duration and decays more severe as the Al_2O_3 aggregates increases.

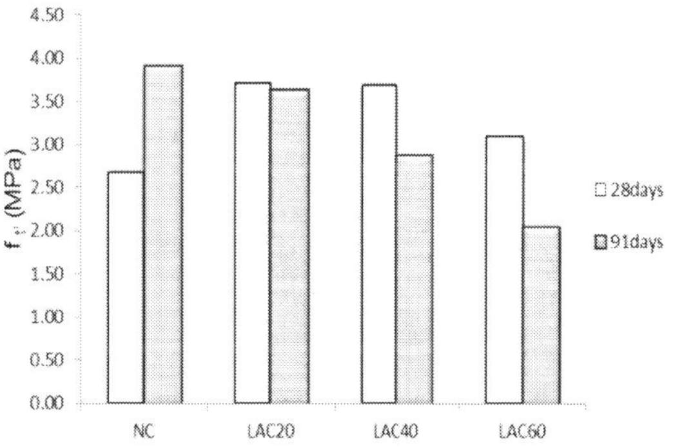

Fig. 4 Tensile strength of concretes

3.4 Elastic Modulus

Test results showed that the modulus of elasticity of LAC exceeds that of normal concrete and increases with Al_2O_3 aggregates for a given water/cemnt ratio. Figure 5 reveals that an increase in Al_2O_3 aggregates causes an increase in the elastic modulus, which appears consistent with that of compressive strength.

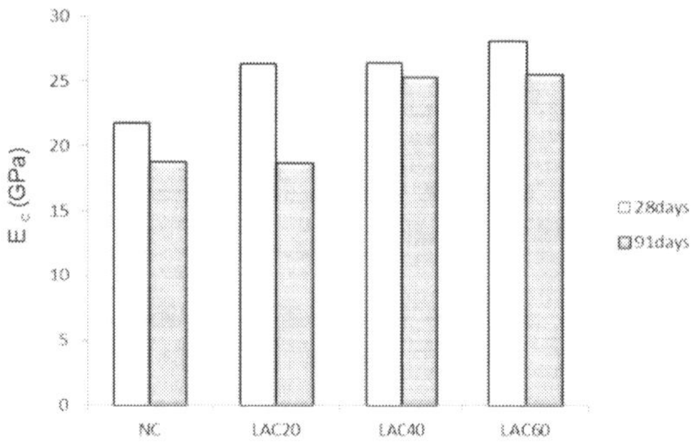

Fig. 5 Elastic moduli of concretes

4 Summary

LAC have higher compressive strength than normal for a given water/cement ratio. The compressive strength seems to be higher in the earlier age and decays after 7 days of curing. Concrete including 40% of Al_2O_3 aggregates as sand replacement performs the highest strength and workability as well. Elastic modulus of concrete appears to increase with the Al_2O_3 inclusion. The tensile strength seems to be decreasing with the increase of Al_2O_3 aggregates in concrete. The voids due to the effect of conversion do affect the mechanical properties of concrete.

Acknowledgement

This research was supported by the Ministry of Science and Technology of Taiwan under the Grant MOST103-2211-E-324-035, which is sincerely gtateful.

References

1. P. Lessing, *Developments in the Formulation and Reinforcement of Concrete* (Sidney Mindess, 2008)
2. Y.-C. Kan, K.- C. Pei and C.- L. Chang, J. NUCL ENG DES, **228**, 119-27(2004)
3. Yu-Cheng Kan, Hsuan-Chih Yang and Kuang-Chih Pei, Adv Mat Res, **512 - 515**, pp. 2908 – 13 (2012)
4. Masaharu Kinno et al., PROG NUCL SCI, **1**, 28-31 (2011)
5. Masaharu Kinno et al., *SMiRT 19 Proceeding*, (2007)

DOI: 10.1051/matecconf/201713004003

Investigation of the rebound number and compressive strength of concrete with quarry dust as fine aggregate

Suppachai Sinthaworn [1-2, a]

[1] Research Unit in Sustainable Innovation in Civil and Environmental Engineering, Srinakharinwirot University, Thailand
[2] Department of Civil Engineering, Faculty of Engineering, Srinakharinwirot University, Thailand

Abstract. This article presents the consideration of relation between compressive strength and rebound number of concrete cooperating with quarry dust as fine aggregate (natural river sand was replaced by quarry dust at the rate of 0%, 25%, 50%, 75% and 100% by weight of fine aggregate). The properties of the tested concrete samples are w/c = 0.6, maximum size of coarse aggregate is 20 mm., cement contents are between 308 and 348 kg/m^3, slumps range from 0 to 100 mm., the 28-day compressive strength from 14 to 30 MPa. It was found that the rebound number results were affected by quarry dust especially the standard deviator of rebound number. The cube compressive strength at 28 days and the supplementary curve from the instruction manual were discussed. Moreover, the prediction equation is proposed to estimate the compressive strength of concrete cooperating with quarry dust as fine aggregate.

1 Introduction

Some parts of Thailand experienced a shortage in natural sand for concrete production while crushed sand (quarry waste) is abundant [1]. Therefore, the quarry waste is researched for serving as fine aggregate [2-3]. However, the engineering properties of fine aggregate affects the physical and mechanical properties of concrete such as surface hardness and compressive strength [4].

Non-destructive Testing (NDT) of concrete by hardness method (i.e., Schmidt hammer) has been continuously using since the late 1940s. Schmidt hammer or rebound hammer is widely used in several standards (e.g., ASTM C805, BS 1881: Part 202). In addition, the strength of concrete is governed by both aggregate characteristics and cement paste. A particular aggregate type also results in difference of the correlation between rebound number and compressive strength [4-5]. Then, using NDT for concrete with the new aggregate type should be checked and calibrated.

The objective of this research is to provide the information of the rebound number and related data in order to investigate the compressive strength of concrete with quarry wastes as fine aggregate.

2 Experimental Procedure

2.1. Concrete Sample

Concrete specimens were made as normal strength concrete. Ordinary Portland cement type I was used while maximum size of coarse aggregate is 20 mm. Coarse aggregate had specific gravities of 2.70 and water absorptions of 0.38%. Two types of fine aggregate are natural river sand and quarry dust. The natural river sand was replaced by quarry dust at the rate of 0%, 25%, 50%, 75% and 100% by weight of fine aggregate. Cement contents are between 308 and 348 kg/m^3. The concrete mixtures are shown in Table 1

After mixing, the initial slump was measured. Then, the concrete specimens (15×15×15 cm) were casted. After casting for 24 hours, all concrete specimens were remolded and cured in water at room temperature. Compressive strength of specimens were tested at the ages of 7 and 28 days.

In summary, the properties of the concrete samples are w/c = 0.6, slumps range from 0 to 100 mm. the 28-day compressive strength from 14 to 30 MPa.

2.2 Rebound hammer

This study used the rebound hammer (mechanical concrete test hammer, MATTEST brand) as shown in Fig.1. The interpretative curve as in the instruction manual is also shown in Fig.1.

Before destructive test (compressive strength test by universal testing machine), concrete specimens were investigated by the rebound hammer to obtain the rebound number from 10 locations on left and right face of the cube specimen as shown in Fig. 2.

[a] Corresponding author: suppachai@g.swu.ac.th

 DOI: 10.1051/matecconf/201713004003

Table 1. Mixture proportion of concrete samples.

Mix. ID	Weight per cubic meter [kg]				
	Cement	Water	Coarse Aggregate	River Sand	Quarry Dust
W0	312	187	1,247	624	0
W25	312	187	1,247	468	156
W50	312	187	1,247	312	312
W75	312	187	1,247	156	468
W100	312	187	1,247	0	624
V0	348	209	1,197	583	0
V25	348	209	1,197	438	146
V50	348	209	1,197	292	292
V75	348	209	1,197	146	438
V100	348	209	1,197	0	583
A0	308	179	1,009	874	0
A25	308	179	1,009	656	219
A50	308	179	1,009	437	437
A75	308	179	1,009	219	656
A100	308	179	1,009	0	874

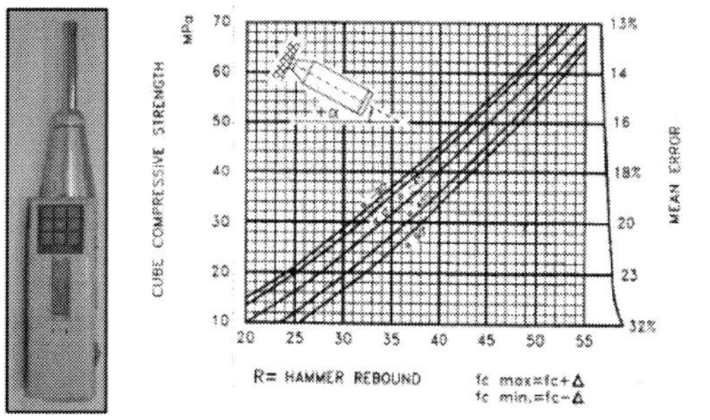

Fig. 1. The rebound hammer and the interpretative curve (from user manual [6]).

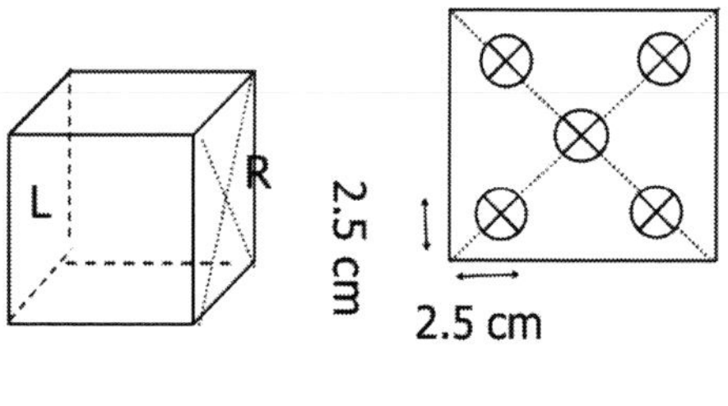

Fig. 2. The tested locations of rebound number on the cube specimen.

3 Results and Discussions

3.1. Median and Average of Rebound Numbers

From the manual, the rebound number can be considered reliable if any readings are not further from the average than the amount specified below:

Rebound Number (H)	15	30	45
Discard reading +/-	2.5	3.0	3.5

The average rebound number should be obtained from the average of the 10 best readings.

However, to discard some readings in practical might results in a biased data because of human error. On the other hand, to discard some readings which are different from average could effect on the final average value of rebound number. Therefore, using of the average value without discard some reading could be more reasonable.

The Median is the "middle" of a sorted list of numbers. The median is quite similar idea with discard the numbers that above and below averages. Fig. 3a shows the relationship between median and average of 10 readings of rebound number. This figure reveals that the average values of 10 readings of rebound number are slightly higher than that of median value. Therefore, to interpret the result by median or average value might differ approximately 2 units of rebound number.

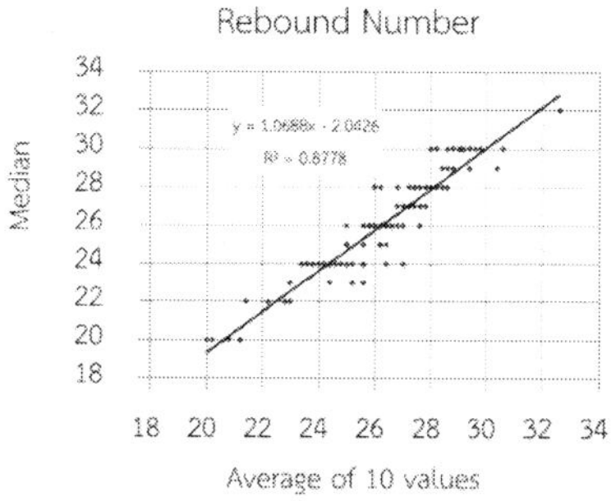

(a)

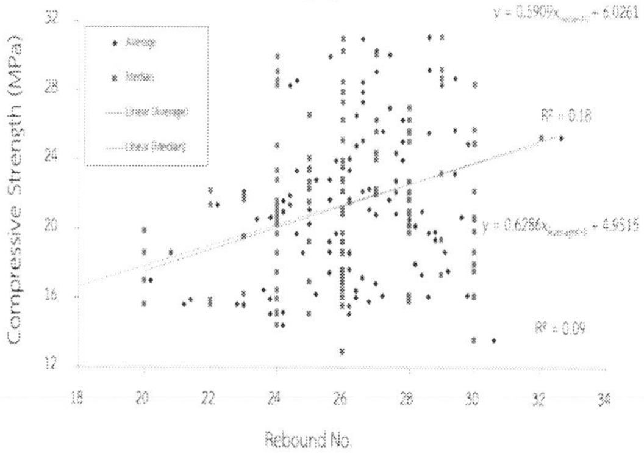

(b)

Fig. 3. (a) Relationship between median and average of rebound number. (b) Compressive strength versus rebound number of all specimens.

3.2 Relationship between Compressive Strength and Rebound Number

The compressive strength of all concrete samplings versus the rebound number are illustrated in Fig.3b. The two linear regressions of the rebound number (average and median) and compressive strength show quite small R^2 value. However, the regression from median value of the obtained rebound number seems more accurate value to predict the cube compressive strength than using the regression from average rebound number.

3.3 Comparative Prediction of Compressive Strength between from the Manual and the Regression

Fig.4 demonstrations the prediction of compressive strength by rebound number from the manual and the linear regression of the relationship between compressive strength and rebound number. The linear regression (blue line) shows approximately ±20% fluctuation from the mean compressive strength while the suggestion value and range from the manual (green line) matches only a small regions of the data. Moreover, it should be noted that compressive strength range from 12 to 32 MPa could be obtained from a narrow range of rebound number from 20 to 32 (12 units).

3.4 The Prediction Equation

Fig.5 displays the plots of the compressive strength of concrete versus the obtained rebound number. Those concretes were separated into three types of fine aggregate which are 1) river sand {red triangle, normal}, 2) quarry dust {dash line, Q}, and 3) mixed of river sand and quarry dust. Mostly results are in the prediction band of the regression line (prediction value ± 20%). However, some results locate above and below the prediction area. It could not find an evidential on the effect of fine aggregates type from Fig.5. However, the proposed equation for concrete with quarry dust as fine aggregate is mentioned in Eq 1.

$$f'c = 0.5909\, H + 6.0261 \qquad (1)$$

where H is the median of the obtained rebound number. $f'c$ is the cube compressive strength of concrete (MPa).

4 Conclusions

Based on the results of this study, the following conclusions could be drawn:

- The average and median value of the obtained rebound number might not equal, and the median of the obtained rebound number seems more reasonable value to predict the cube compressive strength than using average value of the obtained rebound number.

- The proposed equation for concrete with quarry dust as fine aggregate is mentioned as using the median of the obtained rebound number. However, the predicted compressive strength obtains from a narrow range of the rebound number (only 12 units). It is possible to achieve a new relationship between rebound number and compressive strength by the experiment for the new case to get the better prediction result.

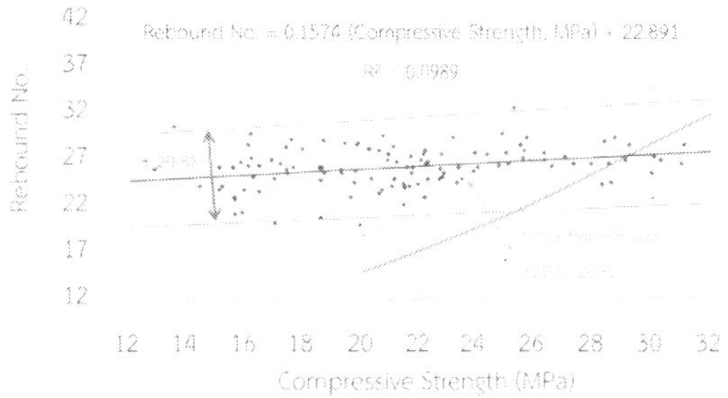

Fig. 4. Comparative prediction of compressive strength from the manual and a regression.

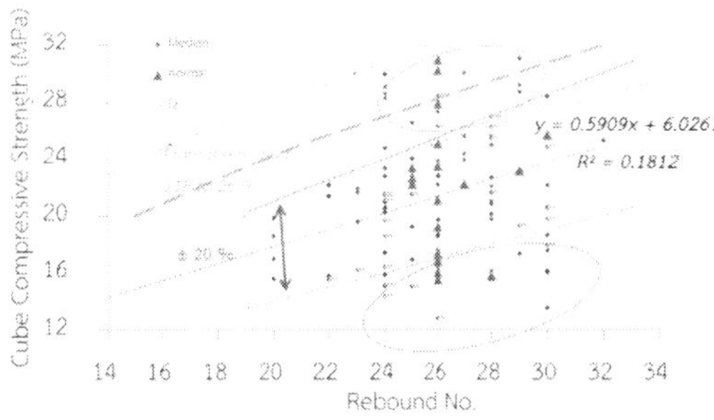

Fig. 5. The plots of compressive strength versus rebound number of the different mixtures.

Acknowledgements

This work was financially supported by Faculty of Engineering, Srinakharinwirot University under contract No. 147/2552 and the authors gratefully acknowledge the financial support to present this work in the conference from Strategic Wisdom and Research Institute, Srinakharinwirot University.

References

1. S. Sinthaworn, T. Koseekageepat, O. Saengmanee, Advanced Materials Research 974, 350-353 (2014),
2. S. Sinthaworn, W. Teerajetgul and A. Sirivachiraporn, Advanced Materials Research 875-877, 619-623 (2014)
3. S. Sinthaworn, Materials Science Forum, 886, 159-163 (2017)
4. J.H. Bungey and S.G. Millard, "Testing of Concrete in Structures", Third edition, Blackie Academic & Professional, Chapman & Hall, 1996.
5. S. Sinthaworn. 2010." Assessment of Strength of Concrete by Rebound Hammer", SWU Engineering Journal **5(2)**, 41-49 (2010). (in Thai).
6. MATTEST, Mechanical Concrete Test Hammer, Manual User Guide, www.matest.com, Italy.

 DOI: 10.1051/matecconf/201713004004

The Synthesis and Characterization of Novel Bis-cationic Asphalt Emulsifier

Xiao-Meng Yu, Lai-Shun Shi[a], Cun-Fei Ma, Lu Hou, Jing-Wen Kong and Ya-Li Sun

School of Chemistry and Chemical Engineering, Shandong University, Jinan 250061, P. R. China

Abstract. Three novel bis-cationic asphalt emulsifiers were synthesized by two steps reaction of N-tallow-1,3-diaminopropane, methyl acrylate, sodium hydroxide and chloroacetic acid. The structure of the product was confirmed by FTIR, ^1H NMR and elemental analysis. The emulsifier exhibits satisfactory emulsification. The prepared bituminous emulsion showed higher storage stability. The emulsifier belongs to medium-set and slow-set asphalt emulsifier. It can be designed to perform in a wide variety of applications including tack coat, fog seal, chip seal and slurry seal in asphalt pavement construction.

1 Introduction

Asphalt has many excellent properties such as good adhesion properties, aging resistance and water resistance. It is widely used in road repair, waterproofing materials, sealing materials, etc [1]. Emulsified asphalt performance depends on the nature of the asphalt emulsifier. According to the different charge of asphalt emulsifier, it is divided into cationic, anionic, nonionic, amphoteric [2]. In recent years, it is widely used and studied due to the good performance, high adhesion of cationic asphalt emulsifier [3]. Negm [4] synthesized a series of cationic surfactants with benzaldehyde, p-aminopyridine, xylene, 1,2-dibromoethane, 1,6-dibromohexane and 1,12-dibromododecane by two-step reaction. Li [5] synthesized an asymmetric Gemini cationic surfactants, characterized its structure and discussed the affect factors of stability.

Depending on the breaking speed of asphalt emulsifier, it is divided into slow, rapid, and medium set type. Slow set asphalt emulsifier [6] is a key material for slurry seal. Slurry seal [7] is solid surface sealing layer which emulsified asphalt, water, aggregates and slag mix by a certain percentage forming a paste and flow of synthetic slurry mixture and then spread evenly on the road. Rapid set emulsifier [8] is mainly used for chip seal layer and spray patching potholes. Fracturing chippings overlay often uses the medium set type emulsifier [9].

In previous papers, we have synthesized different kinds of asphalt emulsifiers [10-14]. Three novel bis-cationic asphalt emulsifiers were synthesized by two steps reaction in this paper. The structure of the emulsifier was confirmed by FTIR, ^1H NMR and elemental analysis.

2 Experimental

2.1 Materials and characterization

N-tallow-1,3-diaminopropane was technical grade and obtained from Zibo HuaRun Chemical Co., Ltd, Zibo, China. All the other reagents were of analytical reagent grade and used without further purification. The asphalt used was AH-90, which was provided by Shengli Refinery of Qilu Petrochemical Company, Zibo, China. The mineral aggregate was the mixture of different size of marble stone. The FTIR spectra were recorded from 400 to 4000 cm^{-1} wavenumber range with averaging 32 scans at a resolution of 4 cm^{-1} on a Bruker Tensor-27 FTIR spectrophotometer. The ^1H NMR spectra were recorded on a Bruker Avance 300M NMR spectrometer (Bruker, Germany), using D_2O as solvent and TMS as internal reference. Chemical shifts (δ) were given in ppm. Elemental analyses were performed by the Elementar vario EL type III element analyzer (Elementar, German).

2.2 Synthesis of asphalt emulsifier

Scheme 1 shows the synthesis route of the bis-cationic asphalt emulsifier (A).

$$C_{18}H_{37}NH(CH_2)_3NH_2 + CH_2{=}CHCOOCH_3 \longrightarrow$$

$$\underset{\text{(Intermediate)}}{C_{18}H_{37}NH(CH_2)_3NH(CH_2)_2COOCH_3} \qquad (1)$$

$$C_{18}H_{37}NH(CH_2)_3NH(CH_2)_2COOCH_3 + 2ClCH_2COONa + H_2O \longrightarrow$$

$$C_{18}H_{37}\underset{\substack{|\\ CH_2COONa}}{NH^+}(CH_2)_3\overset{\substack{CH_2COONa\\ |}}{NH^+}(CH_2)_2COOH \cdot 2Cl^- + CH_3OH \qquad (2)$$

Scheme 1. Synthetic route of asphalt emulsifier (A).

[a] Corresponding author: LSHUNSH@sdu.edu.cn

N-tallow-1,3-diaminopropane (326.0 g, 1.00 mol) and 400 g isopropanol were added in a three-necked flask, and heated to 70 ℃ in a water bath. Methyl acrylate (90.3 g, 1.05 mol) was added dropwise and reacted for 4h. The reaction solution was processed by vacuum distillation to remove the solvent, then recrystallized with anhydrous ethanol for three times to obtained the intermediate.

Sodium chloroacetate aqueous solution was synthesized by the reaction of 196.6 g (2.08 mol) chloroacetic acid, 83.2 g (2.08 mol) NaOH and 560.0 g distilled water. Then the sodium chloroacetate aqueous solution was added dropwise to the reaction solution synthesized above and stirred for 5 h at 65 ℃ in water bath. Finally, the reaction solution was processed by vacuum distillation to remove the solvent, then recrystallized with anhydrous ethanol for three times to obtained the product.

Scheme 2 shows the synthesis route of the bis-cationic asphalt emulsifier (B).

$$C_{18}H_{37}NH(CH_2)_3NH_2 + 2CH_2{=}CHCOOCH_3 \longrightarrow C_{18}H_{37}NH(CH_2)_3N(CH_2CH_2COOCH_3)_2 \quad (3)$$

$$C_{18}H_{37}NH(CH_2)_3N(CH_2CH_2COOCH_3)_2 + 2ClCH_2COONa + 2H_2O \longrightarrow$$

$$\begin{array}{l} \qquad\qquad\qquad CH_2COONa \\ C_{18}H_{37}NH^+(CH_2)_3N^+(CH_2CH_2COOH)_2 \cdot 2Cl^- + 2CH_3OH \\ \qquad\quad CH_2COONa \end{array} \quad (4)$$

Scheme 2. Synthetic route of asphalt emulsifier (B).

N-tallow-1,3-diaminopropane (326.0 g, 1.00 mol) and 400 g isopropanol were added in a three-necked flask, and heated to 70 ℃ in a water bath. Methyl acrylate (175.5 g, 2.04 mol) was added dropwise and reacted for 4h. The reaction solution was processed by vacuum distillation to remove the solvent, then recrystallized with anhydrous ethanol for three times to obtained the intermediate.

Sodium chloroacetate aqueous solution was synthesized by the reaction of 196.6 g (2.08 mol) chloroacetic acid, 83.2 g (2.08 mol) NaOH and 560.0 g distilled water. Then the sodium chloroacetate aqueous solution was added dropwise to the reaction solution synthesized above and stirred for 5 h at 65℃ in water bath. Finally, the reaction solution was processed by vacuum distillation to remove the solvent, then recrystallized with anhydrous ethanol for three times to obtained the product.

Scheme 3 shows the synthesis route of the bis-cationic asphalt emulsifier (C).

$$C_{18}H_{37}NH(CH_2)_3NH_2 + 3CH_2{=}CHCOOCH_3 \longrightarrow$$

$$\begin{array}{l} C_{18}H_{37}N(CH_2)_3N(CH_2CH_2COOCH_3)_2 \\ \qquad\quad CH_2CH_2COOCH_3 \end{array} \quad (5)$$

$$\begin{array}{l} C_{18}H_{37}N(CH_2)_3N(CH_2CH_2COOCH_3)_2 + 2ClCH_2COONa + 3H_2O \longrightarrow \\ \qquad\quad CH_2CH_2COOCH_3 \end{array}$$

$$\begin{array}{l} \qquad\quad CH_2CH_2COOH \\ C_{18}H_{37}N^+CH_2CH_2CH_2N^+(CH_2CH_2COOH)_2 \cdot 2Cl^- + 3CH_3OH \\ \qquad\quad CH_2COONa \qquad CH_2COONa \end{array} \quad (6)$$

Scheme 3. Synthetic route of asphalt emulsifier (C).

N-tallow-1,3-diaminopropane (326.0 g, 1.00 mol) and 400 g isopropanol were added in a three-necked flask, and heated to 70 ℃ in a water bath. Methyl acrylate (261.6 g, 3.04 mol) was added dropwise and reacted for 4h. The reaction solution was processed by vacuum distillation to remove the solvent, then recrystallized with anhydrous ethanol for three times to obtained the intermediate.

Sodium chloroacetate aqueous solution was synthesized by the reaction of 196.6 g (2.08 mol) chloroacetic acid, 83.2 g (2.08 mol) NaOH and 560.0 g distilled water. Then the sodium chloroacetate aqueous solution was added dropwise to the reaction solution synthesized above and stirred for 5 h at 65 ℃ in water bath. Finally, the reaction solution was processed by vacuum distillation to remove the solvent, then recrystallized with anhydrous ethanol for three times to obtained the product.

2.3 Performance test of bituminous emulsion

2.3.1 Preparation of bituminous emulsion

12.5 g emulsifier was dissolved in 200 ml water, then the solution was heated to 60 ℃ . The pH value of the aqueous solution was adjusted to 2.0 by adding hydrochloric acid. 300 g AH-90 asphalt was heated to 120 ℃ . The bituminous emulsion was prepared by mixing the aqueous solution containing emulsifier and asphalt in a colloid mill for 1 min.

2.3.2 Demulsification speed test of bituminous emulsion

Demulsification speed of bituminous emulsion was determined according to T0658-1993 standard. B type mineral aggregate (200 g) and water (30 g) were placed in a 500-mL bowl. Bituminous emulsion (50 g) was then added, and the mixture was stirred for 1 min at 60 revolutions per minute. The mixing time and the appearance were recorded by observing the uniform of the aggregate mixing performance.

3 Results and discussion

3.1 Structure characterization of intermediate

Figure 1 gives the FTIR spectrum of the intermediate in Scheme 1. The absorption at 3275 cm^{-1} (peak 1) is assigned to the N-H stretching. The absorption at 2917 cm^{-1} (peak 2) is assigned to the methyl asymmetrical stretching. The absorption at 2854 cm^{-1} (peak 3) is assigned to the methylene symmetrical stretching. The absorption at 1735 cm^{-1} (peak 4) is assigned to C=O stretching of ester. The absorption at 1569 cm^{-1} (peak 5) is assigned to N-H bending vibration. The absorptions at 1470 cm^{-1} (peak 6) and 1376 cm^{-1} (peak 7) are assigned to the methyl asymmetrical and symmetrical bending vibration, respectively. The absorption at 1207 cm^{-1} (peak 8) is assigned to C-N stretching. The absorption at 1125 cm^{-1} (peak 9) is assigned to C-O-C stretching. The absorption at 724 cm^{-1} (peak 10) is assigned to methylene in-plane swing vibration. Therefore, the FTIR spectrum agrees with the chemical structure of the intermediate to be synthesized.

3.2 Structure characterization of asphalt emulsifier (A)

Figure 2 gives the FTIR spectrum of asphalt emulsifier (A) in Scheme 1. The absorption at 3250-3500 cm^{-1} (peak 1) is assigned to the superimposed peak of O-H and N-H stretching. The absorption at 2918 cm^{-1} (peak 2) is assigned to the methyl asymmetrical stretching. The absorption at 2846 cm^{-1} (peak 3) is assigned to the methylene symmetrical stretching. The absorption at 1737 cm^{-1} (peak 4) is assigned to C=O stretching of carboxylic acid. The absorption at 1634 cm^{-1} (peak 5) is assigned to C=O stretching of carboxylate. The absorptions at 1470 cm^{-1} (peak 6) and 1395 cm^{-1} (peak 7) are assigned to the methyl asymmetrical and symmetrical bending vibration, respectively. The absorption at 1207 cm^{-1} (peak 8) is assigned to C-N stretching. The absorption at 1068 cm^{-1} (peak 9) is assigned to C-OH stretching. The absorption at 926 cm^{-1} (peak 10) is assigned to O-H out-of-plane swing vibration of carboxylic acid. The absorption at 717 cm^{-1} (peak 11) is assigned to methylene in-plane swing vibration. Therefore, the FTIR spectrum agrees with the chemical structure of asphalt emulsifier (A) to be synthesized.

^{1}H NMR (300 MHz, D_2O), δ: 0.81 (t, 3H, $-CH_3$), 1.22(s, 30H, $-CH_2-$), 1.66(s, 2H, $CH_3(CH_2)_{15}C\underline{H}_2-$), 2.16(s, 2H, $-NH^+CH_2C\underline{H}_2CH_2NH^+-$), 2.59(m,2H,$-NH^+CH_2C\underline{H}_2COOH$), 3.12-3.67(m, 8H, $-C\underline{H}_2NH^+C\underline{H}_2CH_2C\underline{H}_2NH^+C\underline{H}_2CH_2COOH$), 3.73-4.24(m, 4H, $-C\underline{H}_2COONa$) ppm.

Anal.: Calcd. for $C_{28}H_{54}N_2O_6Cl_2Na_2$: C, 46.47%; H, 6.92 %; N, 3.87 %. Found: C, 47.02 %; H, 9.43 %; N, 3.98 %.

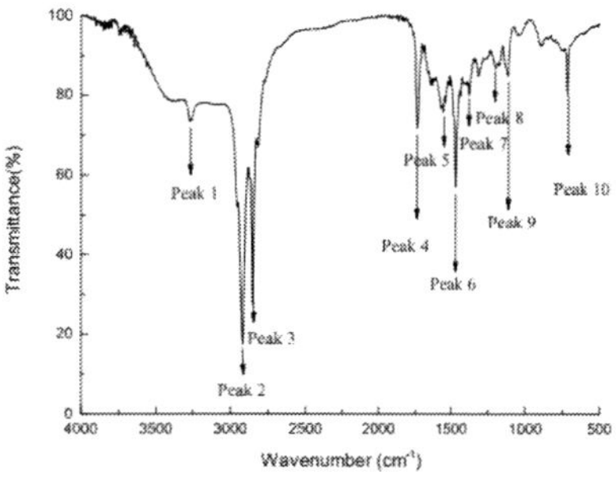

Figure 1. FTIR of intermediate in Scheme 1.

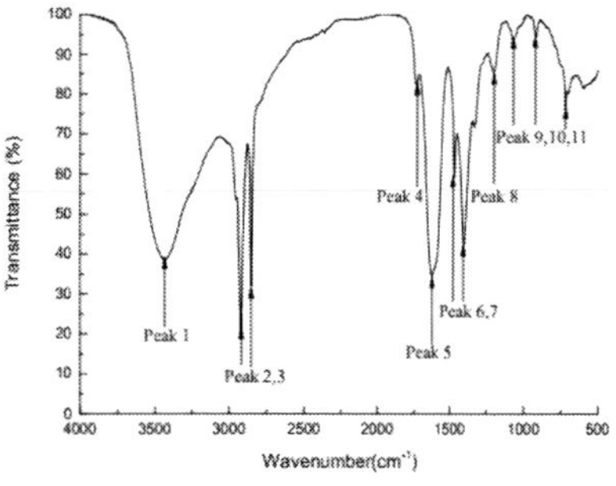

Figure 2. FTIR spectrum of asphalt emulsifier (A).

3.3 Structure characterization of asphalt emulsifier (B)

Figure 3 gives the FTIR spectrum of asphalt emulsifier (B) in Scheme 2. The absorption at 3250-3500 cm^{-1} (peak 1) is assigned to the superimposed peak of O-H and N-H stretching. The absorption at 2925 cm^{-1} (peak 2) is assigned to the methyl asymmetrical stretching. The absorption at 2850 cm^{-1} (peak 3) is assigned to the methylene symmetrical stretching. The absorption at 1735 cm^{-1} (peak 4) is assigned to C=O stretching of carboxylic acid. The absorption at 1605 cm^{-1} (peak 5) is assigned to C=O stretching of carboxylate. The absorptions at 1470 cm^{-1} (peak 6) and 1395 cm^{-1} (peak 7) are assigned to the methyl asymmetrical and symmetrical bending vibration, respectively. The absorption at 1254 cm^{-1} (peak 8) is assigned to C-N stretching. The absorption at 1068 cm^{-1} (peak 9) is assigned to C-OH stretching. The absorption at 917 cm^{-1} (peak 10) is assigned to O-H out-of-plane swing vibration of carboxylic acid. The absorption at 701 cm^{-1} (peak 11) is assigned to methylene in-plane swing vibration. Therefore, the FTIR spectrum agrees with the chemical structure of asphalt emulsifier (B) to be synthesized.

^{1}H NMR (300 MHz, D_2O), δ: 0.85(t, 3H, $-CH_3$), 1.26(s, 30H, $-CH_2-$), 1.70(s, 2H, $CH_3(CH_2)_{15}C\underline{H}_2-$), 2.24(s, 2H, $-NH^+CH_2C\underline{H}_2CH_2N^+-$), 2.57-2.64(m,4H,$-CH_2C\underline{H}_2COOH$), 3.01-3.58(m, 10H, $-C\underline{H}_2NH^+C\underline{H}_2CH_2C\underline{H}_2N^+(C\underline{H}_2CH_2COOH)_2$), 3.83-4.28(s,4H, $-C\underline{H}_2COONa$) ppm.

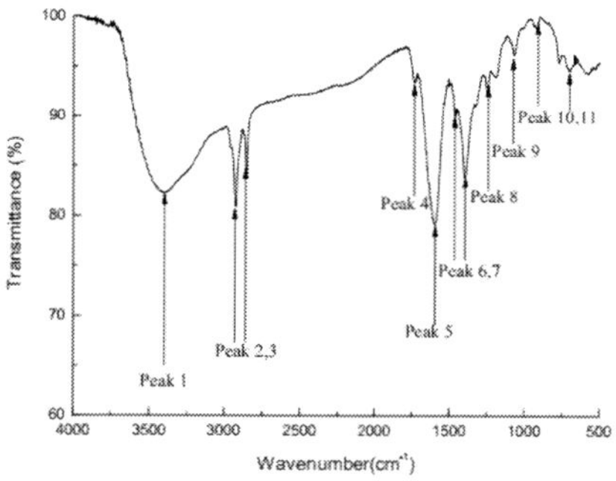

Figure 3. FTIR spectrum of asphalt emulsifier (B).

3.4 Structure characterization of asphalt emulsifier (C)

Figure 4 gives the FTIR spectrum of asphalt emulsifier (C) in Scheme 3. The absorption at 3427 cm^{-1} (peak 1) is assigned to O-H stretching. The absorption at 2925 cm^{-1} (peak 2) is assigned to the methyl asymmetrical stretching. The absorption at 2854 cm^{-1} (peak 3) is assigned to the methylene symmetrical stretching. The absorption at 1733 cm^{-1} (peak 4) is assigned to C=O stretching of carboxylic acid. The absorption at 1606 cm^{-1} (peak 5) is assigned to C=O stretching of carboxylate. The absorptions at 1463 cm^{-1} (peak 6) and 1395 cm^{-1} (peak 7) are assigned to the methyl asymmetrical and symmetrical bending vibration, respectively. The absorptions at 1261 cm^{-1} (peak 8) and 1197 cm^{-1} (peak 9)

DOI: 10.1051/matecconf/201713004004

are assigned to C-N stretching. The absorption at 1072 cm^{-1} (peak 10) is assigned to C-OH stretching. The absorption at 930 cm^{-1} (peak 11) is assigned to O-H out-of-plane swing vibration of carboxylic acid. The absorption at 772 cm^{-1} (peak 12) is assigned to methylene in-plane swing vibration. Therefore, the FTIR spectrum agrees with the chemical structure of asphalt emulsifier (C) to be synthesized.

^{1}H NMR (300 MHz, D_2O), δ: 0.85(t, 3H, $-CH_3$), 1.25(s, 30H, $-CH_2-$), 1.70(s, 2H, $CH_3(CH_2)_{15}C\underline{H}_2-$), 2.24(s, 2H, $-N^+CH_2C\underline{H}_2CH_2N^+-$), 2.63-2.85(m,6H,$-CH_2C\underline{H}_2COOH$), 2.93-3.51(m,12H, $-C\underline{H}_2N^+(C\underline{H}_2CH_2COOH)-C\underline{H}_2CH_2C\underline{H}_2N^+(C\underline{H}_2CH_2COOH)_2$), 3.89-4.25(s,4H, $-C\underline{H}_2COONa$) ppm.

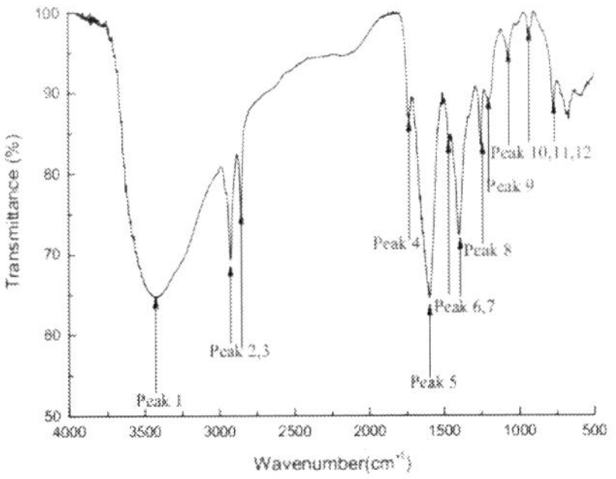

Figure 4. FTIR spectrum of asphalt emulsifier (C).

3.5 Demulsification speed test of bituminous emulsion

The emulsifiers have excellent emulsifying ability to AH-90 asphalt. The prepared bituminous emulsion showed higher storage stability. Table 1 gives the mixing time and the types of the asphalt emulsifiers. It can be designed to perform in a wide variety of applications including tack coat, fog seal, chip seal and slurry seal in asphalt pavement construction.

Table 1. Demulsification speed test of bituminous emulsion.

asphalt emulsifier	mixing time (s)	types of the asphalt emulsifier
A	12	medium-set
B	47	slow-set
C	14	medium-set

4 Summary

Three novel bis-cationic asphalt emulsifiers were synthesized by two steps reaction of N-tallow-1,3-diaminopropane, methyl acrylate, sodium hydroxide and chloroacetic acid. The emulsifier belongs to medium-set and slow-set asphalt emulsifier. It can be designed to perform in a wide variety of applications including tack coat, fog seal, chip seal and slurry seal in asphalt pavement construction.

Acknowledgement

The authors would like to thank the financial support from the Shandong Provincial Science and Technology Project (No. 2015GGX107005).

References

1. F. Yang, G. Li, J. Qi, et al. Synthesis and surface activity properties of alkylphenol polyoxyethylene nonionic trimeric surfactants. Applied Surface Science, 257, 312-318(2010)
2. K. Isobe, R. Tamaki. Nonionic emulsifier for asphalt. U.S. Patent 6,114,418. (2000)
3. P.R.J. Blanpain, J.W. Creak, G.F. Mori, et al. Cationic asphalt emulsions. U.S. Patent 4,997,868. (1991)
4. N.A. Negm, M.A. El-Hashash, D.E. Mohamed, et al. Gemini cationic surfactants: synthesis and influence of chemical structure on the surface activity. Journal of Surfactants and Detergents, 16, 733-738(2013)
5. H.P. Li, H. Zhao, K. Liao. The preparation of asphalt emulsions with dissymmetric Gemini quaternary ammonium salts cationic surfactants. Energy Sources, Part A: Recovery, Utilization, and Environmental Effects, 35, 2285-2293(2013)
6. W.A. McIntosh. Alkali lignin to stabilize slow-break asphalt emulsions. Industrial & Engineering Chemistry, 44, 1656-1659(1952)
7. X. Zhi, W. Wang, Y. Tsai. Cost-benefit timing for applying slurry seal on actual roadway tests in China. Journal of Central South University, 19, 2394-2402(2012)
8. P. Goullet, P. Scotte. Method of making fast-breaking bituminous emulsions. U.S. Patent 4,338,136. (1982)
9. A.C. Ludwig. Emulsification of rock asphalt. U.S. Patent 5,089,052. (1992)
10. C. Huai, L. Shi, N. Li. Synthesis of a novel betaine-type asphalt emulsifier and its investigation by online FTIR spectrophotometry. Research on Chemical Intermediates, 39, 597−614(2013)
11. L. Shi, M. Sun, N. Li, B. Zhang. A novel betaine type asphalt emulsifier synthesized and investigated by online FTIR spectrophotometric method. Chem. Ind. Chem. Eng. Q., 21, 113−121(2015)
12. L. Shi, Y. Zhao. Synthesis and performance evaluation of a Gemini cationic asphalt emulsifier. Petroleum asphalt, 24(6), 66-71(2010)
13. N. Li, L. Shi, X. Gong, Q. Xu, X. Liu, X. Wang. Synthesis of a novel cationic asphalt emulsifier and its investigation by online FTIR spectrophotometry. Res. Chem. Intermed., 41(4), 1935−1950(2015)
14. X. Tao, L. Shi, M. Sun, N. Li. Synthesis of lignin amine asphalt emulsifier and its investigation by

DOI: 10.1051/matecconf/201713004004

online FTIR spectrophotometry. Advanced Materials Research, 909, 72−76(2014)

Study on Performance of Steel Fiber Concrete Bridge Pier Specimens under Horizontal Cyclic Loading

Chen Baiben[1,a], Feng Zhongren[1] and Shi Xiaoji[1]

[1]*School of Civil Engineering and Architecture, Wuhan University of Technology, 430070 Wuhan Hubei, China*

Abstract. Because of that steel fiber can effectively prevent the extension and development of small cracks in the concrete, steel fiber reinforced concrete has good toughness and tensile strength. In the application of building materials, steel fiber reinforced concrete is an ideal elastic-plastic material. For the seismic performance, it has advantages. In order to analyze the seismic performance of steel fiber reinforced concrete, 4 piers of the scale model test under horizontal cyclic loading were done. The results showed that failure mode of steel fiber reinforced concrete is better than that of ordinary concrete, and has a large yield moment under the external loads.

1 Introduction

The steel fiber concrete is different from ordinary concrete, which adds steel fibers into concrete, in which cement and steel fibers shared external force in the early loaded. The cement is the main bearing material for external load. With the increasing of load, the cracks in concrete became increasing and have the development trends in the whole material. In this time, the steel fibers have tension and are continuing into the loading stage, them bear lots of load force. Different from that ordinary concrete cracking is the failure of cement, steel fiber reinforced concrete is mainly due to steel fiber pull-out from the concrete, resulting in the entire concrete cracking, loss of tensile properties.

Due to incorporation of steel fiber, the fiber material and cement in the concrete combine into one in the steel fiber reinforced concrete, the levels of strength, toughness and tensile properties are increasing. It has lots of advantages over traditional materials in building materials. Especially in recent years, the extensive development of bridge construction is vary fast, the steel fiber reinforced concrete can effectively prevent the coupling effect of the pier in horizontal and vertical seismic force, deformation, shear force and cracking can be control, and the structure will not damage in earthquake.

21 steel fiber reinforced high-strength concrete columns were done the eccentric and non eccentric loading test by Foster [1], and the bearing capacity and ductility of steel fiber on concrete columns would be improved. 26 steel fiber reinforced high-strength concrete column bending performance test under low cyclic loading were done by Li Fenglan [2], the axial compression ratio, characteristics value of steel fiber content, stirrup characteristic value, shear span ratio and longitudinal reinforcement and other factors on the influence of the ductility of steel fiber reinforced high strength concrete performance were researched. Zhu Zhao [3] used finite element software to establish the local steel fiber reinforced concrete bridge pier model, and analyzed the reasonable distribution of pier steel fiber length. Zhang Yuye [4-6] through the experiment of steel fiber reinforced concrete pier, proposed a reasonable length of steel fiber reinforced concrete using local formula. The scholars [7-9] at home and abroad have also studied the mechanical properties of steel fiber reinforced concrete and its application in frame columns, the shear resistance and the structural energy dissipation of the steel fiber reinforced concrete are obviously improved.

The horizontal cyclic loading performance tests of four steel fiber reinforced concrete bridge pier specimens were analyzed. In different steel fiber volume, seismic performances of bridge material were compared, which reflect the yield failure patterns of steel fiber reinforced concrete in cyclic load under the condition of the horizontal cyclic loading.

2 Constitutive relation of concrete under cyclic loading

It is well known that the unloading paths and loading paths of concrete are quite different except at low stress levels. When unloading to the initial stress state, the strain does not recover completely, but retains a part of the permanent strain (plastic strain). In some applications, in order to simulate the properties under cyclic loading conditions, the simple model is combined with the loading criterion. The loading criterion is not related to the constitutive relation. For unloading scenarios, it is assumed that the deformation is due to the control of the

[a] Corresponding author: cbb0505@163.com

modulus of elasticity different from the loading conditions. But there are problems such as uniqueness and continuity.

For concrete materials, it is difficult to simulate the cyclic loading capacity and simulate the degradation of strength and stiffness of concrete. Normally, need a similar strain introduced to simulate the deformation history and determine the amount of a single hysteresis behavior. The concept of equivalent uniaxial strain can be used to solve such problems.

By using the incremental constitutive equation and the shear stiffness assumption expressed by other material constants, the incremental constitutive relation expressed in terms of incremental uniaxial strain is given. On this basis, the relationship between equivalent uniaxial strain and stress is introduced to form the basis of incremental elastic modulus derived from stress and equivalent uniaxial strain parameters.

2.1. Incremental constitutive relation

Under biaxial stress conditions, the orthotropic increment stress-strain relation can be written:

$$\begin{Bmatrix} \dot{\sigma}_{11} \\ \dot{\sigma}_{22} \\ \dot{\tau}_{12} \end{Bmatrix} = \frac{1}{1-\nu_1\nu_2} \begin{bmatrix} E_1 & \nu_2 E_2 & 0 \\ \nu_1 E_2 & E_2 & 0 \\ 0 & 0 & (1-\nu_1\nu_2)G \end{bmatrix} \begin{Bmatrix} \dot{\varepsilon}_{11} \\ \dot{\varepsilon}_{22} \\ \dot{\gamma}_{12} \end{Bmatrix} \tag{1}$$

Among them, E_1, ν_1 and E_2, ν_2 are tangent elastic moduli along the orthotropic principal axes 1 and 2, and G is the tangent shear modulus related to the axes 1 and 2.

The equivalent Poisson's ratio μ is defined as follows:

$$\mu^2 = \nu_1\nu_2 \tag{2}$$

Suppose that the shear modulus G is the invariant of the shaft transformation:

$$(1-\mu^2)G = \frac{1}{4}(E_1 + E_2 - 2\mu\sqrt{E_1E_2}) \tag{3}$$

Therefore, the formula (1) can be turned into:

$$\begin{Bmatrix} d\sigma_1 \\ d\sigma_2 \\ d\tau_{12} \end{Bmatrix} = \frac{1}{(1-\mu^2)} \begin{bmatrix} E_1 & \mu\sqrt{E_1E_2} & 0 \\ 0 & E_2 & 0 \\ 0 & \mu\sqrt{E_1E_2} & \frac{1}{4}(E_1+E_2-2\mu\sqrt{E_1E_2}) \end{bmatrix} \begin{Bmatrix} d\varepsilon_1 \\ d\varepsilon_2 \\ d\gamma_{12} \end{Bmatrix} \tag{4}$$

2.2. Equivalent uniaxial strain

The variation of the tangent modulus E_1 and E_2 with the stress is determined by introducing the concept of equivalent uniaxial strain, assuming that the formula (4) is as follows:

$$\begin{Bmatrix} d\sigma_1 \\ d\sigma_2 \\ d\tau_{12} \end{Bmatrix} = \begin{bmatrix} E_1B_{11} & E_1B_{12} & 0 \\ E_2B_{21} & E_2B_{22} & 0 \\ 0 & 0 & G \end{bmatrix} \begin{Bmatrix} d\varepsilon_1 \\ d\varepsilon_2 \\ d\gamma_{12} \end{Bmatrix} \tag{5}$$

Among them, the coefficient B_{ij} (i, j=1, 2) can be determined by making the matrix entries in formula (5) correspond to the corresponding items in formula (4). By calculation:

$$\begin{aligned} d\sigma_1 &= E_1(B_{11}d\varepsilon_1 + B_{12}d\varepsilon_2) \\ d\sigma_2 &= E_2(B_{21}d\varepsilon_1 + B_{22}d\varepsilon_2) \\ d\tau_{12} &= Gd\gamma_{12} \end{aligned} \tag{6}$$

The upper form can be written as follows:

$$\begin{Bmatrix} d\sigma_1 \\ d\sigma_2 \\ d\tau_{12} \end{Bmatrix} = \begin{bmatrix} E_1 & 0 & 0 \\ 0 & E_2 & 0 \\ 0 & 0 & G \end{bmatrix} \begin{Bmatrix} d\varepsilon_{1u} \\ d\varepsilon_{2u} \\ d\gamma_{12} \end{Bmatrix} \tag{7}$$

The strain vector of each formula zip code is defined as the equivalent increment single axis strain vector, and the definition of its element is accomplished by the corresponding increment of the actual strain increment and the formula (6):

$$d\varepsilon_{1u} = B_{i1}d\varepsilon_1 + B_{i2}d\varepsilon_2 \tag{8}$$

The total equivalent effect can be obtained by integrating the equivalent uniaxial strain:

$$\varepsilon_{iu} = \int \frac{d\sigma_i}{E_i} \tag{9}$$

3 Steel fiber reinforced concrete column specimens

The test specimens were made of C40 concrete, which is made of steel fiber with a certain volume proportion in commercial concrete. Among them, every cubic meters of concrete materials: water is 160kg; cement is 240kg; sand is 782kg; stone is 1038kg; admixture is 180kg and additive is 6.0kg. Mass ratio is like that, water: cement: sand: stone: admixture: additive =0.67: 1: 3.26: 4.58: 0.25: 0.50. The ordinary concrete cube compressive strength for 28 days is 50.6MPa, and that of steel fiber reinforced concrete is 51.7MPa in which steel fiber volume ratio is 1.5%.

HRB500 high strength rebar is used for longitudinal bars which diameter is 20mm, and HRB335 steel is used for stirrup which diameter is 8mm.

Specimens are a total of four groups, ordinary concrete specimens is specimen S, and the steel fiber volume ratio is different for steel fiber reinforced concrete specimens, which 1% is specimen A, 1.5% is specimen B, 2% is specimen C. Steel fiber is a kind of corrugated steel fiber, as shown in Figure 1. The test component parameter list are shown in Table 1.

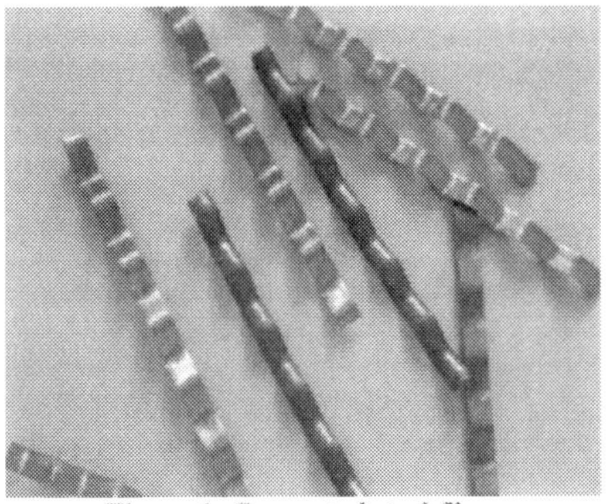

Figure 1. Corrugated steel fiber.

Table 1. Test component parameter list.

No.	S	A	B	C
Concrete	ordinary	steel fiber	steel fiber	steel fiber
Main bar	HRB500	HRB500	HRB500	HRB500
Steel fiber volume ratio (%)	/	1.0	1.5	2.0
Height (mm)	1000	1000	1000	1000
Longitudinal reinforcement diameter (mm)	20	20	20	20
Longitudinal reinforcement ratio (%)	3.14	3.14	3.14	3.14
Stirrup diameter (mm)	8	8	8	8
Spacing between stirrups (mm)	80	80	80	80
Volume stirrup ratio (%)	1.14	1.14	1.14	1.14
Axial compression ratio	0.1	0.1	0.1	0.1

The specimens were made of solid square section, the size of the pier column is 300mm × 300mm, the height is 1000mm, and the bottom was set with a fixed cap. The section size and reinforcement of the pier column model are shown in Figure 2. The forming specimen is shown in Figure 3.

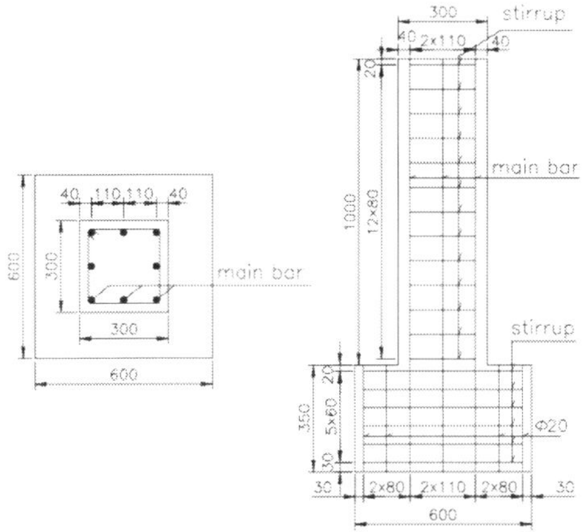

Figure 2. Section size and reinforcement (unit: mm).

Figure 3. The forming specimen for test

4 Horizontal cyclic loading test

The test load is carried out by 10t-100t hydraulic servo static and dynamic loading system, the maximum horizontal load is 1000kN, the maximum horizontal displacement is 1000mm. The vertical force is realized by adding hydraulic jack on the top of the specimen. The horizontal load is controlled by displacement amplitude and the displacement is 3, 6, 9, 12, 15, 18, 21, 24, 27, 30, 35, 40, 45, 50, 55 and 60mm. At the bottom of the specimen, by using hoop frame to fix in the geosynclines. The loading device is shown in Figure 4.

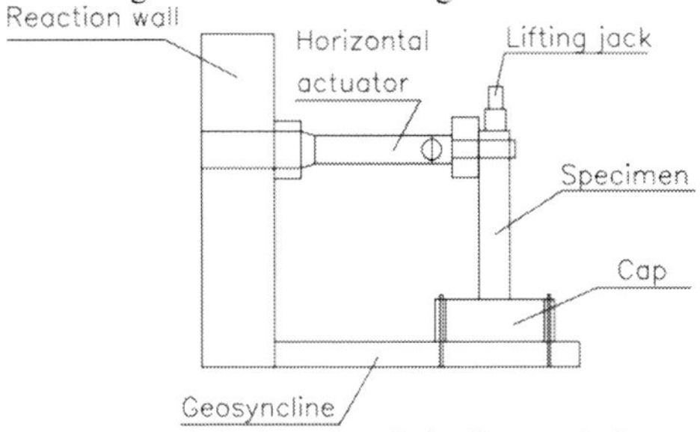

Figure 4. Horizontal cyclic loading test device.

5 Failure patterns

5.1. Failure characteristics of ordinary concrete (specimens S)

Under cyclic loading, the crack of the specimen increases gradually with the increase of displacement. The first crack appears at the time of displacement is 9mm, the position of first crack from the 311mm along the height direction of the specimen extends along the 45 degree angle into the middle lower place. With the increase of displacement, the cracks increase gradually, and the crack width increases. When the displacement is 24mm, through cracks from left to right is formed, and the width is about 3mm. when the displacement reaches 30mm, the surrounding concrete scaling appear because of concrete crushing, confined concrete by stirrup continue to bear force based on confined effect from hoop steels. When the displacement is 45mm, the steel reinforcement of the specimen yield, and the concrete in the cross section falls off, and the load can not be improved. The specimen failure is shown in Figure 5.

Figure 5. Failure characteristics of ordinary concrete specimens.

5.2. Failure characteristics of steel fiber reinforced concrete (specimen B)

Under cyclic loading, when the first crack appears to be 15mm, the crack is small and the development is slow. With the increase of displacement, the crack extends along the 45 degree angle, and the small transverse cracks occur near the main crack. When the displacement reaches 27mm, there is a relative penetration crack, the crack width is 1mm, the main crack width is larger, 2mm. Small cracks occur in the middle of the pier. When the displacement reaches 50mm, the left and right main cracks intersect with each other. When the displacement reaches the maximum value of the test, there is no large area shedding phenomenon in the concrete of the protective layer. The specimen failure is shown in Figure 6. The failure characteristics of specimens A and C are similar to those of B.

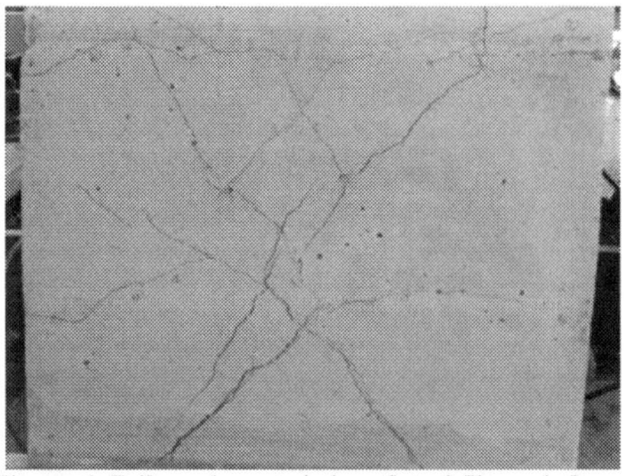

Figure 6. Failure characteristics of steel fiber reinforced concrete specimens.

According to the above characteristics, it can be seen that due to the effect of steel fiber, the internal crack of concrete is slow, the crack width is small, the number of small cracks increases, and the integrity of the components is better compared with ordinary concrete.

6 Hysteretic curve

According to the data of displacement by testing equipment at the top of pier, subtract to horizontal displacement at the bottom of pier, the horizontal load and lateral displacement of pier hysteretic curve of 4 specimens of pier were drawing combined with horizontal seat actuator afterburner. Figure 7 show the hysteretic curve of the specimens.

From the hysteretic curves, all the hysteretic curves have shape of "pinch" bow, the ordinary concrete specimen pinch form is small, on the same level of displacement under the action, it has the tendency to spread. The steel fiber concrete is not diffusion state role in repeated displacement under load, comparison of both lag the curve, steel fiber reinforced concrete specimens are plump, and the ordinary concrete specimen is low.

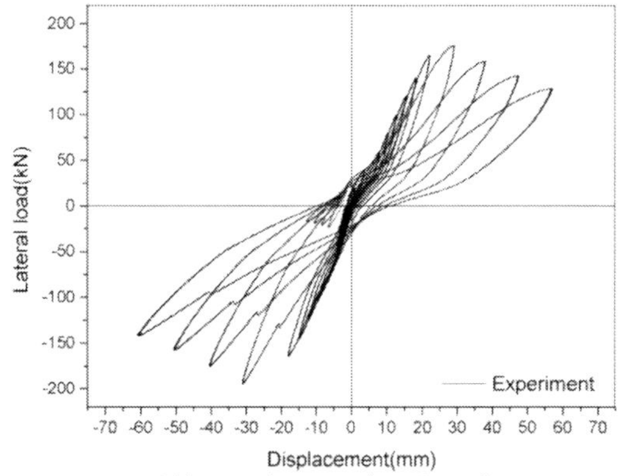

(a) hysteretic curve of specimen S

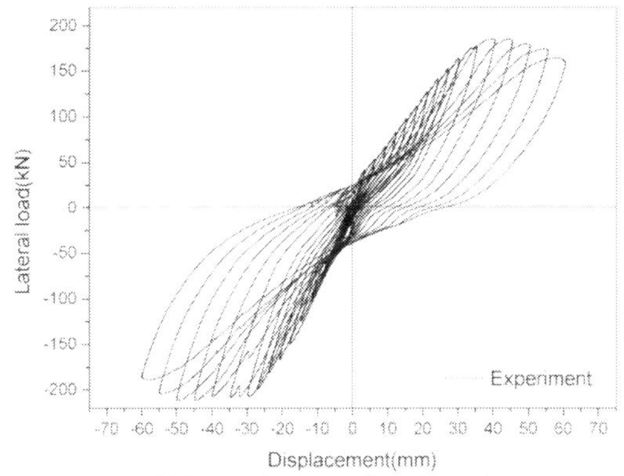

(b) hysteretic curve of specimen A

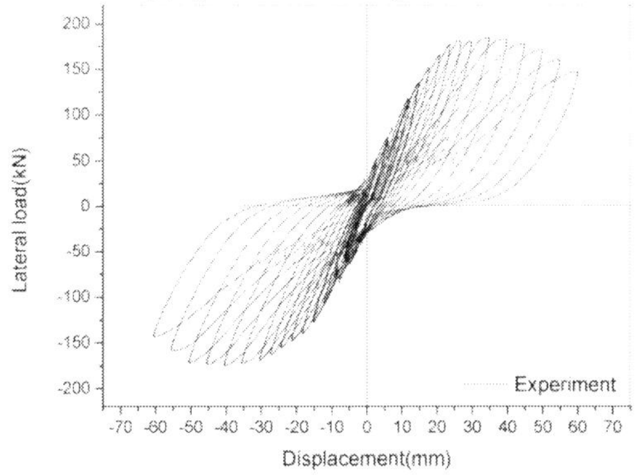

(c) hysteretic curve of specimen B

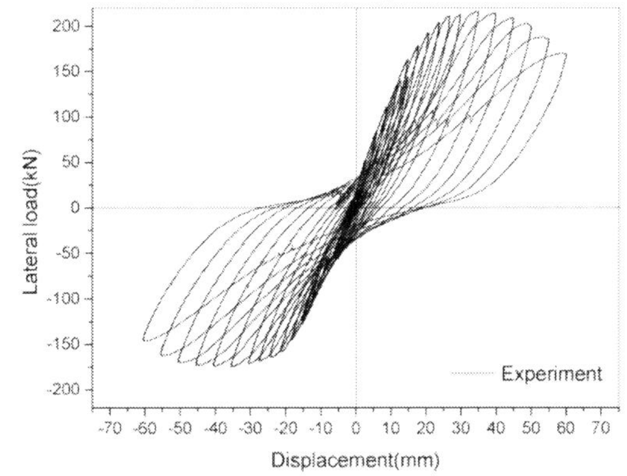

(d) hysteretic curve of specimen C

Figure 7. Hysteretic curves.

In the early stage of displacement loading, ordinary concrete specimen has the same change trend compared with steel fiber reinforced concrete specimens, with the increase of the displacement, the steel fiber concrete of change trend decreased more gently, and ordinary concrete is lower than that, which has larger slope. Comparison of both loading and unloading trend, ordinary concrete has asymmetrical shape, unloading slope is less than loading slope, but the loading and unloading slope of steel fiber concrete is consistent.

Comparing the three hysteretic curves of specimens which have different steel fiber content, the three forms are consistent, and show the ideal loading characteristics of elastic-plastic materials. The greater the amount of steel fiber mixing, the greater horizontal force under same displacement. It has a larger area surrounded by displacement force curve, which shows that the structure energy performance is good.

7 Summary

By comparing the test results of ordinary concrete and steel fiber reinforced concrete specimens with different steel fiber content, the following conclusions can be obtained by the combination of the failure mode and relevant data:

(1) Steel fiber concrete can prevent the development and extension of concrete cracks under the action of external force with the effect of steel fiber. Steel fiber reinforced concrete specimens in comparison with ordinary concrete, the crack of former appear earlier and the width is larger, the failure trend is obvious, and holding point is smaller; the latter under horizontal cyclic loading showed good integrity, the crack width is small and extends slowly, but small cracks are more.

(2) Compared with ordinary concrete specimens, the hysteretic curves of steel fiber reinforced concrete specimens are relatively full, and the displacement force curve has a large area, which has a better energy dissipation effect.

(3) The loading and unloading slope of steel fiber reinforced concrete specimens are more uniform, and the slope of the ordinary concrete specimen is small. The former has obvious elastic-plastic characteristics.

Acknowledgments

Thanks are given for financial support from the National Natural Science Foundation of China (NSFC), grant 51408449.

References

1. S.J. Foster, M.M. Attard. Journal of Structural Engineering, 127, 28-34 (2001)
2. F.L. Li, C.K. Huang, S.C. Wen, etc. ENGINEERING MECHANICS, 22, 159-164 (2005) (in Chinese)
3. Z. Zhu, Y. Fu, P. Guo, etc. STRUCTURAL ENGINEERS, 20, 73-78 (2010) (in Chinese)
4. Y.Y. Zhang, K.A. Harries, W.C. Yuan. Engineering Structures, 48, 255-265 (2013)
5. Y.Y. Zhang, H.Y. Wei, W.C. Yuan. Journal of Vibration and Shock, 31, 102-107 (2012) (in Chinese)
6. Y.Y. Zhang, H.Y. Wei, W.C. Yuan. Journal of Tongji University(Natural Science), 41, 813-820 (2013) (in Chinese)
7. A. Demeke, I.A. Tegos. ACI Materials Journal, 91, 579-584 (1994)
8. R.S. Olivoto, F.A. Zuccarello. Composites Part B: Engineering, 41, 246-255 (2010)
9. P. Robins, S. Austin, J. Chandler, etc. Cement Concrete Research, 31, 719-729 (2001)

Preparation and Characterization of Manganese Slag and Fly Ash-based Geopolymer

Ya-guang Wang, Feng-lan Han[a], Shi-zhen Zhao and Jing-qiu Mu
Beifang University of Nationalities, Yinchuan 750021, China

Abstract. In this study, a series of manganese slag and fly ash-based geopolymers were prepared though alkali activation by varing the amount of manganese slag. The 3-day, 7-day and 14-day compressive strengths of these samples were tested. The maximum strength of 42.78 MPa was obtained at 14th days of testing when 455 g of fly ash, 195g of manganese slag, 20% of the alkali content , the curing temperature of 100°C, the curing time of 12h were used. XRD and FTIR characterization results shown that the polymerization reaction occurs between the glassiness in the manganese slag and the fly ash while adding alkali activator, and the main structure formed was Ca-A-S-H, which contributed the major strength in manganese slag and fly ash-based geopolymer.

1 Introduction

Fly ash is a slag obtained from the flue gas of coal-fired power plant. Electrolytic manganese slag is a by-product of the industrial production of manganese dioxide. At present, China's annual output of electrolytic manganese metal accounted for 98% of the global [1]. As a by-product of electrolytic manganese metal, the electrolytic manganese slag production ratio was 1:7–10 manganese ore with agrade of 12–14%, resulting in an annual electrolytic manganese slag production rate of approximately 8–10 Mt [2]. The current allowable disposal method for electrolytic manganese slag is landfill, which can cause soil contamination, river or groundwater pollution and some other serious environmental pollution [3,4]. Duan et al determined that the main pollutants in the residue were manganese sulfate, ammonia nitrogen and heavy metals, such as cadmium and nickel. They also found that 13.7% Mn, 17% Se, 44.09% ammonia and 55.3% SO_4^{2-} from raw materials entered residues [5]. Thus, for environmental protection and sustainable development purpose, how to re-utilize electrolytic manganese solid waste has been attracted extensive attention. One of the important way of recycling of manganese slag and fly ash is to use them to prepare the geopolymer material.

Geopolymer, as new green cementing material in recent years, is semi-crystalline aluminosilicate polymer based on Si-O and Al-O tetrahedron converted from amorphous structure. Geopolymer is not only showing excellent properties of organic polymer, ceramic and cement, but also can be prepared by some simple process with low cost [6]. Palomo and Hardjito et al indicated that the fly ash based geopolymer will be the key development way of cement for its wonderful mechanical properties and durability [7,8]. Kumar et.al used the slag to substitute a part of preparation of fly ash geopolymer in their study , the results pointed that the C-S-H gel and N-A-S-H gel were coexist in the reaction product and the addition of slag not only reduced the setting time but also high compressive strength of products [9]. D.Panias et.al investigated the effects of water, NaOH and sodium silicate on compressive strength of prepared fly ash based polymer [10]. Zhang Yunsheng et al studied the synthesis procedure of slag base material and the effect of heavy metal [11]. J.S.J.van Deventer et al analyzed the reaction mechanism of geopolymer materials prepared by solid waste [12].

In this study, we prepared manganese slag and fly ash-based geopolymer using manganese slag and fly ash as raw material, sodium silicate and sodium hydroxide as activator. On the effect of the added amount of manganese slag added on the compressive strength of geopolymer was studied. X- ray diffraction (XRD) and Fourier transform infrared spectroscopy (FIIR) were used to characterize the product of the geopolymer materials.

2 Materials and experimental details

2.1 Raw materials

Fly ash was obtained from a power plant in Ningxia. Manganese slag was obtained from Ningxia Tianyuan Manganese Industry Group. The alkaline activators used in this study were sodium silicate (Na_2SiO_3) and sodium hydroxide (NaOH). NaOH (AR) was obtained from Tianjin Kermel Chemical Reagent Co. Ltd. Na_2SiO_3 (SiO_2 / Na_2O = 3.1) was obtained from Panyu chemical plant. The fly ash and manganese slag were placed the

[a] Corresponding author: 625477897@qq.com

oven (DHG-9070A), baking 2 hours at 105 °C, grinding time was 2 hours in ball mill (SYMΦ500).

Laser particle size instrument (Microtrac-X-100) to determine the size and distribution of fly ash and manganese slag, and the fly ash d_{50} was 18.47μm and the manganese slag d_{50} was 21.48μm. Main constituents of fly ash and manganese slag were analyzed by chemical analysis. The chemical constituents of the fly ashes and manganese slag are given in Table 1.

Table1. Chemical component of raw materials(wt.%).

Materials	Fly ash	Manganese slag
CaO	5.37	9.24
SiO_2	42.62	26.19
Al_2O_3	19.85	4.66
MgO	0.58	1.86
Fe_2O_3	3.43	2.59
TiO_2	0.54	0.12
Lost on ignition	2.90	28.92

Figure 1. shows the XRD of the fly ash. There is a diffuse hump at about 20-35° meaning that the fly ash contains a certain amount of amorphous aluminosilicate minerals, and the main mineral phases are SiO_2 and $Al_2O_3{\cdot}SiO_2$. Figure 2. shows the XRD of the manganese slag. It show that the main mineral phases of the manganese slag are $CaSO_4{\cdot}2H_2O$ and SiO_2.

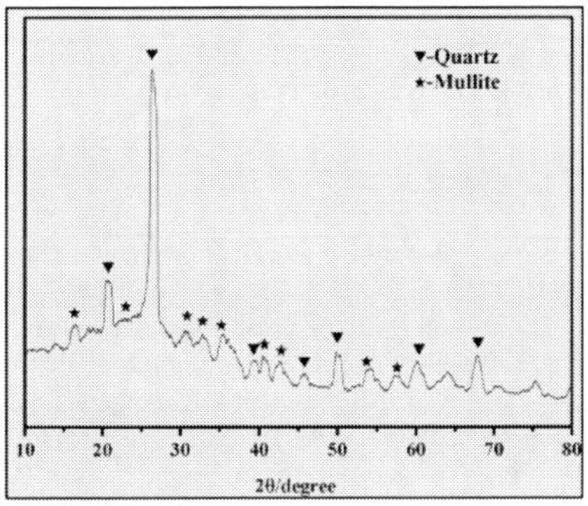

Figure 1. XRD of fly ash.

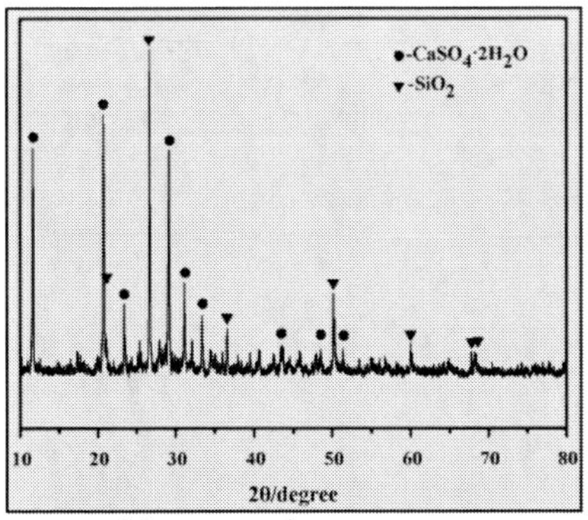

Figure 2. XRD of Manganese slag.

2.2 Casting

First, water, NaOH and Na_2SiO_3, were mixed throughly, and then cooled down to room temperature. The raw powder materials including fly ash and Manganese slag were weighed, and mixed together homogeneously for 1-2 min in a mixer at low speed. The mixed alkali solution was poured into a mixer and blended with the dry powder for 1-2 min at low speed, then for 3-4 min at high speed. Finally, the mixed paste was cast into 40×40×40 mm molds. The samples were cured at a temperature of 100°C for 12 hours. After that, the samples were kept at room temperature until the third day, seventh day and fourteenth day, when the tests were performed. A minimum of 3-6 samples was produced for each mix design to determine the average value of the compressive strength.

2.3 Mole ratio of geopolymer

The amount of alkaline activator used in this study was 20% of the total mass (except for the quantility of water). The used five different ratios of manganese slag / total solids were 0.1, 0.2, 0.3, 0.4 and 0.5. The details of the mixtures are shown in Table 2.

Table 2. Mix design details.

Fly ash(g)	Manganese slag(g)	Alkaline activator	NaOH /Na_2SiO_3
585	65	20%	2:1
520	130	20%	2:1
455	195	20%	2:1
390	260	20%	2:1
325	325	20%	2:1

2.4 Sample Characterization

Pressure tester to test the compressive strength of the sample. Powder samples were used for X-ray diffraction (XRD) and Fourier transform infrared spectroscopy (FTIR) analysis.

The XRD analysis for the powder specimens was conducted using a shimazu XRD-6000 X-ray diffractometer, with a scanning rate of 4° 2θ/min .

The FTIR spectra in the 1600–400cm^{-1} range were recorded at 20°C using Spectrum100 infrared analyzer.

3 Results and discussions

3.1 Results of Compressive Strength

The test results of compressive strength are shown in Figure 3. Figure 3. shows compressive strengths of

 DOI: 10.1051/matecconf/201713004006

manganese slag and fly ash-based geopolymers mortars up to 3, 7, 14 days of curing. Results indicate that as the compressive strength increases, age of geopolymer sample increases, and as the content of manganese slag increases, strength of geopolymer increases and later decreases. When the amount of manganese slag was 30%, the compressive strength was the highest. The decrease of the activator efficiency may be a result of some components of manganese slag reacted with activator while the addition of manganese slag increasing. Therefore, the compressive strength decreased.

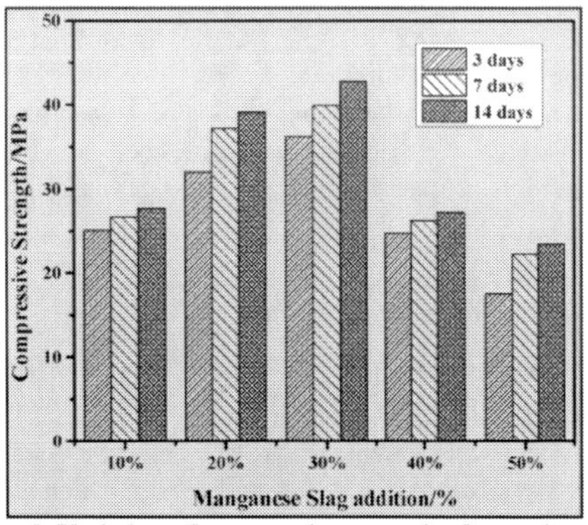

Figure 3. Variation of compressive strength of geopolymer specimens as function of manganese slag content.

3.2 Results of XRD analysis

X-ray diffractograms for different addition amounts of manganese slag and fly ash based geopolymer at 14 curing days are shown in Figure 4. It demonstrated that the crystal in the geopolymer were SiO_2, $NaAl(SO_4)_2 \cdot 6H_2O$, $CaAl_2Si_2O_8 \cdot 4H_2O$ and Na_2SO_4. Compared with XRD of fly ash(Figure 1), the diffuse hump decreased and shifted to the right, illustrating that amorphous aluminosilicate minerals reacted with activator(25-40°). Part of quartz and mullite crystal phase convert to amorphous structure with the reaction processing, and Ca-A-S-H gel was formed in the alkaline environment. Ca-A-S-H gel was the main source of the compressive strength of manganese slag and fly ash-based geopolymer.

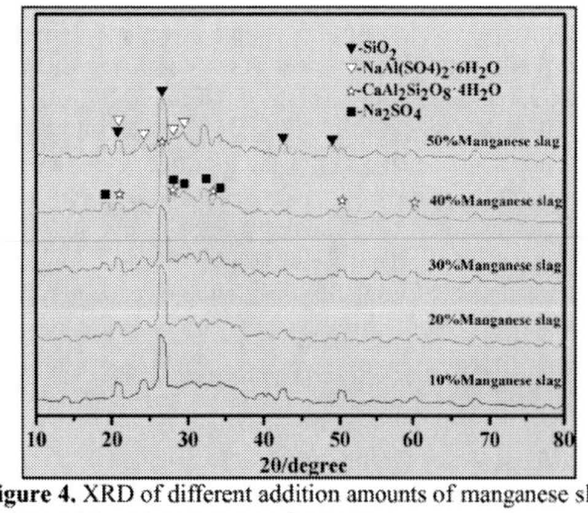

Figure 4. XRD of different addition amounts of manganese slag and fly ash based geopolymer for 14 days.

3.3 Results of FTIR analysis

Figure 5. is the FTIR spectra of different addition amounts of manganese slag and fly ash based geopolymer samples at 14 curing days . The asymmetric stretching vibration of Si-O-Al or Si-O-Si in the prepared geopolymer appears at 1122 cm^{-1}. The peaks at 982 cm^{-1} and 668 cm^{-1} corresponded to TO4 (T=Si or Al) tetrahedral asymmetric and symmetric stretch vibration, respectively. The peak at 798cm^{-1} was the Si-O-Si bending vibration bond. The absorption peak near 615cm^{-1} corresponds to the symmetric stretch vibration of Al-O-Si. The peak around 456 cm^{-1} was assigned to in-plane bending of Al-O and Si-O linkage. The appeared asymmetric stretch vibration absorption peak of Si-O-T(Si or Al) indicated the formation of silicon oxygen tetrahedron and tetrahedral structure in the prepared geopolymer.

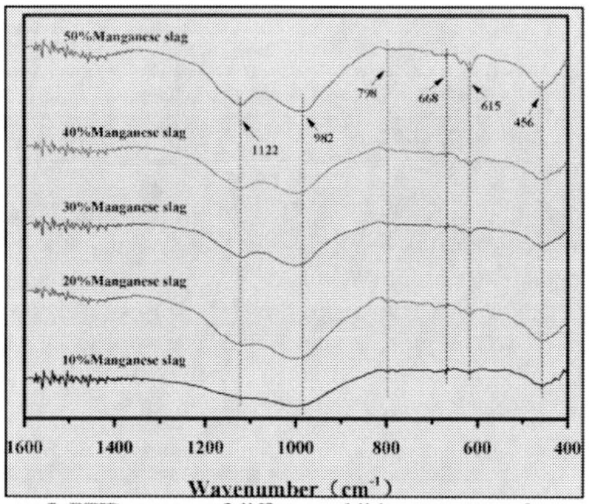

Figure 5. FTIR spectra of different addition amounts of manganese slag and fly ash based geopolymer for 14 days.

4 Conclusion

Based on this study, the following conclusions were drawn:

1. The optimum adding amount of manganese slag in fly ash is 30%, and the strength is highest at this time, up to 42.78 Mpa. With the content of manganese slag further increasing, the activity of manganese slag decrease, which leads to decrease of the polymerization degree and the strength.
2. The XRD of the sample had Ca-A-S-H crystal, which indicates that the polymerization reaction is occurred. The new appeared asymmetric stretch vibration absorption peak of Si-O-T(Si or Al) in the FIIR spectra indicated the formation of silicon oxygen tetrahedron and tetrahedral structure in the prepared geopolymer.

Acknowledgments

Thanks to China EU International Cooperation Project (SQ2013ZOG300003) and Ningxia Hui Autonomous Region science and technology support project (2014ZYH50).

References

1. Tan Z..*In:The 5th international form of electrolytic Mn products market.* (Changsha,Hunan,2013)
2. Bing Du, Changbo Zhou ,Zhigang Dan. Construction and Building Materials. **50**,291(2014)
3. Duan N, Zhou CB, Yu XL. Resour Environ Yangtze Basin. **16**,764(2007)
4. Su Y, Yuan XZ, Zeng GM, Li HM, Li L. Environ Prot Chem Ind. **29**,300(2009)
5. Duan N, Wang F, Zhou CB, Zhu C, Yu H. Resour Conserv Recycl. **54**,506(2010)
6. Xu. Hu. J. S. J. Van Deventer. International Journal of Mineral Processing. **3**,247(2000)
7. Palomo A, Grutzeck M W, Blanco M T. Cement and Concrete Research. **29**,1323(1999)
8. Hardjito D, Wallah S E, Sumajouw D M J. *The Seventh CANMET ACI International Conference on Recent Advances in Concrete Technology.* (Las Vegas, USA, 2004)
9. Kumar S, Kumar R,Mehrotra S P. Jounal of materials science. **45**,607(2010)
10. Panias D, Giannopoulou I P, Perraki T. Colloids Surf, (Physicochem Eng Aspects), 1,(2007)
11. Zhang Yunsheng, Sun Wei, Chen Qianli. Journal of Hazardous Materials. **143** ,206(2007)
12. Van Deventer J S J, Provis J L, Duxson P. Journal of Hazardous Materials. **139**,506(2007)

Effect of Strain Rate on Microscopic Deformation Behavior of High-density Polyethylene under Uniaxial Stretching.

Takumitsu Kida, Yusuke Hiejima* and Koh-hei Nitta

Department of Chemical and Materials Science, Kanazawa University, Kakuma, Kanazawa 920-1192, Japan

Abstract. The microscopic deformation behaviors such as the load sharing and the molecular orientation of high-density polyethylene under uniaxial stretching at various strain rates were investigated by using *in-situ* Raman spectroscopy. The chains within crystalline phase began to orient toward the stretching direction beyond the yielding region and the orientation behavior was not affected by the strain rate. While the stretching stress along the crystalline chains was also not affected by the strain rate, the peak shifts of the Raman bands at 1130, 1418, 1440 and 1460 cm^{-1}, which are sensitive to the interchain interactions obviously, depended on the strain rate; the higher strain rates lead to the stronger stretching stress or negative pressure on the crystalline and amorphous chains. These effects of the strain rate on the microscopic deformation was associated with the cavitation and the void formation leading to the release of the internal pressure.

1 Introduction

Semi-crystalline polymers such as polyethylene (PE) and polypropylene (PP) have been used in many applications owing to the excellent mechanical properties such as high drawability, toughness, workability, lightweight and so on[1–3]. One of the origins of these mechanical properties is the interminglement of the crystalline and amorphous phase[4–6]. Therefore, the correlations between these complicated morphological and mechanical properties have been widely studied to improve the mechanical properties of semi-crystalline polymers[3,5,7–10].

Although uniaxial stretching is one of the most fundamental deformation process, PE and PP show complicated mechanical response and a long scale structural transformation during deformation[11,12]. In particular, specimens shows inhomogeneous deformation called as the necking in the post-yielding[13,14]. As the specimen is further stretched, necked portion propagates to the entire specimen. During these deformation, microscopic structures transform from the spherulitic structure to the fibrillar structure[5]. It has been established that these structural changes are directly related to the fracture mechanism of semi-crystalline polymers.

For *in-situ* study of the structural changes during deformation, one of the most powerful tools is the rheo-optics, which is the simultaneous measurements of mechanical testing and optical measurements such as birefringence, infrared (IR) spectroscopy, X-ray scattering, or Raman spectroscopy[9,15–28]. Such a technique enables us to detect bulk properties (sress, strain, and strain rate) and coresponding local structural parameters (molecular orientation, microscopic stress, crystallinity, chain conformation) at same time during deformation. Especially, *in-situ* Raman spectroscopy have been useful to evaluate the molecular orientation distribution[22,24] and the microsocpic stress applied to the skeletal chains for PE and PP under deformation[22,25,29].

It has been observed that the mechanical properties of semi-crystalline polymers strongly depend on the strain rates [30–32], though the microscopic origin is unclear at this stage. Although the rheo-Raman spectroscopy was applied, the strain rate effect was not clearly observed [18], presumably due to the low strain rate of ~10^{-3} s^{-1}. Recent development of optical apparatus enable us to measure Raman spectra at sufficiently short exposure time of ~1 s. We have developed custom-made *in-situ* Raman spectroscopy[22,24] with the minimum exposure time during deformation is about 0.1 s. In this study, we performed rheo-Raman spectroscopy of high-density polyethylene (HDPE) to evaluate the effects of strain rate on the microscopic deformation behavior. The molecular orientation and the load sharing of molecular chains of HDPE were evaluated during uniaxial stretching process.

2 Experimental

2.1 Sample preparation

Ziegler-Natta catalyzed HDPE (M_w = 10×10^4, M_w/M_n = 5.9) supplied from TOSOH corporation was used. The

* hiejima@se.kanazawa-u.ac.jp

pellets were melted by a hot press machine for 5 min at 210°C and 20 MPa followed by quenched in iced water. After the quenching, sample sheets were annealed at 110°C for 5 h. The volumetric crystallinity was about 66% calculated from the density obtained by Archimedes method, where the crystalline and amorphous density were assumed to be ρ_c = 1000 and ρ_a = 855 kg/m^3, respectively[33]. The notch-shaped specimen (2 mm gauge length and 4 mm width) was used as tensile specimen, which cut out of the sample sheets.

2.2 *In-situ* Raman spectroscopy

To evaluate the real time changes of the microscopic deformation, we developed the *in situ* Raman spectroscopic system by installing a custam made tensile tester to the Raman spectroscopy as shown in Fig.1. The details of *in situ* Raman spectroscopy have been described in the previous papers[22–25]. A DPSS laser (LASOS, Jena, Germany, 639.7 nm in the wavelength, 200 mW in laser power) was used as the incident laser light and a monochromator (PIXIS 100 and Spectra Pro 2300i, Princeton Instruments, Trenton, USA, NJ) was used as the detector. The polarization directions of the incident and scattered light were controlled by a 1/2 wavelength plate and a wire-grid polarizer, respectively. Then, three kinds of polarized Raman spectra (*zz*, *yy*, *zy*) were obtained to evaluate the molecular orientation under uniaxial deformation. In addition, for measuring the changes of the peak position of Raman bands, a 1/4 wavelength plate was used to eliminate the effect of polarization. The laser light was irradiated into the notched portion of the specimen as shonw in Fig.1 (b). Since the two grips of the tensile machine move toward the opposite directions at the same rate, the center of the specimen was irradiated during elongation. The elongation rate was set to 0.5 – 10.0 min^{-1} at 20°C. The *in-situ* Raman spectra were accumulated 10 times with an exposure time of 500 ms.

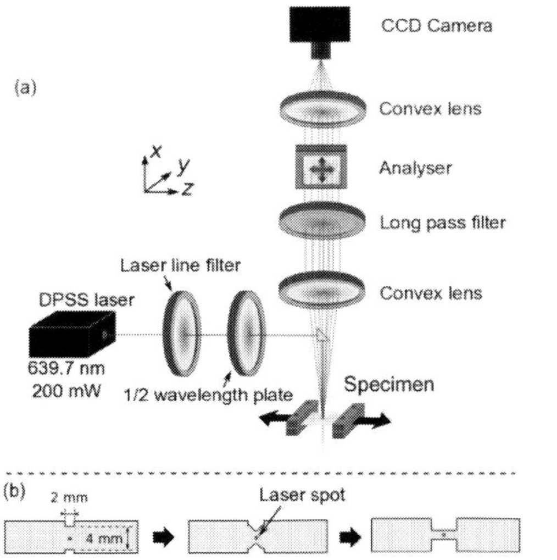

Figure 1. (a) Schematic illustration of *in-situ* Raman spectroscopy system and (b) macroscopic deformatoin behavior of notch-shaped specimen under *in-situ* Raman spectroscopy.

2.2.1 Orientation parameter

The intensity of Raman bands reflects the molecular orientation, which enables us to obtain the orientation parameters[34]. For PE sample, the Raman band at 1130 cm^{-1} assigned to the C-C symmetric stretching mode of crystalline chains has been used to obtain the orientation parameters[22,35]. In this study, two orientation parameters were determined from the intensities of three polarized Raman spectra as follows: [36–39]

$$\langle P_2 \rangle = \frac{3\langle \cos^2\theta \rangle - 1}{2}, \tag{1}$$

$$\langle P_4 \rangle = \frac{35\langle \cos^4\theta \rangle - 30\langle \cos^2\theta \rangle + 3}{8}. \tag{2}$$

Here, θ is the angle between the stretching direction and the principal axis of the Raman tensor. Moreover, the orientation distribution function (ODF) defined as the following equations was determined from the orientation parameters as [38,39]

$$N(\theta) = \sum_{l=0}^{\infty}\left(\frac{2l+1}{2}\right)\langle P_l \rangle P_l(\cos\theta). \tag{3}$$

For the Raman spectroscopy, although only two orientation parameters of $\langle P_2 \rangle$ and $\langle P_4 \rangle$ were obtained, the higher ranks of orientation parameters ($l \geq 6$) can be calculated by maximizing the information entropy. Then, the ODF is written as [24,38,39]

$$N(\theta) = \frac{\exp[\lambda_2 P_2(\cos\theta) + \lambda_4 P_4(\cos\theta)]}{\int_0^{\pi} \sin\theta d\theta \exp[\lambda_2 P_2(\cos\theta) + \lambda_4 P_4(\cos\theta)]}, \tag{4}$$

where λ_2 and λ_4 are the Lagrange multipliers determined to satisfy the following two constraints.

$$\langle P_2 \rangle = \int_0^{\pi} \sin\theta d\theta P_2(\cos\theta) N(\theta), \tag{5}$$

$$\langle P_4 \rangle = \int_0^{\pi} \sin\theta d\theta P_4(\cos\theta) N(\theta). \tag{6}$$

2.2.2 Peak shift

The peak shift $\Delta\nu$ for the Raman bands under uniaxial deformation is defined as the following equation[40,41]:

$$\Delta\nu = \nu - \nu_0, \tag{7}$$

where ν is the peak position at a certain strain and ν_0 is the peak position for the undrawn sample. When the stretching and compression stresses are applied on the molecular chains, the peak shifts have negative and positive values, respectively. The assignments of the Raman bands used in this work is listed in Table 1[42,43]. The anti-symmetric (1063 cm^{-1}) C–C stretching and symmetric (1130 cm^{-1}) C–C stretching vibrational modes correspond to the skeletal chain vibration in the parallel and perpendicular to the molecular chain axis, respectively. These two Raman bands have been used to evaluate the microscopic stress applied to the skeletal chains of the crystalline phase[22,25]. The Raman bands at 1440 and 1460 cm^{-1} corresponding to the CH_2 bending mode of the amorphous chains, are useful to detect the microscopic pressure on the amorphous phase because CH_2 bending mode is sensitive to the microscopic pressure[25,44]. The Raman band at 1418 cm^{-1} assigned to the CH_2 bending modes of the crystalline chains is also strongly affected by the stress, though the stress sensitivity of 1418 cm^{-1} band is opposite to the other Raman bands; $\Delta\nu_{1418}$ shows positive shifts under stretching stress or negative pressure [44,45].

Table 1. Vibrational and phase assignments for the Raman bands of PE[42,43].

Peak position /cm^{-1}	Vibrational mode*	Phase
1063	ν_{as} (C–C)	Crystalline
1130	ν_s (C–C)	Crystalline
1418	δ(CH_2)	Crystalline
1440	δ(CH_2)	Amorphous
1460	δ(CH_2)	Amorphous

* ν_{as}: anti-symmetric stretching, ν_s: symmetric stretching
δ: bending

3 Results

3.1 Stress-strain curve

Fig.2 shows the stress-strain curves of HDPE at various strain rates. Before the first yielding point ($\varepsilon \sim 0.4$) corresponding to the maximum stress point, the specimen is deformed homogeneously. Beyond the first yield point, the stress decreases with the strain, and an inhomogeneous deformation called necking initiates around the second yielding point. The yielding stress of these stress-strain curves become higher at higher strain rates. In addition, the shape of stress-strain curve in the yielding region becomes narrower with increasing the strain rates. In the strain-hardening region after yielding, stress gradually increase with increasing the strain at all strain rates.

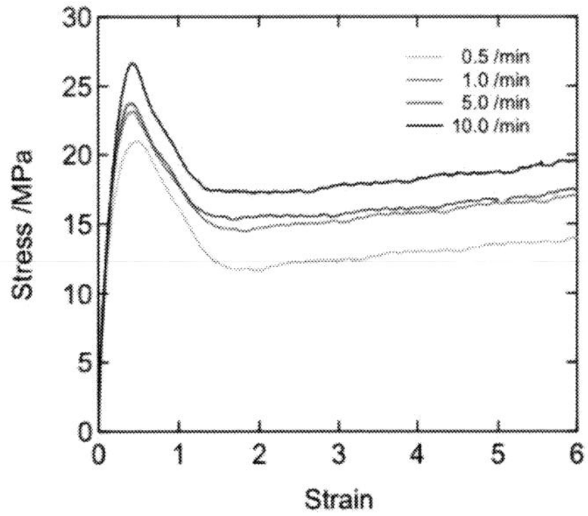

Figure 2. Stress-strain curves elongated at various strain rates.

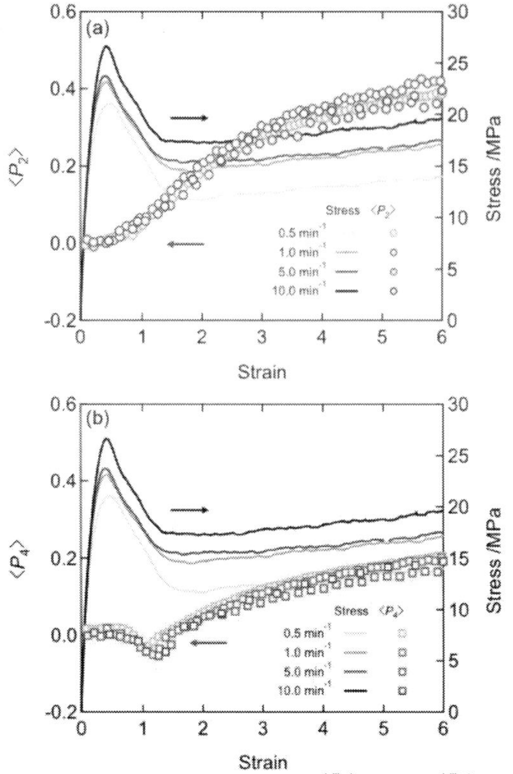

Figure 3. Strain dependences of (a) $\langle P_2 \rangle$ and (b) $\langle P_4 \rangle$ and stress-strain curves elongated at various strain rates.

3.2 Molecular orientation

Fig.3 shows the stress-strain curves and strain dependences of orientation parameters $\langle P_2 \rangle$ and $\langle P_4 \rangle$. In addition, the ODFs at various strain stretched at a strain rate of 0.5 min^{-1} are shonw in Fig.4. The arrows in the figure indicate the vertical axes corresponding to each plot. In the elastic region, the molecular orientation of the crystalline chains remains random since both $\langle P_2 \rangle$ and $\langle P_4 \rangle$ were almost zero and ODF was constant irrespective of the polar angle. After the first yielding point, $\langle P_2 \rangle$ began to increase with increasing the strain, indicating that the crystalline chains begin to orient toward the stretching direction. On the contrary, $\langle P_4 \rangle$ decreased after the first yield point and the ODF showed a broad peak at an intermediate direction (30-50°). Very similar results have been reported by *in-situ* WAXD[9,46] and Raman spectroscopy of HDPE[22,23] and iPP[24]. This oblique orientation in the yielding region is explained by the plastic deformation such as the intracrystalline slip[9,46] or the fragmentatioin process into the lamellar cluster units[22]. In the strain-hardening region, both orientation parameters drastically increased with the strain, and reached asymptotic values in the highly-strained region (ε > 4). The ODFs showed sharp peak at $\theta = 0$, suggesting that the molecular chains within the crystalline phase simply orient in the stretching direction. These molecular orientation behavior of the crystalline chains is unchanged with the strain rates, indicating that the microscopic structural transformation are little affected by the strain rate, though the macroscopic stress-strain behavior is obviously affected.

3.3 Load sharing

Fig.5 shows the strain dependences of the peak shifts for the crystalline chains. In the elastic region, it was found that almost no stress is applied to the crystalline chains because peak shifts of crystalline chains remain zero. These results indicate that in the elastic region the stress mainly concentrates on the amorphous chains, which is in good agreement with the previous data of SAXS measurements [26,27] and the microscopic observation [47]. $\Delta\nu_{1063}$ drastically decreased in the yielding region followed by a weak red shifts, indicating that the stretching stress is applied to the crystalline chains beyond the first yielding region. $\Delta\nu_{1130}$ increased beyond the first yield point and showed a maximum around the second yield point, where $\Delta\nu_{1418}$ showed a minimum. These results indicate that the compression stress is applied to the intermolecular direction of the crystalline chains [25,44,45] due to the densification caused by the necking. Beyond the yielding region, $\Delta\nu_{1130}$ rapidly decreased with strain, approaching an asymptotic value in the highly-strained region, suggesting that the stretching stress is applied to the crystalline chains. The sharp increase of $\Delta\nu_{1418}$ is explained by the expansion of the orthorhombic crystalline structure due to the stretching stress [25,45]. While the peak shifts depended little on the strain rate up to the yielding region, $\Delta\nu_{1130}$ and $\Delta\nu_{1418}$ in the strain-hardening region appreciably depended on the strain rate. Because these two bands reflect the microscopic structural transformation perpendicular to the molecular chain, the strain rate dependence of the peak shifts is explained by the microscopic deformation in the perpendicular to the molecular chain axis. It is noteworthy that the strain rate dependence of the $\Delta\nu_{1063}$ corresponding to the stress along the molecular chain axis is negligible during the entire strain range.

Fig.6 shows the strain dependences of the peak shifts for the amorphous CH_2 bending modes. In the elastic region, both peak shifts $\Delta\nu_{1440}$ and $\Delta\nu_{1460}$ were almost zero, suggesting that no microscopic pressure is applied to the amorphous phase. After the first yield point,

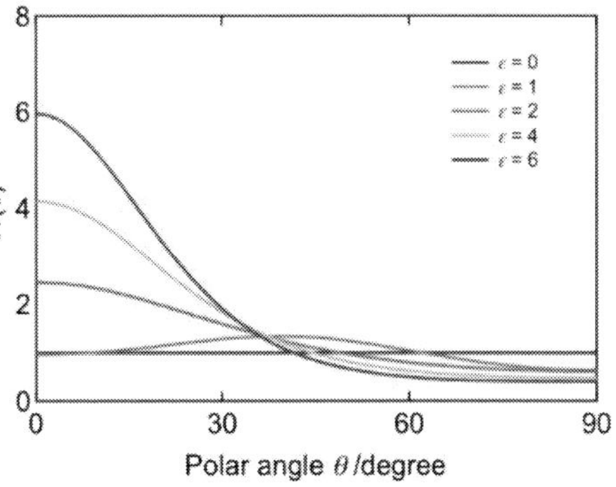

Figure 4. ODFs at various strains during elongation at various strain rates.

$\Delta\nu_{1440}$ and $\Delta\nu_{1460}$ began to increase and show maximum points around the second yielding points. The maximum values around the second yielding point of $\Delta\nu_{1440}$ and $\Delta\nu_{1460}$ were obviously larger than those of the crystalline phase ($\Delta\nu_{1130}$ and $\Delta\nu_{1418}$), indicating that the compression stress under necking is more concentrated on the amorphous phase than in the crystalline chains. Beyond the yielding region, stretching stress applied initially to the amorphous phase because both peak shifts of amorphous chains showed a decrease with the strain. In the highly-strained region, the microscopic load sharing of the amorphous phase remained essentially constant, suggesting that the microscopic structural transformation has been practically completed in the yielding region.

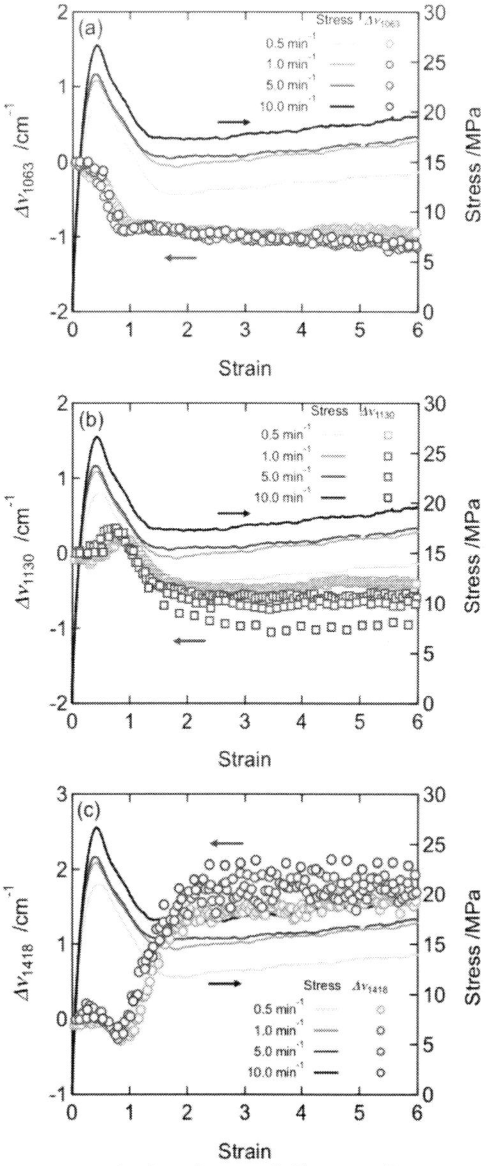

Figure 5. Strain dependences of (a)$\Delta\nu_{1063}$, (b)$\Delta\nu_{1130}$ and (c)$\Delta\nu_{1418}$ along with stress-strain curves elongated at various strain rates.

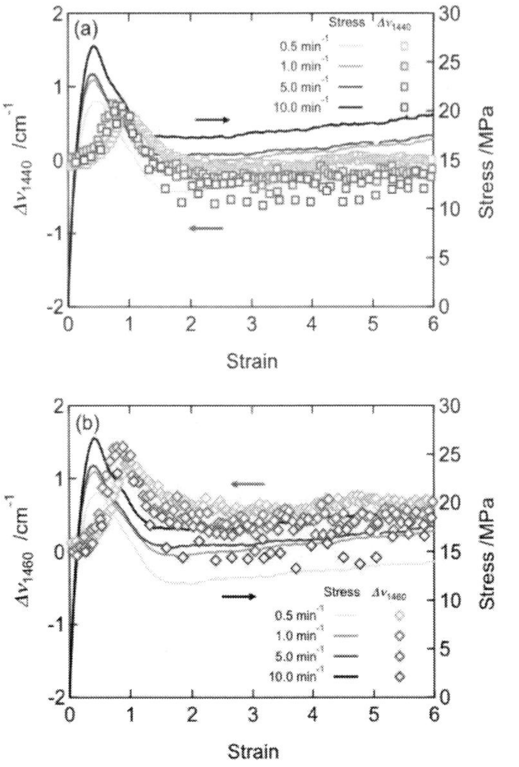

Figure 6. Strain dependences of (a)$\Delta\nu_{1440}$ and (b)$\Delta\nu_{1460}$ along with stress-strain curves elongated at various strain rates.

The strain rate dependences of $\Delta\nu_{1440}$ and $\Delta\nu_{1460}$ were observed in the strain-hardening region, as observed for the crystalline bands at 1130 and 1418 cm^{-1}. Then, it is suggestd that a higher strain rate leads to a stronger stretching loading on the amorphous chains.

4 Discussion

While the molecular orientation are independent of strain rate, the peak shifts of the Raman bands except for the 1063 cm^{-1} band obviously depend on strain rate. This is because they are sensitive to the microscopic deformation in the perpendicular to the molecular chain axis. Then, while the overall microscopic deformation remains essentially the same, the strain rate is likely to affect the interchain interactions.

It has been proposed that the increase of the strain rate promotes the cavitation and the void formation with necking[10,32], resulting in the release of the internal negative pressure which is induced by the Poisson contraction during the uniaxial stretching[48,49]. Under hydrostatic pressure [44], the CH_2 bending modes at 1440 and 1460 cm^{-1} and the C–C symmetric stretching mode at 1130 cm^{-1} show blue shifts and the CH_2 bending modes at 1418 cm^{-1} shows red shift. Then, the large shift beyond the yielding region (red shift for the 1130, 1440 and 1460 cm^{-1} and blue shift for the 1418 cm^{-1}) can be considered to result from the formation of cavitation and voids. The data of the load sharing and the molecular orientation of the amorphous chains were indispensable to reveal the effect of strain rate on the microscopic deformation mechanism.

5 Conclusions

In-situ Raman spectroscopy was performed to investigate the microscopic deformation during uniaxial stretching and to clarify the effects of the strain rate. Beyond the yielding region, the crystalline chains were stretched and compressed along and perpendicular to the molecular chain axis, respectively. In addition, the compression stress was found to be also applied to the amorphous phase due to the inhomogeneous deformation of the specimen under necking. Moreover, the crystalline chains oriented in the intermediate direction accompanied with the plastic deformation of the crystalline phase. In the strain-hardening region, the crystalline chains simply oriented in the stretching direction, and the stretching stress was applied to both crystalline and amorphous chains. Although the overall deformation mechanism of the microscopic structure was independent of the strain rates, interchain interactions of the crystalline and amorphous chains were affected by the strain rate. These effects of strain rate were caused by the formation of cavitation and voids during the yielding.

References

[1] I.M. Ward, J. Sweeney, *Mechanical Properties of Solid Polymers: 3rd edition* (Wiley&Sons, London, 2013)

[2] I.M. Ward, *Structure and properties of oriented polymers* (Chapman&Hall, London, 1997)

[3] G. Strobl, *The Physics of Polymers, 2nd ed.* (Springer, Verlag Berlin Heidelberg, 1996)

[4] R. Seguela, J. Polym. Sci. Part B Polym. Phys. **43**, 1729 (2005)

[5] A. Peterlin, J. Mater. Sci. **6**, 490 (1971)

[6] E.F. Oleinik, Polym. Sci. Ser. C. **45**, 17 (2003)

[7] K.-H. Nitta, M. Takayanagi, J. Macromol. Sci. Part B Phys. **42**, 107 (2003)

[8] K.-H. Nitta, M. Takayanagi, J. Polym. Sci. Part B Polym. Phys. **37**, 357 (1999)

[9] Z. Jiang, Y. Tang, J. Rieger, H.-F. Enderle, D. Lilge, S. V. Roth, R. Gehrke, Z. Wu, Z. Li, Y. Men, Polymer **50**, 4101 (2009)

[10] M. Kuriyagawa, K.-H. Nitta, Polymer **52**, 3469 (2011)

[11] K.-H. Nitta, T. Ishiburo, J. Polym. Sci. Part B Polym. Phys. **40**, 2018 (2002)

[12] K. Hong, A. Rastogi, G. Strobl, Macromolecules **37**, 10165 (2004)

[13] P.I. Vincent, Polymer **1**, 7 (1960)

[14] C. G'sell, N. Aly-Helal, J. Jonas, J. Mater. Sci. **18**, 1731 (1983)

[15] M. Plass, R. Streck, J. Nieto, H.W. Siesler, Macromol. Symp. **265**, 166 (2008)

[16] H.W. Siesler, Polym. Bull. **12**, 481 (1984)

[17] S. Onogi, A. Tanaka, Y. Ishikawa, T. Igarashi, Polym. J. **7**, 467 (1975)

[18] P. Tarantili, A. Andreopoulos, C. Galiotis, Macromolecules **31**, 6964 (1998)

[19] H. Shinzawa, W. Kanematsu, I. Noda, Vib. Spectrosc. **70**, 53 (2014)

[20] M. Mizushima, T. Kawamura, K. Takahashi, K.-H. Nitta, Polym. Test. **38**, 81 (2014)

[20] M. Mizushima, T. Kawamura, K. Takahashi, K.-H. Nitta, Polym. Test. **38**, 81 (2014)

[21] Y. Tang, Z. Jiang, Y. Men, L. An, H.F. Enderle, D. Lilge, S. V. Roth, R. Gehrke, J. Rieger, Polymer **48**, 5125 (2007)

[22] T. Kida, T. Oku, Y. Hiejima, K.-H. Nitta, Polymer **58**, 88 (2015)

[23] T. Kida, Y. Hiejima, K.-H. Nitta, Polym. Test. **44**, 30 (2015)

[24] T. Kida, Y. Hiejima, K.-H. Nitta, Express Polym. Lett. **10**, 701 (2016)

[25] T. Kida, Y. Hiejima, K.-H. Nitta, Int. J. Exp. Spectrosc. Tech. **1**, 1 (2016)

[26] B. Xiong, O. Lame, J.-M. Chenal, C. Rochas, R. Seguela, G. Vigier, Polymer **55**, 1223 (2014)

[27] B. Xiong, O. Lame, J.-M. Chenal, C. Rochas, R. Seguela, G. Vigier, Macromolecules **48**, 2149 (2015)

[28] B. Xiong, O. Lame, J.-M. Chenal, C. Rochas, R. Seguela, Express Polym. Lett. **10**, 311 (2016)

[29] J.C. Rodríguez-Cabello, J.C. Merino, T. Jawhari, J.M. Pastor, J. Raman Spectrosc. **27**, 463 (1996)

[30] S. Hobeika, Y. Men, G. Strobl, Macromolecules **33**, 1827 (2000)

[31] T. Liu, I.R. Harrison, Polymer **29**, 233 (1988)

[32] P.-Y. Ben Jar, Polym. Eng. Sci. **54**, 1871 (2014)

[33] J. Brandrup, E. Immergut, E.A. Grulke, *Polymer Handbook* (Wiley&Sons, New York, 1990)

[34] M. Tanaka, R.J. Young, J. Mater. Sci. **41**, 963 (2006)

[35] M. Pigeon, R.E. Prud'homme, M. Pezolet, Macromolecules **24**, 5687 (1991)

[36] S. Frisk, R.M. Ikeda, D.B. Chase, J.F. Rabolt, Appl. Spectrosc. **58**, 279 (2004)

[37] M. Richard-Lacroix, C. Pellerin, Macromolecules **46**, 5561 (2013)

[38] D.I. Bower, J. Polym. Sci. Polym. Phys. Ed. **10**, 2135 (1972)

[39] D.I. Bower, J. Polym. Sci. Polym. Phys. Ed. **19**, 93 (1981)

[40] R.P. Wool, Polym. Eng. Sci. **20**, 805 (1980)

[41] R.P. Wool, R.S. Bretzlaff, B.Y. Li, C.H. Wang, R.H. Boyd, J. Polym. Sci. Part B Polym. Phys. **24**, 1039 (1986)

[42] M. Gall, P. Hendra, C. Peacock, Polymer **13**, 104 (1972)

[43] M. Gall, P. Hendra, O. Peacock, M.E.A. Cudby, H.A. Willis, Spectrochim. Acta Part A Mol. Spectrosc. **28A**, 1485 (1972)

[44] Y. Zhao, J. Wang, Q. Cui, Z. Liu, M. Yang, J. Shen, Polymer **31**, 1425 (1990)

[45] J. Otegui, J.F. Vega, S. Martin, V. Cruz, A. Flores, C. Domingo, J. Martinez-Salazar, J. Mater. Sci. **42**, 1046 (2007)

[46] R. Hiss, S. Hobeika, C. Lynn, G. Strobl, Macromolecules **32**, 4390 (1999)

[47] K.-H. Nitta, M. Takayanagi, J. Mater. Sci. **38**, 4889 (2003)

[48] B. Xiong, O. Lame, J.M. Chenal, C. Rochas, R. Seguela, G. Vigier, Polymer **54**, 5408 (2013)

[49] C. Fond, J. Polym. Sci. Part B Polym. Phys. **39**, 2081 (2001)

Damage behavior of honeycomb sandwich structure under low-energy impact

Jialin Cheng[1], Zejin Li[1], Chaolin Shuai[1], Shuchang Long[2], Xiaohu Yao[2,a]

[1] Chengdu Aircraft Industrial (Group) Co., Ltd, PR China
[2] Department of Engineering Mechanics, School of Civil Engineering and Transportation, South China University of Technology, Guangzhou, 510640, P. R. China

Abstract. Honeycomb sandwich structure is widely used in aircrafts and ships to absorb impact energy. Damage caused by low-energy impact is difficult to investigate, but will significantly reduce the strength of the sandwich structure. This paper presented a systematical experimental study on the damage behavior of honeycomb sandwich structure under different configurations. Drop weight tests were carried out to investigate the effect of impact energy, core material and face plate structure on the dynamic response of sandwich plate. Delamination regions were obtained through ultrasonic scanning. Conclusions were made that the damage behavior of the face plate was similar with composite laminates while the energy absorption capacity was related to the parameters of the honeycomb core and the mismatch angle of the face plate.

1 Introduction

A sandwich structure consists of three parts: two stiff face plates and a light weighted core. Since the structure can absorb a large amount of energy with small deformation, it is widely used in aircrafts, ships and other structure that may suffer impact loads. For a sandwich structure, the damage caused by a low-energy impact is difficult to obtained, but will significantly reduce the strength of the structure, especially the interface strength. As a result, it is important to summarize the damage regulation of sandwich structure under low-energy impact.

The sandwich structure studied in this paper was called honeycomb sandwich structure. The face plate was stacked by unidirectional composite laminates, while the core was composed by honeycomb structure made of Nomex. Composite laminates were widely used in aircrafts because of the high strength-weight ratio. Honeycomb structure was stable and light weighted, furthermore it can absorb a large amount of energy with the buckling of honeycomb cells. The combination of two materials made the structure stiff and light. However, the delamination within face plate, the debonding between face plate and core, and the crush of honeycomb may seriously reduce the strength of the whole structure.

Studies were carried out all over the world on the impact response of sandwich structure. Mines [1] carried out low-velocity impact tests on sandwich panels composed by a specific series of material and geometry. Results have shown that the density of the core influenced the progression of failure and that higher impact velocities increased the energy absorption capabilities of the panels. Dharmasena [2] performed explosive tests on honeycomb core sandwich panels. The sandwich structure was found suffered significantly smaller back face deflections than solid plates of identical mass. Schubel [3] compared the low velocity impact test with quasi-static test on composite sandwich structure. It was proved that results of the two tests were similar except the localized damage while quasi-static test was easier to perform and analyze. Zhao [4] used the viscoelastic split bars and the two-strain measurement method to measure the crushing behavior of honeycombs. The experimental results show that only the out-of-plane crushing behavior was affected by the loading rate. In the study of Gustin [5], impact compression after impact, and tensile stiffness properties of carbon fiber and Kevlar combination sandwich composites were investigated. They found that the addition of Kevlar and hybrid to the face sheet improved the maximum absorbed energy of the structure. Herup [6] characterized the damage initiation of Nomex honeycomb sandwich panels as a function of facesheet thickness and loading rate. Liu [7] conducted tensile, stabilized compressive and step-by-step compressive tests to study the material properties and mechanical response of the Nomex honeycomb core under transverse loading. Anderson [8] presented the results of experimental investigation on impact response of sandwich structure with honeycomb core and foam core. The extent of damage within the structure was provided by the correlation of the residual indentation and cross-sectional views.

In this paper, a systematical experimental study was carried out on the damage behavior of honeycomb

[a] Corresponding author: yaoxh@scut.edu.cn

sandwich structure under low-energy impact. The effect of impact energy, core material and face plate stacking sequence were investigated by three groups of tests. The dynamic response was recorded as impact force curves, and the delamination region was obtained through C-scan. In conclusion, the damage regulation of sandwich structure was summarized and some future works were listed.

2 Material and Method

Drop weight tests were carried out to investigate the damage behavior of honeycomb sandwich structures under impact loads. The test procedure was based on ASTM-D7766. Specimens were cut into plates with a size of 150mm×100mm. The face plates of the sandwich plate were stacked by T700/3234 unidirectional carbon/epoxy composite while the honeycomb core was made of Nomex. The structure of the specimen was presented in Figure 1. The mechanical parameters of composite layers were listed in Table 1.

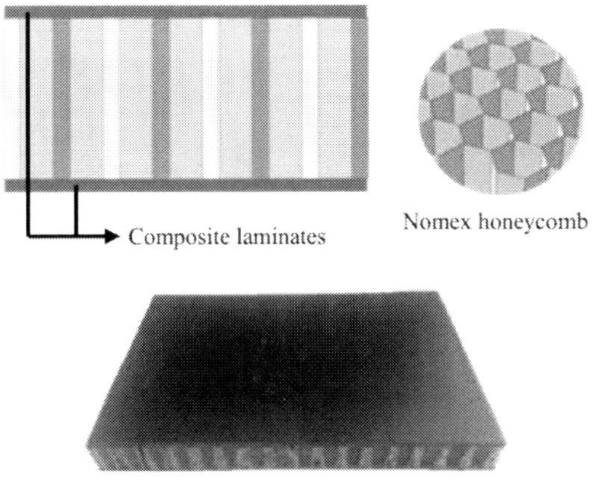

Figure 1. Structure of honeycomb sandwich plates

Table 1. Mechanical parameters of T700/3234 UD composite

Density		1700 kg/m^3
Intra-laminar	**Elastic properties**	E_1=110GPa, E_2 =7.8GPa, ν_{12}=0.32,G_{12}=G_{13}=G_{23}=4GPa
	Strength	X^T=2093MPa, X^C=870MPa, Y^T=50MPa, Y^C =198MPa, S^L= 104MPa
	Fracture energy	G_{ft}=G_{fc}=10N/mm, G_{mt}= G_{mc}=1 N/mm
Inter-laminar	**Elastic properties**	K_n=K_s=K_t=850MPa
	Strength	T_n=3.3MPa T_s=T_t=7MPa
	Fracture energy	G_n=0.306N/mm G_s=G_t=0.632N/mm

The drop weight machine used in the tests was InstronDynatup 9250HV made in USA. The specimen was clipped on a rigid supporting structure with a rectangular cut of 125mm×75mm in the center. The center of the specimen was impacted by a hemisphere punch head weighted 18.89kg with a diameter of 16mm. The impact force and the displacement of the punch were recorded during the test procedure. After the drop weight tests, specimens with visible damage were examined by C-scan. The SAM-300 ultrasonic scanning microscope made in Germany was used in the nondestructive tests.

3 Result and discussion

The test results were divided into three groups in order to reveal the damage regular of honeycomb sandwich structures under different configurations. The damage behavior was characterized by the visible failure of material, the dynamic response of specimen and the delamination area inside the plate.

3.1. Effect of impact energy

Three specimens with the same components were tested under different impact energy. The stacking sequence of face plates is [45/-45/90/0/0/90/-45/45], which represents that the fiber direction in top layer of the face plate is 45°, while the second layer is -45°, and so on. Fiber direction is defined as the extension direction of carbon fibers during the forming process. Curves of impact force versus time and displacement were plotted in Figure 2. All curves share the same trend with two peak values. The first peak value represented the ultimate load of the top face plate. After the penetration of the top plate, the honeycomb core began to bear the main load. Since the core material was much weaker than the face plate, the load level of this stage was much lower than the peak value. When the core was crushed by the punch head, the bottom face plate began to bear the main load, shown as the second peak value. With the load level of 17.21J, the impact energy was absorbed completely by the damage of top face plate and the crush of core. As a result, the second peak value was very small since the bottom face plate did not bear the impact load.

The area enclosed by the curve of impact force versus displacement equals to the energy absorbed by the structure. Although the strength of the core material was low, the buckling of the honeycomb cells contributed a large displacement in the vertical direction. As a result, the honeycomb core absorbed a large amount of impact energy and prevented the bottom face plate from any damage and failure.

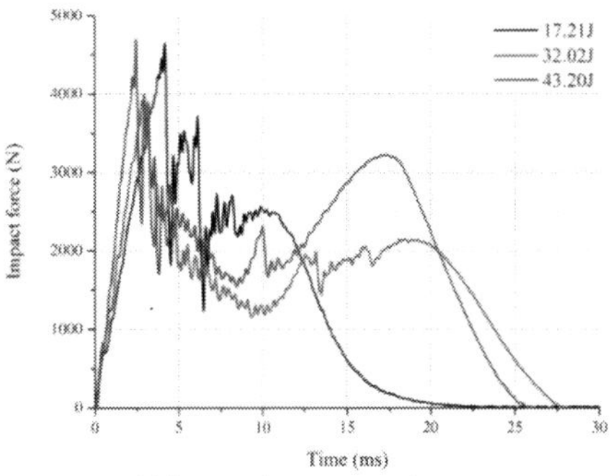

(a) Impact force versus time curves

 DOI: 10.1051/matecconf/201713005002

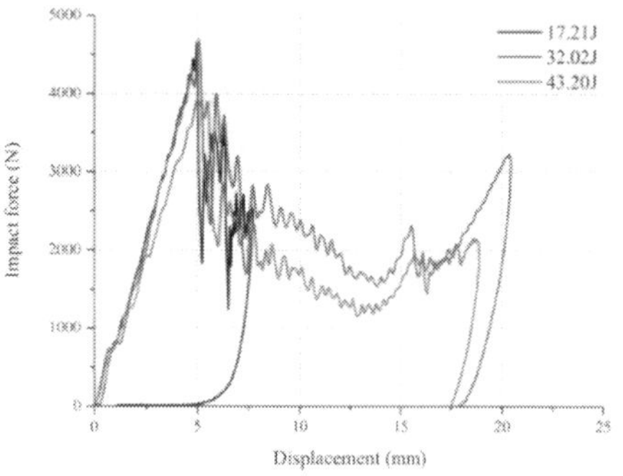

(b) Impact force versus displacement

Figure 2. Curves of specimens under different impact energy

The C-scan results of the three specimens were shown in Table 2. The region in white represented the delamination area. Delamination distributed around the impact zone and along the fiber direction of the sub layer below the interface. This phenomenon was similar with the regulation of composite laminates [9]. On the contrary, the debonding area between top face plate and core was found along the sub layer above the interface.

Table 2. Delamination area of top face plate under different impact energy

Impact Energy	17.21J	32.02J	43.20J
Interface I :-45/90		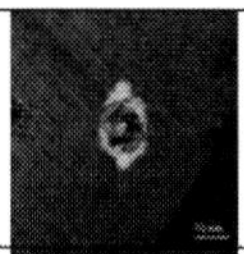	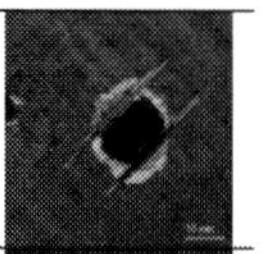
Interface II :90/0			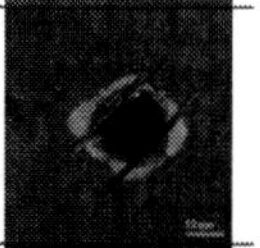
Interface III :45/core			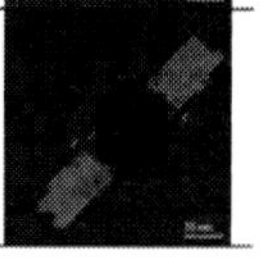

3.2 Effect of honeycomb core

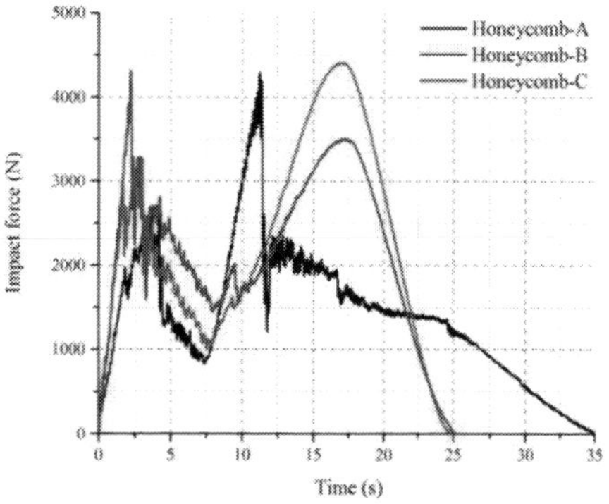

(a) Impact force versus time curves

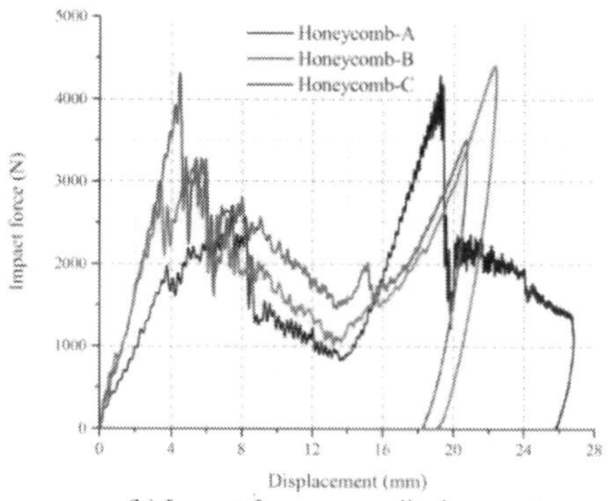

(b) Impact force versus displacement

Figure 3. Curves of specimens with different core parameters

Three specimens with different core parameters were impact under 40J. The cores, made of Nomex honeycomb, were different in cell length (L) and density (ρ). Detailed description can be found in Table 3.

Table 3. Delamination area of top face plate with different core parameters

Honeycomb type	A	L=2.75mm ρ=48kg/m^3	B	L=2.75mm ρ=80kg/m^3	C	L=1.83mm ρ=80kg/m^3
Interface I :45/90	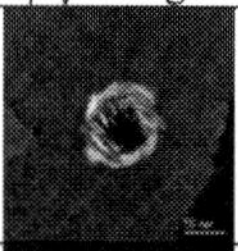		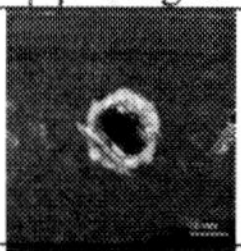		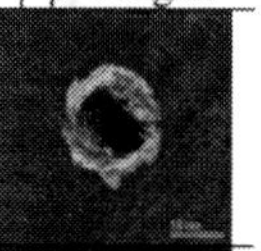	
Interface II :45/0	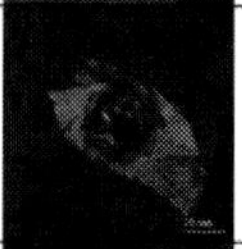					
Interface III : -45/core						

According to the curves in Figure 3, the slops of rising steps for B and C were almost the same, which means honeycomb core with the same density share the same stiffness. As for specimen A, whose core density was much lower, the slope of the curves was low while the first peak value reduced. In the impact process of specimen A, after the damage of top plate and the crush of core, there was a mount of impact energy remains, which leads to the damage of bottom plate. Conclusion can be made that a weak core material can reduce the damage of top face plate but will lead to serious failure of bottom face plate.

Delamination areas of the three specimens were shown in Table 3. The regulation was similar with the first group of tests, especially the shape of debonding area between top face plate and core.

3.3 Effect of face plate stacking sequence

The stacking sequence of composite laminates was proved to affect the delamination area between sub layers.

In order to study the effect of face plate structure on the sandwich plate, two specimens with different stacking sequence were tested under 30J.

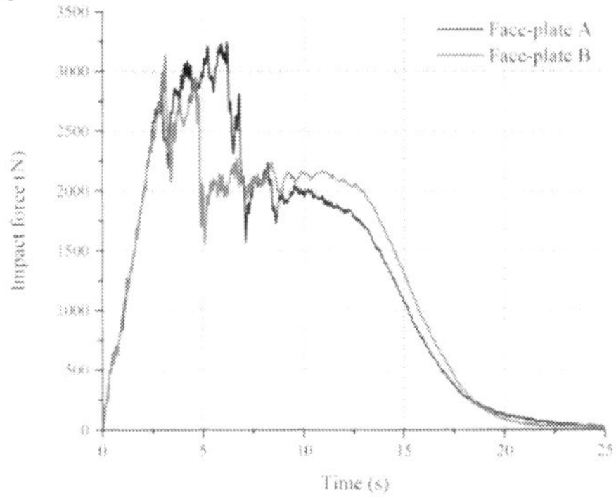

(a) Impact force versus time curves

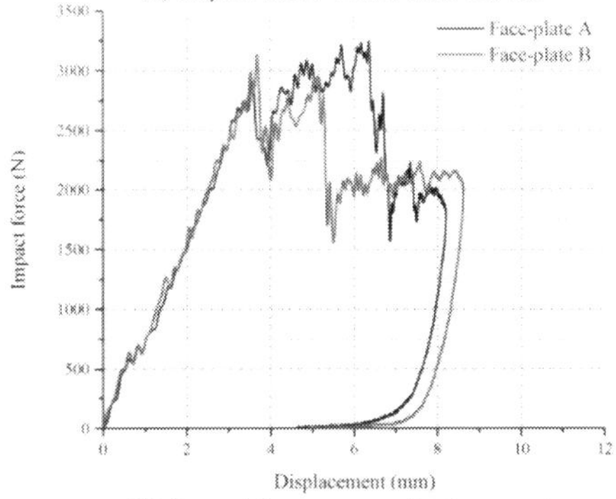

(b) Impact force versus displacement

Figure 4. Curves of specimens with different core parameters

Table 4. Delamination area of top face plate with different core parameters

Ply of face plate	45/90/-45/0 /0/-45/90/45		45/-45/90/0 /0/90/-45/45	
Interface I	90 /-45	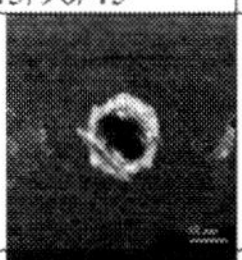	-45 /90	
Interface II	-45 /0	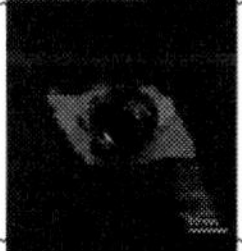	90 /0	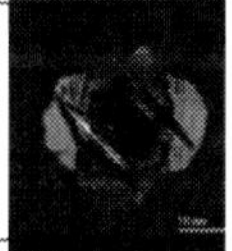
Interface III	45 /core		45 /core	

The dynamic responses of the two specimens were almost the same, except the time of the load drop. The sudden change of the impact force can be regarded as the characterization of the penetration. Later penetration will lead to more damage in top face plate, and at last improve the energy absorption capacity of sandwich structure. As can be seen, the mismatch angle of the stacking sequence of specimen A was smaller than that of specimen B. It is proved that a face plate with big mismatch angles was easy to be penetrated. Delamination areas were also presented in Table 4. The shape of debonding zone was controlled by the ply angle as discussed in above sections.

4 Conclusion

This paper presented the study on the damage behavior of honeycomb sandwich structure under low energy impact. Damage of the structure under different configurations was characterized and some regulations were summarized. Conclusions can be made as follow:

(1) The curves of impact force versus time and displacement can be divided into three steps. At first, the top face plate bears the main load. After the ultimate strength is met, the top face was penetrated and the core begins to bear main load. When the honeycomb was completely crushed, the major impact load was applied on the bottom face plate.

(2) A weak core can reduce the damage of the top face plate but will lead to serious failure of the bottom face plate.

(3) A face plate with big mismatch angles is easy to be penetrated. Reducing the mismatch angles of the face plate can increase the energy absorption capacity of the whole structure.

(4) The delamination area within a face plate share the same regulation of composite laminates. While the debonding area between top face plate and core was found along the fiber direction of the sub layer above the interface.

Future works will focus on numerical studies of honeycomb sandwich structure. Proper institutive model will be chosen for the Nomex honeycomb material. The damage process of face plate and the crush of core will be discussed in detail. Residual strength of sandwich structure will also be considered in follow-up works.

Acknowledgement

The financial sponsorship and support from the Natural Science Foundation of China (11472110, 11372113), the Fundamental Research Funds for the Central Universities (2014ZG0033), and New Century Excellent Talents(NCET-13-0218). This paper is also supported by the opening project of State Key Laboratory of Explosion Science and Technology (Beijing Institute of Technology) (KFJJ14-2M).

References

1. Mines R A W, Worrall C M, Gibson A G. Low velocity perforation behavior of polymer composite sandwich panels [J]. International Journal of Impact Engineering, **21** (1998).
2. Dharmasena K P, Wadley H N G, Xue Z, et al. Mechanical response of metallic honeycomb sandwich panel structures to high-intensity dynamic loading [J]. International Journal of Impact Engineering, **35** (2008).

DOI: 10.1051/matecconf/201713005002

3. Schubel P M, Luo J J, Daniel I M. Low velocity impact behavior of composite sandwich panels [J]. Composites Part A Applied Science & Manufacturing, **36** (2005).
4. Zhao H, Gary G. Crushing behavior of aluminum honeycombs under impact loading [J]. International Journal of Impact Engineering, **21** (1998).
5. Gustin J, Joneson A, Mahinfalah M, et al. Low velocity impact of combination Kevlar/carbon fiber sandwich composites [J]. Composite Structures, **69** (2005).
6. Herup E J, Palazotto A N. Low-velocity impact damage initiation in graphite/epoxy/Nomex honeycomb-sandwich plates [J]. Composites Science & Technology, **12** (2013).
7. Liu L, Wang H, Guan Z. Experimental and numerical study on the mechanical response Of Nomex honeycomb core under transverse loading [J]. Composite Structures, **121** (2015).
8. Anderson T, Madenci E. Experimental investigation of low-velocity impact characteristics of sandwich composites [J]. Composite Structures, **50** (2000).
9. Long S, Yao X, Zhang X. Delamination prediction in composite laminates under low-velocity impact [J]. Composite Structures, **132** (2015).

Research on process and simulation of the isothermal forging of a spur bevel gear

Zhao Xinhai [1,2,a], Zhao Liangang [1], Tian Zhirui [1], Liu Chuanjin [1], and Wu Xianghong [1]
[1] *Key Laboratory of Liquid–Solid Structural Evolution & Processing of Materials, Ministry of Education, Shandong University, Jinan, Shandong, 250061, PR China*
[2] *State Key Laboratory of Materials Processing and Die and Mould Technology, Huazhong University of Science and Technology, Wuhan Hubei, 430074,PR China*

Abstract. This paper took the lead in applying isothermal forging technology to form a spur bevel gear in a single operation. According to the isothermal forging process, the forging part drawing of the spur bevel gear was defined. The 20CrMnTi was chosen as the gear materials. A closed-die structure was used in the forging process. Then, the model of the dies and workpiece were built by UG software. After the models was imported into DEFORM-3D, the simulation of the isothermal forging process of the bevel gear was executed. Through simulation, the shape of the gear, the distributions of effective stress and strain, the forming load of the dies, the laws of metal flow and the filling of the metal during the forging process were gained and thoroughly discussed. The results revealed that the full forging process can be divided into four periods and precision spur bevel gear can be obtained without causing under-filling and folding defect by using the isothermal forging with closed-die forging structure, which provides guidance to manufacturing the spur bevel gear.

1 Introduction

Straight tooth bevel gears have been widely applied in automobile, agricultural machinery, engineering machinery, etc. According to the difference in deformation temperature of billet, the precision forging technology can be divided into hot precision forging, cold precision forging, warm forging and isothermal forging[1].

Since the precision forging gear process was derived from Germany in the 1950s, many gears have been manufactured with less or no cutting processing and the precision of gears can reach 0.02~0.05 mm[2]. However, when it comes to the current precision forging, there are still a great number of problems in actual production[3-4].

1,Resistance of deformation is so large that it is necessary to carry out intermediate heat treatment and superficial treatment in the cold forging process;

2,It is easy to result in the reduction of accurate dimension and surface quality due to the strong oxidation in hot forging;

3,Although warm forging overcomes the drawback of cold forging and hot forging, requirement for equipment is higher.

In addition, it is easy to cause a small amount of shrinkage after cooling and there are difficulties in forming gears in a single forging operation.

Based on the above problems, experimental and theoretical analysis in the forming of spur bevel gears have been carried out. Summarily, the forming technology for spur bevel gears were more considered in the research. Lin et al.[5] studied and analyzed the cold extrusion process and fabrication technology for a differential bevel gear. Grzegorz Samołyk[6] and Deng et al [7] studied both a theoretical and experimental investigation of the cold rotary forging process of a spur bevel gear. W. Bochniak et al[8] investigated the forging process by the KOBO method and employed this method in forming a spur bevel gear to obtain forge piece with complete formation of the teeth under less forming force. Song[9] carried out the process design for closed-die forging of bevel gears using 3-D FEM simulations. Nevertheless, flash in the gear and fracture on the punch were occurred due to the over punch stroke during forming process.

Isothermal forging, which is conducted under the condition of lower strain rate and the constant temperature between the die and billet[10], has many advantages such as small working allowance, high precision and low residual stress. But few researchers have carried out investigation on applying isothermal forging to form spur bevel gears because it is difficult to ensure the die and billet to remaining high deformation temperature for a long time and the die trend to soften under high temperature. However, in recent years, with the rapid development of isothermal forging technology of titanium alloy, the heat insulation device and die material[11] can be employed in isothermal forging of gear steel 20CrMnTi. To ensure the precision of the simulation, the mechanical properties under low temperature of 20CrMnti was gained by a series of

[a] Corresponding authors：xhzhao@sdu.edu.cn

 DOI: 10.1051/matecconf/201713005003

experiments. Moreover, several researchers constantly developed die materials for isothermal forging, such as N3[12] and DM02[13], which provides a possibility to form bevel gears by isothermal forging.

This paper carried out the research on the isothermal forging process of the spur bevel gear by using the FEM simulation method. The metal flow, forming load and the deforming field valuables during the forging process were obtained.

2 FEM simulation models

The part drawing of a straight tooth bevel gear is shown in Figure 1.

Figure 1. Part drawing of the gear

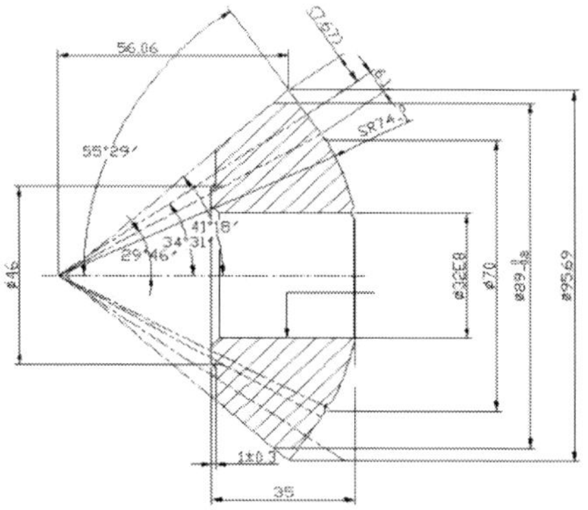

The die structure used in this current study is closed-die forging. According to the final forging drawing, three dimension model of die and billet at the proper position was established using UG software. Then, the STL file of the models were transferred by UG and introduced into Deform-3D to carry out numerical simulation. Meanwhile, the material property of workpiece was defined , which is based on the constitutive equation obtained before and mechanical properties of 20CrMnTi under isothermal spheroidisation(shown in table 1)[7]. The finite element model is shown in Figure 2, which takes one tenth of the entity to calculate for the sake of improving simulation accuracy and save calculation time. In this model, both the upper die and the lower die were made by two parts. At the beginning, the upper punch and the top die move down in the same velocity until the upper die comes into contact with the bottom die, therefore, a closed cavity has been formed. Then, the upper punch continues moving down and the bottom punch begins to move upwards at the same speed as the upper punch, which contributes to making metals flow into the cavity.

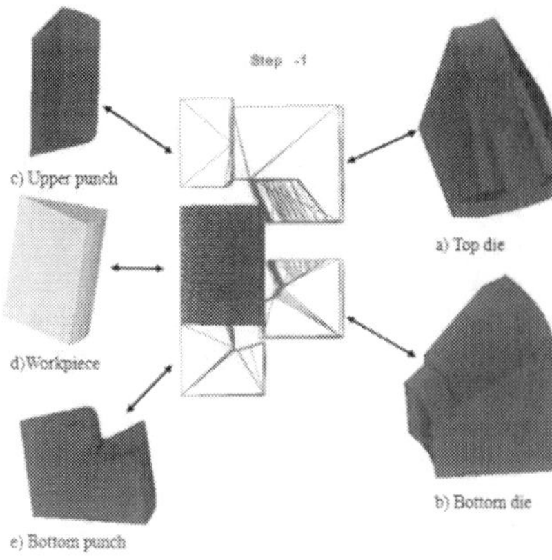

Figure 2. The FEM model of the spur bevel gear

Table 1. Mechanical properties of 20CrMnTi under isothermal spheroidisation.

Heat treatment mode	Isothermal spheroidisation
Poisson' s ratio	0.3
Density(kg/m^3)	7800
Yield strength (MPa)	325.68
Ultimate strength(MPa)	521.17
Young' s modulus(GPa)	225.5

In addition, the general conditions for the finite element model are summarized in Table 2.

Table 2. The general conditions for the finite element model

Condition	Description
Deformation temperature	900 ℃
Workpiece	plastic
Dies	rigid
Velocity of dies	0.05 mm/s
Friction model	Shear friction
Friction between dies and workpiece	0.3
Number of elements	50000

3 Simulation results and discussion

3.1 Analysis of defects and load prediction

Figure 3.presents the spur bevel gear obtained by numerical simulation by Deform-3D. From Figure 3., it can be observed that the spur bevel gear can be formed without causing under-filling because of the closed-die forging manner, meanwhile folding defect does not appear, which can be explained for lower feed rate. Though there are burrs appeared at the clearance between dies and punches, it's size was thin and small and can be removed easily.

Figure 3. The bevel gear obtained by simulation

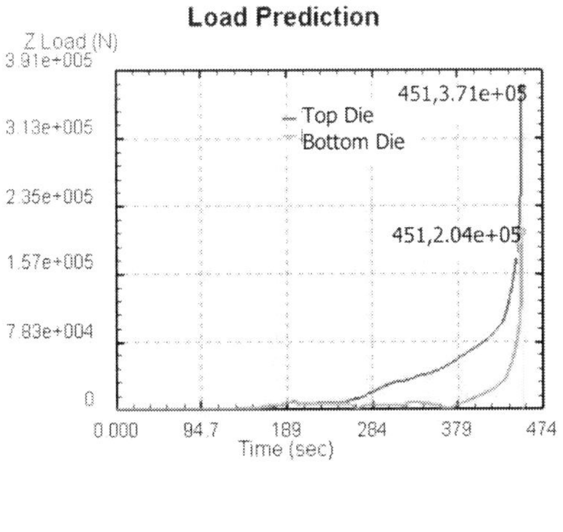

Figure 4. Forming load of top die and bottom die

Figure 4. gives the forming load of top die and bottom die during the isothermal forging. Since the beginning of forging process, the forming load generally keep in lower value with slow-growing. Until the metal begins to flowing into the cavity of big end teeth, the axial forging force sharply increases in the same exponential due to the great metal flow resistance of complete filling of die cavity. In general, compared with the other forging process, the forming force of isothermal forging may be more advantageous. This can be owing to the fact that lower feed rate contributes to the flow and dynamic recrystallization of metal in the process of forming, which leads to less deformation resistance and less forming force. When it comes to the top die whose forming load is higher than other die and punches, its maximum forming load is only 371 KN.

3.2 Analysis of metal flow

The velocity field obtained by DEFORM -3D is shown in Figure 5. According to the difference in the direction of metal flow, there are four periods during the whole forging process. At the first period(Figure 5a), namely in the very beginning of the forging process, most of metal move downward and the billet presents upsetting deformation due to downward pressure from the upper die and punch. Then, after the top die comes into contract with bottom die, the metal begins to flow inside the cavity and metal flow is mainly in radial direction, and this is the second period(Figure.5b). At the third period(Figure. 5c), with the increase of reduction and contract area, when the billet has contract with the small tooth section of top die, metal begins to flow in the upper right direction. In the next, as a result of the resistance from the contract surface, flow direction of the metal in the center of billet happens to offset lengthways while flow direction of the teeth offsets clockwise. At the final finishing period(Figure. 5d), the metal mainly flows along the normal direction of the cavity, which helps to cause the complete filling of die cavity.

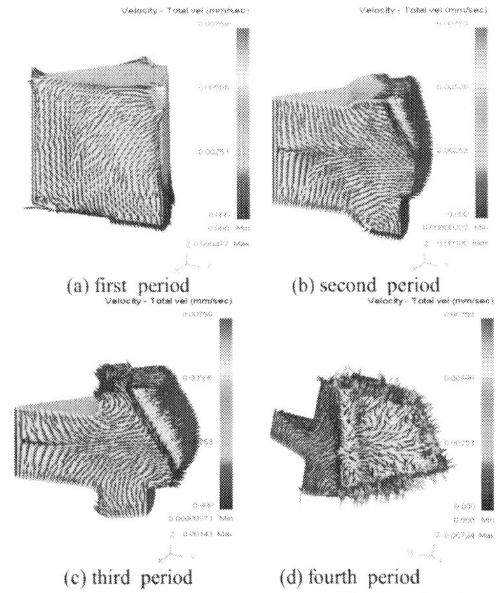

(a) first period (b) second period

(c) third period (d) fourth period

Figure 5. The distribution of velocity field during the isothermal forging

3.3 The distribution of stress and strain

The distribution of effective strain and stress at different time during the isothermal forging process is shown in Figure 6. It can be seen from Figure 6. that, the evolution of effective strain and stress is similar with those of the forming load in the whole isothermal forging, which also can be described as the following two periods. At the end of the first period, as shown in Figure 6.(a) and Figure 6.(c), the effective strain and stress (20~55 Mpa) basically maintain in lower value with a little change. When the metal begins to flow into the cavity of big end teeth, the second period started. At the end of the second period, as shown in Figure 6.(b) and Figure 6.(d), the strain and stress rapidly increase up to 4.91 and 282 Mpa respectively since the metal is pressed against the wall of the die cavities. Moreover, the larger strain and stress locate in the edge areas of end face and tooth crest in the forged gear, which reveals the metal of this area undergoes the highest degree of deformation.

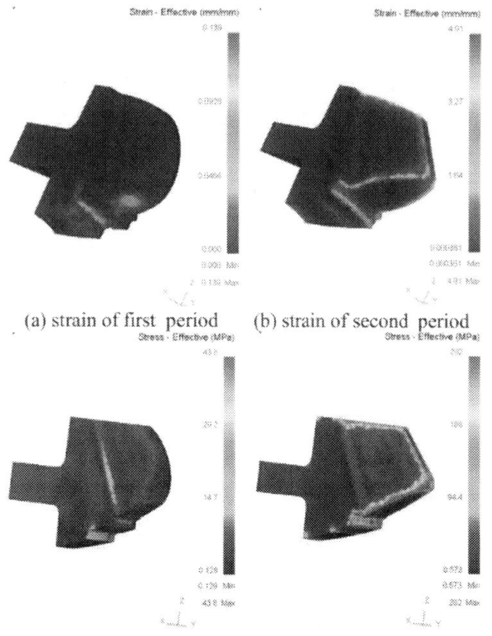

(a) strain of first period (b) strain of second period

(c) stress of first period (d) stress of second period

Figure 6. The distribution of effective strain and stress during the isothermal forging

4 Summary

This paper conducted the simulation of the isothermal forging process of spur bevel gear by using the FEM method. The following conclusions can be achieved:

1,Using the isothermal forging with closed-die structure, the spur bevel gear can be formed without causing under-filling and folding.

2,The forming load of the dies during the isothermal forging process were obtained. The load of the top die is larger than those of the lower die.

3,According to the simulation result, the forging progress can be divided into four periods. The velocity, effective strain and effective stress of the every periods during the process were obtained.

Acknowledgements

This work was financially supported by the fund of State Key Laboratory of Materials Processing and Die & Mould Technology, Huazhong University of Science and Technology (P2014-11) and The Interdisciplinary training project of Shandong University (2014JC024).

References

1. Wang X.Y. Metal Proc.;7(2012)20-21.
2. Yuan X. Net shape Forming Engineer. 6(2010)51-59.
3. Wang J.,Li P. J. of Plasticity Engineer. 1(2013) 6-10.
4. Zhao J.H. Metal Proc. 17(2008)14-17.
5. Lin C.,Xu P..Modern Manu. Engineer. 6(2011) 69-72.
6. Samołyk G. J.Mater.Proc. Technol. 213(2013)1692-1702.
7. Deng X.B., H. Lin.Mater Des. 32(2011)1376 -89.
8. Bochniak W., A. Korbel, R. Szyndler. J. Mater. Proc. Technol. 173(2006)75-83.
9. Song J.H.,Im Y.T.J. Mater. Proc. Technol. 192-193(2007)1-7.
10. Zhang Z.L.,Li H.R..Hot Working Technol. (2014):12-15.
11. Guo L.F.,Zhu Y.C.Chinese J. of Rare Material. 2(2012)357-362.
12. ZHAO H.B. ,WU.C.X. Mater. Engineer. 1(2009)18-21.
13. Xiao C.B.,Song.J.X. Mater. Engineer. (2005)44-47.

Ultrasonic dissimilar joining of aluminum alloy and polymer with the composite material of ABS polymer doping carbonized rice husk

Chin-Pao, Cheng*, Chun-Hu Cheng, I-Wen Wang Yu-Chen Chen, Yang-Sheng You

Dept. of Mechatronic Engineering, National Taiwan Normal University, Taiwan, R.O.C.

Abstract. The metal housing is typically jointed with plastic fittings by conventional gluing method or embedding injection molding to produce this type of devices. We propose to improve this new technique with more practical approach. In plastic-aluminum substrate dissimilar joining, the 5052 aluminum plate coarsening process was performed to increase the porosity of the permeable dissimilar phase. The ABS polymer plus carbonized rice husk powder was later induced or deposited on the microstructure to improve the bonding effect. The plastic -aluminum substrate dissimilar joining is completed by the final step of ultrasonic welding. The finished substrate will be tested on the properties of tensile strength to ensure its quality. According to the simulation analysis and measuring results, the maximum temperature between the interface of ABS polymer and 5052 aluminum alloy is about 400~450℃ during ultrasonic welding, which can make the surface of ABS polymer to be melted. Furthermore, after drilling micro-hole array and covering ABS plus carbonized rice husk powder, the 5052 aluminum alloy shows better joining effect with ABS polymer sheet by ultrasonic welding. This improved approach does not require mold or injection molding machinery to produce the high quality plastic -aluminum bonding parts.

1 INTRODUCTION

Because of features such as excellent texture, anti-electromagnetic waves, heat dissipation, and high strength, metal cases can be used to mobile devices to make them shatterproof and shockproof as well as display a luxurious texture【1】. However, metal cases must be assembled with the internal plastic parts of mobile devices【2】. Light metals and plastics are commonly used together because of the demand for lightweight transportation, and the bonding between them is receiving increasing attention. Past conventional bonding methods include mechanical fastening and adhesive bonding, which possess some limitations such as stress concentration, a requirement for large bonding areas, differences in the nature of dissimilar materials, and harmful gas emissions【3】. In addition, when metals are bonded to polymers, their interfaces do not produce interfacial reaction layers similar to intermetallic compounds, and generally require intermolecular bonding forces such as van der Waals or electrostatic forces to achieve bonding effects. However, these bonding forces are relatively weak and possess stress concentration problems, with the plastic parts prone to falling off under external forces. New bonding technology for metal and plastic undergo constant innovation to enhance bonding quality. The Japanese company Taisei Plas proposed the Nano Molding Technology (NMT) for metal–plastic bonding, wherein a metal sheet is first processed using a special solution to form nano-holes on its surface【4,5】. It is then placed into a mold and subjected to injection molding for plastics to enter the nano-holes and solidify, thereby achieving the bonding of metal and plastic. The bonding can be applied to consumer electronics such as mobile phone cases or the internal mechanism designs of battery back covers. However, the process involves high temperature and high pressure effects that can deform the thin metal plates, whereas the use of plastic molds and injection molding machines also increases production costs.

Welding is another viable bonding method, with the early application of ultrasonic welding being mainly used for the bonding of plastic materials; it has also been used to bond dissimilar aluminum alloy and plastic materials in recent years【6,7】. Balle *et al.* used ultrasonic welding to bond aluminum plates and carbon fiber reinforced polymer, achieving satisfactory bonding effect. In this study, a micro-hole array was first fabricated on the aluminum alloy surface, after which the processed aluminum and plastic was bonded using ultrasonic welding, which resulted in occlusion between the plastic and the coarsened metal surface【7】.

To strengthen the bonding interface, rice husks were subjected to high temperature firing and pulverized, after which charcoal was removed, and remnants were mixed with plastic powder and placed in the bonding interface between the plastic and aluminum alloy. The joint was then subjected to tensile strength analysis after ultrasonic welding. The aim was to achieve a simplified process

[a] Corresponding author: cpcheng@ntnu.edu.tw

with reduced production costs and enhanced bonding strength. Rice husks are low-priced agricultural waste from rice grain production, and mixing them with plastic following their carbonization to form composite materials can improve part strength.

2 Experimental Process

The aluminum alloy 5052 and the plastic acrylonitrile butadiene styrene (ABS) were adopted as the experimental materials for this study. Both were first processed into 120 mm × 40 mm × 3 mm sheets before welding, with a joint area of 30 mm × 30 mm, after which they were bonded in an overlapping fashion. A micro-hole array was adopted in this study to stabilize and quantify the coarsening of the aluminum alloy surface. Within the 30 × 30 mm2 range of the aluminum alloy bonding area, 20 holes with a diameter of 1.0 mm were each drilled along the length and width of the alloy using a computer numerical control (CNC) machine at a fixed distance. The array contained 400 holes in total, each having a depth of 2 mm (Fig. 1), and the aluminum alloy was then bonded with the ABS using ultrasonic welding. For the plastic to be fully filled into the holes, ABS powder was added to the bonding interface. In addition, to strengthen the bonding interface, carbonized rice husk powder and ABS powder were further mixed to a specific proportion (Fig. 2) and added to the interface area, after which ultrasonic welding was performed (Fig. 3) and its effectiveness was evaluated.

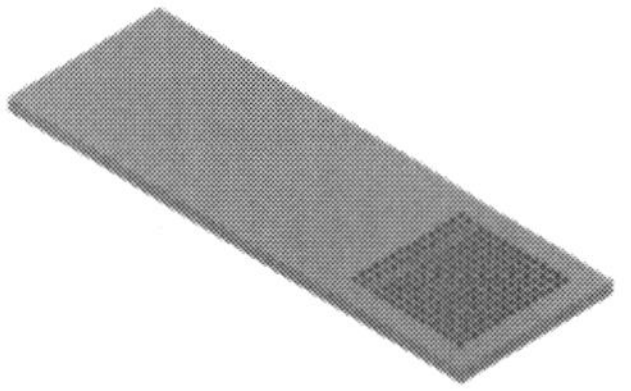

Fig.1 Micro-hole array on the aluminum alloy surface

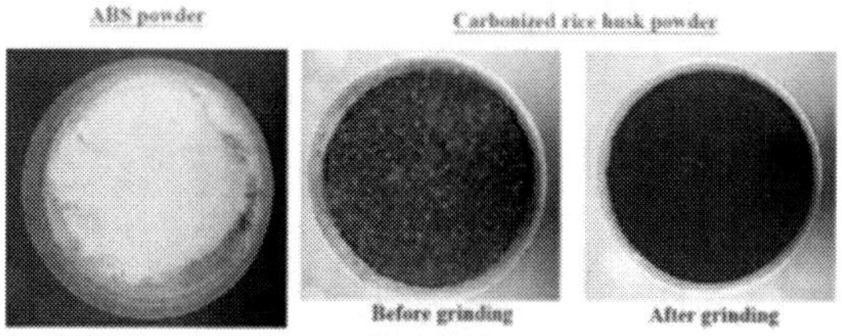

Fig.2 ABS and carbonized rice husk powder

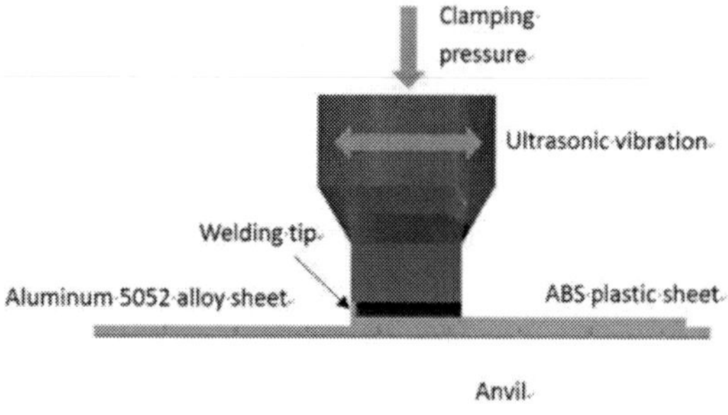

Fig. 3 Ultrasonic welding of two overlapping test pieces

First, an aluminum alloy sheet with a processed surface was placed at the bottom of a fixture, and an overlapping plastic sheet was placed on top of it (Fig. 3). Ultrasonic energy was transmitted to the plastic through the horn, and a high temperature was produced in its internal molecules because of friction. When the temperature exceeded the melting point of the plastic, the ABS underwent local melting and was squeezed through the holes of the coarsened metal where mechanical interlocking was achieved after it cooled down. The fixed parameters for the ultrasonic welding in this experiment were a frequency of 20 025 Hz, a vibration amplitude of 20 μ, a delay time of 0.8 s, a curing time of 1.5 s, and a welding time of 2~4 s. To explore the quality of the dissimilar bonding interface, the bonding strength of aluminum alloy and ABS was tested under various conditions using a tensile testing machine. Two pieces of 3 mm material were adhered to the end of both test pieces during the experiment for the tensile force to be exerted on the centerline of the test pieces during the tensile test (Fig. 4).

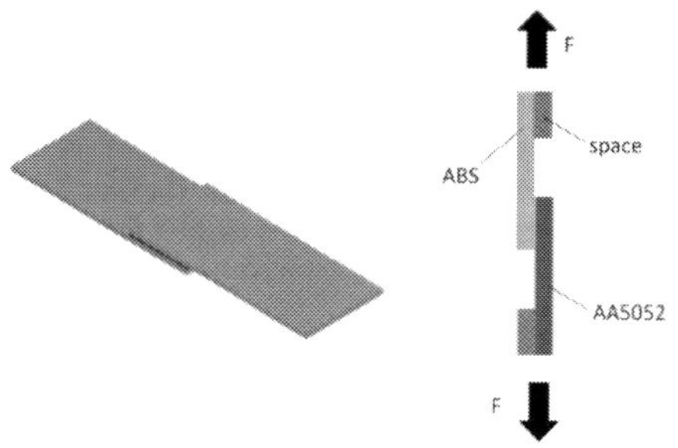

Fig. 4 Schematic diagram of the tensile test

3. Results and Discussion

3.1 Temperature Distribution

To understand the interfacial temperature changes of the aluminum alloy and ABS using ultrasonic welding, the temperature measurement was conducted directly using a thermocouple and then the change during the bonding process was recorded. Fig. 5 shows the laying position of the thermocouple. First, a groove 1.1 mm wide and 20 mm long was added to the top of the aluminum alloy, wherein a K-type thermocouple with a 1 mm diameter was placed every 6 mm starting from the centerline. A temperature recorder was connected to measure the temperature distribution at different positions to analyze the possible changes in the material properties during the bonding process.

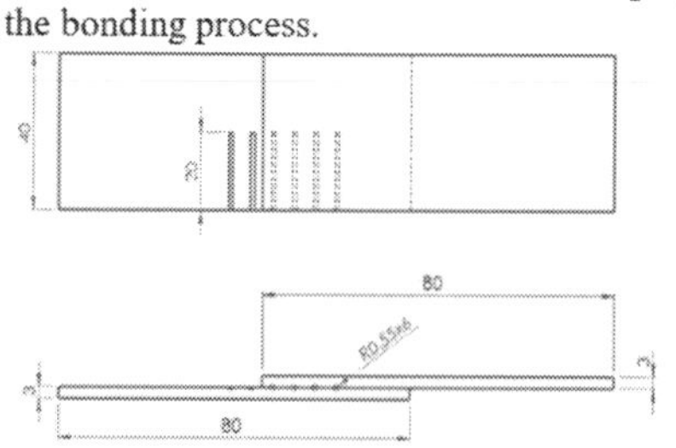

Fig. 5 Schematic diagram of the temperature measurement position

Fig. 6 shows the temperature change curves for each part after the measurement; the melting time was 2 s, and the instantaneous maximum temperature at the center reached approximately 450 °C. The maximum temperature 18 mm from the center was also about 400 °C, which is higher than the melting point of the ABS and was sufficient for the melting of the plastic surface despite the very short time.

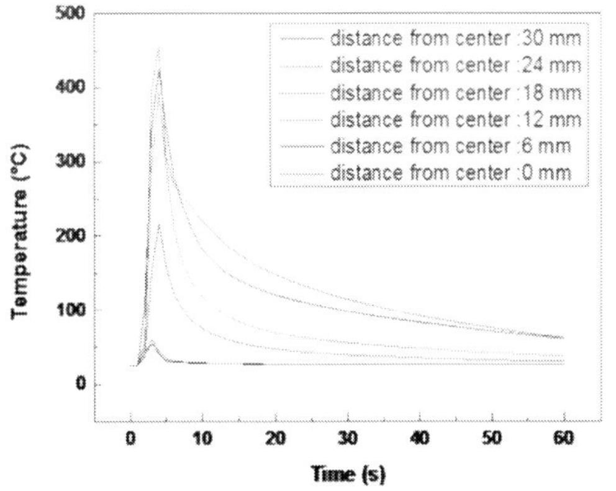

Fig. 6 Temperature change curves for multiple positions

The second part of this study conducted a simulation analysis using the finite element analysis (FEA) software Abaqus/CAE student edition. The value of the heat source generated during the ultrasonic welding process had to be obtained before the analysis. According to Elangovan *et al.*, the heat source causing the temperature change during the ultrasonic welding process can be divided into two categories: that generated by deformation, and that by the friction between the two test piece surfaces. At the early stage of welding, the temperature rapidly spread to the deformation zone because of the high thermal conductivity of the aluminum alloy. This can be regarded as a uniform distribution of the total power produced in the deformation zone【8】.

The heat flux due to deformation form Eq. (1)

$$Q_w = \frac{\sqrt{(Y_T/2)^2 - ((F_N/A_{DZ})/2)^2} \times A_{DZ} \times 4 \times \xi_0 \times fw}{A_W} \quad (1)$$

The area involved in friction is given

$$A_{FR} = A_R - A_W \quad (2)$$

Heat flux by friction will become

$$Q_{FR} = \frac{\mu \times F_N \times 4 \times \xi_0 \times fw}{A_{FR}} \quad (3)$$

Table I List of symbols

A_{DZ}	area of deformation zone (mm^2)
f_W	welding frequency (Hz)
F_N	normal force (N)
A_W	heat flux due to deformation (W/mm^2)
Y_T	temperature dependent yield strength(N/mm^2)
ξ_0	sonotrode amplitude (μm)
μ	static friction
A_{FR}	area of friction zone (mm^2)
A_R	Area involved in friction(mm^2)
Q_w	heat flux due to deformation (W/mm^2)
Q_{FR}	heat flux due to friction (W/mm^2)

According to the simulation analysis results (Fig. 7), the maximum temperature in the central region of the contact surface was more than 400 °C, which is close to the result of the previous thermocouple measurement. In addition, because the thermal conductivity coefficients of the ABS and aluminum alloy were vastly distinct, and the base of the fixed test piece was also of aluminum alloy material, an obvious temperature gradient change was observed in the lower aluminum alloy plate. A high temperature was only produced on the surface of the upper ABS plate, and the melting effect was produced on the plastic surface during ultrasonic welding. In the thermal analysis model, the interface contact of the horn and ABS, as well as that of the ABS and aluminum alloy, was in an ideal condition. However, under actual circumstances, full contact cannot be achieved because of thickness changes and surface roughness. A complete bond can only be obtained by increasing the welding time, as well as the molten plastic used to fill the gap between the two adherents.

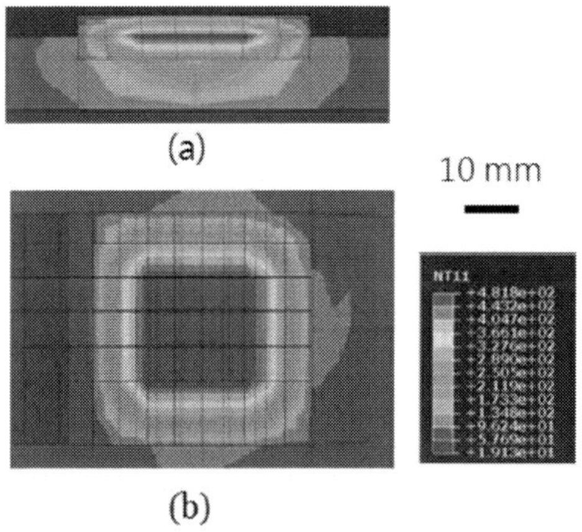

Fig. 7 Predicted temperature distribution for the 2 s bonding of the weld symmetry plane: (a) Z-axis profile; (b) X-Y plane

3.2 Ultrasonic Welding

The carbonized rice husk used in this study was produced by introducing high temperature and dry distillation at 900 °C to attain high silicon dioxide (SiO2) content and low carbon content. Based on the electron microscope images of the carbonized rice husk in Fig. 8, the rice husks were mainly irregular particles, identified as SiO2 under X-ray diffraction (XRD) analysis after the removal of the charcoal (Fig. 9). This study hopes to encourage the use of this waste resource because of its relatively low cost. In subsequent experiments, the high-temperature-fired carbonized rice husks were added to the bonding interface of the ABS with the hope of achieving a strengthening effect.

Fig. 8 Electron microscope images of carbonized rice husks

An aluminum alloy surface with a micro-hole array was placed into mixed powder of ABS and carbonized rice husk, after which it was subjected to ultrasonic welding. Fig. 10 shows that the aluminum alloy and the plastic achieved a bonding effect after the mixed powder was melted and filled into the holes. The bonded test pieces were then subjected to tensile testing, and the tensile strength changes following distinct bonding times are shown in Fig. 11. The test pieces had a lower bonding strength with a bonding time of 2 s, mainly because of the insufficient time, during which the mixed powder was incapable of fully melting and effectively filling the holes. The bonding strength of the test pieces decreased at a long bonding time of 4.5 s, a result that could be caused by the excessively long bonding time that resulted in the softening and deformation of the ABS. The bonding strength of the specimens with bonding time 3~4 s was close to the ABS base material, and the addition of mixed powder with carbonized rice hull particles also enhanced the bonding strength of the test pieces. These results indicate that under sufficient bonding conditions, the fracture positions of the stretched test pieces were all located at the ABS base material instead of on the bonding interface, thus demonstrating a satisfactory bonding effect.

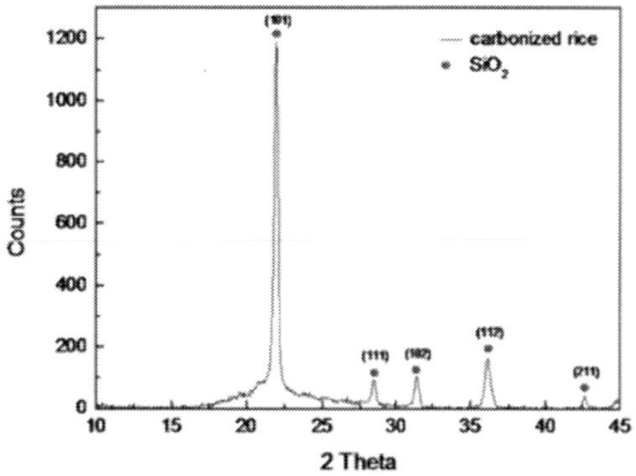

Fig. 9 XRD analysis of carbonized rice husks

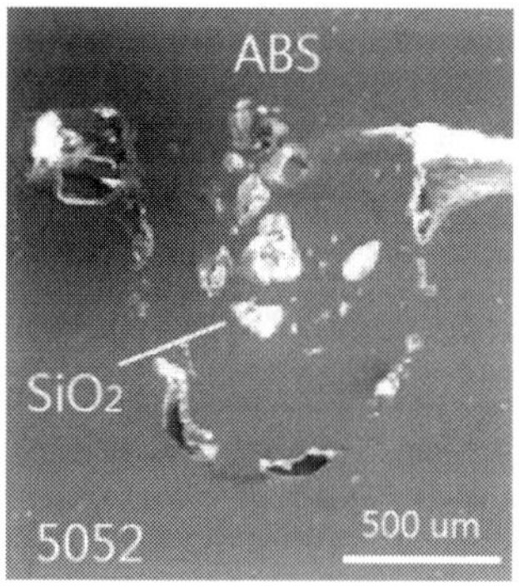

Fig. 10 Mixed powder of ABS and carbonized rice husk, melted and filled into the hole

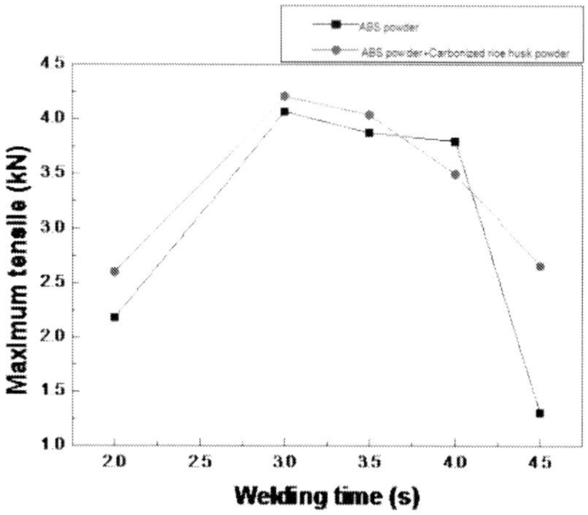

Fig. 11 Relationship between the tensile loading and bonding time after the bonding of the test pieces

4 Conclusions

This study conducted a dissimilar bonding of 5052 aluminum alloy and ABS using ultrasonic welding. A micro-hole array was produced at the junction of the aluminum alloy to increase the rate of dissimilar agents entering the holes, and the mixed powder consisting of ABS and carbonized rice husk was added at the interface to improve the bonding effect. The experiment obtained the following results:

1. The bonding interfacial temperature changes for the ultrasonic welding of the aluminum alloy and ABS were measured using a thermocouple, which showed that the maximum temperature at the bonding region was approximately 400~450°C when the welding time was 2 s. A similar result was also obtained using finite element simulation analysis, and a more obvious heat transfer effect was produced on the aluminum alloy side.
2. Regular holes were produced on the aluminum alloy surface using a micro-hole array, and ABS mixed with SiO_2-based carbonized rice husk was added to the bonding interface, after which it was subjected to ultrasonic welding with ABS. The tensile test results indicated that the fracture position of the test pieces

was on the ABS base material when the bonding time was 3~4 s, and a maximum load of approximately 4 kN was obtained, indicating a satisfactory bonding effect. A bonding time exceeding 4 s led to the deformation of the ABS, resulting in diminished bonding strength.

Acknowledgments

The authors would like to thank Ministry of Science and Technology of the Republic of China, for their financial support to this research under Contract no. MOST 104-2221-E-003-006.

References

1. C. O. Annerfors, S. Petersson, Nano molding technology on cosmetic aluminum parts in mobile phones, Sweden, School of Mechanical Engineering Lund University, 2007, 1-116.
2. G. Ramarathnam, M. Libertucci, M. M. Sadowski and T. H. No, Joining of polymers to metal, Welding Research Supplement, December 1992, 483-s~ 490-s.
3. P. Kah, R. Suoranta, J. Martikainen, C. Magnus, Techniques for joining dissimilar materials: metals and polymers, Rev. Adv. Mater. Sci. 36, 2014, 152-164.
4. http://www.uneec.com/tw/04_technology/01_list.aspx?SID=91.
5. http://taiseiplas.lekumo.biz/e/nmt01.html.
6. S. Katayama, Y. Kawahito, Laser direct joining of metal and plastic, Scripta Materialia, Volume 59, Issue 12, December 2008, Pages 1247–1250.
7. F. Balle, G. Wagner and D. Eifler, Ultrasonic metal welding of aluminium sheets to carbon fibre reinforced thermoplastic composites, Advanced Engineering Materials, 2009, 11, 35-39.
8. S. Elangovan, S. Semeer and K. Prakasan, Temperature and stress distribution in ultrasonic metal welding—An FEA-based study, Journal of materials processing technology, 209, 2009, 1143–1150.

Numerical analysis of the adherends with similar thickness on weld-bonded single lap aluminium joint

Jianli Li[1], Peng Wang[2] and Min You[1, 3,a]

[1]Hubei Three Gorges Polytechnic, Yichang 443000, China
[2]Wuhan Tower Works, Power Construction Corporation of China, Wuhan 430011, China
[3]Hubei Key Laboratory of Hydroelectric Machinery Design & Maintenance, China Three Gorges University, Yichang 443002, China

Abstract. The effect of the adherend with similar thickness varied from 1 mm to 3 mm on the stress distribution in weld-bonded single lap aluminium joint was investigated using elasto-plastic finite element method (FEM). The results from the numerical simulation show that all the values of the peak stresses along the mid-bondline at the points near the both ends of the lap zone as well as the ones in the region of the nugget are increased when the adherend thickness increased. It is suggested that the adherend thickness of 2 mm to 2.5 mm be appropriate to optimize the stress distribution in the weld-bonded single lap aluminium joint.

1 Introduction

Single lap joint is one kind of the most applied adhesively bonded joints used in the aerospace and automotive industries. Although there has been a considerable amount of research on the single lap joints, only few studies have been carried out on the effect of the adherend thickness. [1-3]. Aydin et al [2] investigated the effect of adherend thickness on the failure of adhesively-bonded single-lap joints. Darwish and Al-Samhan [3] analyzed the effect of adherend thickness dissimilarity on the peel and shear strength of spot-welded dissimilar thickness joints. The experimental study of the effect of adhesive thickness on the impact toughness of adhesively bonded steel butt-joints was carried out [4] and the effect of the adherend thickness with asymmetric rigidity as well as the nugget location on the stress distribution along the mid-bondline and adherend in weld-bonded aluminium single lap joints was investigated in author's work [5-6]. This work is aimed to study the effect of the adherend with similar thickness on the stress distribution in the weld-bonded single lap aluminium joint.

2 Finite element model

The model and the meshes were built using the ANSYS finite element software as shown in Fig.1 and Fig.2. The properties of the materials used in this study are listed in Table 1. The load applied was taken as 40 MPa and the dimensions of the aluminium alloy adherend were made in accordance with the Chinese standard GB 7124 and the width of the joint is 25 mm. The triangular element was used for both bondline and the nugget and the quadrilateral element for adherend (Fig.2). The thickness of the bondline was set as 0.2 mm and divided into 4 layers. To investigate the effect of the adherend thickness thoroughly, four different adherend thicknesses were taken into account as 1 mm, 2 mm, 2.5 mm and 3 mm respectively. The nugget was assumed in a shape of ellipsoid.

Table 1. Materials properties.

Materials	Elastic Modulus (MPa)	Poisson's Ratio	Yield Strength (MPa)	Tang Modulus (MPa)
Aluminum alloy LY-2	71,000	0.32	400	240
Nugget	102,000	0.29	800	210
Phenolic resin adhesive	2,875	0.42	90	500

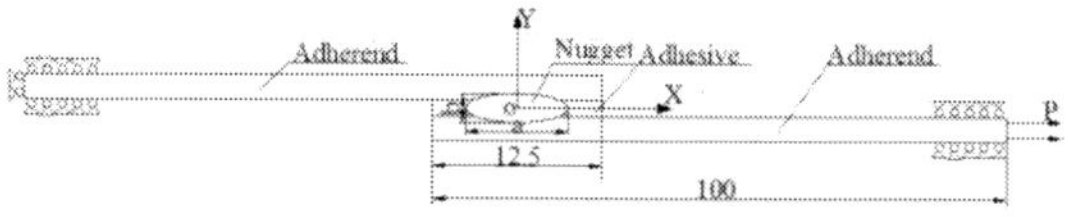

Figure 1. Finite element model (unit: mm) [6].

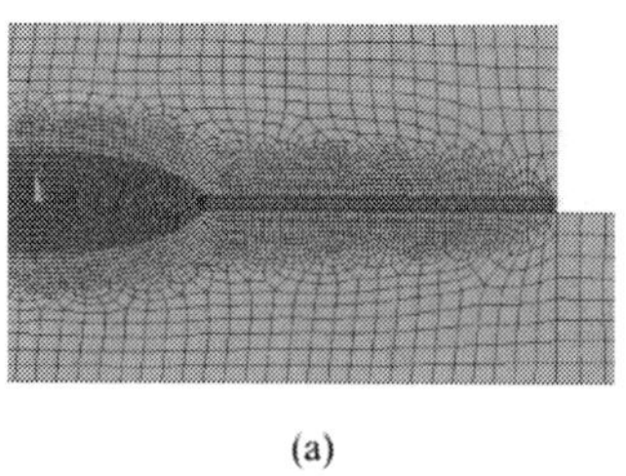

(a)

[a] Corresponding author: youmin@ctgu.edu.cn

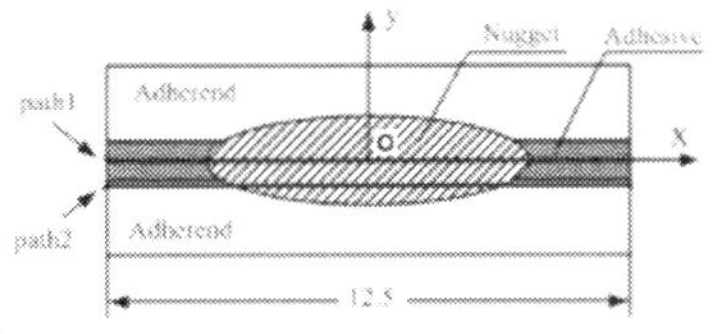

(b)

Figure 2. Finite element meshes for right half of over lap zone (a) and the paths (b) [6].

3 Results and discussion

3.1 Stress distribution along the mid-bondline

The effect of the adherend with similar thickness on the stress distribution along x axis (path 1) is presented in Fig.3a to Fig.3d. The results from the simulation show that all the peak values of the stress components Sx, Sy, Sxy and the von Mises equivalent stress Seqv occurred at the points near the both ends of the over lap zone are significantly increased when the adherend thickness is increased. For instance, the peak value of the longitudinal stress Sx is increased 80.3 % from 7.1 MPa to about 12.8 MPa when the adherend thickness increased from 1 mm to 3 mm (Fig. 3a) meanwhile the peak value of the peel stress Sy is increased 86.4 % from 17.6 MPa to about 32.8 MPa (Fig. 3b). And all the stresses distributed along the x-axis are symmetrical to the centre of the overlap zone. The peak value of the stress Sy, Sxy and Seqv occurred at the points near the edges of the nugget is increased significant as the adherend thickness increased. For instance, the absolute value of the valley stress Sy is increased about 286.4 % from 4.4 MPa to about 12.6 MPa when the adherend thickness increased from 1 mm to 3 mm (Fig. 3b). But it is nearly the same for the peak value of the stress Sx occurred in the region of the nugget (17.8 MPa to 17.9 MPa) as the adherend thickness is increased from 1 mm to 2.5 mm and then it is increased to 19.8 MPa when the thickness is 3 mm. Comparing this figure with the Fig. 3 in Ref. [5], it is found that the stress distribution tendency along the mid-bondline in weld-bonded aluminium single lap joint is similar to it but the effect of adherend thickness is more significant presented in the Ref. [5]. For instance, the effect of the adherend thickness on the longitudinal stress Sx (Fig. 3e) is more significant than that one presented in Fig. 3a. The differences in the study conditions are the thickness of the upper plate kept as 2 mm while the lower one's was varied from 1 mm to 3 mm and the load was set as 3 kN [5]. From Fig. 3a, it is clear that the load is mainly subjected by the nugget and the stress distributed in this region is relative higher than that in Fig.3e and it is almost no increment with the adherend thickness increase when the thickness is not great than 2.5 mm. In other words, the adherends with same thickness is more beneficial than that one with dissimilar thickness. The stress distribution tendency of Sxy (Fig. 3c) is similar to that of Seqv (Fig. 3d).

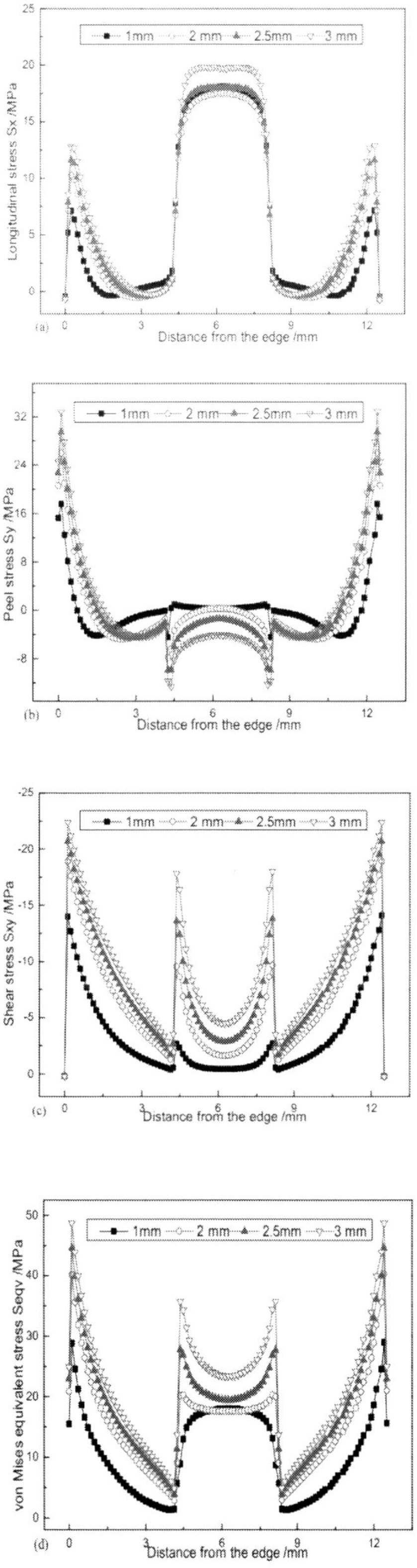

 DOI: 10.1051/matecconf/201713006002

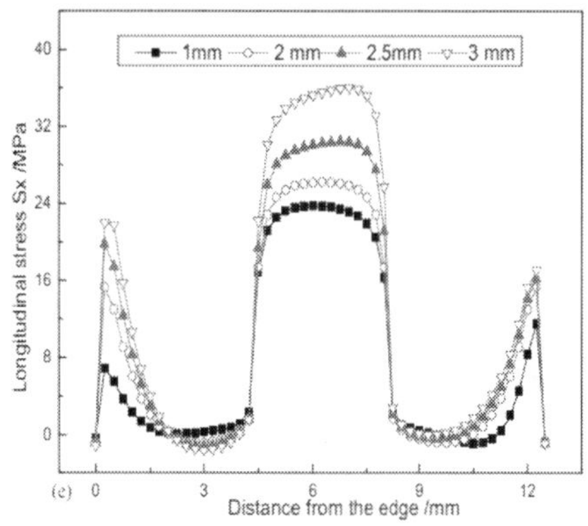

Figure 3. Effect of adherend thickness on the stress distribution along the mid-bondline: (a) longitudinal stress Sx, (b) peel stress Sy, (c) shear stress Sxy and (d) von Mises equivalent stress Seqv; (e) longitudinal stress Sx [5].

3.2 Stress distribution in the adherend near the interface

When the conditions of the research was kept as the same, the effect of the adherend thickness on the stress distributed in the adherend near the interface along the path2 ($y = -0.15$ mm) is shown in Fig.4. It is clear that the distribution tendency of the longitudinal stress Sx (Fig.4a) is similar to that of the von Mises equivalent stress Seqv (Fig.4d) and the values of the peak stresses at the point corresponding to the right end of the over lap zone are all over 125 MPa when the adherend thickness of the joint varied from 1 mm to 3 mm. All the absolute values of the peel stress Sy are increased as the thickness of the adherend increased and the same tendency is presented in the region closed to the nugget (Fig.4b). The highest absolute value of the peel stress Sy is occurred at the point corresponding to the left edge of the nugget (about 23.5 MPa, with 3 mm thickness adherend, which is nearly 11 times to the lowest one, about 2.1 MPa with 1 mm thickness). The negative peel stress Sy is beneficial to the strength and load bearing capacity of the joint [7].

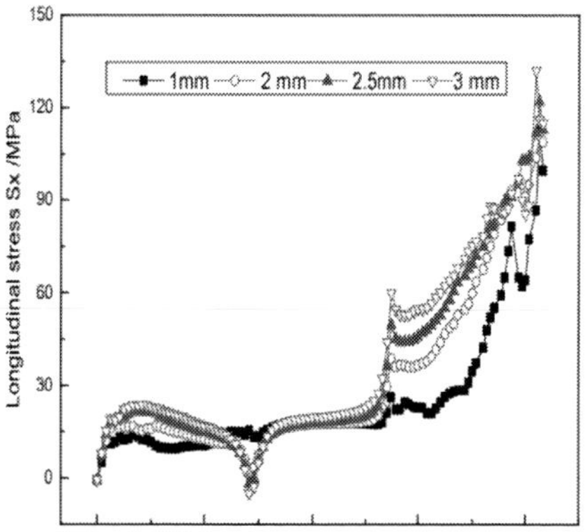

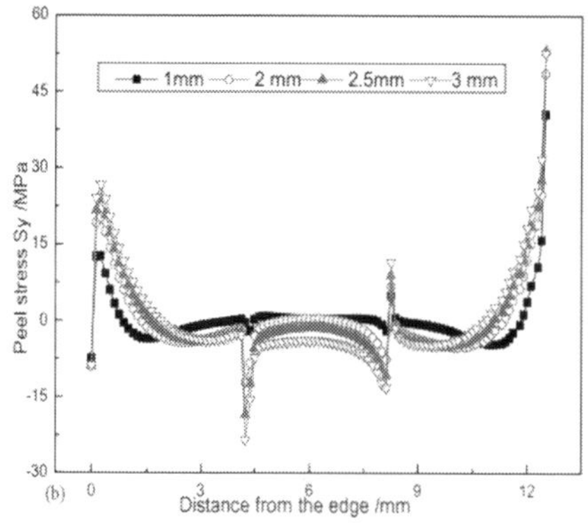

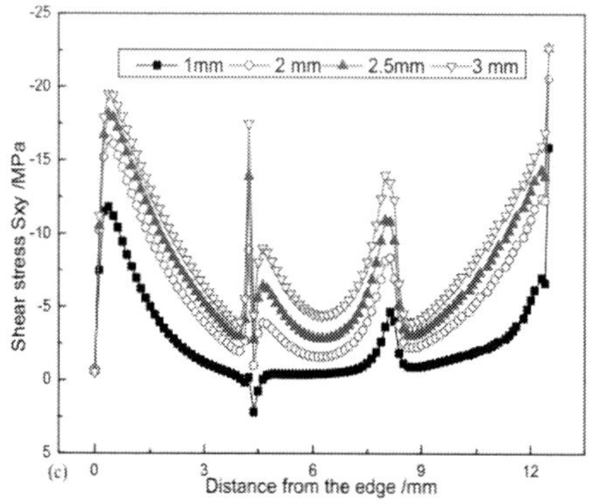

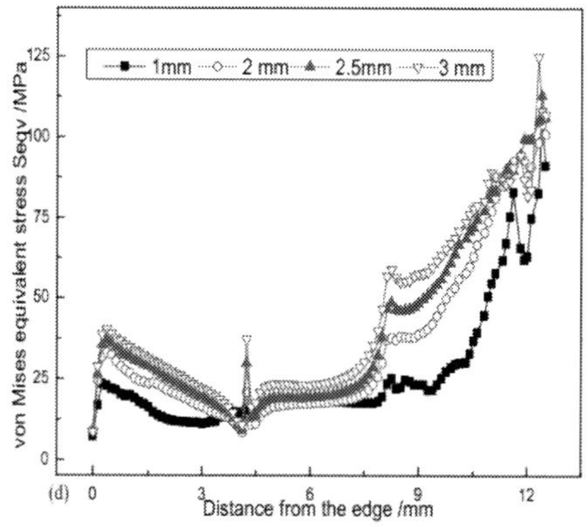

Figure 4. Effect of the adherend thickness on the stress distribution in adherend near the interface: (a) longitudinal stress Sx, (b) peel stress Sy, (c) shear stress Sxy and (d) von Mises equivalent stress Seqv.

4 Summary

The results obtained show that all the peak values of the stress Sx, Sy, Sxy and Seqv along the mid-bondline (x-axis) at the points near both ends of the over lap zone are increased significant when the adherend thickness is increased from 1 mm to 3 mm. The peak values of the stress Sy, Sxy and Seqv occurred at the points near the edges of the nugget are increased significant when the adherend thickness was increased. And it is nearly the same for the peak value of the stress Sx occurred in the region of the nugget as the adherend thickness is increased from 1 mm to 2.5 mm and then it is increased again with the 3 mm adherend thickness. At the points corresponding to the right end of lap zone and in the region close to the nugget, all the absolute values of the peak or valley stresses along the adherend 0.05 mm below the interface increased as the thickness of the adherend increases. It is recommended that the suitable adherend thickness be 2 mm to 2.5 mm for weld-bonded single lap aluminium joint according to its synthetic effects.

Acknowledgements

The authors would like to acknowledge the financial supported by the Hubei Province Natural Science Foundation of China under project no. 2014CFA123.

References

1. A. Ozel, M. D. Aydin, S. Temiz, J Adhesion Sci. Technol. **18**, 313 (2004)
2. M. D. Aydin, A. Ozel, S. Temiz, J Adhesion Sci. Technol. **19**, 705 (2005)
3. S. M. Darwish, A. Al-Samhan, J Mater. Proces. Technol. **147**, 51 (2004).
4. M. You, J. Hu, X. Zheng, A. He, C. Chen, Advanced Mater. Research, **230-232**, 1350 (2011)
5. M. You, J. Yan, X. Zheng, J. Zhang, D. Zhu, Proc. 2010 Int. Conf. Computer Eng. & Technol. **5**, 587 (2010)
6. J. Li, P. Wang, M. You, H. Yu, Advances in Eng. Research, **101**, 404(2016).
7. M. You, Z. Li, X. Zheng, S. Yu, G. Li, D. Sun, Int. J Adhesion & Adhesives, **29**, 280 (2009).

MATEC Web of Conferences 130, 06003 (2017)
CMPSE2017 DOI: 10.1051/matecconf/201713006003

Effect of main layer thickness on the response of multi layer bonded steel butt joint under Charpy impact test

Jianli Li[1], Yalan Zhao[2], Yingying Li[3] and Min You[1,3,a]

[1]Hubei Three Gorges Polytechnic, Yichang 443000, China
[2]Wuhan HBR Technology Co., Ltd, Wuhan 430025, China
[3]Hubei Key Laboratory of Hydroelectric Machinery Design & Maintenance, China Three Gorges University, Yichang 443002, China

Abstract. The numerical analysis of the main layer thickness on the impact response of the steel butt joint bonded by multi-layer under the Charpy impact test is investigated using the elasto-plastic finite element method (FEM). The results obtained from numerical simulation show that both the elastic strain and plastic strain occurred at the point near the upper or lower surface decreased significantly when the main layer thickness increased from 0.1 mm to 0.4 mm. The value of the normal stress Sx and the von Mises equivalent stress Seqv is increased first when the main layer thickness increased and then it decreased significantly when the main layer thickness is greater than 0.2 mm. There is a strong value fluctuation for stress Seqv at the point 0.5 mm away from the upper surface when the main layer thickness is greater than 0.2 mm.

1 Introduction

The influence of the adhesive thickness is significant on the adhesively bonded metal joints [1-2].In recent years, some investigations related to the impact properties of the adhesively bonded metal joint have been done [3-5]. The effect of the adhesives thickness, elastic modulus of the adhesives on the butt joint was investigated by the finite element method [3]. The FEM method was also applied to study the failure procedure of the adhesives [4]. The effect of notch depth [5] as well as the multi-layer [6] under Izod impact testing on the impact response of the adhesively bonded steel butt-joint was investigated and it was found that the impact energy absorbed by unit area of joint is increased as the notch depth increased. The effect of multi-layer on the stress distribution in adhesively bonded single lap aluminium joint [7]was significant. The aim of this work is to study the effect of multi-layer on the impact response of the steel butt-joint under the Charpy impact testing.

2 Establishment of the model

The diagram of Charpy impact test and the multi layer is shown in Fig. 1 and the properties of the materials used in the study are listed in the Table 1. The specimen for impact test modelling is in accordance with GB/T 19748 was bonded by a sandwich (primer Phenolic-main layer Epoxy- primer Phenolic, A-B-A for short) adhesive multi-layer as shown in Fig. 1b. The adherends were made from Q235 structural steel and the two halves of specimen (length 27.3 mm, width 10 mm and thickness 10 mm) was bonded with a multi adhesive layer consisted of two primer with the thickness of 0.1mm and a main layer with the thickness of 0.1 mm, 0.2 mm, 0.3 mm and 0.4 mm respectively. The impact time was set as 0.2 ms with an initial velocity of 3.2 m/s. The finite element model was built using the ANSYS/ LS-DYNA software as shown in Fig.2. The eight-node hexahedral element Solid164 was used for both adhesive layer and adherend and it was restrained in Y direction by two supports. Two typical nodes are marked as P (x=0, y=4.5 mm, z=0) and Q (x=0, y=- 4.5 mm, z=0) in Fig. 2c (0.5 mm away from the upper or lower surface respectively) for analyzing the response of the bondline under the Charpy impact test.

Table 1. Materials properties.

Materials	Elastic Modulus (MPa)	Poisson's Ratio	Yield Strength (MPa)	Tang Modulus (MPa)
Steel Q235	203,000	0.27	235	6100
A (Phenolic Resin)	2,875	0.42	90	500
B (Epoxy Resin)	1,888	0.33	50	50

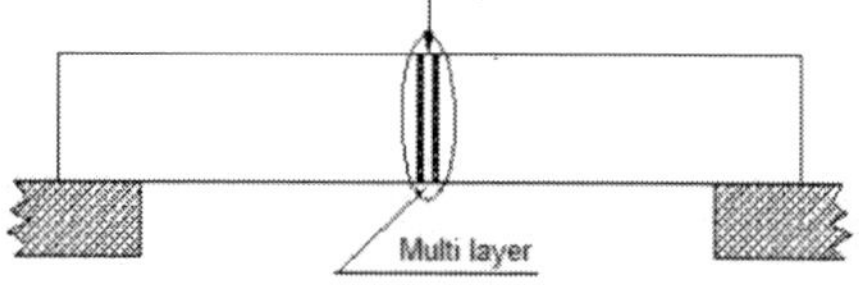

(a)

[a] Corresponding author: youmin@ctgu.edu.cn

 DOI: 10.1051/matecconf/201713006003

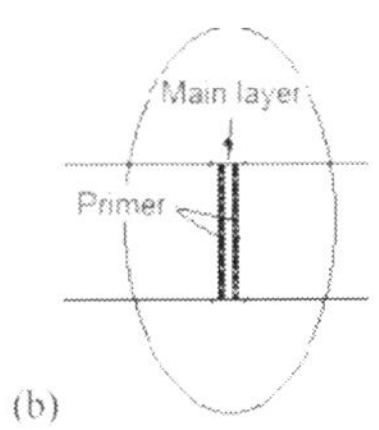

Figure 1. Diagram of the Charpy impact testing (a) and the multi layer (b).

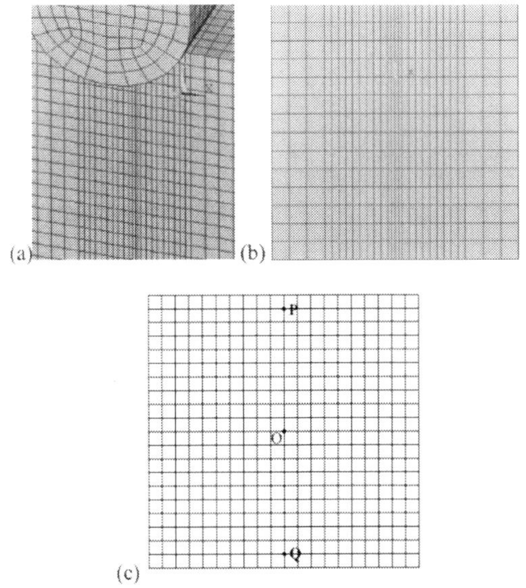

Figure 2. (a) Model of FEM, (b)mesh for the multi-layer bonded butt joint and (c) the location of the node P and Q [6].

3 Results and discussion

3.1 The response of strain after impact

The effect of the main layer thickness on the impact response with time at the point P (0.5 mm away from the upper surface) and Q (0.5 mm away from the lower surface) under Charpy impact test is shown in Fig.3. It can be seen from the figure that both the elastic strain occurred at P or Q is increased rapidly after about 0.04 ms of time lap to a certain value (about 0.072 ms the elastic strain reached to 0.0364) and then the increase becomes slowly (Fig. 3a and 3c) but the increase of the plastic strain is steadily vs. time (Fig. 3b and Fig. 3d). In general, both the elastic strain and plastic strain at the node P is increased first and then it is decreased as the main layer thickness increased from 0.1 mm to 0.4 mm. And the strain of the multi layer with a 0.2 mm thick main layer is the greatest after 0.1 ms. The delay time of the elastic strain response is longer when the main layer thickness is increased. At node Q the highest elastic or plastic strain occurred in the multi layer with a 0.3 mm thickness after 0.12 ms meanwhile the lowest strain occurred in the bondline of the multi layer with a 0.1 mm thick main layer.

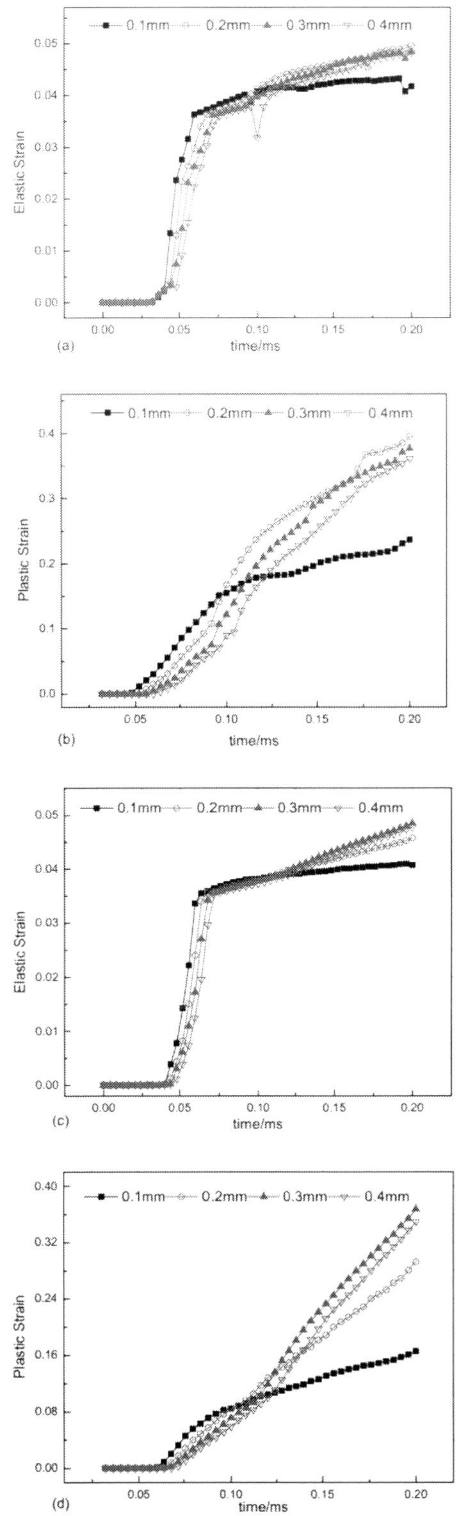

Figure 3. The effect of main layer thickness and the node on the strain vs. time: node P (a) & (b), node Q (c) & (d).

3.2 The response of stress after impact

The effect of the main layer thickness on the normal stress Sx and von Mises equivalent stress Seqv response with the time at the point P and Q under Charpy impact test is shown in Fig.4. It can be seen from the figure that the response of stress Sx and Seqv at the point P and Q is also significantly affected by the main layer thickness. When the main layer thickness increased from 0.1 mm to 0.4 mm, the absolute values of the stress Sx response at node P and Q decreased greatly and the response time became longer (Fig. 4a and 4c). When the time is longer than 0.1 ms, the von Mises equivalent stress Seqv at node P present a strong stress fluctuation especially for the main layer is thicker than 0.2 mm (Fig. 4b). The von Mises equivalent stress Seqv occurred at node Q (Fig. 4d) is much different from the stress Seqv occurred at node P. The value of stress Seqv is decreased first and then increased again as the thickness increase to a value of 0.4 mm. It is increased evidently when the main layer thickness reached 0.4 mm to a value greater than 50 MPa (Fig. 4d).

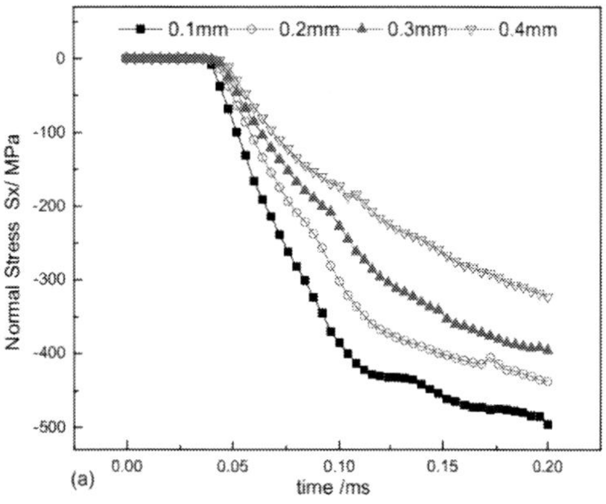

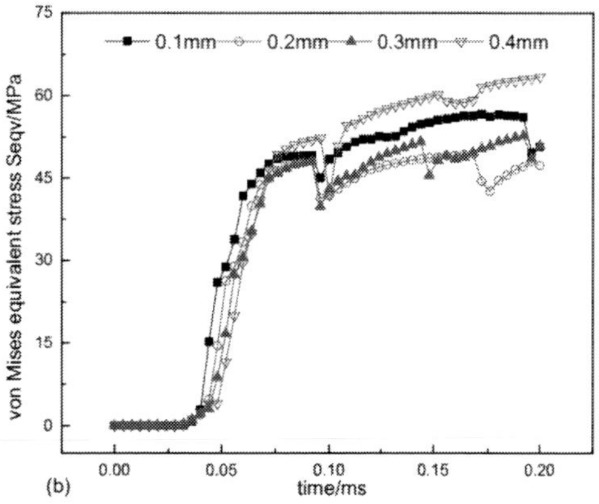

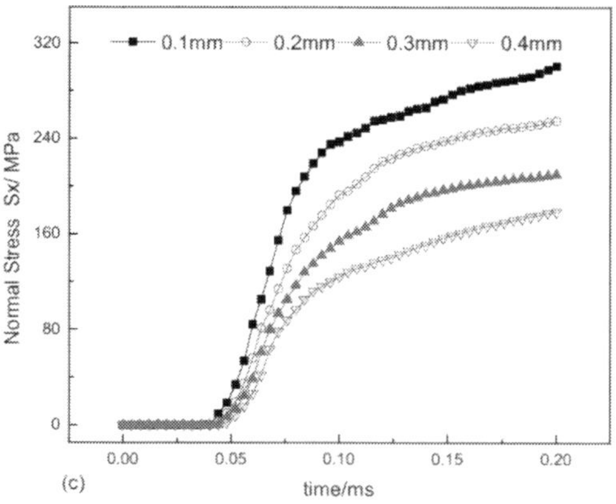

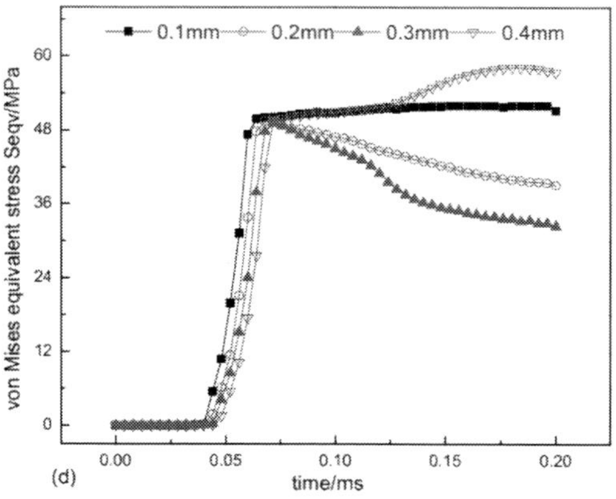

Figure 4. The effect of main layer thickness and the node on the normal stress and von Mises equivalent stress vs. time: node P (a) & (b), node Q (c) & (d).

4 Summary

Under the study condition of this work, the results obtained from the numerical modelling show that the impact response of the butt-joint bonded by multi-layer bondline is evidently affected by the main layer thickness under the Charpy impact test. The absolute values of the stress Sx response at node P and Q decreased greatly when the main layer thickness increased from 0.1 mm to 0.4 mm. There is a value fluctuation for the stress Seqv at the point 0.5 mm away from the upper surface when the main layer thickness is greater than 0.2 mm meanwhile the value of the stress Seqv at the point 0.5 mm away from the lower surface is decreased significantly except the 0.4 mm thick layer.

Acknowledgements

The authors would like to acknowledge the financial supported by the Hubei Province Natural Science Foundation of China under project no. 2014CFA123.

References

1. L.F. Kawashita, A.J. Kinloch, D.R. Moore, J.G. Williams, Int. J Adhesion & Adhesives, **28**, 199 (2008)
2. J. M. Arenas, J. J. Narbon, C. Alia, Int. J Adhesion & Adhesives, **30**, 160 (2010)
3. M. You, J. Hu, X. Zheng, A. He, C. Chen, Advanced Materials Research, **230-232,** 1350 (2011).
4. X. Zheng, L. Wu, M. You, K. Liu, M. Li, Advanced Materials Research, **602-604,** 2279 (2013)
5. X. Zheng, M. Li, M. You, W. Liu, K. Liu, Advanced Materials Research, **644,** 197 (2013)
6. J. Li, Y. Zhao, Y. Li, M. You, Materials Science Forum, **904**, 68 (2017)
7. M. You, Y. Zhao, J. Li, Y. Li, Advanced Materials Research, **1061-1062,** 471 (2015).

Simulations about the effect of heat treatment temperatures on the properties of diffusion bonded Mg/Al joints

Yunlong Ding[1], *Jiangang Wang*[2] *and Dongying Ju*[3, 4,a]

[1] *Graduate School of Saitama Institute of Technology, Fusaiji 1690, Fukaya, Saitama 369-0293, Japan, e5002lmf@sit.ac.jp*
[2] *Hebei Key Laboratory of Material Near-Net Forming Technology, Hebe University of Science and Technology, Hebei Shijiazhuang 050018, China, wm094212@163.com*
[3] *Department of Material Science and Engineering, University of Science and Technology Liaoning, Anshan 114051, China*
[4] *Department of Material Science and Engineering, Saitama Institute of Technology, Fusaiji 1690, Fukaya, Saitama 369-0293, Japan,*
dyju@sit.ac.jp

Abstract. In this study, Finite elements method (FEM) was used for simulations on the annealing treatment. And the distributions of residual stress were obtained from the resualts of simulitions. Besides, in order to study the mechanical properties. Composite plates of AZ31B magnesium alloys and 6061 aluminum alloys were fabricated by vacuum diffusion bonding method. Then experiments about annealing, measurement of residual stresses and tension test were also carried out. The results of residual stresses obtained by X-ray diffraction (XRD) were compared with the results of simulations. It can be thought that the results of residual stress obtained from simulations and that from experiments are in a good agreement. What's more, The residual stress after annealing at 250℃ is the least of all, while the tensile strength is the strongest. On the basis of results of simulations and experiments, the conclusion can be obtained that 250℃ is the most appropriate annealing temperature.

1 Introduction

Magnesium is the lightest metal [1]. As the use of magnesium alloys, the weight of components can be reduced significantly. However, it's corroded easily. While aluminum alloys have attractive mechanical and metallurgical properties, including high strength and excellent corrosion resistance [2]. So wide and cross applications of these two metals make the connection of Mg/Al dissimilar metals inevitable[3]. For example, the Mg-Al laminated armor is used in tanks and armored vehicles[4], Mg/Al complex structure is used in aerospace engines and components[5,6,7]. So it is significant to achieve reliable connection of Mg/Al dissimilar metals. Presently, there are many studies about welding of magnesium and aluminium. And there are many welding methods that have been used to join Mg alloys and Al alloys. However, no matter which technique is used, if the brittle and hard intermetallic compounds, such as Al_3Mg_2 and $Mg_{17}Al_{12}$ were formed in the joints. The performances of welded joints will be deteriorated seriously[8].

In this study, for the purpose of improving the tensile strength, simulations on annealing treatment were applied to search for the apposite annealing temperature. Besides, experiments on annealing treatment were also carried out. What's more, residual stress and tensile strength were investigated in this work.

2 Simulations

During the processes of diffusion bonding and heat treatment, along with the change of temperature, microstructure will transform, and at the same time, thermal stress emerges. If the stress exceeds the elastic limit, plastic deformation will result from it. A series of varieties don't emerge individually but interact on each other. The theory system related is constructed and called metallo-thermo-mechanics, which is the foundation of thermal treatment analysis.

[a] Corresponding author: *Prof. Dongying Ju, E-mail: dyju@sit.ac.jp, Tel: 81-48-5856826, Fax: 81-48-5855928*

When AZ31 magnesium alloy and 6061 aluminum alloy is joined by the method of diffusion bonding, diffusion between the elements Mg and Al should be considered. Diffusion phenomenon can be analyzed by Fick's law and expressed as the following equation:

$$\frac{\partial C}{\partial t} = D\frac{\partial^2 C}{\partial x^2} \tag{1}$$

Where C is the concentration of element. D which is a function of C, is diffusion coefficient standing for the diffusion property of material, and can be express using the following equation:

$$D = K_0 \exp(K_1 C) \tag{2}$$

Where K_0 and K_1 are coefficients. Generally speaking, if the influence resulted from the variety of microstructure is ignored, the diffusion equation can be expressed with the element's concentration C :

$$\dot{C} = div(DgradC) \tag{3}$$

The boundary condition is represented as the following equation:

$$-D_c gradC \cdot n = \beta_c (C - C_w) \tag{4}$$

Where β_c is the concentration of gas around, C_w is the concentration of elements around.

If the energy in objects is represented as $e = g + T\eta + tr(\sigma\varepsilon^e)$, then the first law of thermodynamics can be wrote as :

$$\rho\dot{e} - tr(\sigma\dot{\varepsilon}) + div\boldsymbol{h} = 0 \tag{5}$$

If Fourier law ($\boldsymbol{h} = kgradT$) is used, plastic work and transformation latent heat is not considered, items that relate elastic strain and hardening coefficient are ignored, then equation (5) can be wrote as the following equation:

$$\rho c\dot{T} - kdiv(gradT) = 0 \tag{6}$$

Where ρ is the density of material, c is specific heat, k is thermal conductivity. When the coefficient of heat transfer and the temperature of fluid that is in touch with the objects are changeless, boundary condition is expressed using the following equation:

$$-kgradT \cdot \boldsymbol{n} = h(T)(T - T_\omega) \tag{7}$$

Where $\boldsymbol{n}$ is a vector whose direction is outward from the surface of objects. T_ω is temperature around. $h(T)$ is the coefficient of heat transfer between object and environment. Generally, coefficient of heat transfer is a function of temperature T, which can be obtained from the experiment value of cooling curve [9].

When plastic materials are with load, elastic or plastic deformation will occur. The generalized Hooke's law is applicable to three-dimensional stress and strain, and can be expressed using the equation (8):

$$\dot{\varepsilon}_{ij} = \dot{\varepsilon}^e + \dot{\varepsilon}^p_{ij} + \dot{\varepsilon}^T_{ij} \tag{8}$$

Where $\dot{\varepsilon}_{ij}$ is total strain rate, $\dot{\varepsilon}^e$ is elastic strain rate, $\dot{\varepsilon}^p_{ij}$ is plastic strain rate, and $\dot{\varepsilon}^T_{ij}$ is thermal strain rate. Especially, elastic strain rate and thermal strain rate are expressed with the following equation.

$$\begin{cases} \dot{\varepsilon}^e_{ij} + \dot{\varepsilon}^T_{ij} = E^e_{ijkl}\dot{\sigma}_{kl} + \beta_{ij}\dot{T} \\ \beta_{ij} = \dfrac{\partial E^e_{ijkl}}{\partial T}\sigma_{kl} + \alpha_{ij}\dfrac{\partial \alpha_{ij}}{\partial T}(T - T_0) \end{cases} \tag{9}$$

Where α is linear expansion coefficient, and is temperature difference. E^e_{ijkl} is a coefficient that is represented as equation (10):

$$E^e_{ijkl} = \frac{1}{2G}\left\{(\delta_{ik}\delta_{jl} + \delta_{il}\delta_{jk}) - \frac{\nu}{1+\nu}\delta_{ij}\delta_{kl}\right\} \tag{10}$$

Where ν and G are Poisson's ratio and shear modulus [19]. Plastic strain rate is expressed as equation (11):

$$\begin{cases} \dot{\varepsilon}^p_{ij} = \dfrac{1}{\hat{G}}\left(\dfrac{\partial F}{\partial \sigma_{mn}}\dot{\sigma}_{mn} + \dfrac{\partial F}{\partial T}\dot{T}\right)\dfrac{\partial F}{\partial \sigma_{ij}} \\ \dfrac{1}{\hat{G}} = -\dfrac{1}{\dfrac{\partial F}{\partial \kappa}\dfrac{\partial F}{\partial \sigma_{kl}}\sigma_{kl}} \\ F = F(\sigma, T) \end{cases} \tag{11}$$

Where F is Mises yield function.

First, the definitions of mixture and maxing rule are explained. Intermetallic compounds, such as Al_3Mg_2 and $Mg_{17}Al_{12}$ will generate during the diffusion process of magnesium alloy and aluminum alloy. It is thought that there are many micro-structures mix together. Mixture contains N compositions whose volume fraction are $\xi_I (I = 1, 2, \cdots\cdots, N)$, and if all the volume fractions are together, then 1 will be obtained. It can be expressed as:

$$\sum_{I=1}^{N} \xi_I = 1 \tag{12}$$

If the property of composition I is represented by χ_I, then the property χ contains all the properties of N compositions. Therefore, the property χ can be expressed as equation (13), which is called maxing rule [10].

$$\chi = \sum_{I=1}^{N} \chi_I \xi_I, \quad \xi_I (I = 1, 2, \cdots\cdots, N) \tag{13}$$

According to the maxing rule, the coefficients of AZ31 magnesium alloy and 6061 aluminum alloy, and all the coefficients that relate the Intermetallic compounds of Mg and Al are calculated. Then simulations were carried out. Temperature conditions were 200℃, 250℃, 300℃ and room temperature 20℃. It was thought that the interface can't move during diffusion process. Therefore,

 DOI: 10.1051/matecconf/201713006004

restrict was only applied to the contact surface, so that it can't move any directions. Then simulations were carried out.

3 Experimental

AZ31B magnesium alloy sheets and 6061 aluminum alloy sheets were successfully joined using vacuum diffusion bonding method in the equipment with atmosphere of argon. The size of sample was shown as Fig.1. The joining temperature was 440℃ and the flowing rate of gas was set as 150ml/min. Before diffusion bonding, the oxide layers on the surfaces of substrates were cleaned using abrasive papers (GRIT 240, 600, 800, 1200) and polished with polishing cloth and polishing compound. Then the grinded sheets were wiped with acetone. After vacuum diffusion bonding, the samples were used for the annealing treatment experiments. The heat treatment temperatures were 200℃、250℃、300℃, and the holding time was 60min. Then cooled down to the room temperature in the electric furnace.

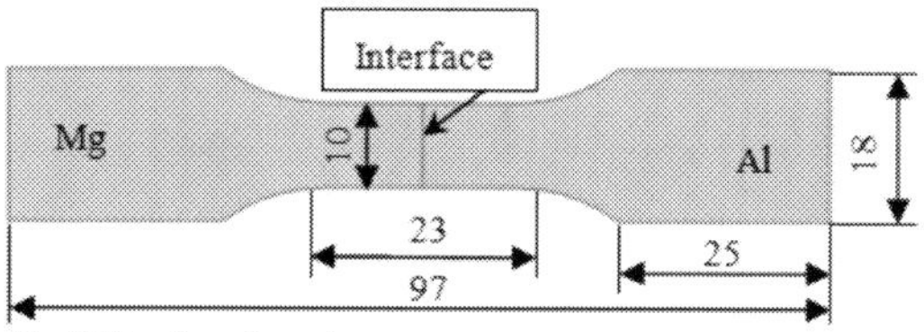

Fig.1 The size of specimen

In order to investigate the distribution of residual stress of the specimens annealed at different temperatures, residual stress was measured by XRD (X-Ray Diffraction). Based on residual stress testing principle, and used x-ray whose wavelength was λ. First, radiate the specimen with different incident angles, and obtained the corresponding diffraction angle 2θ, then the slope M of 2θ-$\sin^2\psi$ (generally, ψ was set as 0°、15°、30°、45°) can be calculated。In addition, the relationship between 2θ and $\sin^2\psi$ can be obtained，then residual stress σ can be calculated according to the following equation:

$$\sigma = K \cdot \Delta 2\theta / \Delta \sin 2\psi = K \cdot M \tag{14}$$

Where K is the stress constant of x-ray diffraction analysis, which can be expressed as:

$$K = -\frac{1}{2} \cdot \cot\theta_0 \cdot \frac{E}{1+\upsilon} \cdot \frac{\pi}{180} \tag{15}$$

Where E is the elastic modulus of material, θ_0 is the diffraction angle without stress, υ is Poisson's ratio [11].

When the side of 6061 Al alloy was measured, 2θ was set as 140°, stress constant was -163.32MPa/°, tube type was Cr, wavelength was kα, and the size of collimator was φ0.5mm. While 2θ was set as 155°, stress constant was -79.14MPa/° at the side of AZ31Mg alloy. What's more, tube type, wavelength and the size of collimator were the same with 6061Al alloy. In addition, the 2θ of $Mg_{17}Al_{12}$ and Al_3Mg_2 were set as 150° and 145°, and stress constant were respectively -98.97MPa/° and -126.22MPa/°. Then the results of residual stress were calculated at last.

Besides, tensile strength was tested by using tensile machine, and there are three specimens that were annealed at the same temperature were measured in every group. During the experiment, first, the parts were regulated into the same vertical plane, and the specimen was clamped, then loaded gradually until the specimen was fractured. What's more, all the specimens were tested respectively, and the tensile strength was obtained.

4 Results and discussions

The results of simulations are shown in Fig.2 a, b, c and d. The distributions of residual stress obtained by simulations at different temperatures are nearly the same, but the stress values are different from each other. Because of the phenomenon of stress concentration emerges at the edge of interface, in another words, the place will fracture easily.

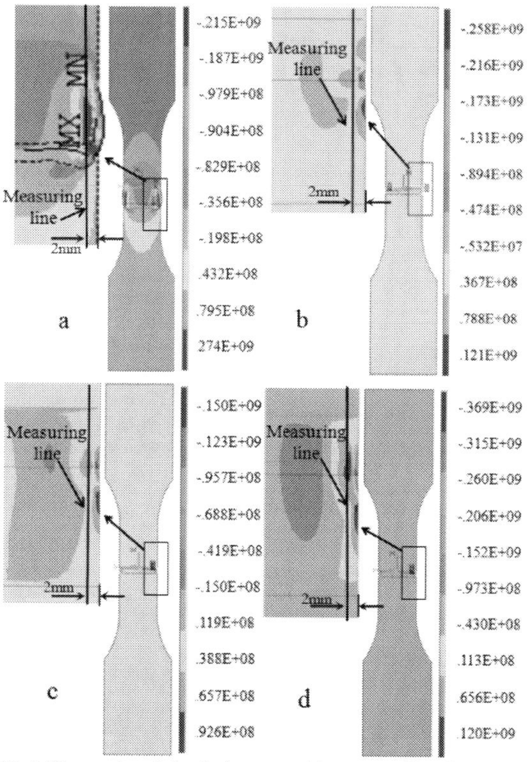

Fig.2 The results of simulations, a : without annealing, b : annealing at 200℃, c : annealing at 250℃, d : annealing at 300℃

In this paper, the line crossing the edge of interface and the place where stress concentration emerges was named as measuring line. The line is 2mm distance from the edge(the measuring line is shown in Fig.2). Then the stress distribution along the line crossing the edge of interface and the place where stress concentration emerges was investigated. Based on the stress values at the places of nodes on the measuring line and the distance between nodes, Fig.3 can be got, and it shows the distribution of residual stresses that were got by simulations. It can be thought that residual stresses in the interface are tensile stress, and the values of stresses near the interface are larger than other places. The stresses near interface at the treated conditions 200℃, 250℃, and 300℃ are respectively 66MPa, 61MPa and 63MPa. When without annealing, the stress is about 71 Mpa. In another words, The further from interface the smaller it will be. What's more, the residual stress turn to be compressive stress, when the place is far from interface and it nearly to be 0MPa in the further place.

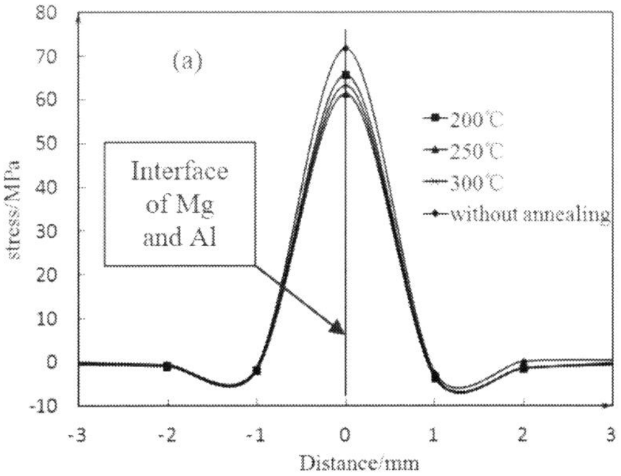

Figure.3 The distribution of residual stress, obtained by simulation

The result of residual stress measured by XRD is shown in Fig.4. the diffraction peak of Al emerges in the place where 2θ is near 139°, and indices of crystal face is (311). While the diffraction peak of Mg occurs when the value of 2θ is near 152°, and indices of crystal face is (104), However, the diffraction peaks of intermetallic compounds near Al and Mg are respectively 139.4° and 143°[12]. It can be known that residual stress in the interface is tensile stress, and the value is the largest. The stress turns to be lower along with the increasing of the distance from interface. The residual stress nearly to be 0MPa in the further place.

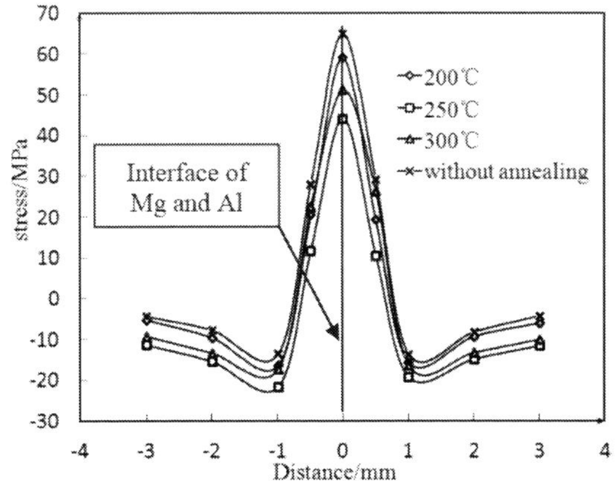

Figure.4 The distribution of residual stress, measured by XRD

In addition, when the annealing condition is 300℃, the largest stress is about 51MPa, when annealing temperature is 200℃, the stress near interface is about 60MPa, while treated at 250℃, the largest stress is approximately 44Mpa. However, when without annealing, the stress is about 65 Mpa, which emerges at the place called interface. According to the annealing theories, The stress in the place where is far from the welded joints, turn to be smaller and smaller, and gradually close to 0 Mpa. So it could be thought that the results of residual stress obtained from simulations and experiments are in a good agreement with the welding theory.

Stress is a vector, and the tensile stress is as plus, while compressive stress is as minus in this study. When stress emerges in materials, the mechanical behavior of materials will be effected. Such as fatigue strength, tensile strength and so on. What's more, fracture will also be result in. As the study above, the residual stress in the interface of Mg/Al alloy will effect the mechanical behavior of composite material. The bigger the residual stress is, the weaker the tensile strength will be. For the purpose of verifying the results on residual stress, the analysis about tensile strength is carried out as following.

The average tensile strengths of composite sheets treated at different annealing treatment temperature were obtained by tensile experiments. When the annealing temperature is 250℃, the tensile strength is the strongest, which is about 47MPa. However, when annealed at 200℃ and 300℃, tensile strength are approximately 41MPa and 39MPa. While without annealing, the strength is about 37 Mpa, which is nearly the same with what has been reported in other study[13]. As it is known that residual stress has important impact on the strength of materials. If the residual stress is tensile tress, then it will decrease the tensile strongth of the materials and increase it's compressive strongth. In contrtast, the compressive redisual stress will decrease the compressive strongth and increase the tensile strongth of the materials. So it is apparent that the results of the experiments about tensile strength is consistent with that of residual stress.

Conclusions

In this paper, simulations about annealing of bonded Mg/Al composite sheets, and experiments on residual stress, tensile strength were carried out. According to the outcomes of simulations and experiments, It can be thaught that the residual stress after annealing at 250℃ is 44MPa, which is the least of all. While the tensile strength after annealing at 250℃ is 47Mpa, which is the strongest. And the results got by simulations are in a good agreement with that of experiments. So the conclusion can be obtained that 250℃ is the most appropriate annealing temperature.

Acknowledgments

This research work has been partially supported by the grant subsidy of the "Nano Project" for Private

Universities:2011-2014 from MEXT, Japan. This study was also supported by the "Advanced Science Research Laboratory" in Saitama Institute of Technology.

References

1. L.M. Zhao, Z.D. Zhang. Scripta Materialia. **58**, 283-286 (2008)
2. Y.B. Yan, Z.W. Zhang, W. Shen, J.H. Wang, L.K. Zhang, B.A. Chin. Materials Science and engineering A. **527**, 2241-2245 (2010)
3. J.C. Feng, Y.R. Wang, Z.D. Zhang. The Chinese Journal of Nonferrous Metals. **15,** 2 (2005)
4. H. Li, M. Qian, D. Li. Laser Journal. **28**, 5 (2007)
5. J. Wang, J.C. Feng, Y.X. Wang. Mater Sci Technol. **24**, 7 (2008)
6. X.R. Li, W. Liang, X.G. Zhao, Y. Zhang, X.P. Fu, F.C. Liu. Journal of Alloys and Compounds. **471**, 408-411 (2009)
7. Fei Liu, Daxin Ren, Liming Liu. Materials and Design. **46** , 419-425 (2013)
8. Liming Liu, Fei Liu, Meili Zhu. Materials. **7**, 2 (2014)
9. D.Y. Ju. Journal of The Society of Materials Science. **55**, 7(2006)
10. T. Inoue. JSME international Journal. Ser. 1. **31**, 4 (1988)
11. J.G. Wang, D.Y. Ju. Study on evolution technology of anisotropic mechanical properties and microstructure evolution of metal thin plate under complex stress condition[D], (Saitama Institute Of Technology，Japan, 2011)
12. D.Q. Zhang, J.W. He, Residual stress Analysis by x-ray diffraction and It's Functions, (Xi'an Jiaotong University, China,1999)
13. J. SHANG, K.H. WANG, Q. ZHOU, D.K. ZHANG, J.HUANG, J.Q. GE. Trans. Nonferrous Met. Soc.China. **22**, 1961-1966 (2012)

FEM model-based optimal control synthesis for curing a large composite structure with CAD imported geometry

Sergey Shevtsov[1,a], Ilya Tarasov[2], Vladimir Axenov[3], Igor Zhilyaev[1], Jiing-Kae Wu[4] and Natalia Snezhina[5]

[1]*Southern Center of Russian Academy, Aircraft Technologies Department, 344006 Rostov on Don, Russia*
[2] *AP Group LLC, Department of New Materials, 192236 Saint Petersburg, Russia*
[3] *LANIT Group, CAD Department, 129075 Moscow, Russia*
[4] *National Kaohsiung Marine University, Mechanical Engineering Department, 81157 Kaohsiung city, Taiwan R.O.C.*
[5] *Don State Technical University, Aircraft Engineering Department, 344000 Rostov on Don, Russia*

Abstract. The paper studies a problem of optimal control synthesis for curing a large composite structure using open die moulding in an autoclave. The necessary control should eliminate early hard skin formation, emergence of resin-rich or resin-dry areas, insufficient consolidation, and uneven cure. This purpose is achieved through providing uniform distribution of degree of cure and temperature within the cured part volume. The used approach is illustrated on the example of the large CFRP aircraft panel manufactured by means of two-stage curing cycle. The forward problem combines the heat transfer and autocatalytic kinetic equations, which are linked by the specific heat capacity, thermal conductivity coefficient and by the exothermal heat source that depend on the prepreg's degree of cure. The control synthesis problem is formulated as a multi-objective optimization problem where minimized objectives are deviations of temperature and degree of cure within a cured part, considering constraints imposed by thermal-kinetic properties of prepreg and manufacturing requirements. The Pareto-based optimization algorithm, which executes the mapping of a subset of design space to the objectives space by the cyclic calling the forward problem implemented as a finite element model, allows estimating the best achievable quality indicators of manufactured composite parts, finding satisfactory parameters of the control law, and decision-making considering all imposed constraints.

1 Introduction

The landscape of composite part manufacturing has been spreading rapidly over the recent years. The growth in use of composites is the most intensive in aircraft and rotorcraft industries due to increased specific and fatigue strength or stiffness and better weight saving of these materials, which significantly exceed the same properties of metal lightweight alloys. Most typical polymeric composites reinforced with glass / carbon fibers with the thermoset resin matrix are used in manufacturing of such parts as wing and tail skins, spars, stringers, ribs, fuselage sections, helicopters main and tail rotor blades, pressure vessels, radomes, cowlings, etc. [1].

Components based on long fibers, tows, yarn, or fabrics are manufactured using some kind of laminating procedure. In this process, sheets of reinforcement precoated with resin (prepreg) or freshly applied resin are forced against the surface of a mould under required conditions of pressure, temperature, and time. Many different technologies have been developed to produce aircraft components of different shape, dimensions, load-carrying, and wall thickness. A wide variety of components that have the open-shell geometry or smooth panel shape are widespread in aircraft structures. The open-mould forming is most suitable for components with such types of geometry. According to this technology involving the use of only one mould surface, over which layers of reinforcing fibres or fabric are tightly placed, the laminate with orthotropic or transversely isotropic structural symmetry is most often formed. Then a mould packed with a raw material is placed into a vacuum bag, which is a flexible plastic film embracing the surface of the lay-up, and a vacuum-tight bag together with a mould is disposed in an autoclave [1, 2].

The autoclave system includes a vessel containing a bag mould with laid-up prepreg, sources for heating the gas stream and circulating it uniformly inside the vessel, a subsystem for pressurizing the gas stream, a subsystem for

[a] Corresponding author: sergnshevtsov@gmail.com

applying vacuum to the part covered with a vacuum bag, and a subsystem for controlling operating parameters. Gas circulation inside an autoclave is essential to provide mass flow for temperature uniformity and heat transfer to the part load. The vacuum is applied to remove air and volatiles from the polymerized prepreg, while the pressure is required to consolidate all layers together into a laminate. At the first stage of heating the prepreg, resin viscosity in prepreg plies decreases, attains a minimum, and then increases rapidly (gelation stage) as the curing (cross-linking) reaction begins and proceeds toward completion. The temperature up to 150°C – 200°C may be required to complete the polymerization process of epoxy resin systems is most often applied to the open mould using hot-air blowers and electric elements installed in an autoclave [1, 2]. Figure 1 shows a typical two-stage cure cycle for a glass / carbon fiber–epoxy prepreg.

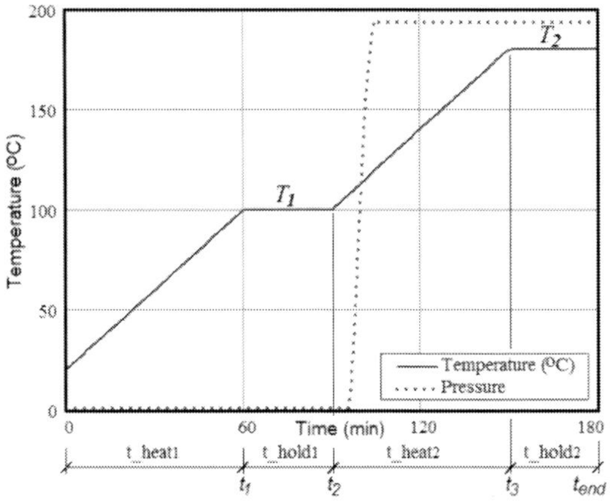

Figure 1. An example of two-stage cure cycle for a carbon fiber–epoxy prepreg.

The first stage of this cure cycle consists of increasing the temperature at a controlled rate up to the start of resin fluidization and dwelling at this temperature until the minimum resin viscosity is reached. Such an isothermal hold is applied at the stages of resin fluidization as well as at the process completion to allow the temperature distribution to become more uniform, especially in large components with thickness variations, to prolong the time allowed for volatiles removal, and to allow pre-reactions and consolidation of the resin. At the end of cure cycle after resin vitrification, the temperature is slowly reduced while the laminate is still under pressure. The ready part is removed from the mould and, sometimes, postcured at an elevated temperature.

In order to eliminate problems such as overheating, early hard skin formation (which prevents the release of gases), porosity, resin-rich areas, resin-dry areas, poor surface finish, insufficient consolidation, and uneven cure, the correct timing of applications of temperature and pressure should be used. All means of process monitoring that support traditional autoclave curing can provide information on temperature, pressure, and sometimes resin viscosity (with DEA sensors placed in some points of the cured part). Unfortunately, these means cannot indicate the state of cure throughout the cured component [3]. Due to this circumstance, the composite technology most often uses the trial-and-error approach to the tooling design and process development that causes the high cost of advanced composites. Such a difficulty is most significant for the large composite structures and can be overcome by using methods of process modelling or simulation that can predict a desired tooling and process design window, within which a reduced number of trials can be conducted [1].

A good process model should correctly describe a number of coupled physical and chemical phenomena, including heat transfer, thermal, kinetic and rheological properties of material transformation at all stages of cure, and should take into account different constraints imposed by the process equipment, material properties, etc. This requires preliminary resin testing using DSC, DMA, DEA techniques to identify all resin properties during the full cure cycle. Due to the complex mathematical description of the problem and complexity of the components geometry, the finite element method is the most efficient and has been used recently by many authors to model a variety of polymeric composite cure processes [4 - 7]. Such a finite element implementation of the cure process allows formulating an optimization problem where the degree of cure at the final stage of the curing process and variations of the degree of cure and temperature within the cured body during the whole process can be accepted as objectives, whereas time dependencies of applied heat flow, temperature and pressure are design variables. Such an approach, which has been implemented in some recent works [8 - 15], allows synthesizing the most appropriate process control that provides the best quality of the cured part [16 - 18] for the acceptable processing time as well as determining some process parameters, which are related to the state of the controlled plant and can be effectively monitored, e.g., temperature at some accessible points on the mould or cured part surface. The multi-objective approach to the problem optimization of such a system is the most efficient in a practice when mapping of the area in the design space to the objective space is considered, and a pseudo optimal solution is selected, taking into account various constraints, including trade-offs between the full processing time, maximum allowable temperature gradient and curing uniformity.

Such a multi-objective optimization model based approach to the synthesis of the pseudo optimal process control is illustrated in this paper using an example of the cure process of carbon fiber reinforced epoxy-based large composite panel with many longitudinally oriented reinforcing ribs (see Fig. 2). The part with open-shell geometry and wall thickness varied in the range of 5 – 8 mm should be manufactured by means of the open mould curing technology.

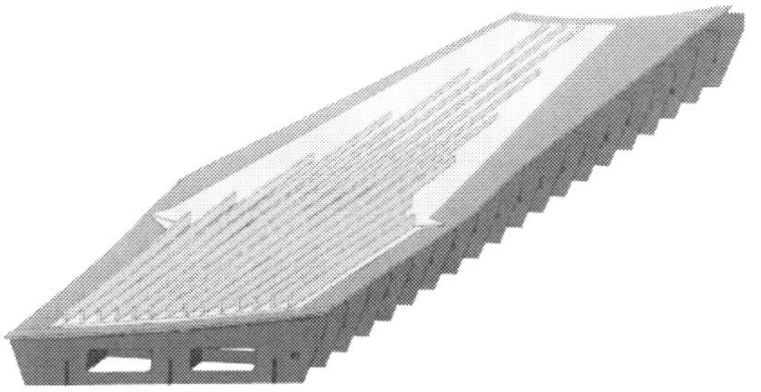

Figure 2. Geometry of the cured CFRP panel placed on the open mould, which is supported by the numerous footings that prevent the mould deformations during cure

The present study consisted of the following phases. By using the results of DSC studies we have established that the autocatalytic equation is the most appropriate to describe the resin kinetic behaviour when cured at varying heating rates. All six parameters of this equation, the value of reaction enthalpy, and heat conduction dependence on degree of cure have been identified using numerical technique of evolutionary computing [19]. DEA testing results of the prepreg that describe the transformation of the resin rheology during cure have been analyzed to determine the temperatures of both isothermal holds, which are present in Fig.1.

The geometry of the cured part together with a mould made of thick-walled CFRP has been redesigned to exclude some topological imperfection, and exported to the finite element soft package. The governing system of coupled heat transfer and kinetic equations for the prepreg is similar to that used in [15], but it has different boundary conditions and dependencies of thermal, kinetic and rheological properties on the temperature and degree of cure. Assuming a two-stage cure cycle, we varied the duration of the first and second heating up and two dwelling sections to provide minimum variations of the degree of cure and temperature gradient within the cured part, while maximum duration of each cure stage and the whole cure cycle and temperatures at the first and second isothermal holds are constrained. Optimum area in a four-dimensional design space has been visualized by the set of its two-dimensional projections with the contour lines that determine the values of the objectives.

2 Identification of cure kinetics and prepreg properties

In order to determine the thermal and kinetic properties of the used prepreg Toray T800, all dynamic DSC tests have been carried out using NETZSCH equipment at the heating rates 0.8…1.5 deg/min. The time histories of the differential heat flow have been numerically processed to fit them by the analytical dependencies given by different kinetic models. The main criterion for choosing the most appropriate kinetic model was the ability to successfully describe the behaviour of experimental data obtained at the different heating rate. The quality indicator of these data approximation by each studied kinetic equation was the discrepancy between experimental and calculated cure rate (see Fig. 3) and was averaged on the used temperature interval. The principal types of the kinetic equations are listed in [20]. Among these types the Kamal & Sourour equation (below simply named as Kamal)

$$\partial\alpha/dt = A_2\exp(-E_2/RT)\cdot\alpha^m(1-\alpha)^n \qquad (1)$$

and autocatalytic-1 equation (below named as autocatalytic)

$$\partial\alpha/dt = \left[A_1\exp\left(-\frac{E_1}{RT}\right) + A_2\exp\left(-\frac{E_2}{RT}\right)\cdot\alpha^m\right](1-\alpha)^n, \qquad (2)$$

where degree of cure α is defined by the relationship $Q_{exo} = Q_{tot}\cdot\partial\alpha/\partial t$ between actual value of exothermal heat Q_{exo} and enthalpy Q_{tot} of cure reaction; E_1, E_2 are the activation energies, R is the universal gas constant, T is the Calvin temperature, m and n are the reaction orders, and t is time.

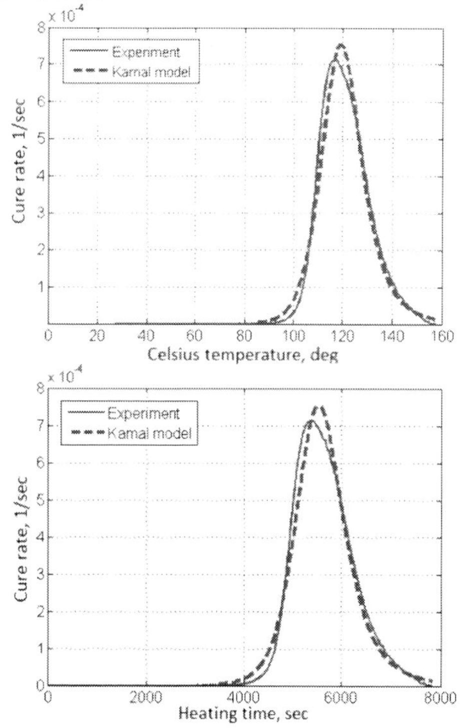

a)

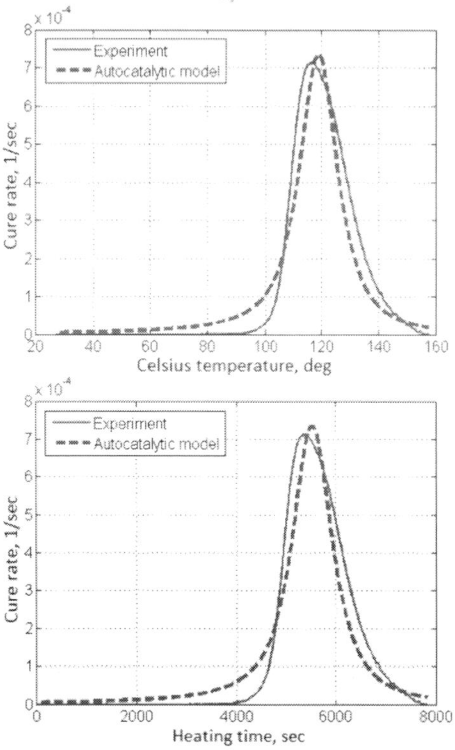

b)

Figure 3. Dependencies of cure rate obtained during the dynamic DSC test at the heating rate 1 deg/min combined with the theoretical dependencies given by the simple Kamal (a) and autocatalytic (b) models

The main difference between equations (1) and (2) is that Kamal equation can describe each polymerization process if initial value of $\alpha > 0$ only, whereas autocatalytic equation (2) provides a proper description of the cure evolution even at its zero initial condition. Figure 3 demonstrates, when parameters of equations (1) and (2) are identified using the experimental data obtained from the DSC test at the identical heating rate, Kamal model gives better consistency with experimental data. But this model is valid for only one temperature law – ramp with the heating rate at which was obtained the initial DSC data that have been used for the model identification. Indeed, figure 4a demonstrates an apparent dependency of the cure rate on the temperature. But this dependence is due to various heating rate. Figure 4b shows the absolute overlap of three curves of the cure rates, which have been calculated from equation (1). The main inference from this careful study is that cure evolution according to Kamal model is controlled by time, instead by the temperature.

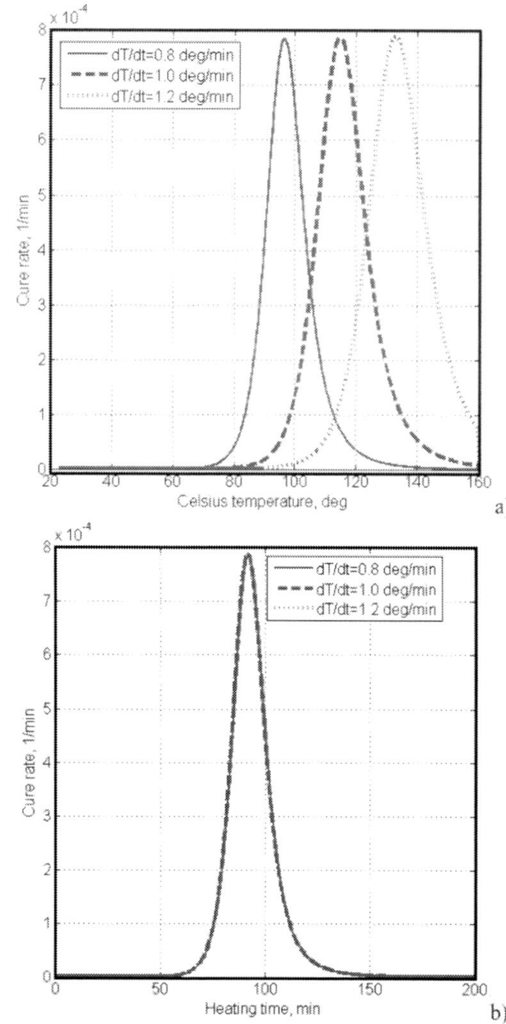

Figure 4. Temperature and time dependencies of cure rate calculated from the simple Kamal equation (1) at the different heating rate

Moreover, the choice of initial value for cure in (1) is arbitrary, and it is carried out only to satisfy the requirement of better fitting the experimental data, without any reasonable justification. Meanwhile, many authors [4, 6, 14-16, 27] use this model in their studies without due justification. Hence, use the simple Kamal equation in models involving heating rate variation is improperly.

Autocatalytic model (2) doesn't have of this shortcoming. When it has been identified so, that its parameters provide the best fitting data for the dynamic DSC implemented at the boundary values of interval for the heating rate variation, this model gives the correct quantitative description of cure evolution at the heating rates inside this interval. The ability of the autocatalytic model has been confirmed in [22, 29] and successfully used in [11] to optimize the cure cycle of thick composite parts. Such a model has been identified and used in our study to formulate the coupled thermal-kinetic problem of composite part cure.

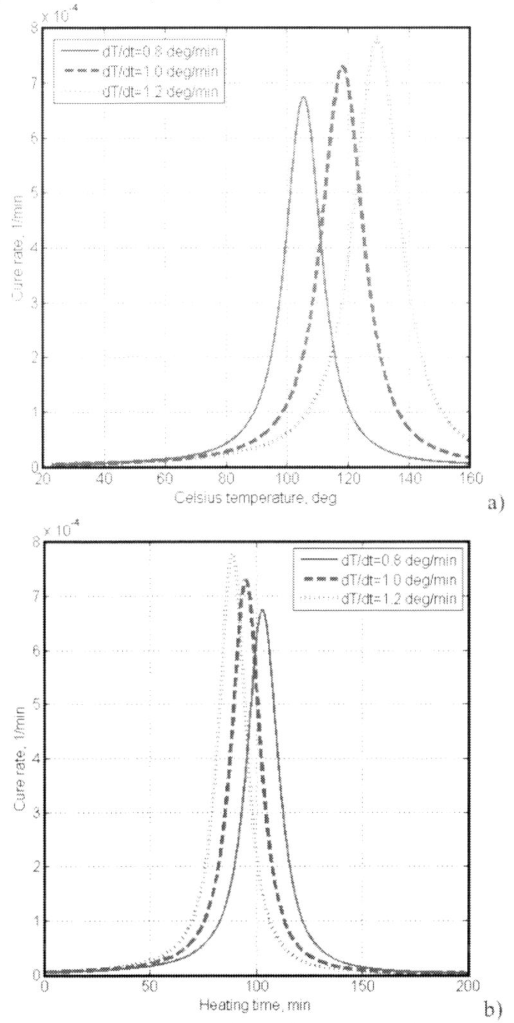

Figure 5. Temperature and time dependencies of cure rate calculated from the autocatalytic equation (2) at the different heating rate

As the further results of kinetic model identification, the enthalpy Q_{tot} =0.105 MJ/kg of cure reaction and dependence of specific heat capacity (in J/(kg•K)) on degree of cure were found

$$C_p = 2300 - 550 \cdot \alpha \,. \tag{3}$$

3 Rheology transformations during cure

The rheology transformations of the prepreg during cure have been studied by DEA technique with the use NETZSCH equipment at different frequencies and varied heating laws. To prevent electrical contact of sensor Idex115 surface with conductive carbon prepreg, a special insulating layer was used to separate them. After preliminary testing, the temperature program, as present in Fig. 6, was used to determine the critical points for the cured prepreg (minimum viscosity and start of gelation – inflection point).

Figures 7a and 7b show the time and temperature dependencies of ion viscosity of the prepreg during cure and the critical points that have been used to assign the temperature of isothermal held at the finite element modelling of the part cure. Inflection points of curves coincide to the temperatures of cure rate peaks found from DSC testing. Positions of these points allow assigning instants for applying air pressure in autoclave.

4 Forward cure modelling: problem formulation

The full problem statement combines the heat transfer equation

$$\rho_{c/m} C_{c/m} \partial T/\partial t + \nabla \cdot \left(-k_{c/m} \nabla T\right) = \begin{cases} Q_{exo} \\ 0 \end{cases} \tag{4}$$

with the kinetic equation (2). Subscripts c and m of material density $\rho_{c/m}$, specific heat capacity $C_{c/m}$, and thermal conductivities $k_{c/m}$ in equation (4) indicate the domains of composite prepreg and mould respectively. Heat sources intensity Q_{exo} acts inside the composite body, while there are no any internal heat sources inside the mould. Equation (2) shows how the local value of the degree of cure depends on the temperature T, whose distribution is described by heat transfer equation (4). In turn, coefficients C_c, k_c and internal heat sources intensity Q_{exo} in equation (4) depend on the state of prepreg materials, which is a function of the degree of cure $\alpha(\mathbf{r})$. All thermal properties of the prepreg material during cure have been identified using results of experimental studies.

The initial value of the absolute temperature T was taken equal to 20°C everywhere in simulated domains.

All surfaces of the mould and outer surfaces of the cured part were exposed to the same temperature $T(t)$ whose time history is represented in Figure 1.

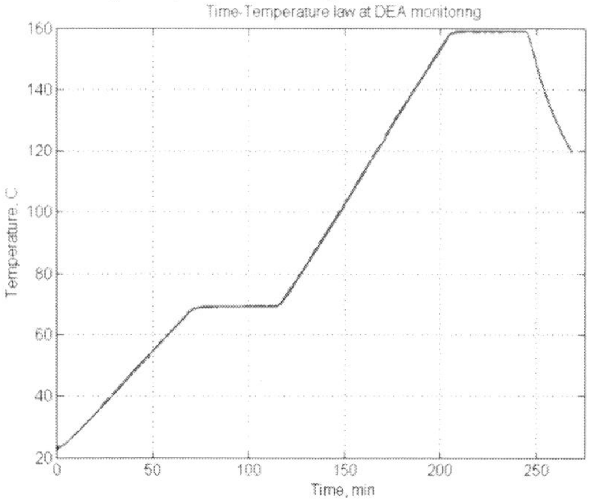

Figure 6. Temperature schedule used at DEA testing of the prepreg

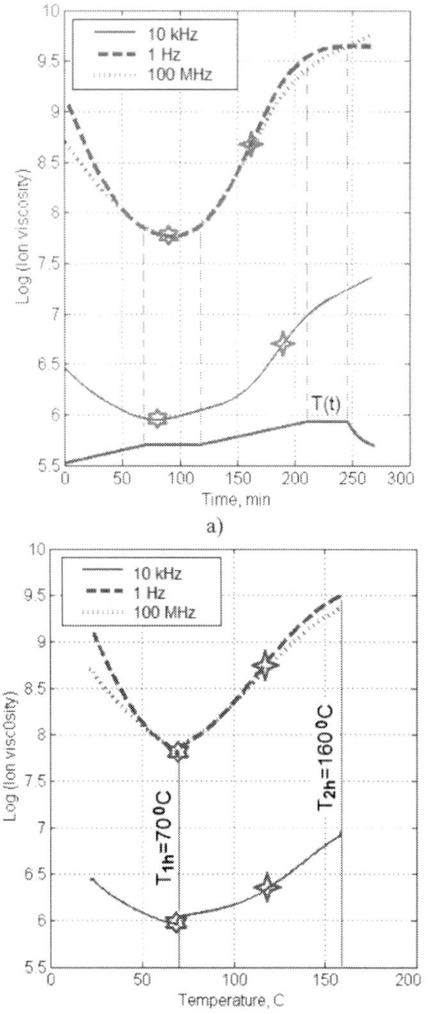

a)

b)

Figure 7. Time (a) and temperature (b) dependencies of Log ion viscosity for the prepreg DSC testing at the heating law, which is present in Fig. 6

Four-pointed stars and hexagrams depict the critical points

To facilitate the finite element meshing the thermal resistance and thus the presence of the vacuum bag thin film were ignored, whereas release film separating prepreg and upper surface of mould was substituted by the special boundary condition

$$-\mathbf{n}_d \cdot \left(-k\left(\nabla_t T_s + \nabla_n T_s\right)\right) = -\frac{T_u - T_d}{R_s} + \frac{1}{2} d_s \cdot Q_s$$
$$-\mathbf{n}_u \cdot \left(-k\left(\nabla_t T_s + \nabla_n T_s\right)\right) = -\frac{T_d - T_u}{R_s} + \frac{1}{2} d_s \cdot Q_s, \quad (5)$$

where

$$R_s = -\frac{d_s}{k}; \quad T_u = T_s \,|_{L=0}; \quad T_d = T_s \,|_{L=d_s}, \quad (6)$$

$\mathbf{n}_u$ and $\mathbf{n}_d$ are the normal vectors to the upper and lower boundaries of the film layer, d_s is the film thickness, $Q_s = 0$ is the internal heat source intensity that acts in the film (no exothermal heat) and T_s is the auxiliary variable, which is determined inside virtual layer only.

Our numerical experiments that have been implemented with meshed thin layer (thickness 0.1 mm) and virtual layer with boundary conditions (5) and (6) showed that the temperatures discrepancy in all used ranges didn't exceed 0.02 ^{0}C.

The values of materials properties for CFRP mould and release film are present in Table 1.

Table 1. Material properties for mould and release film.

Material property	CFRP mould	Release film
Density, kg/m^3	1500	1200
Thermal conductivity, W/(m*K)	0.75+0.1(T-20)/180	0.25
Specific heat capacity, J/(kg*K)	850+550(T-20)/200	1170

In order to find an appropriate finite element meshing and indicative values of the time integration step, several segments were cut from a whole panel (see Fig. 8).

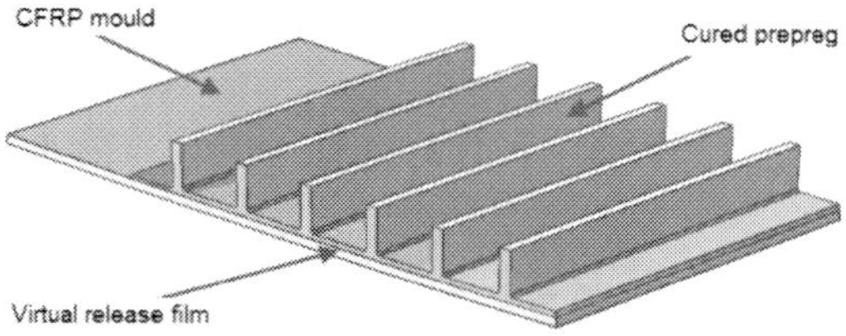

Figure 8. An example of segment that was cut from a whole panel for the preliminary testing FEM model.

All boundary conditions for these segments are illustrated in Fig. 9. The results of these preliminary simulations then have been used at the finite element implementation of forward problem of whole panel curing process in Comsol Multiphysics 5.2 environment.

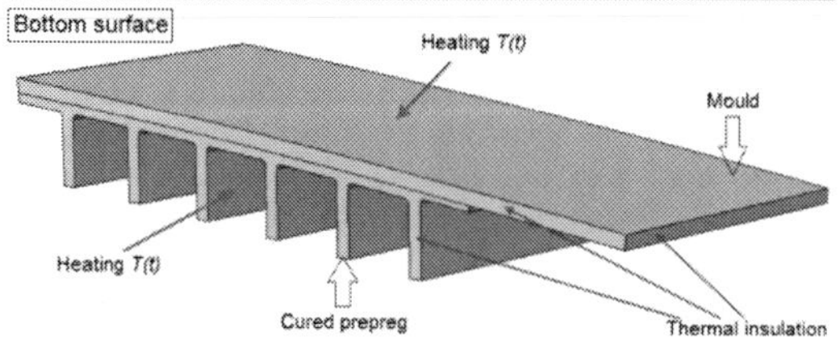

Figure 9. External boundary condition for the segments of part at the finite element transient analysis cure process

5 Optimization problem statement and implementation

The optimization procedure was organized as follows. The FE model of part placed onto a mould performed a parametric sweep of design variables inside pre-assigned intervals. Full cure cycle has been performed for each set of design variables, and these results were stored in text file. So, the operating algorithm fills the area in the design space with points, mapping them to the areas in the objective space. The calculation results then were numerically analyzed to satisfy all imposed constraints and reconstruct the optimum areas. Four timings of the two-stage cure cycle (4D design space) were constrained by the inequalities

$$\begin{cases} t\_heat_1 \le 75\,\text{min} \\ t\_hold_1 \le 45\,\text{min} \\ t\_heat_2 \le 90\,\text{min} \\ t\_hold_2 \le 40\,\text{min} \\ \sum t_i \le 240\,\text{min} \end{cases} \qquad (7)$$

Full cycle duration was limited by the value 240 min due to manufacturing conditions.

Conversion at the end of cure cycle was calculated by integration and averaging over the cured composite volume Ω ; it was assumed to be no less than 0.98

$$\langle\alpha\rangle = \int_\Omega \alpha(\mathbf{r}, t_{end}) \cdot dv \Big/ V \le 0.98 \qquad (8)$$

Cure process objectives included two variables that should be minimized. First objective is the maximum of the mean deviation of conversion during the cure cycle

$$\langle\Delta\alpha\rangle = \max_{t\in[0;t_{end}]} \int_\Omega \left|\alpha(\mathbf{r},t) - \langle\alpha\rangle\right| \cdot dv \Big/ V . \qquad (9)$$

The second objective is the mean deviation of the temperature at the first and second stages of heating-up. This objective, which allows to estimate the temperature uniformity within the cured prepreg, was calculated according to the following relationships

$$\langle\Delta T\rangle_{1,2} = \max_{t\in\left\{\begin{smallmatrix}[0;t_1]\\ [t_2;t_3]\end{smallmatrix}\right\}} \int_\Omega \left|T(\mathbf{r},t) - \langle T(t)\rangle_{1,2}\right| \cdot dv \Big/ V , \qquad (10)$$

where mean temperature at the heating stages of cure cycle

$$\langle T(t)\rangle_{1,2} = \int_\Omega T(\mathbf{r},t)\cdot dv \Big/ V ; \quad at \quad t \in \begin{cases}[0;t_1]\\ [t_2;t_3]\end{cases} . \qquad (11)$$

These values of both objectives calculated at the different control parameters are the points in the objective space that were analyzed to find the reasonable minimum of criteria (9) and (10). Most attractive cases can be analyzed also not only for averaged, but local distribution of both objectives.

A multidimensional visualization techniques used here transforms a multidimensional problem so that it can be mapped to a 2D visual space. In the considered case when a control law is parameterized by four parameters whose values completely characterize the thermal action on the cure system, techniques based on multiple 2D displays is the most appropriate for providing insight into complex cure phenomena. We use some projections of the subset in the objective space to the 2D design subspace where these projections are drawn as level lines.

Some optimization results are present in Figs. 10a and 10b. Four and six-pointed stars denote the temperature control laws, parameters of which are present in Table 2.

Table 2. Design variables for the studied control schedules.

Schedule	Marker	Timings, min (see Fig.1)			
		t_ heat₁	t_ hold₁	t_ heat₂	t_ hold₂
1st		70	42	90	35
2nd		64	36	82	35

The plots in Fig. 10 demonstrate that durations of first heating and first isothermal hold have the greatest impact on the spatial uniformity of degree of cure. It can be explained by the fact that these timings provide a uniform spatial distribution of α directly before the start of the second heating, when cross-linking reaction evolves very fast and this reaction is controlled not only by the temperature, but by time too. This is due to the increased influence of the multiplier α^m at the end of the first isothermal hold (see Eq. 2). The highest variation in the degree of cure occurs at the initial period of the second heating stage as resin gelation begins as shown in Figure 12. The highest cure quality can be attained through the acceptable deceleration of the heating rate. Figures 11 and 12 demonstrate time histories for the temperature in autoclave; averaged temperature and degree of cure deviations for two studied control schedules (see Table 2). It should be noted that the correct quantitative conclusions can only be reached when taking into account the actual geometry of the composite structure and the mould as well as their material properties. This confirms the need for use CAD model of cured parts.

DOI: 10.1051/matecconf/201713007001

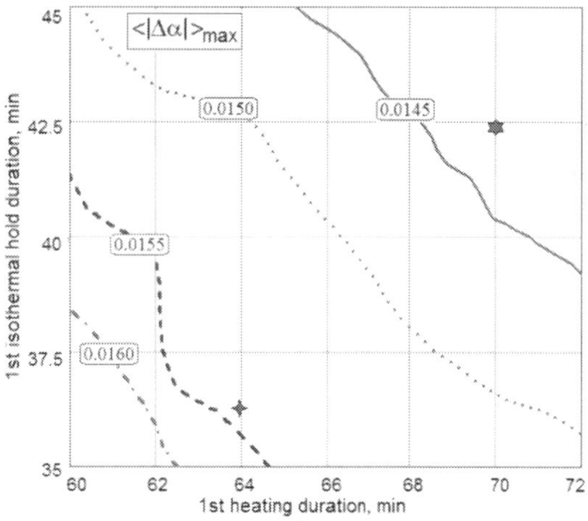

a)

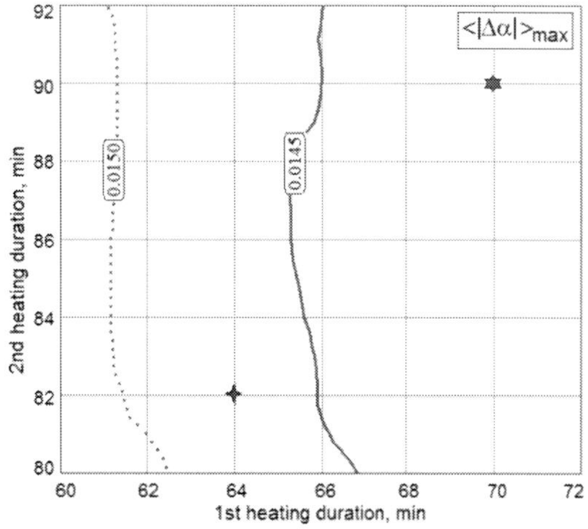

b)

Figure 10. Contour lines for the maximum variation of the degree of cure within the prepreg body througout the cycle.
a) Projection on the plane *(t_heat₁, t_hold₁)*;
b) projection on the plane *(t_heat₁, t_heat₂)*

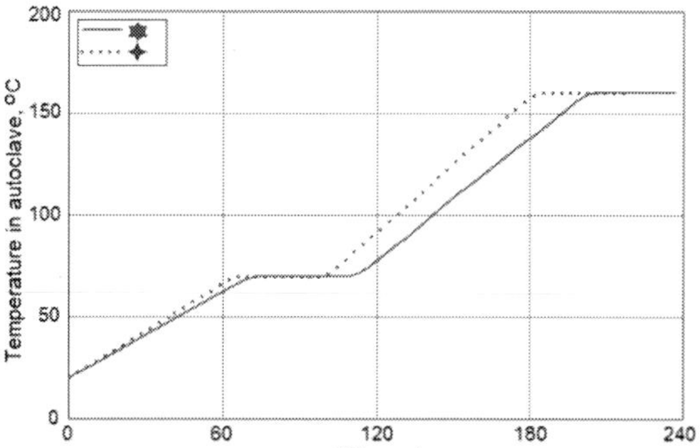

Figure 11. Two time histories of the temperature in atoclave

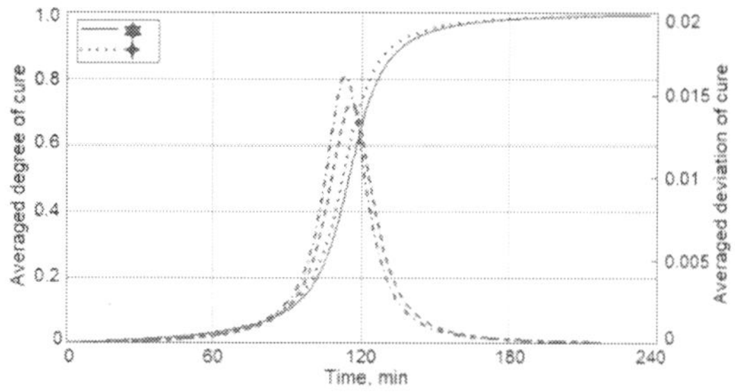

a)

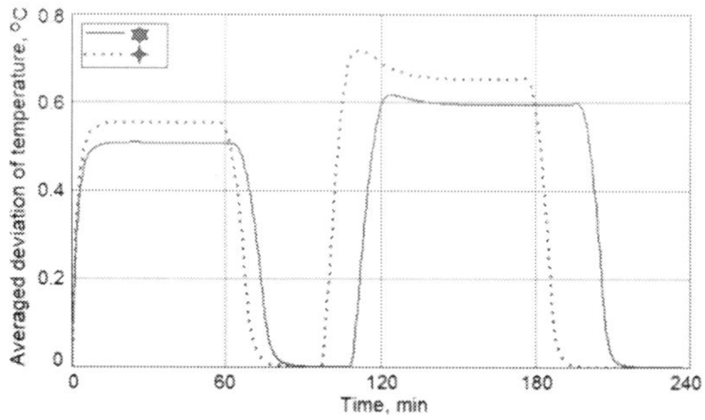

b)

Figure 12. The time histories of degree of cure $\alpha(t)$, its deviation (a) and deviation of temperature (b) averaged over prepreg volume at the curing schedules, which are present in Figure 11 (see Table 2)

Conclusions

We have defined and solved the problem of the optimal control synthesis for the cure process of large CFRP composite panel cured under vacuum in an autoclave as a multi-objective optimization problem to ensure minimum variations of the degree of cure and temperature inside the prepreg body throughout the cure cycle, taking into account all constraints imposed by the prepreg thermal-kinetic properties and other operating conditions. The forward modeling problem for the transient cure process of the composite structure includes coupled heat transfer and kinetic phenomena in the prepreg throughout the two-stage cure cycle using the part geometry, which is exported from the CAD model to the finite element soft package. The input parameters of the process model implemented by using Comsol Multiphysics software are timings that form the control law for the temperature in an autoclave and appear as points in the 4D design space. The process quality criteria computed by integration over the cured prepreg volume at each time step are the averaged deviations of degree of cure and temperature determined at the completion of the transient analysis covering the entire cure cycle time. The suggested approach allows understanding the complex nature of curing phenomena,

 DOI: 10.1051/matecconf/201713007001

measuring the maximum achievable process quality, finding values satisfying the constraints imposed by the control law parameters, and making a reasonable decision on the choice of such values.

References

1. A. Baker, S. Dutton and D. Kelly, *Composite Materials for Aircraft Structures*, 2nd ed., (AIAA ltd., Reston, Virginia, 2004)
2. P.K. Mallick, *Fiber-Reinforced Composites: Materials, Manufacturing, and Design*, 3rd ed. (CRC Press, Boca Raton, FL, 2007)
3. G. Chiesura et al. Sensors. **16**, 6 (2016)
4. H.C. Park, and S.W. Lee, J. Compos. Mater. **35** (2001)
5. Y. Gu, M. Li, Z. Zhang and Z. Sun, J. Compos. Mater. **40** (2006)
6. T.W. Capehart, and H.G. Kia, J. Compos. Mater. **41** (2007)
7. L. Sorrentino, W. Polini and C. Bellini, Adv. Compos. Mater. **23**, 3, (2014)
8. S. Johnson, C.D. Rudd and D.J. Hill, P. I. Mech. Eng. B-J. Eng. **209** (1995)
9. J.-H. Oh and D.-G. Lee, J. Compos. Mater. **36**, 1 (2002)
10. A. Mamoune, A. Saouab, T. Ouahbi and C.-H. Park, J. Reinf. Plast. Comp. **30** (2011)
11. P.E. Jahromi, A. Shojaei and S.M.R. Pishvaie, J. Reinf. Plast. Comp. **31** (2012)
12. A. Saad, A. Echchelh, M.Hattabi and M. El Ganaoui, J. Reinf. Plast. Comp. **31** (2012)
13. J. Xu, M. Qiao, B. You, and Y. Yu, J. Reinf. Plast. Comp. **31** (2012)
14. L. Sorrentino, C. Bellini, Polym.-Plast. Technol. **54** (2015)
15. I.V. Tarasov, S.N. Shevtsov, A.V. Evlanov and E.E. Orozaliev, IFAC P. Ser. **28**, 11 (2016)
16. R.B. Williams, B.W.Grimsley, D.J. Inman, and W.K. Wilkie, J. Reinf. Plast. Comp. **23** (2004)
17. L. Esposito, L. Sorrentino, F. Penta, C. Bellini, Int. J. Adv. Manuf. Tech. **83**, 5-8 (2016)
18. C. Brauner, T. French and A.S. Hermann, J. Compos. Mater. **51**, 7 (2017)
19. J.-K. Wu et al. Proc. Int. Conf. on Information Science, Electronics and Electrical Engineering - ISEEE 2014. (2014)
20. R.V.N. Melnik, Model. Simul. Mater. Sc. **10** (2002)
21. M.R. Kamal, S. Sourour, Polym. Eng. Sci. **13** (1973)
22. M.-K. Um, I.M. Daniel and C.-S. Hwang, Compos. Sci. Technol. **62** (2002)
23. J. Kim, T.J. Moon, and J.R. Howell, J. Compos. Mater. **36**, 21 (2002)
24. B.-C. Chern, T.J. Moon, J.R. Howell and W. Tan, J. Compos. Mater. **36** (2002)
25. Wen, X., et al. High Perform. Polym. **23** (2011)
26. M.J. Yoo, et al. Eur. Polym. J. **46** (2010)
27. J. Zhang et al. Thermochim. Acta. **549** (2012)
28. R. Hardis, J.L.P. Jessop, F.E. Peters and M.R. Kessler, Compos. Part A-Appl. S. **49** (2013)
29. G. Van Assche, S. Awier, and B. Van Mele, Thermochim. Acta **388** (2002)
30. M. Vafayan, M.H. Beheshty, M.H.R. Ghoreishy and H. Abedini, J. Compos. Mater. **48** (2014)

Effect of polymer concentration on the lifetime and transmittance behavior of a self-oscillating polymer chain with a high lower critical solution temperature

Tomoka Nakazumi[1] and Yusuke Hara[1,a]

[1]*Research Institute for Sustainable Chemistry, ISC, National Institute of Advanced Industrial Science and Technology, AIST, Central 5-2, 1-1-1 Higashi, Tsukuba 305-8565, Japan.*

Abstract. In this study, we investigated the lifetime and self-oscillating behavior of a polymer chain with a high lower critical solution temperature (LCST). The polymer chain comprised 4-vinyl-4′-methyl-2,2′-bipyridinebis(2,2′-bipyridine)bis(hexafluorophosphate) ruthenium as catalyst for the Belousov–Zhabotinsky reaction and N-ethylacrylamide as the polymer main-chain with a high LCST when compared to N-isopropylacrylamide. We demonstrated that the self-oscillating behavior was significantly affected by the polymer concentration and measuring temperature. Moreover, we established that the lifetime of the transmittance self-oscillation can be predicted from the polymer concentration and measuring temperature.

1 Introduction

The Belousov-Zhabotinsky (BZ) reaction is a famous self-oscillating reaction that occurs under constant temperature conditions [1–8]. The behavior of this reaction differs under unstirred and stirred conditions. When unstirred, the reaction period (time scale: dozens of seconds) and space scale (several millimeters) allow complex patterns originating from the transition-metal catalyst to be observed with the naked eye. Therefore, the chemical waves and patterns of the BZ reaction have been much investigated as a model for spatiotemporal structure formation by scientists in various fields including mathematics, physics, and chemistry. Under strong acid conditions comprising nitric acid (HNO_3) or sulfuric acid (H_2SO_4), the organic substrate [malonic acid (MA)] in the BZ reaction is oxidized by an oxidizing agent [sodium bromate ($NaBrO_3$)] in the presence of a transition-metal catalyst.

In this study, we synthesized a self-oscillating polymer chain, covalently bonded to the transition-metal catalyst moiety [$Ru(bpy)_3$], to undergo a periodical transmittance change induced by the BZ reaction [9–21]. In this latter reaction, a periodical change in the solubility of the polymer chain was synchronized with the periodical redox change in the $Ru(bpy)_3$ moiety. This occurred because the $Ru(bpy)_3$ moiety exhibits different solubilities in its reduced and oxidized states. Thus, the reduced $Ru(bpy)_3$ moiety in the self-oscillating polymer chain is hydrophobic compared to the oxidized $Ru(bpy)_3$ moiety. This difference in solubility originates from the transmittance self-oscillation induced by the BZ reaction. No color changes were observed between the reduced and oxidized $Ru(bpy)_3$ during transmittance self-oscillation measurements. Thus, the isosbestic points of the polymer solution in the reduced and oxidized states were detected by a spectrophotometer. We concluded that transmittance self-oscillation is due to the soluble-insoluble change in the polymer solution in the reduced and oxidized states induced by the BZ reaction.

In this investigation, we adopted poly(N-ethylacrylamide) [poly(NEAAm), Figure 1] as a self-oscillating polymer main-chain with a high lower critical solution temperature (LCST). This polymer chain undergoes transmittance self-oscillation in a wide temperature range when compared to the conventional self-oscillating polymer chain comprising a poly(N-isopropylacrylamide) [poly(NIPAAm)] main-chain [22-23]. Previous investigations have employed the latter self-oscillating main-chain with an LCST of ~32 °C [10-21]. Moreover, for the NIPAAm-based self-oscillating polymer chain {poly[NIPAAm-*co*-$Ru(bpy)_3$]}, the LCST in the reduced state is lower than that of poly(NIPAAm) because of the low solubility of the reduced $Ru(bpy)_3$ moiety [19]. Therefore, the self-oscillating polymer chain cannot undergo self-oscillation at high temperatures. This occurs because the NIPAAm-based self-oscillating polymer chain is easily aggregated in the reduced state that originates from the hydrophobic main-chain and interactions between the hydrophobic reduced $Ru(bpy)_3$ moiety. At high temperatures, the aggregation of the self-oscillating polymer chain is much faster because of the lower LCST of the polymer chain. Therefore, the self-oscillating polymer solution cannot disaggregate as the self-oscillating polymer solution changes from the reduced state to the oxidized state; this latter state

[a] Corresponding author: y-hara@aist.go.jp

DOI: 10.1051/matecconf/201713007002

originates from the higher hydrophobic nature of the polymer main-chain.

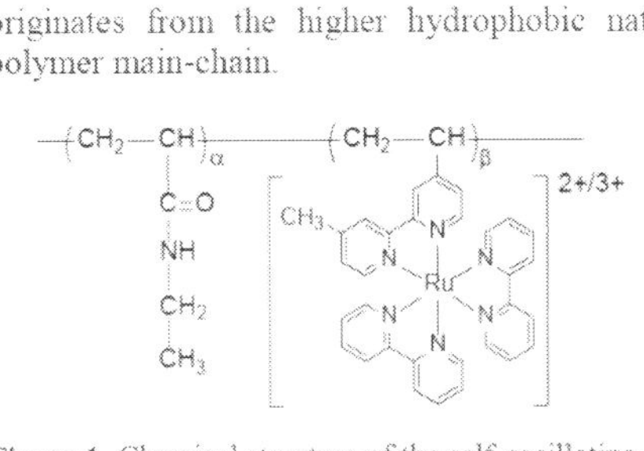

Figure 1. Chemical structure of the self-oscillating polymer chain {poly[NEAAm-*co*-Ru(bpy)$_3$]}. NEAAm = N-ethylacrylamide; Ru(bpy)$_3$ = 4-vinyl-4′-methyl-2,2′-bipyridinebis(2,2′ bipyridine)bis(hexafluorophosphate) ruthenium.

In this study, to determine the activation energy, we measured the transmittance self-oscillation of a poly[NEAAm-*co*-Ru(bpy)$_3$] aqueous solution comprising the three BZ substrates (sulfuric acid, malonic acid, and sodium bromate) at high temperatures. We also examined the effect of the polymer concentration on the self-oscillating behavior and the lifetime of the transmittance self-oscillation.

2 Experimental section

2.1 Synthesis of poly[NEAAm-*co*-Ru(bpy)3]

Poly[NEAAm-*co*-Ru(bpy)$_3$] (Figure 1) was synthesized by free-radical polymerization at 83 °C for 5 h under N_2 bubbling. 4-vinyl-4′-methyl-2,2′-bipyridinebis(2,2′-bipyridine)bis(hexafluorophosphate) ruthenium [Ru(bpy)$_3$; 1.25 g], NEAAm (10.7 g), and 2,2'-azobis(2-methylbutyronitrile) (0.63 g) were dissolved in an ethanol/water (1:1; 125.3 g) mixture. The resultant poly[NEAAm-*co*-Ru(bpy)$_3$] aqueous solution was subsequently dialyzed against ethanol for 10 d to remove any unreacted monomers and the initiator.

2.2 Measurement of transmittance self-oscillation

Polymer solutions comprising the three BZ substrates [malonic acid (0.09 M), sodium bromate (0.5 M), and sulfuric acid (0.4 M)] were prepared. The self-oscillating behavior of the polymer solution was measured using a spectrophotometer at 570 nm; this is the isosbestic point of the polymer solutions in the reduced and oxidized states. The transmittance self-oscillation was detected by the spectrophotometer while stirring.

3 Results and Discussion

Figure 2 illustrates the transmittance self-oscillating behavior of the poly[NEAAm-*co*-Ru(bpy)$_3$] solutions with a fixed concentration of the three BZ substrates at (A) 23 °C and (B) 28 °C. The self-oscillating behavior of the polymer solutions was significantly affected by the measuring temperature. In previous studies, Hara measured the LCST of the poly[NEAAm-*co*-Ru(bpy)$_3$] solution in the reduced and oxidized states [22-23]. The LCST values of these solutions in the reduced and oxidized states were determined as 16 °C and 44 °C, respectively. This difference in solubility originated from the transmittance self-oscillation induced by the BZ reaction. In the oxidized state, the LCST was higher than that in the reduced state. The higher LCST in the oxidized Ru(bpy)$_3$ moiety allowed self-oscillation at higher temperatures when compared to the NIPAAm-based self-oscillating polymer solution.

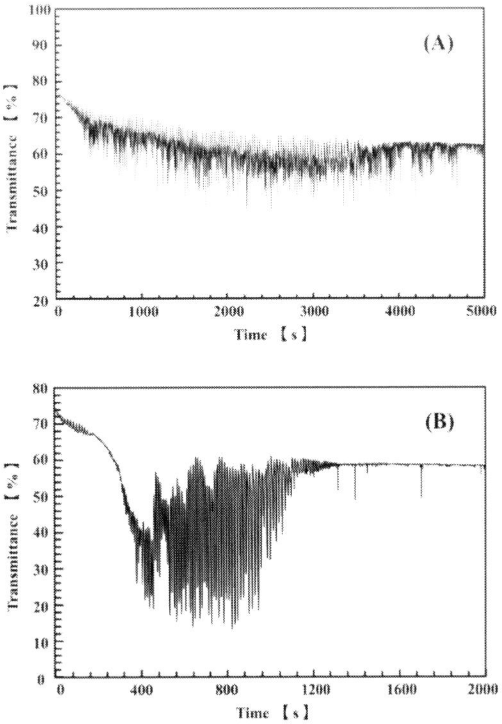

Figure 2. Transmittance self-oscillation for a 0.5 wt.% poly[NEAAm-*co*-Ru(bpy)$_3$] aqueous solution with a fixed initial concentration of sulfuric acid (0.4 M), sodium bromate (0.5 M), and malonic acid (0.09 M) under fixed temperatures of (A) 23 °C and (B) 28 °C. NEAAm = N-ethylacrylamide; Ru(bpy)$_3$ = 4-vinyl-4′-methyl-2,2′-bipyridinebis(2,2′bipyridine)bis(hexafluorophosphate) ruthenium.

Figure 3 demonstrates the Arrhenius dependence of the transmittance self-oscillation of the NEAAm-based polymer solutions with fixed concentrations of the three BZ substrates; the period of self-oscillation decreased with an increase in temperature. The activation energy of the polymer solutions was calculated as 86.7 kJ/mol. This value was higher than that of the poly[NEAAm-*co*-Ru(bpy)$_3$] solution including nitric acid as the strong acid for the BZ reaction. Thus, since sulfuric acid exhibits a smaller oxidizing power than nitric acid, we attributed this difference in activation energy to the different oxidizing power of the strong acid employed. The

DOI: 10.1051/matecconf/201713007002

different oxidizing power greatly affected the BZ reaction because the chemical process of the oxidation in the BZ reaction is significantly important and affects the ease of the elementary reaction.

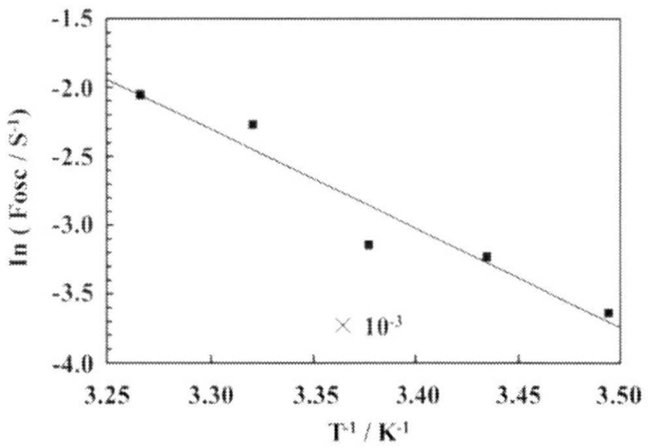

Figure 3. Arrhenius dependence of the self-oscillating frequency (*Fosc*) on temperature for a poly[NEAAm-*co*-Ru(bpy)$_3$] solution with fixed concentrations of the three BZ substrates [sulfuric acid (0.4 M), malonic acid (0.09 M), and sodium bromate (0.5 M)]. NEAAm = N-ethylacrylamide; Ru(bpy)$_3$ = 4-vinyl-4′-methyl-2,2′-bipyridinebis(2,2′ bipyridine)bis(hexafluorophosphate) ruthenium.

Figure 4 represents the relationship between the lifetime of the transmittance self-oscillation and the measuring temperature. The lifetime of the transmittance self-oscillation elongated as the measuring temperature decreased. At higher temperatures, the number of self-oscillations was larger at shorter times because the period of self-oscillation was shorter. Therefore, we considered that the duration of the transmittance self-oscillation for poly[NEAAm-*co*-Ru(bpy)$_3$] was shorter at higher temperatures. This occurred because malonic acid, the energy source of the BZ reaction, was consumed after shorter periods of time at higher temperatures.

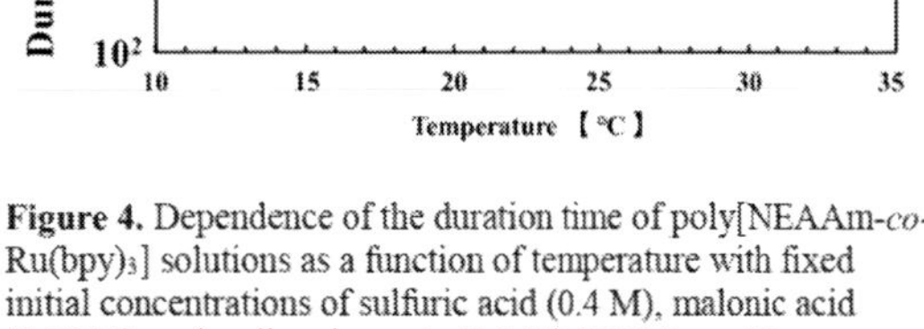

Figure 4. Dependence of the duration time of poly[NEAAm-*co*-Ru(bpy)$_3$] solutions as a function of temperature with fixed initial concentrations of sulfuric acid (0.4 M), malonic acid (0.09 M), and sodium bromate (0.5 M). NEAAm = N-ethylacrylamide; Ru(bpy)$_3$ = 4-vinyl-4′-methyl-2,2′-bipyridinebis(2,2′ bipyridine)bis(hexafluorophosphate) ruthenium.

Figure 5 displays the effect of the polymer concentration on the self-oscillating behavior. The period of self-oscillation was greatly affected by the polymer concentration. The large noise observed in Figure 5 (A was derived from the carbon dioxide (CO_2) generated in the BZ reaction.

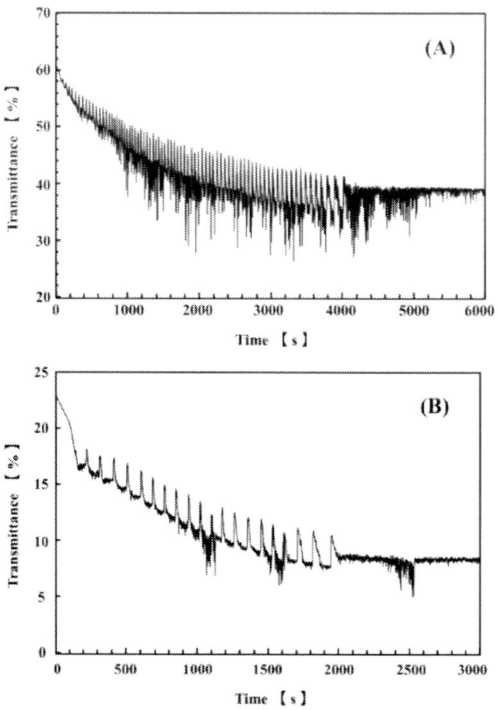

Figure 5. Transmittance self-oscillation for poly[NEAAm-*co*-Ru(bpy)$_3$] solutions [(A) 1.0 wt.% and (B) 3.0 wt. %] with fixed initial concentrations of sodium bromate (0.5 M), sulfuric acid (0.4 M), and malonic acid (0.09 M) at 13 °C. NEAAm = N-ethylacrylamide; Ru(bpy)$_3$ = 4-vinyl-4′-methyl-2,2′-bipyridinebis(2,2′ bipyridine)bis(hexafluorophosphate) ruthenium.

Figure 6 illustrates the relationship between the period of the transmittance self-oscillation and polymer concentration. The period of self-oscillation increased with an increase in polymer concentration. Moreover, when the polymer concentration increased, the viscosity of the polymer solution also increased while the diffusion coefficient of the BZ substrates decreased. Therefore, the period of self-oscillation decreased with an increase in polymer concentration. On the other hand, as the polymer concentration increased, the concentration of the Ru(bpy)$_3$ BZ reaction catalyst also increased. This result indicated that the Ru(bpy)$_3$ concentration exhibited less influence on the period of the transmittance self-oscillation of the poly[NEAAm-*co*-Ru(bpy)$_3$] aqueous solution.

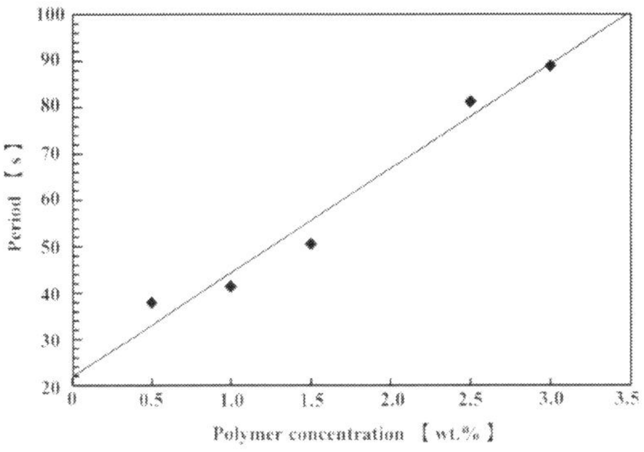

Figure 6. Dependence of the period of the poly[NEAAm-*co*-Ru(bpy)$_3$] solutions as a function of polymer concentration with fixed initial concentrations of sulfuric acid (0.4 M), malonic acid (0.09 M), and sodium bromate (0.5 M) at 13 °C. NEAAm = N-ethylacrylamide; Ru(bpy)$_3$ = 4-vinyl-4′-methyl-2,2′-bipyridinebis(2,2′ bipyridine)bis(hexafluorophosphate) ruthenium.

Figure 7 represents the relationship between the concentration of the poly[NEAAm-*co*-Ru(bpy)$_3$] solution and the lifetime of the transmittance self-oscillation. The duration time of the self-oscillation was significantly affected by the polymer concentration. As the polymer concentration decreased, the duration time increased. On the other hand, when the polymer concentration increased, the period of self-oscillation also increased. For the 3.0 wt. % polymer concentration, the period of self-oscillation was longer while the lifetime was shorter. Thus, as the number of self-oscillations decreased, the lifetime of the self-oscillation became shorter. This tendency is the opposite of the trend observed for the plot of the lifetime against measuring temperature (Figures 3 and 4).

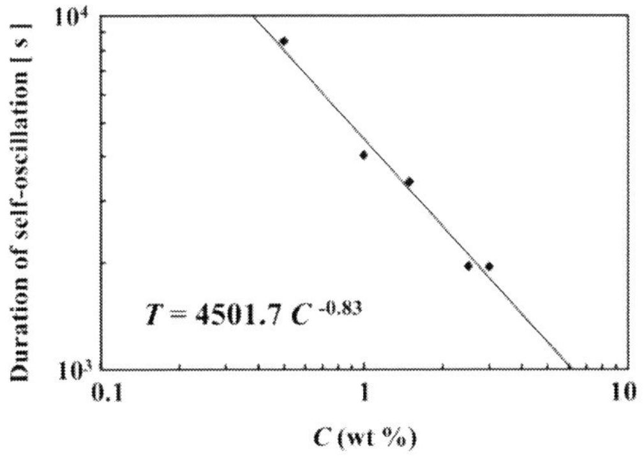

Figure 7. Dependence of the duration time of the poly[NEAAm-*co*-Ru(bpy)$_3$] solution as a function of polymer concentration, with fixed initial concentrations of sulfuric acid (0.4 M), malonic acid (0.09 M), and sodium bromate (0.5 M) at 13 °C.

4 Conclusion

In this study, we demonstrated the effect of the measuring temperature on the transmittance self-oscillating behavior of the poly[NEAAm-*co*-Ru(bpy)$_3$] solution. We established that the period of self-oscillation follows the Arrhenius equation. The activation energy of the polymer solution was determined as 86.7 kJ/mol. Moreover, the lifetime of the self-oscillation became shorter with an increase in temperature because the malonic acid in the polymer solution was consumed in a shorter time. In addition, we confirmed the influence of the polymer concentration on the self-oscillating behavior and lifetime by changing the polymer concentration. The period of self-oscillation increased when the polymer concentration increased. This result indicated that the diffusion coefficients of the three BZ substrates were significantly affected by the polymer concentration. On the other hand, as the polymer concentration increased, the lifetime of the transmittance self-oscillation became shorter. This occurred because the cycle number and the duration of the self-oscillation decreased as the polymer concentration increased. This trend was the reverse of the trend observed for the change in temperature.

5 Acknowledgements

This work was supported by Grants-in-Aid (KAKENHI) for Young Scientists (B) (16K17493, 15K18014) and Scientific Research (B) (17H03209, 15H03827). This work was also carried out under the auspices of the New Energy and Industrial Technology Development Organization (NEDO) of Japan under the Industrial Technology Research Grant Program in 2011.

References

1. B.P. Belousov, Sb. Ref. Radiat. Med. (Collections of Abstracts on Radiation Medicine, Medgiz, Moscow) **1**, 145 (1959)
2. A.N. Zaikin, A.M. Zhabotinsky, Nature **225**, 535 (1970)
3. J. Lindsay *Oscillations and traveling waves in chemical systems* (Eds. R.J. Field, M. Burger, Wiley, New York, 1985)
4. S.K. Scott, *Chemical chaos, 1st ed.* (Oxford University Press, Oxford, 1991)
5. E.J. Reusser, R.J. Field, J. Am. Chem. Soc. **101**, 1063 (1979)
6. G. Nicolis, I. Prigogine, *Self-organization in nonequilibrium systems* (Wiley, New York, 1977)
7. G. Nicolis, I. Prigogine, *Self-Orgainization in nonequilibrium Systems* Wiley, New York 27, 1977)
8. J.D. Murray, *Mathematical Biology* (Springer-Verlag, Berlin, 1990)
9. T. Ishiwatari, M. Kawagishi and M. Mitsuishi, J. Polym. Sci.: Polym. Chem. **22**, 2699 (1984)
10. R. Yoshida, T. Sakai, S. Ito, T. Yamaguchi, J. Am. Chem. Soc. **124**, 8095 (2002)
11. R. Yoshida, T. Takahashi, T. Yamaguchi and H. Ichijo, J. Am. Chem. Soc. **118**, 5134 (1996)

12. Y. Hara, R. Yoshida, J. Phys. Chem. B **109**, 9451 (2005).
13. Y. Hara, R. Yoshida, Langmuir **21**, 9773 (2005).
14. Y. Hara, R. Yoshida, J. Chem. Phys. **128**, 224904, (2008).
15. Y. Hara, R. Yoshida, J. Phys. Chem. B **112**, 8427 (2008).
16. Y. Hara, R. Yoshida, Macromol. Chem. Phys. **210**, 2160 (2009).
17. Y. Hara, R. Yoshida, Macromol. Rapid Commun. **30**, 1656 (2009).
18. Y. Hara, H. Mayama, Y. Yamaguchi, Y. Takenaka, R. Fukuda, J. Phys. Chem. B **117**, 14351 (2013).
19. Y. Hara, Y. Takenaka, Sens. **14**, 1497 (2014).
20. Y. Hara, Y. Yamaguchi, H. Mayama, J. Phys. Chem. B **118**, 634 (2014).
21. Y. Hara, K. Fujimoto, H. Mayama, J. Phys. Chem. B **118**, 608 (2014).
22. Y. Hara, Adv. Mater. Res. **941**, 1208 (2014).
23. Y. Hara, Adv. Mater. Res. **941**, 1212 (2014).

Influence of the Belousov–Zhabotinsky substrate concentration on the lifetime and self-oscillating behavior of a polymer solution

Tomoka Nakazumi[1] and Yusuke Hara[1,a]

[1]Research Institute for Sustainable Chemistry, ISC, National Institute of Advanced Industrial Science and Technology, AIST, Central 5-2, 1-1-1 Higashi, Tsukuba 305-8565, Japan.

Abstract. In this study, we synthesized a self-oscillating polymer chain with N-ethylacrylamide (NEAAm) as the main-chain. This exhibits a higher lower critical solution temperature than an N-isopropylacrylamide main chain. We subsequently measured the transmittance self-oscillation of poly[NEAAm-*co*-Ru(bpy)$_3$] [Ru(bpy)$_3$ = 4-vinyl-4′-methyl-2,2′-bipyridinebis(2,2′ bipyridine)bis(hexafluorophosphate) ruthenium] solutions comprising three Belousov–Zhabotinsky (BZ) substrates (malonic acid, sodium bromate, and sulfuric acid), under stirring at constant temperature. The soluble-insoluble self-oscillation of the polymer solution originated from the different solubility of the Ru(bpy)$_3$ moiety as a catalyst of the BZ reaction in the reduced and oxidized states. We demonstrated that the self-oscillating behavior was significantly affected by the initial concentration of the sodium bromate. Moreover, we clarified that the lifetime of the transmittance self-oscillation can be predicted from the initial concentration of malonic acid.

1 Introduction

The use of the self-oscillating polymer chain copolymerized with 4-vinyl-4′-methyl-2,2′-bipyridinebis(2,2′bipyridine)bis(hexafluorophosphate) ruthenium [Ru(bpy)$_3$] as a catalyst for the Belousov-Zhabotinsky (BZ) reaction [1-12] has been much investigated for application to soft actuators and soft robots with autonomous motion. The polymer chain is driven while consuming malonic acid, the energy source of the BZ reaction. This occurs because self-oscillating behavior is induced by a BZ reaction originating from the Ru(bpy)$_3$ moiety in the polymer chain [13-22]. The BZ reaction has been much investigated and is a well-known oscillating reaction that causes spontaneous redox oscillation and a wide variety of nonlinear phenomena. In this reaction, an oxidant (sodium bromate) oxidizes an organic acid (malonic acid) in the presence of a metal catalyst [Ru(bpy)$_3$] that in turn undergoes periodical redox change. When the redox state of the metal catalyst changes, its solubility also changes simultaneously. In a previous study, polymer chains covalently bonded to Ru(bpy)$_3$ were driven by making use of the periodical solubility change of the metal catalyst induced by the BZ reaction. This polymer solution, comprising three BZ substrates (malonic acid, sodium bromate, and sulfuric acid), undergoes transmittance self-oscillation. Thus, soluble-insoluble self-oscillations originate from the different solubilities of the Ru(bpy)$_3$ moiety in the reduced and oxidized states. In a previous investigation, the main-chain of the self-oscillating polymer selected was poly(N-isopropylacrylamide) [poly(NIPAAm)] [15] which exhibits a lower critical solution temperature (LCST) of ~32 °C. As the Ru(bpy)$_3$ was covalently bonded to the poly(NIPAAm), the LCST of poly[NIPAAm-co-Ru(bpy)$_3$] in the reduced state was lower than that in the oxidized state because the reduced Ru(bpy)$_3$ moiety exhibited lower solubility compared to the oxidized Ru(bpy)$_3$ moiety [15]. This difference in solubility originates from the transmittance self-oscillation induced by the BZ reaction. Hara *et al.* selected the self-oscillating polymer main-chain with the higher LCST to undergo self-oscillation in a wide range of temperature conditions [23, 24]. In these studies, N-ethylacrylamide (NEAAm) was adopted because of its high LCST compared to the poly(NIPAAm) chain; the LCST of poly[NEAAm-*co*-Ru(bpy)$_3$] was 44 °C in the oxidized state and 16 °C in the reduced state. The difference in solubility of the poly[NEAAm-*co*-Ru(bpy)$_3$] in the reduced and oxidized states is attributed to the transmittance self-oscillation under constant temperature conditions.

Notably, it is important to establish the control factor of the period of the transmittance self-oscillation for application to soft actuators and soft robots. Thus, in this study, we examined the influence of the BZ substrate (sodium bromate and malonic acid) concentration in the polymer solutions on the period and transmittance self-oscillating behavior. Moreover, the duration of self-oscillation is also an important factor in the development of autonomous soft actuators and soft robots with a long life. Thus, we evaluated the influence of malonic acid concentration on the lifetime of the transmittance self-oscillation.

[a] Corresponding author: y-hara@aist.go.jp

DOI: 10.1051/matecconf/201713007003

2 Experimental section

2.1 Synthesis of poly[NEAAm-*co*-Ru(bpy)$_3$]

Poly[NEAAm-co-Ru(bpy)$_3$] (Figure 1) was synthesized by random polymerization at 83 °C for 5 h under nitrogen bubbling. Ru(bpy)$_3$ (1.25 g) and NEAAm (10.7 g) monomers together with 2,2'-azobis(2-methylbutyronitrile) (initiator, 0.63 g) were dissolved in an ethanol/water (1:1; 125.3 g) mixture. The resultant polymer mixture was subsequently dialyzed against ethanol for 10 d.

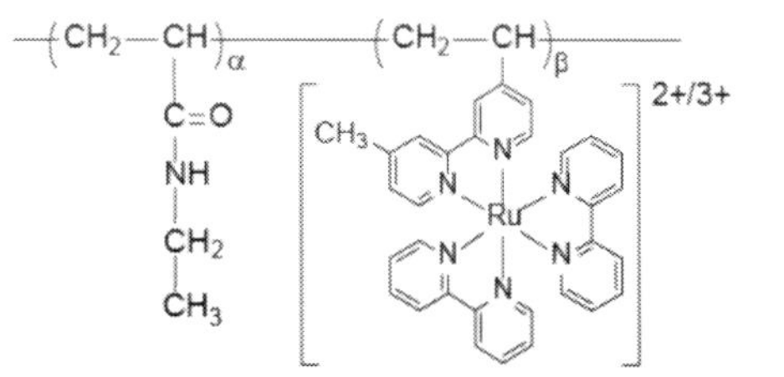

Figure 1. Chemical structure of the self-oscillating polymer chain {poly[NEAAm-*co*-Ru(bpy)$_3$]}. NEAAm = N-ethylacrylamide; Ru(bpy)$_3$ = 4-vinyl-4′-methyl-2,2′-bipyridinebis(2,2′ bipyridine)bis(hexafluorophosphate) ruthenium.

2.2 Measurement of transmittance self-oscillation of poly[NEAAm-*co*-Ru(bpy)$_3$]

0.3 wt.% of poly[NEAAM-*co*-Ru(bpy)$_3$] aqueous solution comprising the three BZ substrates (malonic acid, sodium bromate, and sulfuric acid) were prepared to measure the transmittance self-oscillation. The self-oscillating behavior was detected by a spectrophotometer at 570 nm under constant stirring. This wavelength is the isosbestic point of the polymer solution in the reduced and oxidized states. The temperature of the polymer solution was fixed at 11 °C to detect stable self-oscillating behavior.

3 Results and Discussion

Figures 2 and 3 represent the transmittance self-oscillations of the poly[NEAAm-co-Ru(bpy)$_3$] aqueous solution comprising fixed concentrations of the three BZ substrates at constant temperature (11 °C) under stirring.The large noise observed in these self-oscillating graphs was attributed to the carbon dioxide (CO_2) generated in the BZ reaction. As illustrated in Figure 2, the self-oscillating behavior was significantly affected by the initial concentration of the sodium bromate. The period of self-oscillation was much slower in Figure 2(A) than in 2(B), even though the difference in concentration was only 0.2 M. This trend indicated that the amount of sodium bromate is significantly important to control the transmittance self-oscillating behavior. Thus, we concluded that the oxidation process of Ru(bpy)$_3$ in the BZ reaction is significantly affected by the concentration of the sodium bromate. Conversely, the influence of the initial concentration of malonic acid was smaller, at a fixed concentration (0.5 M) of sodium bromate, [Figures 3(A) and 3(B)].

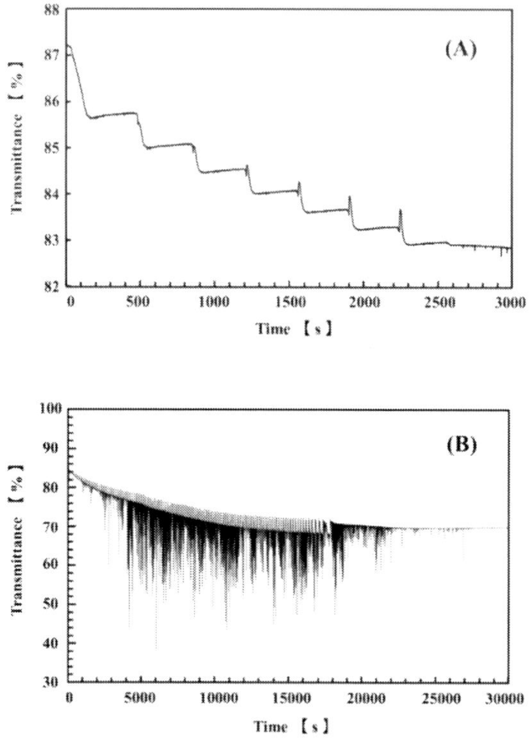

Figure 2. Transmittance self-oscillation profiles for a 0.3 wt. % poly[NEAAm-*co*-Ru(bpy)$_3$] solution with fixed initial concentrations of malonic acid (0.06 M) and sulfuric acid (0.3 M): concentration of sodium bromate at (A) 0.1 M and (B) 0.3 M, at 11 °C. NEAAm = N-ethylacrylamide; Ru(bpy)$_3$ = 4-vinyl-4′-methyl-2,2′-bipyridinebis(2,2′bipyridine)bis(hexafluorophosphate) ruthenium.

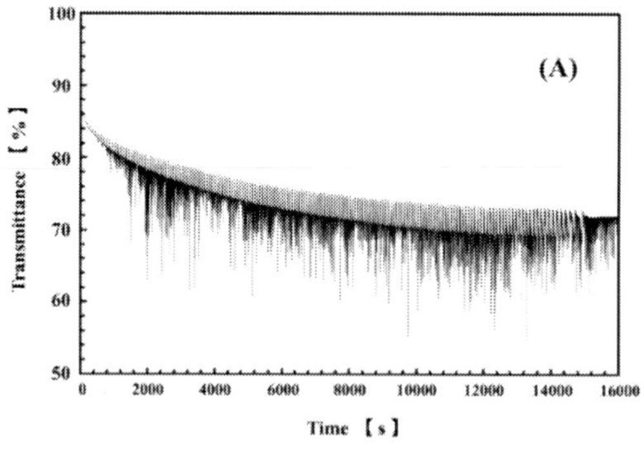

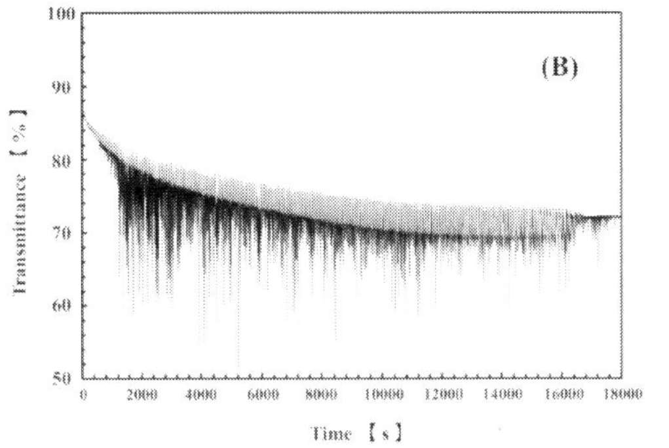

Figure 3. Transmittance self-oscillation profiles for a 0.3 wt. % poly([NEAAm-*co*-Ru(bpy)$_3$] aqueous solution with fixed initial concentrations of sodium bromate (0.5 M) and sulfuric acid (0.3 M); concentrations of malonic acid at (A) 0.08 M and (B) 0.1 M, at 11 °C. NEAAm = N-ethylacrylamide; Ru(bpy)$_3$ = 4-vinyl-4′-methyl-2,2′-bipyridinebis(2,2′bipyridine)bis(hexafluorophosphate) ruthenium.

Figure 4 illustrates the relationship between the initial substrate concentrations and the period of the transmittance self-oscillation of the poly[NEAAm-co-Ru(bpy)$_3$] aqueous solution. The period of the transmittance self-oscillation at an initial concentration of sodium bromate of 0.9 M was approximately ten-fold smaller than that observed when the concentration was reduced to 0.1 M. This trend demonstrated that the period of self-oscillation could be highly accelerated by changing the sodium bromate concentration. On the other hand, as the initial concentration of malonic acid was changed, the changing range of the period was comparatively smaller. These results demonstrated that the initial concentration of sodium bromate is significantly effective in accelerating the period of the transmittance self-oscillation. An increase in the solution temperature has also been reported to have a significant effect [15].

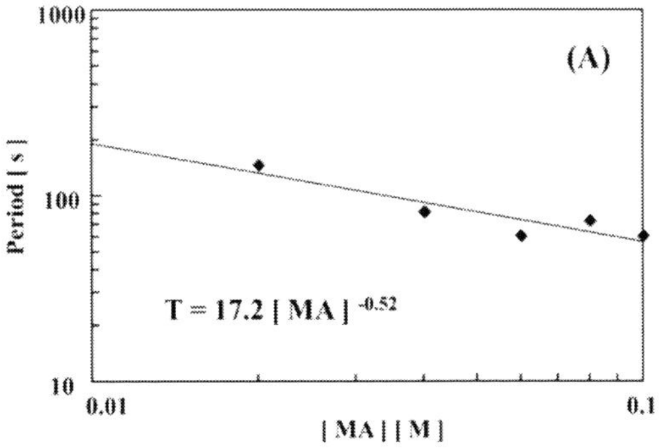

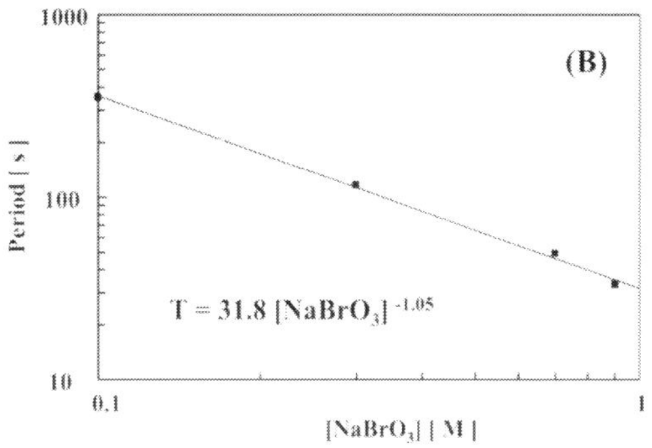

Figure 4. Logarithmic plots of period T (s) *vs.* initial molar concentration of one Belousov–Zhabotinsky (BZ) substrate at constant temperature (11 °C) under fixed concentrations for the other two BZ substrates: (A) sodium bromate (0.5 M) and sulfuric acid (0.3 M); (B) malonic acid (0.06 M) and sulfuric acid (0.3 M).

Figure 5 displays the lifetime of the transmittance self-oscillation of the poly[NEAAm-co-Ru(bpy)$_3$] aqueous solution at 11 °C. The lifetime of the self-oscillation was found to be linearly related to the initial concentration of malonic acid. This demonstrated that the lifetime of the self-oscillation can be predicted from the initial concentration of malonic acid because the malonic acid is the energy source of the BZ reaction.

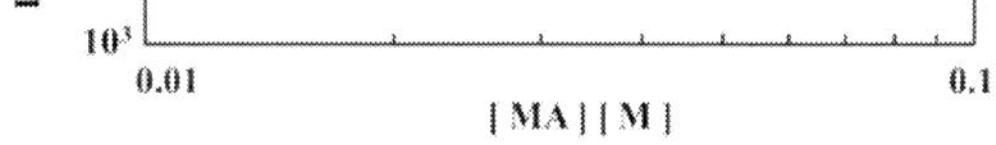

Figure 5. Lifetime of the transmittance self-oscillation for 0.3 wt.% poly[NEAAm-*co*-Ru(bpy)$_3$] solutions as a function of malonic acid concentration, with fixed initial concentrations of sulfuric acid (0.3 M) and sodium bromate (0.5 M), at 11 °C. NEAAm = N-ethylacrylamide; Ru(bpy)$_3$ = 4-vinyl-4′-methyl-2,2′-bipyridinebis(2,2′ bipyridine)bis(hexafluorophosphate) ruthenium.

4 Conclusion

In this study, transmittance self-oscillations of a polymer chain with a high LCST were measured by changing the initial concentration of malonic acid and sodium bromate to clarify the control factor of the period. This study demonstrated that the self-oscillating behavior of poly[NEAAm-co-Ru(bpy)$_3$] was significantly affected by the initial concentration of sodium bromate. We also established that the period of self-oscillation for the poly[NEAAm-co-Ru(bpy)$_3$] aqueous solution can be determined from the initial concentrations of malonic acid and sodium bromate. In addition, we proved that the lifetime of the self-oscillation of the polymer chain can be predicted from the initial concentration of malonic acid.

5 Acknowledgements

This work was supported by Grants-in-Aid (KAKENHI) for Young Scientists (B) (16K17493, 15K18014) and Scientific Research (B) (17H03209, 15H03827). This work was also carried out under the auspices of the New Energy and Industrial Technology Development Organization (NEDO) of Japan under the Industrial Technology Research Grant Program in 2011.

References

1. B.P. Belousov, Sb. Ref. Radiat. Med. (Collections of Abstracts on Radiation Medicine, Medgiz, Moscow) **1**, 145 (1959)
2. A.N. Zaikin, A.M. Zhabotinsky, Nature **225**, 535 (1970)
3. R.J. Field, E. Koros, R.M. Noyes, J. Am. Chem. Soc. **94**, 8649 (1972)
4. R.J. Field, R.M. Noyes, J. Chem. Phys. **60**, 1877 (1974)
5. R.J. Field, R.M. Noyes, J. Am. Chem. Soc. **96**, 2001 (1974)
6. G. Nicolis, I. Prigogine, *Self-Organization in nonequilibrium systems* (Wiley, New York, 27, 1977)
7. E.J. Reusser, R.J. Field, J. Am. Chem. Soc. **101**, 1063 (1979)
8. J. Lindsay *Oscillations and traveling waves in chemical systems* (Eds. R.J. Field, M. Burger, Wiley, New York, 1985)
9. J.D. Murray, *Mathematical biology* (Springer-Verlag, Berlin, 1990)
10. S.K. Scott, *Chemical chaos, 1st ed.* (Oxford University Press, Oxford, 1991)
11. I.R. Epstein, K. Showalter, J. Phys. Chem. **100**, 13132 (1996)
12. I.R. Epstein, J.A. Pojman, *An introduction to nonlinear chemical dynamics: oscillations, waves, patterns, and chaos* (Oxford University Press, New York, 1998)
13. T. Ishiwatari, M. Kawagishi, M. Mitsuishi, J. Polym. Sci.: Polym. Chem. **22**, 2699 (1984)
14. R. Yoshida, T. Takahashi, T. Yamaguchi, H. Ichijo, J. Am. Chem. Soc. **118**, 5134 (1996)
15. Y. Hara, Y. Takenaka, Sens. **14**, 1497 (2014)
16. Y. Hara, R. Yoshida, J. Phys. Chem. B **109**, 9451 (2005)
17. Y. Hara, R. Yoshida, J. Chem. Phys. **128**, 224904, (2008)
18. Y. Hara, R. Yoshida, J. Phys. Chem. B **112**, 8427 (2008)
19. Y. Hara, R. Yoshida, Macromol. Rapid Commun. **30**, 1656 (2009)
20. Y. Hara, R.A. Jahan, Polym. **3**, 330 (2011)
21. Y. Hara, H. Mayama, Y. Yamaguchi, Y. Takenaka, R. Fukuda, J. Phys. Chem. B **117**, 14351 (2013)
22. Y. Hara, Chemosens. **1**, 3 (2013)
23. Y. Hara, Adv. Mater. Res. **941**,1208 (2014)
24. Y. Hara, Adv. Mater. Res. **941**, 1212 (2014)

The theoretical interpretation of the Linear/circularly polarized light-induced molecular orientation of the azo group-containing trans-Achiral Schiff base dinuclear complex composite material

Nobumitsu Sunaga[1, a], Takashiro Akitsu1[1, b], Taro Konomi[2] and Masahiro Katoh[2]

[1] Tokyo University of Science, Faculty of Science, Department of Chemistry, 162-8601 1-3 Kagurazaka, Shinjuku-ku, Tokyo, Japan
[2] Institute for Molecular Science, UVSOR, 444-8585 38 Nishigonaka, Myodaiji, Okazaki, Aichi, Japan
[a] Author to whom correspondence should be addressed; E-Mail: 1317707@ed.tus.ac.jp; Tel.: +81-3-5228-8271.
[b] Author to whom correspondence should be addressed; E-Mail: akitsu@rs.kagu.tus.ac.jp; Tel.: +81-3-5228-8271; Fax: +81-3-5261-4631.

Abstract. We have investigated linearly polarized UV and circularly polarized UV light-induced molecular orientation, namely increasing optical anisotropy of newly synthesized AZ containing chiral Schiff base Ni (II), Cu (II) and Zn (II) complexes, as organic/inorganic hybrid materials in polymethylmethacrylate (PMMA) cast films. The driving force and reason for polarized UV light-induced molecular orientation of AZ is so-called Weigert effect. The increase of the dichroic was measured with polarized absorption electronic spectra. In addition, we study on newly synthesized composites of AZ containing chiral Schiff base metal complexes with AZ to irradiate with not only linearly but also circularly polarized UV light (Wavelength selective (260, 318(π - π * of AZ) and 380 nm) and continuous wavelength). In addition, we comparison between simulated by TD-DFT and experimental spectra of electronic and CD spectra, and we calculated dipole moments by TD-DFT. Furthermore, we are using the CD spectrum after circularly polarized light UV irradiation, tried to observation change of supramolecular chiral order of the complexes.

1 Introduction

In our previous study, by using typical Mn12 single-molecule magnet[1] and AZ in PMMA cast films, we could observe switching of AC susceptibility in association with photoisomerization of AZ at 3 K .[2] C1s and O1s peaks of PMMA could be observed in XPS of the hybrid material, while Mn2p1/2,3/2 peaks, those were shown as bulk Mn12, could not be observed at the surface of PMMA. Which suggests that all components are contained inside the PMMA films (not surface) .[3] Therefore, we employed polarized electronic spectroscopy to measure aggregation and molecular orientation of the components of hybrid materials. We have investigated polarized electronic spectra of hybrid material of Mn12 and AZ in PMMA cast film after alternate UV and visible light irradiation for 3 min π - π * and n- π * bands suggested increasing of optical anisotropy accompanying with reversible photoisomerizaion of AZ, which may have resulted in previously reported reversible increasing and decreasing of ac magnetic susceptibility. Furthermore, increasing optical anisotropy could be also observed for hybrid materials composed of chiral trans-type Schiff base Co(II), Ni(II), Cu(II), and Zn(II) complexes and AZ in PMMA cast films after alternate irradiation of polarized UV and visible light .[4] In order to elucidate intermolecular interactions by chiral recognition between components effects of conformation of ligands, we prepared and measured hybrid materials composed of enantiomers or diasteromers of Schiff base Mn(III) complexes and AZ in PMMA cast films. The results suggested that molecular recognition and intermolecular interactions depend on matrix in which Mn(III) complexes are dispersed or co-existing solutes in the matrix .[5] Organic/inorganic hybrid materials composed of photochromic dyes such as AZ and inorganic coordination compounds have been studied widely as a key technology for optical (magnetic) data recording. One of the ad-vantages of organic/inorganic hybrid materials may be facility of designing supramolecular multifunctional materials, for example metal complexes incorporating azo-ligands with large dichromic ratios.[6-8]

We also focused on photo-tuning of optical anisotropy of AZ, namely polarized UV light-induced alignment caused by selected photoisomerization and molecular reorientation of AZ (Weigert effect)[9-11]. We measured a CD spectrum each of organic/inorganic hybrid materials and invested photo-induced dichroism.[12-14] In addition, we discussed the relations between degree of molecular orientation and electric transition dipole moments.

Recently we have investigated linearly polarized UV light induced molecular orientation of chiral Schiff base nickel(II), copper(II) and zinc(II) diunclear complexes in

[a,b] Corresponding author: 1317707@ed.tus.ac.jp, akitsu@rs.kagu.tus.ac.jp

organic/inorganic hybrid materials containing azo-compounds in a PMMA matrix. Weigert effect makes it parallel orientation of solutes and which was observed optical dichroism with polarized spectroscopy and was discussed with TD-DFT.[15-21] However, we don't deeply understand the research still organic/inorganic hybrid materials of AZ chiral Schiff base metal complexes for emitting circularly polarized UV light.

Here, we synthesized AZ containing achiral Schiff base Ni(II), Cu(II) and Zn(II) diunclear complexes (Fig. 1). So, we produced the organic/inorganic hybrid materials by placing the complexes in PMMA film. In addition, we study on newly synthesized composites of achiral Schiff base metal complexes with AZ to irradiate with not only linearly but also circularly polarized UV light (Wave-length selective (280 nm, 318 nm (π - π * of AZ) and 380 nm) and continuous wavelength), and examine the di-pole moments and different molecular orientation induced. These complexes effectively contain an azo group to cause the Weigert effect, and hydroxyl group is introduced to enhance the intermolecular interactions PMMA. Furthermore, we attempted to observe conformational changes of metal complexes and supramolecular chiral molecular arrangement for the system by means of CD spectra before and after circularly polarized UV light irradiation.

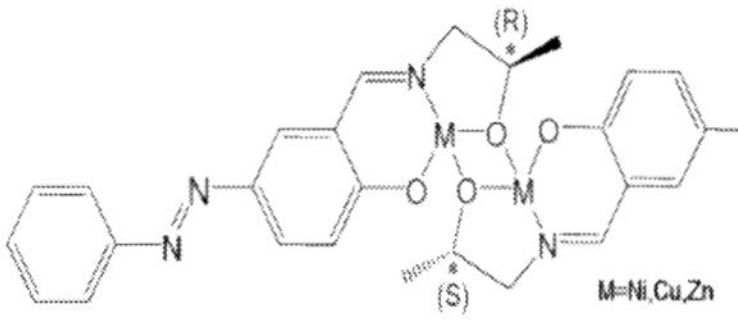

Figure 1. Structures of chiral Schiff base Ni(II)(**1**), Cu(II)(**2**), and Zn(II)(**3**) diunclear complexes.

2 Experimental

2.1 General procedure

Chemicals of the highest commercial grade available (solvents from Kanto Chemical, organic compounds from Tokyo Chemical Industry and metal sources from Wako) were used as received with-out further purification.

2.2 Preparation of the organic/inorganic hybrid materials composes

We synthesized AZ containing achiral Schiff base Ni(II), Cu(II) and Zn(II) diunclear complexes to prepare organic/inorganic hybrid materials composes. Then Acetone solution (1 mL) of **1**-**3** and acetone solution (2 mL) of PMMA (10 %) were cast onto a slide glass to give rise to PMMA film of hybrid materials **1**-**3**+PMMA.

*2.2.1 Preparation of (di- μ -oxo-bis[2-[[(2-hydroxypropyl)imino- κ N]methyl]-4-2-phenyldiazenylphenolato- κ O]2-)nickel(II) (**1**)*

aldehyde (0.5651 g,2.5 mmol) dissolved in methanol (60mL), 1-amino-2-propanol (0.1878 g, 2.5 mmol) was added dropwise and stirred at 373 K for 0.5 h to give orange solution of ligand. Nickel(II) acetate tetrahydrate (0.6822 g 2.5 mmol) was added to the resulting solution to give brown solution of the complex. After stirring for 2h, this crude brown compound was filtered. Yield 0.2068 g (12%). Anal. Calc. for C32H30Ni2N6O2: C, 50.92; H, 4.43; N, 9.73. Found: C, 50.52; H, 4.45; N, 10.36%. IR (KBr (cm-1)): 668, 1117, 1313, 1539, 1603 (N=N), 1652 (C=N), 2922, 3440.

*2.2.2 Preparation of (di- μ -oxo-bis[2-[[(2-hydroxypropyl)imino- κ N]methyl]-4-2-phenyldiazenylphenolato- κ O]2-)copper(II)(**2**)*

aldehyde (0.5651 g,2.5 mmol) dissolved in methanol (60mL), 1-amino-2-propanol (0.1878 g, 2.5 mmol) was added dropwise and stirred at 373 K for 0.5 h to give orange solution of ligand. Copper(II) acetate dihydrate (0.4993 g 2.5 mmol) was added to the resulting solution to give green solution of the complex. After stirring for 2h, this crude green compound was filtered. Yield 0.1053 g (15%). Anal. Calc. for C32H30Cu2N6O2; C, 54.96; H, 3.83; N, 11.7. Found: C, 55.73; H, 4.38; N, 12.12%. IR (KBr (cm-1)): 668, 1112, 1386, 1558, 1607 (N=N), 1635 (C=N), 2923, 3446.

*2.2.3 Preparation of (di- μ -oxo-bis[2-[[(2-hydroxypropyl)imino- κ N]methyl]-4-2-phenyldiazenylphenolato- κ O]2-)Zinc(II)(**3**)*

aldehyde (0.5651 g,2.5 mmol) dissolved in methanol (60mL), 1-amino-2-propanol (0.1878 g, 2.5 mmol) was added dropwise and stirred at 373 K for 0.5 h to give orange solution of ligand. Zinc(II) acetate dihydrate (0.54875g 2.5 mmol) and sodium hydrogen carbonate ware added to the resulting solution to give orange solution of the complex. After stirring for 2h, this crude orange compound was filtered. Yield 0.0717g (10%). Anal. Calc. for C32H30N6O4Zn2 ; C, 57.41; H, 4.31; N, 12.12. Found: C, 57.45; H, 4.36; N, 12.12%. IR (KBr (cm-1)): 688, 1114, 1390, 1472, 1593 (N=N), 1635 (C=N), 2965, 3445.

2.3 Physical measurement

Elemental analyses were carried out with a Perkin-Elmer 2400 II CHNS/O analyzer at Tokyo University of Sience.IR spectra were recorded on a JASCO FT-IR 4200 plus spectrophotometer. UV-vis absorption spectra were recorded on a JASCO V-570 or V-570 UV/VIS/NIR spectrophotometer equipped with a polarizer at 298K. CD spectra were recorded on s JASCO J-820 spectrophotometer at 298K.Photo-irradiation was carried out with D2 light source for 200-350nm (with a visible cut filter) and Xe light source 350-800nm (with a UV cut filter) with a polarizer. Photo-illumination were carried out using a lamp with optical filters (UV λ = 200-400 nm) leading to a sample by using optical fibers, quarter wave plate (SIGMA KOKI) and polarizer through optical filters.

2.4 Physical measurement in UVSOR

Elemental analyses were carried out with a Perkin-Elmer 2400 II CHNS/O analyzer at Tokyo University of Sience.IR spectra were recorded on a JASCO FT-IR 4200 plus spectrophotometer. UV-vis absorption spectra were recorded on a JASCO V-570 or V-570 UV/VIS/NIR spectrophotometer equipped with a polarizer at 298K. CD spectra were recorded on s JASCO J-820 spectrophotometer at 298K.Photo-irradiation was carried out with D2 light source for 200-350nm (with a visible cut filter) and Xe light source 350-800nm (with a UV cut filter) with a polarizer. Photoillumination were carried out using a lamp with optical filters (UV λ = 200-400 nm) leading to a sample by using optical fibers, quarter wave plate (SIGMA KOKI) and polarizer through optical filters.

2.5 Computational methods

All calculations were perfumed using the Gaussian 09W software Revision D.10 (Gaussian, Inc.).[15] The gas phase geometry optimizations were carried out using TD-DFT with B3LYP functional. The vertical excitation energy was calculated with the Lanl2dz for Ni and Cu and with the 6-31+G (d) basis set for H, C, N, O, and method based on the singlet ground state geometry.

3 Results and Discussion

3.1. Electronic and CD spectra of complexes

Fig.2. shows comparison between simulated and experimental spectra of electronic and CD spectra to confirm reliable band assignment. In the case of experimental electronic spectra, d-d bands appear at 689 nm and 684 nm for achiral Ni(II) and Cu(II) diunclear complexes respectively. Also, π - π * bands appear at 285 nm, 282 nm and 286 nm and n- π * bands appear at 369 nm, 342 nm and 352 nm for achiral Ni(II), Cu(II) and Zn(II) diunclear complexes, respectively. In the case of simulated spectra, d-d bands appear at 692 nm and 701 nm and transition electric dipole moment is evaluated to be (-0.0836, -0.3985, -0.6970) and (-1.0799, 0.3828, -0.0029) with oscillator strength of 0.8072 D and 1.1458 D for Ni(II) and Cu(II) complexes, respectively. Also, π - π * bands appear at 385 nm, 392 nm and 381 nm and n- π * bands appear at 469 nm, 442 nm and 453 nm for achiral Ni(II), Cu(II) and Zn(II) diunclear complexes, respectively. Moreover, n- π * and π - π * bands appear at 486 nm and 329 nm and transition electric dipole moment is evaluated to be (0, 0, 0) and (-0.1230, -2.8902, 0) with oscillator strength of 0.0000 and 0.7726 for the trans-form of AZ. The corresponding n- π * and π - π * bands appear at 482 nm and 301 nm and transition electric dipole moment is evaluated to be (-0.7441, 0, -0.2159) and (0.7979, 0, 0.2324) with oscillator strength of 0.0378 and 0.0696 for the cis-form of AZ.[24] According to preceding studies on the related compounds, the optimized structures and assignments of transitions for **1**, **2** and **3** are reasonable.

Selected predominant peaks are assigned to transitions between molecular orbitals: for **1**, 662 nm (d-d band)(HOMO (166) to LUMO+2 (169)), 385 nm (π - π * band)(HOMO-1 (154) to LUMO+1 (168)), and 469 nm (n- π * band)(HOMO-11 (150) to LUMO+5 (172)); for **2**, 701 nm (d-d band)(HOMO-14 (153) to LUMO (168)), 392nm (π - π * band)(HOMO-8 (159) to LUMO (162)), and 442 nm (n- π * band)(HOMO (167) to LUMO+5 (172)); for **3**, 299 nm (π - π * band)(HOMO-5 (155) to LUMO (161)), 381 nm (n- π * band) (HOMO (160) to LUMO+1 (162)).

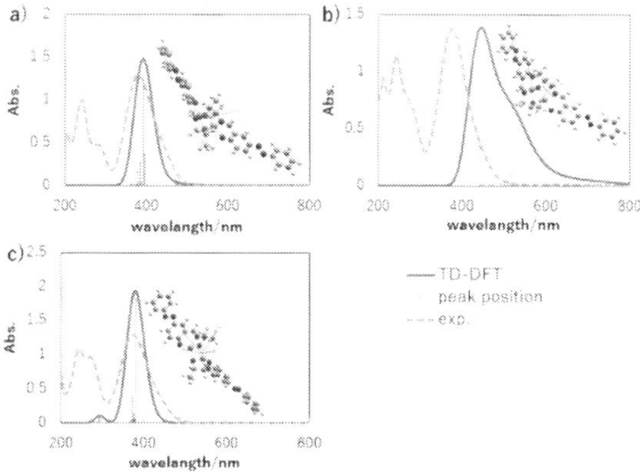

Figure 2. Simulated (blue(TD-DFT)) and experimental (grey) electronic spectra of **1** (**a**), **2** (**b**) and **3** (**c**) with optimized structures.

3.2 Polarized absorption spectra of azo-containing hybrid materials after linearly polarized UV light irradiation

Fig. 3 shows polarized absorption spectra and circular diagrams of angular dependence of absorbance of π - π * band at 280 nm and n- π * band at 380 nm of AZ group and d-d band at 689 nm of 1 for **1**+PMMA. The decrease in absorbance of the π - π * and n- π * at initial is due to photoisomerization. Optical anisotropy increased as time proceed. The circular diagrams suggest that reorientation of AZ effects on relatively rich optical anisotropy observed by π - π * and n- π * bands. Polarized UV light irradiation induced optical anisotropy with de-creasing intensity of π - π * band significantly. Fig. 4 shows polarized absorption spectra and circular diagrams of angular dependence of absorbance of π - π * band at 280 nm and n- π * band at 380 nm of AZ and d-d band at 684 nm of 2 for **2**+PMMA. Initial change of π - π * band feature is also observed because of AZ occurs cis-trans photoisomerization. Fig. 5 shows polarized absorption spectra and circular diagrams of angular dependence of absorbance of π - π * band at 280 nm and n- π * band at 380 nm of AZ. The order of enhancing dichroism is **2**+PMMA>**3**+PMMA>**1**+AZ+PMMA.

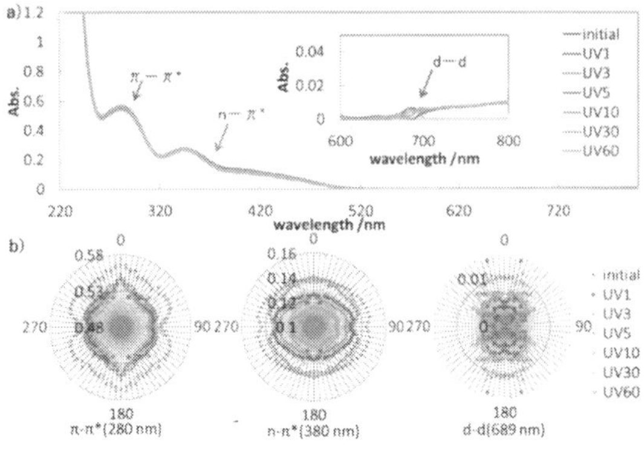

Figure 3. (**a**) Polarized absorption spectra of 1+PMMA with 0 degree showing predominant peaks before and after UV light irradiation for 60 min. (**b**) Circular diagram of angular dependence of absorbance of the predominant peaks polarized absorption spectra of 1+PMMA after UV light irradiation for 0,

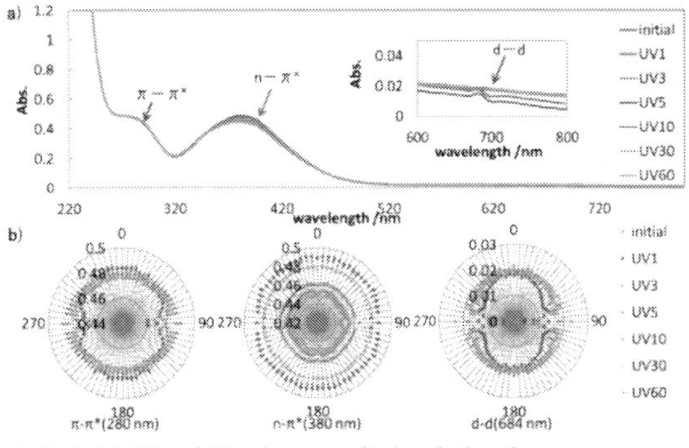

1, 3, 5, 10, 30 and 60 min respectively ads. bands.

Figure 4. (**a**) Polarized absorption spectra of 2+PMMA with 0 degree showing predominant peaks before and after UV light irradiation for 60 min. (**b**) Circular diagram of angular dependence of absorbance of the predominant peaks polarized absorption spectra of 2+PMMA after UV light irradiation for 0, 1, 3, 5, 10, 30 and 60 min respectively ads. bands.

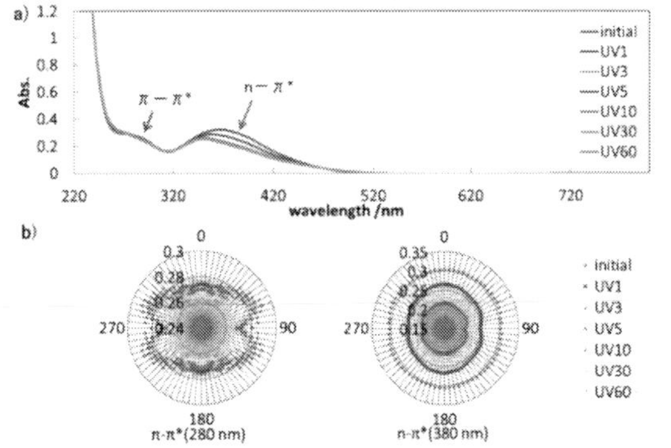

Figure 5. (**a**) Polarized absorption spectra of 3+PMMA with 0 degree showing predominant peaks before and after UV light irradiation for 60 min. (**b**) Circular diagram of angular dependence of absorbance of the predominant peaks polarized absorption spectra of 3+PMMA after UV light irradiation for 0, 1, 3, 5, 10, 30 and 60 min respectively ads. bands.

3.3 Parameters for optical anisotropy

The degree of photoinduced optical anisotropy (spectral dichroism) of absorption electronic spectra can be described commonly by these two parameters:

$$S = (Aparrallel - Aperpendicular) / (2Aperpendicular + Aparrallel) \quad (1)$$

$$R = Aperpendicular / Aparrallel \quad (2)$$

Where Aperpendicular and Aparrallel values denote absorbance measured with the measuring polarizers perpendicular or parallel to electric vector of irradiation polarized light. Ideal isotropic systems of S = 0 and R = 1 and both S and R parameters are changed as increasing dichroism by molecular alignment. General treatment has been explained in our previous articles for cis-trans isomerizaiton as well as Weigert effect. The S values are summarized in Tables respectively. Hereafter we discuss them in a qualitative level.

Table 1. The S values of anisotropic parameters polarized absorption spectra for **1**, **2** or **3**+PMMA after linearly polarized UV light irradiation.

UV irradiation (min.)		initial	1	3	5	10	30
1+PMMA	280 nm, π-π* (AZ)	0	-0.00069	0.00886	0.000206	0.002015	0.005864
	380 nm, n-π* (AZ)	0	-0.00277	0.023518	0.01895	0.009389	0.034243
	689 nm, n-π* (1)	0	-0.09636	-0.52621	-0.55414	-0.49733	-0.16811
2+PMMA	280 nm, π-π* (AZ)	0	0.00535	-0.00899	-0.01197	-0.01036	-0.01004
	380 nm, n-π* (AZ)	0	0.002101	-0.0051	-0.00453	-0.00546	-0.00589
	684 nm, n-π* (2)	0	0.019426	-0.10287	-0.04259	-0.02426	-0.04989
3+PMMA	280 nm, π-π* (AZ)	0	0.008322	0.009576	0.00836	0.013522	0.00445
	380 nm, n-π* (AZ)	0	0.018025	0.008556	0.012543	0.010613	0.009471
	No d-d band						

3.4 Discuss about linear polarized light irradiation

For the change of the absorption spectrum of after linearly polarized UV light irradiation, knowledge regarding molecular orientation will be obtained conformation, from S value of π - π * band, a change in the coordination structure from d-d band. For the azo-containing systems (**1**, **2**, or **3** +PMMA), azo-containing for complexes can be influenced by Weigert effect of AZ. Metal ions providing rigid coordination environment are effective to increase molecular orientation caused by Weigert effect.[22-24] According to Table 1, we can derive the following information associated with molecular orientations with intermolecular interactions in these organic/inorganic hybrid materials. The order of the effect, **2** +PMMA>**3** +PMMA>**1** +PMMA for π - π * bands and n- π * bands, **2** +PMMA> **1** +PMMA for d-d bands Result in the smallest degree of increasing optical anisotropy in both of n- π * and π - π * accompanying with Weigert effect of AZ. Reducing direct transmission of molecular alignment from AZ by rigidity of metal complexes due to Zn(II) ions is suggested.[23, 25-30] The degree of increasing optical anisotropy accompanying with Weigert effect of AZ depends on not shapes of chiral ligands but central Ni(II), Cu(II) or Zn(II) metal ions.

3.5 Discussion about CD spectra after circularly polarized UV light irradiation

Figure 6-8 show the CD spectra of **1**, **2**, or **3**+PMMA, after wavelength selective circularly polarized UV light irradiation 5min. Figure 6. shows the CD spectra of 1+PMMA around 380 nm and 280 nm exhibited changes before and after wavelength selective circularly polarized UV light irradiation (280 nm, 318 nm and 380nm) for **1**+PMMA. In the 280 nm UV irradiation, was induced CD is the largest in two steps of 3, 8 min. The 318 nm UV irradiation, significant induced CD could not be confirmed. In the 380 nm UV irradiation, induced CD became the largest in 8 min. In Figure 7. shows the CD spectra of **2**+PMMA, after the 280 nm UV irradiation, induced CD also increased in proportion to the irradiation time. In the 318 nm UV irradiation, induced CD be-came the largest in the irradiated for 3 min. In the 380 nm UV irradiation, was induced CD is the largest in two steps of 3, 8 min. Figure 8. shows the CD spectra of **3**+PMMA around 380 nm and 280 nm exhibited changes before and after wavelength selective circularly polarized UV light irradiation (260 nm, 318 nm and 380nm) for 3+PMMA. In the 280 nm UV irradiation, induced CD became the largest in two stages for 8 min. In the 318 nm UV irradiation, significant induced CD could not be confirmed. In the 380 nm UV irradiation, it was confirmed that similar induced CD in any irradiation time.

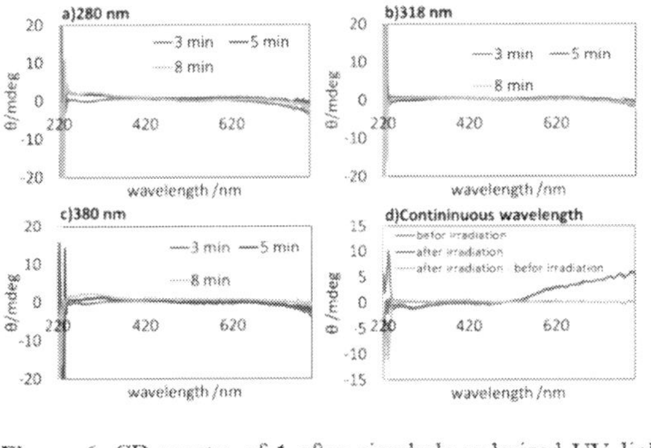

Figure 6. CD spectra of **1** after circularly polarized UV light irradiation with respectively 260, 318, 380 nm and continuous.

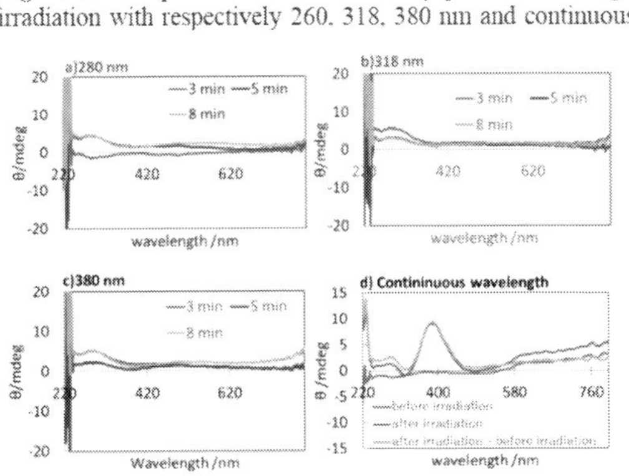

Figure 7. CD spectra of **2** after circularly polarized UV light irradiation with respectively 260, 318, 380 nm and continuous.

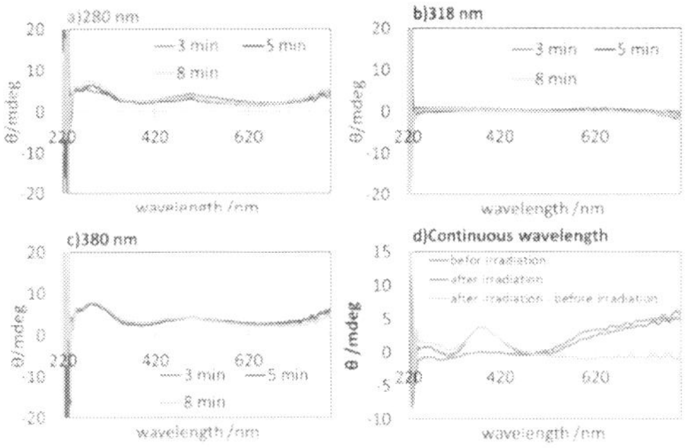

Figure 8. CD spectra of **3** after circularly polarized UV light irradiation with respectively 260, 318, 380 nm and continuous.

Continuous wavelength circularly polarized UV light irradiation, the difference between the CD spectra be-fore and after irradiation of circular polarization, is that peak was observed at around 400nm. This peak was different from peak of the CD spectra due to the chiral complex molecule attributed in TD-DFT calculation. It is considered from this fact, that peak emerging from it is the peak of supramolecular chiral (helical) order and is not derived from the complex itself. There was change in the peak position in the chiral and achiral, and chiral strength had become stronger. On the other hand, using a light source having a narrow wavelength width in UVSOR (Wavelength selective (280, 318(π - π * of AZ) and 380 nm), induced chiral order is caused absorption wavelength selective. Changes in the observed shape of the CD spectra, the induction of supramolecular chirality or conformational change of the molecule can be considered as the cause. Changes in the CD spectra of this time, appeared only change in absorption intensity with increasing the irradiation time, which suggested that molecules of supramolecular chirality induction by Weigert effect of AZ by circularly polarized light irradiation, rather than the changes in molecular shape. In addition, induced chiral order is caused absorption wavelength selective, which suggests that it is a new knowledge of the radiation source only. Peak shift of appearing by changing the wavelength selective is seen. There was change in the peak position in the chiral and achiral, and chiral strength had become stronger. This is suggested that it is something that those which have a chiral place is easily oriented supramolecular chiral (helical) order is well. The order of this effect is **2**+PMMA>**3**+PMMA>**1**+PMMA, and these results suggests that optical anisotropy depends not only on flexibility of coordination environment but also coordination geometries.

3.6 Discussion about Theoretical interpretation of the optical rotation strength using coupled oscillator model

For interpretation of supramolecular chirality, we attempt to discuss spatial relationship between AZ molecules assuming dipole-dipole interaction. For this reason, we will apply coupled oscillator model hereaf-

DOI: 10.1051/matecconf/201713007004

ter.[31-40] If only the Coulomb interaction of the chromo-spheres A between B two can be considered, we assume system dipole interactions between paired the two ones. Now, the Hamiltonian of the system is represented by the equation (3).

$$H = H_A(r_A) + H_B(r_B) + V_{AB} \quad (3)$$

Where Hamiltonian *HA* (*rA*) or *HB* (*rB*) is of the chromophores A or B, *VAB* is Coulomb interaction potential between the chromophore A and B. By the wave function of the ground states and excited states of the chromophores A or B, to introduce electric transition dipole moment, optical rotation strength of the formula is derived as the equation (4).

$$R_{0a} = -\left(\frac{2\pi v_a v_b}{hc(v_a^2 - v_b^2)}\right)(\mu_{0a} \cdot T \cdot \mu_{0b})(R_{AB} \times \mu_{0a} \cdot \mu_{0b}) \quad (4)$$

Where Va or Vb are absorption frequency, μ 0a or μ 0b is electric transition dipole moment, RAB is a distance of A and B, T is parameters. To simplify this, con-sider that both chromospheres are identical, and hereafter assume connecting both two, is orthogonal (Figure 9).

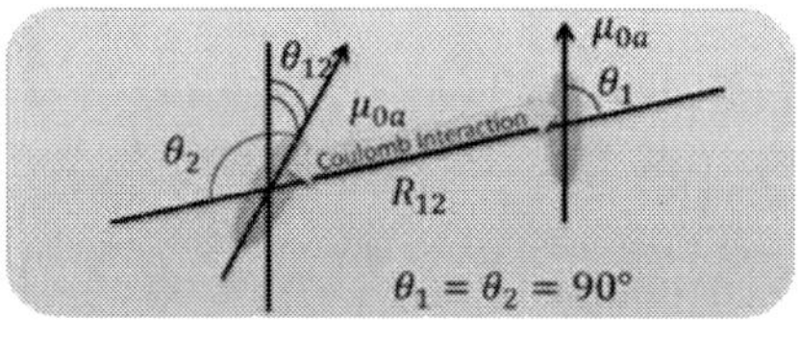

Figure 9. Coulomb interaction between the same chromospheres

Now, from equation (3), the optical rotation strength becomes the following equation (4). Furthermore, we considered as follows. The intensity of CD bands can be represented by optical rotation strength R. From the CD spectra measured, R is deter-mined according to the following equation (6).

$$R_{\pm} = \mp\left(\pi v_0 / 2c\right) R_{12} \mu_{0a}^{2} \sin\theta_{12} \quad (5)$$

$$R = 2.296 \times 10^{-39} \int_0^{\infty} \left\{\Delta\varepsilon(\sigma)/\sigma\right\} d\sigma$$

$$\approx \left(2.296 \times 10^{-39}/\sigma_0\right) \int_0^{\infty} \Delta\varepsilon(\sigma) d\sigma \ \text{(c.g.s)} \quad (6)$$

Where σ is the wave number and is the σ_0 wavenumber of the peaks (the maximum) of the CD spectra. As relational expressions, we calculated from equation (6) optical rotation strength of the induced CD spectra observed. Substituting equation (5), the degree of inclination (θ) of the dye molecules of interest was evaluated.

Table 4. The angle of optical rotation and strength to transition electric dipole moment of each complex in the π - π * or n- π * transition states. For simplicity, R12 was assumed to be 0.5 nm all.

	1+PMMA	2+PMMA	3+PMMA
R (n-π*) /×10^{-42} c.g.s	1.64	2.26	5.7116
θ / ×10^{-40} deg	38.8	60.7	3.1234
μ_{0a}(n-π*) /Debye	3.49503972	3.657174	22.55621
R (n-π*) / ×10^{-42} c.g.s	1.58	1.85	1.57291
θ / ×10^{-40} deg	121	174	215.239
μ_{0a}(π-π*) /Debye	1.89947416	1.562521	1.417491
R (π-π*) / ×10^{-42} c.g.s	2.76	10.5	-
θ / ×10^{-40} deg	457	15200	-
μ_{0a}(π-π*) /Debye	0.0054	0.188345	-

The calculated results indicated that transition di-poles of AZ resulting in large electric dipole transition of n- π * and π - π * transitions are significantly inclined each other, which was attributed to mutual molecular orientation induced by circularly polarized UV light irradiation. In addition, comparison all **1**, **2**, or **3** +PMMA suggested chiral molecular orientation derived from AZ group in complex. Respect to the circularly polarized light irradiation, **2**+PMMA was considered a large light-responsive overall. Further, from the comparison of n- π * transition corresponding to 380 nm circularly polarized light irradiation, **2**+PMMA is suggested to have a relatively large photo responsive. From the comparison of theoretical optical rotation strength, optical response of the magnitude Respect to circularly polarized light ultraviolet light was found to be **2**+PMMA> **3**+PMMA> **1**+PMMA.

4 Conclusion

In summary, we have prepared organic/inorganic hybrid materials of azo-containing achiral Schiff base Ni(II), Cu(II), and Zn(II) complexes in PMMA. Linearly polarized UV light irradiation resulted in increasing optical anisotropy, which was observed with polarized UV-vis spectra (overlapped bands) of π - π *, n- π * and d-d bands. We discussed the assignment and the origin of CD bands of hybrid materials of chiral metal complexes and AZ after irradiation of polarized light by means of TD-DFT calculations. In the PMMA film, the order of this effect is **2**+PMMA>**3**+PMMA>**1**+PMMA. The results suggested that differences in optical anisotropy is ascribed to intermolecular interaction with AZ groups due to flexibility of coordination environment of metal centres of the complexes. Circularly polarized UV light irradiation resulted in increasing intensity of CD bands around 380 nm, which suggested induced chiral molecular orientation. Continuous wavelength circularly polarized UV light irradiation resulted in increasing intensity of CD bands, which suggested induced chiral molecular orientation according to calculated dipole moment. On the other hand, using a light source having a narrow wavelength width in UVSOR (Wavelength selective (280, 318(π - π * of AZ) and 380 nm), induced chiral order is caused absorption wavelength selective, which suggests that it is a new knowledge of the radiated light. This effect is **2**+PMMA>**3**+PMMA>**1**+PMMA, and these results are suggesting that optical anisotropy depends not only on flexibility of coordination environment but also stable coordination geometries. Linearly polarized UV light irradiation resulted in increasing optical anisotropy. It is attributed to the arrangement (not magnitude) of transition electric dipole moments. Ni(II) is prefer a planar structure and Cu(II) is taking a flexible structure. but Zn(II) is prefer a tetrahedral structure for the rigid, which suggests that it is less susceptible to orientation by AZ. We discussed the assignment and the origin of CD bands of hybrid materials of chiral metal complexes and AZ after polarized light irradiation by TD-DFT calculations. As a result, Respect to the circularly polarized light irradiation, **2**+PMMA was considered a large light-responsive overall. Further, from the comparison of n- π * transition corresponding to 380 nm circularly polarized light irradiation, **2**+PMMA is suggested to have a relatively large photo responsive. From the comparison of theoretical optical rotation strength, optical response of the magnitude Respect to circularly polarized light ultraviolet light was found to be **2**+PMMA> **3**+PMMA> **1**+PMMA.

5 Acknowledgment

This work was supported by Research Foundation for Opto-Science and Technology. The authors thank Institute for Molecular Science for the use of UVSOR facility (Prof. Hiromasa. Kato and Dr. Taro. Konomi) and CD spectrometer (Mr. Seiji Makita). The Computations were performed using Research Centre for Computational Science, Okazaki, Japan.

6 References

1. O. A. Blackburn, B. J. Coe, J. Fielden, M. Helliwell, J. J. W. McDounall and M. G. Hutchings, Inorg. Chem., **49**, 9136(2010).
2. A. A. Khandar and Z. Revani, Polyhedron, **18**, 129(1998).
3. S. Leclair, L. Mathew, M. Giguere, S. Motallebi, and Y. Zhao, Macromolecules, **36**, 9024(2003).
4. Y. Einaga, M. Taguchi, G. Li, T. Akitsu, T. Sugai and O. Sato, Chem. Mater., **15**, 8(2003).
5. R. Mikami, M. Taguchi, K. Yamada, K. Suzuki, O. Sato and Y. Einaga, Angew. Chem. Int. Ed., **43**, 6135(2004).
6. T. Yamamoto, Y. Umemura, O. Sato and Y. Einaga, Chem. Mater., **16**, 1195(2004).
7. Y. Einaga, R. Mikami, T. Akitsu and G. Li, Thin Solid Films, **493**, 230(2005).
8. T. Akitsu and A. Yoshida, Curr. Phys. Chem., **1**, 76(2011).
9. T. Akitsu and Y. Miura, J. Chem. Chem Eng., **5**, 443(2011).
10. Y. Aritake, T. Takanashi, A. Yamazaki and T. Akitsu, Polyhedron, **30**, 886(2011).
11. T. Akitsu and T. Itoh, Polyhedron, 2007, **29**, 477(2010).
12. Y. Aritake, T. Takanashi, A. Yamazaki and T. Akitsu, Polyhedron, **30**, 886(2011).
13. Y. Aritake, T Akitsu, Polyhedron, **31**, 278(2012).
14. A. Yamazaki, T. Akitsu, RSC Advances, **2**, 2975(2012).
15. M. J. Frisch, G. W. Trucks, H. B. Schlegel, and D. J. Fox, et al., Gaussian, Inc., Wallingford CT, 2009.
16. S. Yamada, Coord. Chem. Rev., **537**, 190(1999).
17. A. A. Khandar and Z. Revani, Polyhedron, **18**, 129(1998).
18. T. Akitsu, Y. Einaga, Acta Cryst., **C60**, m640(2004).
19. T. Akitsu, Y. Einaga, Polyhedron, **25**, 1089(2006).
20. C. Evans, D. Luneau, J. Chem. Soc., Dalton Trans., 83(2002).
21. H. Sakiyama, H. Okawa, N. Matsumoto and S. Kida, J. Chem. Soc., Dalton Trans., **2935**(1990).
22. H. Sakiyama, H. Okawa, N. Matsumoto and S. Kida, Bull. Chem. Soc. Jpn., **64**, 2644(1991).
23. M. Ulusoy, H. Karabiyik, R. Kilincarslan, M. Aygun, B. Cetinkaya and S. Garcia-Granda, Struct. Chem., **19**, 749(2008).
24. R. S. Forgan , J. E. Davidson, S. G. Galbraith, D. K. Henderson, S. Parsons, P. A. Tasker and F. J. White, Chem.commun., **4049**(2008).
25. T. Akitsu and Y. Einaga, Acta. crystallogr., **E60**, m1555(2004).
26. T. Akitsu and Y. Einaga, Acta. crystallogr., **E60**, m640(2004).
27. S. Banerjee, A. K. Mukherjee, I. Banerjee, R. L. De Neumann, and L. Louer, Cryst.res.technol., **3**, 4815(2005).
28. P. C. Chia, D. P. Freyberg, G. M. Mockler, and E. Sinn, Inorg.chem., **16**, 254(1977).
29. J. M. Fernandez-G, C. Ausbun-Valdes and R. A. Toscano Gonzalez-Guerrero, Z.anorg.allg.chem., **633**, 1251(2007).
30. D. Hall, S. V. Sheat and TN T. Waters, J. chem.soc.A, 460(1968).
31. T. Akitsu and Y. Einaga, Acta. crystallogr., **E60**, m1605(2004).
32. T. Akitsu and C. Kominato, InTech, **515**(2013).
33. L. Nikolova, T. Todorov, M. Ivanov, F. Andruzzi, S. Hvilsted and P. S. Ramanujam, Opt. Mater., **8**, 255(1997).
34. G. Iftime, F. L. Labarthet, A. Natansohn, and Paul Rochon, J. Am. Chem. Soc., **122**, 12646(2000).
35. A. Natansohn and P. Rochon, AdV. Mater., **11**, 1387(1999).
36. S. Banerjee, A. K. Mukherjee, I. Banerjee, R. L. De Neumann, and L. Louer, Cryst.res.technol., **3**, 4815(2005).
37. D. Hall, S. V. Sheat and TN T. Waters, J. chem.soc.A, **460**(1968).
38. S. Matile, N. Berova, K. Nakanishi, J. Fleischhauer,R. W. Woody, J. Am. Chem. Soc., **118**, 5198 (1996).
39. D. J. Caldwell, H. Eyring, "Theory of Optical Activity", Wiley-Interscience, New York (1971).
40. N.Sunaga, S.Furuya, T.Akitsu, et al., Computational Chemistry: Theory, Methods and Applications, Nova Science Publishers, Inc., **85** (2014).

Single-component and fast-curing epoxy resin for liquid composite molding processes

Yiru Wang[1,2,a], Wangshuang Liu[2], Yiping Qiu[1] and Yi Wei[2]
1 Donghua University, College of Textiles, Shanghai 201620, China
2 Donghua University, Center for Civil Aviation Composites, Shanghai, 201620, China.

Abstract. Development of single-component and fast-curing epoxy resins is highly desired for many industry applications. In this work, we report an epoxy system based on diglycidyl ether of bisphenol A (DGEBA) and 1-(2-cyanoethyl)-2-ethyl-4- methylimidazole (1C2E4MIM). The inductive effect of electron-withdrawing cyano group distinctly increases the latency of 1C2E4MIM without sacrificing the curing rate. The results of differential scanning calorimeter (DSC) and dynamic mechanical analysis (DMA) measurements indicate DGEBA/1C2E4MIM epoxy system can be fully cured in 15 min. The rheological, thermal and mechanical properties of DGEBA/1C2E4MIM epoxy system were studied in detail. The results show that the shelf life of this epoxy system is more than 4 days at room temperature and more than 6 months at -18 °C. The cured epoxy resins show high glass transition temperature (>155 °C), tensile strength (>80 MPa) as well as excellent moist heat resistance. Finally, carbon fiber-reinforced polymer composites (CFRPs) were fabricated using this epoxy system as matrix via vacuum assisted resin infusion (VARI) process. The mechanical properties of CFRPs, including tensile, flexural, compressive and interlaminar shear properties, were investigated.

1 INTRODUCTION

Carbon fiber-reinforced polymer composites (CFRPs) are increasingly used in many applications that require high-performance light-weight structural materials. For example, CFRPs are regarded as an ideal light-weight substitute to mental in manufacturing electric car body. Likewise, in commercial aircraft industry, the content of advanced composites in Boeing 787 Dreamliner and Airbus A350 amazingly reaches over 50 wt%. For modern high-performance CFRPs, epoxy resins are the most widely utilized thermosetting matrix materials owing to their superior mechanical performance, dimensional stability, adhesion, heat and chemical resistance.[2-4] The traditional epoxy formulation for CFRPs commonly includes epoxy monomer and hardener (amines or andydrides). Owing to the limited shelf life, the two components have to be stored separately and mixed immediately just prior to use. Compared with two-component epoxy system, single-component epoxy resin can be conveniently operated without weighting and blending processes. In addition, epoxy curing often accounts for a large proportion of the manufacturing cycle of CFRPs, and so fast-curing epoxy system is desired to improve the production efficiency, which is particularly significant in automobile industry.[5,6]

In order to reduce the molding time, short curing time and low viscosity are active demand. Many fast curing epoxy resin systems were studied in past years, which can be roughly divided into five parts, thiol curing agent, amine curing agent, anhydride curing agent, cationic curing agent and anion curing agent. Thiol curing agent can not be used in liquid molding system due to low mechanical properties, and very short pot time, and most importantly low glass transition temperature, therefore they are generally used in adhesion or coating[7-9]. Anhydride curing agent has to take at least 30 min to complete the curing at 120 °C[10], which is too long to meet the industrial production line requirements. Except that, the curing temperature of anhydride is higher. Amine curing agent, cationic curing agent and anion curing agent are relatively suitable to liquid fast molding system due to their short curing time and better mechanical properties and thermo properties. Until now, all of epoxy curing agent used in automobile epoxy product line are Amine curing agents[11,12], which including aliphatic amines, alicyclic amines, and aromatic amines. Alicyclic amines and aromatic amines are much more expensive to aliphatic amines. Most of aromatic amines are solid, and have much higher viscosity when mixed with epoxy resin, which is disadvantage to liquid molding. Except that, aromatic amines have higher curing temperature. Aliphatic amine curing agent is currently used in liquid molding automobile industrial line, but there are still many challenges, the biggest challenge is the balance of viscosity and curing time.[13]

Imidazoles are a class of typical epoxy hardener. Only a catalytic amount of imidazoles can initiate the curing reaction through nucleophilic and anionic polymerization processes.[14,15] Due to the high cross-linking density and low content of polar groups, imidazole-cured epoxy resins have an excellent combination of high modulus, good heat and moisture resistance. In general, imidazoles have intrinsic high curing reactivity towards epoxy resins, and the curing behaviors can be facilely tailored by introducing various substituent groups on the imidazole rings. So far, a large amount of imidazole hardeners with different structures has been commercialized, which provides a generous toolbox to develop single-component and fast-curing epoxy system with reasonable shelf life.

In this work, 2E4MIM and 1C2E4MIM were compared as hardener of single-component and fast-curing epoxy resin for liquid composite molding processes.Furthermore, 1C2E4MIM was investigated in curing behaviour, glass transition temperature (T_g), tensile property and hot-wet properties. Finally, the mechanical properties of carbon fiber reinforced composite by 1C2E4MIM-epoxy resin was tested.

2 EXPERIMENTS

2.1 Materials

Diglycidyl ether of bisphenol-A (DGEBA) with an epoxy equivalent weight of 182~192 (NANYAEPOXY RESIN). 1-(2-Cyanoethyl)-2-ethyl-4-methylimidazole (1C2E4MIM, cas:23996-25-0) purity 99% from Huaweiruike company. 2-ethyl-4-methylimidazole (2E4MIM, cas:931-36-2) purity 99% from Huaweiruike company. T700 carbon fiber plain weave cloth (Zhongfu Shenying Carbon Fiber Co.,Ltd). Peel ply, infusion net, vacuum bag, seal salve, sealing tape, ventilated felt, pipe thread, guid pipe.

2.2 Sample preparing

2.2.1 Preparing of the resin system

Mix the hardener and epoxy resin by parts with speed 200 r/min at ambient temperature for 5 min. Then degas the mixture in vacuum drying oven for 15 min at 45 °C.

2.2.2 Preparing of the neat resin sample

Pure the disgassed mixture into the aluminum mold cured 20 min at 120 °C for level of 4 wt % 2E4MIM and 4 wt %, 6 wt % 1C2E4MIMIM as the hardener. And cured 40 min at 120 °C for level of 2 wt % 1C2E4MIM as hardener. The water absorption sample was obtained by putting the cured resin into deionized water for 80 days at 70 °C. The hot-wet property sample was obtained via putting the cured sample into a deionized water at 100 °C for 48 h.

2.2.3 Preparing of the carbon fiber composites

Lay up the plain wave carbon fiber cloth (30 cm ×30 cm) on a glass plate, put the infusion net (29 cm ×29 cm) on the clothes at the centre, the inlet and outlet pipe

 DOI: 10.1051/matecconf/201713007005

respectively in each side of the carbon fiber cloth. The ventilated felt was put on the outlet side. The outlet pipe was connected with vacuum pump, with a safety resin trap between them. The vacuum bag was put on the top sealed by sealing tape. Heating the degassed mixture to different temperature inject into the carbon fiber preform. Subsequently cured 30 min at 120 °C.

2.3 Measurements

2.3.1 Pot life

The viscosity was obtained by Dynamic Shear Rheometer (DSR) (Rheometer: MARS60) at 25 °C, 1.592 Hz.

2.3.2 Curing behaviour

The curing time was tested by DSR and Differential Scanning Calorimeter (DSC) (NETZSCH DSC214). Isothermal DSC studies were performed by rapidly ramping the sample from 0 °C to 120 °C, with the rate of 100 °C/min, holding for various periods of time. Dynamic scans at 10 °C/min over a temperature range from 25 °C to 220 °C was used to determine the total heat of reaction.

2.3.3 Mechanical properties

The Tg and hot-wet glass transition temperature (HW-Tg) were measured by dynamic mechanical analysis (DMA)(TA Q800) at a heating rate of 10 °C/min. The neat resin mechanical properties were tested according to ASTM by Instron 2360. The tensile properties and flexural properties were tested per ASTM D 638 and ASTM D 790. To evaluate the CFRPs mechanical properties, the T700 plain weave carbon fabric reinforced composite mechanical properties were tested according to ASTM. The casting time and mechanical properties at difference casting temperature have been measured.

3 RESULTS AND DISCUSSIONS

3.1 Comparison of 1C2E4MIM and 2E4MIM as hardener

3.1.1 Pot life of 1C2E4MIM and 2E4MIM as the hardener

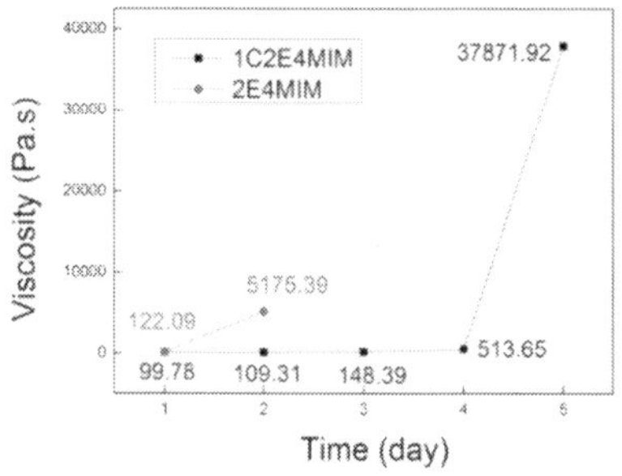

Figure 1. The viscosity change of 1C2E4MIM and 2E4MIM at 25 °C for five days.

The viscosity of 2E4MIM climbed steeply in the next day, while the 1C2E4MIM increased slightly (figure 1). The 2E4MIM was cured at the third day. The viscosity of 1C2E4MIM climbed much slower than 2E4MIM, which indicate that the thermal latency of 1C2E4MIM was better than 2E4MIM own to the cyano group not only has steric effect, but also has coordination effect with imidazole ring.

3.1.2 The DMA result of 2E4MIM and 1C2E4MIM as the hardener

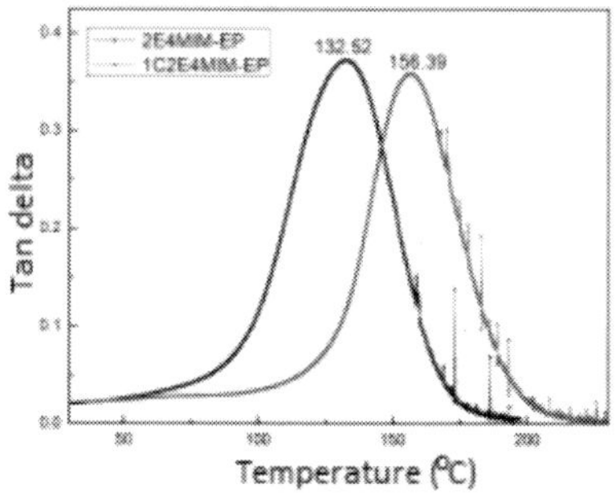

Figure 2. The T_g of 2E4MIM and 1C2E4MIM as hardener individually at same content on 4%.

1C2E4MIM had higher T_g than 2E4MIM which could attribute to the cyano group reacted with epoxy group, therefore increased the cross-linking density (figure 2).

3.1.3 Flexural properties of 2E4MIM and 1C2E4MIM as the hardener

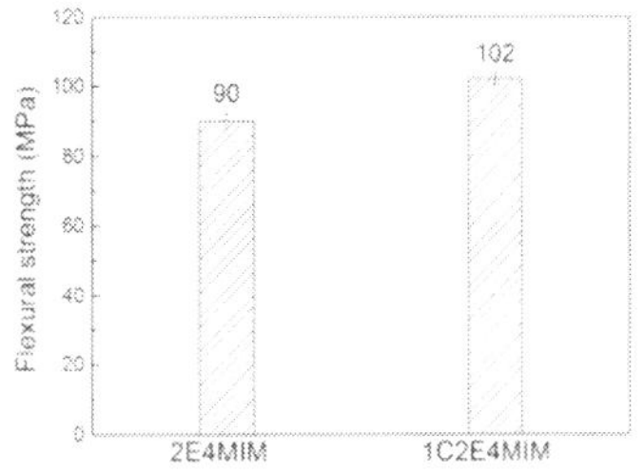

Figure 3. The flexural properties of 2E4MIM and 1C2E4MIM as hardener individually at same content on 4 wt %.

The 1C2E4MIM has higher flexural strength **compares** to 2E4MIM. The higher flexural strength of 1C2E4MIM also own to higher cross-linking of presence of cyano group (figure 3).

Considering the pot life, T_g and bending properties, the 1C2E4MIM has better performance as a RTM resin of fiber reinforced composite. The properties of 1C2E4MIM as hardener were investigated further.

3.2 Comparison of different 1C2E4MIM level

3.2.1 Curing behaviour of different 1C2E4MIM level

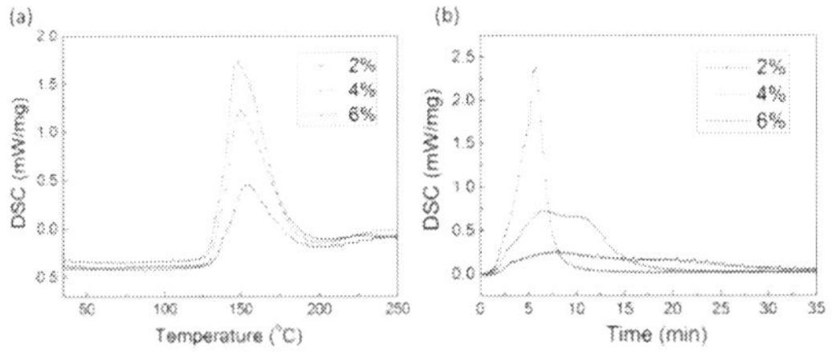

Figure 4. a. DSC curves of the DGEBA/1C2E4MIM system with the different level of 1C2E4MIM at 2 wt %, 4 wt %, 6 wt %: a. on heating rate of 10 °C/min, b. on constant temperature on 120 °C.

Table 1. Curing parameters for epoxy compositions on different level of 1C2E4MIM .

1C2E4MIM content (pbw)	**T_i (°C)**	**T_{max} (°C)**	**ΔH (J/g)**	**Curing time at 120 °C (min)**	**T_g (°C) by DMA**
2%	135.1	155.1	128.8	27	71.61
4%	135.0	149.4	267.5	15	156.39
6%	134.3	147.5	334.4	8	165.06

The increasing of 1C2E4MIM level have no impact on initial curing temperature, but significant impact on heat release and curing time (table 1). The increase of 1C2E4MIM level sharply increased the heat release and decreased curing time (figure 4).

3.2.2 The DMA result of different 1C2E4MIM level

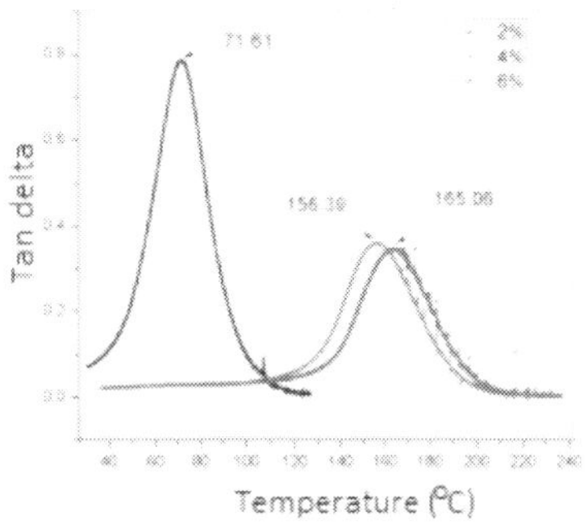

Figure 5. DMA result of different level of 1C2E4MIM at 2 wt %, 4 wt %, 6 wt %

When increased the 1C2E4MIM load, the T_g improved significantly while 1C2E4MIM concentration under 4 wt %, but not that remarkable while 1C2E4MIM concentration above 4 wt % (figure 5), therefore the 4 wt % was investigated as the typical concentration.

3.2.3 Tensile properties of different 1C2E4MIM level

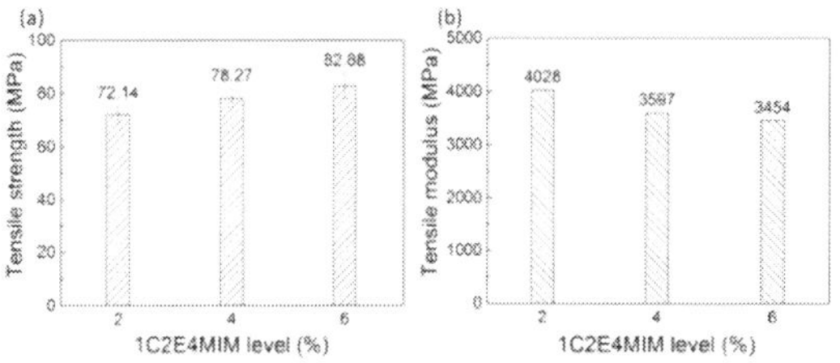

Figure 6. Comparison of neat resin with different level of 1C2E4MIM at 2 wt %, 4 wt %, 6 wt %,and PTPA (a) tensile strength, (b) tensile modular.

The tensile strength did not increase much when 1C2E4MIM content was above 4 wt % , while the tensile modular decreased when 1C2E4MIM increasing (figure 6).

3.2.4 CFRPs mechanical properties

The CFRPs mechanical properties with 4 wt % level of 1C2E4MIM based and T700 plain weave carbon fiber reinforced composite showed high mechanical properties, which could meet the load bearing structure requirements.

Table 2. Carbon fiber reinforced 1C2E4MIM-epoxy resin composites mechanical properties.

Tensile property		Flexural property		Compress property	
Strength (M Pa)	Modular (G Pa)	Strength (M Pa)	Modular (G Pa)	Strength (M Pa)	Modular (G Pa)
632	69	662	55	310	60
Short beam shear strength (M Pa)					
43					

4 CONCLUSIONS

The 2E4MIM and 1C2E4MIM were investigated as fast cure thermal latent epoxy resins, which could be used in automobile production line. The Stern effect and coordination effect on formidable ring of 1C2E4MIM bring better thermal latency and longer pot life. The higher cross-linking with epoxy resin created higher T_g, bending strength. It could be deduced that the latency of the electron deficiency group on formidable ring which could react with epoxy group can both improved the latency and properties. Then the properties of different content of 1C2E4MIM were investigated. The when the content below the 4 wt %, the T_g sharp fell, and curing time last long, when the content above 4 wt %, the Tg climbed very slow, but the heat of reaction were increased and got too high. The hot-wet properties of 1C2E4MIM-Epoxy resin were excellent. And the mechanical properties of T700 carbon fiber reinforced 1C2E4MIM-Epoxy composites result to be auspicious.

Reference

1. Fuchs, E., et al. (2008). "Strategic materials selection in the automobile body: Economic opportunities for polymer composite design." Composites Science and Technology **68**(9): 1989-2002.
2. Ellis, B. (1993). "Chemistry and technology of epoxy resins." Springer-Science+Business Media, B.V.
3. Ellis, B. (1993). "Chemistry and technology of epoxy resins." Springer-Science+Business Media, B.V.
4. Bauer, R. S. (1985). "Epoxy Resins." ACS SYMPOSIUM SERIES 285: 931-961.
5. Global and China Epoxy Resin Industry Report, 2013-2016,http://www.researchandmarkets.com/reports/2669198/global_and_china_epoxy_resin_industry_report#pos-3
6. Friedrich, K. and A. A. Almajid (2012). "Manufacturing Aspects of Advanced Polymer Composites for Automotive Applications." Applied Composite Materials 20(2): 107-128.
7. Guzmán, D., et al. (2015). "Enhancement in the Glass Transition Temperature in Latent Thiol-Epoxy Click Cured Thermosets." Polymers 7(4): 680-694.
8. Parker, S., et al. (2016). "High-Tg Thiol-Click Thermoset Networks via the Thiol-Maleimide Michael Addition." Macromol Rapid Commun 37(13): 1027-1032.
9. Parker, S., et al. (2016). "High-Tg Thiol-Click Thermoset Networks via the Thiol-Maleimide Michael Addition." Macromol Rapid Commun 37(13): 1027-1032.
10. Mahendran, A. R., et al. (2011). "Thermal cure kinetics of epoxidized linseed oil with anhydride hardener." Journal of Thermal Analysis and Calorimetry 107(3): 989-998.
11. Fast-curing epoxy system, Automotive Engineering Magazine[EB/OL],01-Oct-2012 08:45 EDT.Available: http://articles.sae.org/11435/
12. DOW，Solutions from Dow Automotive Systems VORAFORCE 5300 ultra-fast cure composite epoxy system.2015.3.25.
13. CompositesWorld，North America is fastest-growing market for automotive lightweight materials [EB/OL]，2015.5.4, Available: http://www.composi-tesworld.com/news/north-america-is-fastest-growing-market-for-automotive-lightweight-materials.
14. S.K. Ooia, et al. (2000). "DSC studies of the curing mechanisms and kinetics of DGEBA using imidazole curing agents." Polymer 41: 3639–3649.
15. Heise, M. S. and G. C. Martin (1987). "Curing Mechanism and Thermal Properties of Epoxy-Imidazole Systems." Macromolecules.

Morphology and Properties of Poly(Lactic Acid) and Ethylene-Methyl Acrylate Copolymer Blends with Organoclay

Sirirat Wacharawichanant[1,a], Sirinan Ratchawong[1], Phakjira Hoysang[1] and Manop Phankokkruad[2]

[1]*Department of Chemical Engineering, Faculty of Engineering and Industrial Technology, Silpakorn University, Nakhon Pathom 73000, Thailand*

[2]*Faculty of Information Technology, King Mongkut's Institute of Technology Ladkrabang, Bangkok 10520, Thailand*

Abstract. The effects of nanoclay with surface modified by 25-30 wt% of octadecylamine (Clay-ODA) on mechanical, thermal and morphological properties of poly(lactic acid) (PLA)/ethylene-methyl acrylate copolymer (EMAC) blends with various Clay-ODA contents were investigated. The EMAC with 19.60 wt% of methyl acrylate (or EMAC1820) was used, and the ratio of PLA and EMAC1820 was 80/20 by weight, and the Clay-ODA contents were 1, 3, 5 and 7 phr. The morphology analysis showed that the addition of Clay-ODA could improve the miscibility of PLA/EMAC1820 blends due to the decrease the domain sizes of dispersed EMAC1820 phase in the PLA matrix. The mechanical properties showed Young's modulus, stress at break and storage modulus of PLA/EMAC1820 blends was improved after adding Clay-ODA. The addition of Clay-ODA did not effect on the melting temperature and glass transition temperature of PLA and PLA blends. The thermal stability of PLA could improve by adding EMAC1820 and Clay-ODA.

1 Introduction

Poly(lactic acid) (PLA) is biodegradable polymer and a linear aliphatic thermoplastic polyester, which can be synthesized by condensation of lactic acid or ring opening polymerization of lactide [1-2]. PLA can processes similarly to polyolefins by conventional processing methods such as extrusion, tubular blown film, injection molding, and thermoforming and PLA has high strength, high modulus and transparency [3]. However, the main weak points of PLA properties are too brittle for many applications, low heat distortion temperature (HDT) and low thermal stability compared to commercial polymers [4-7]. To more applications of PLA need to improve the mechanical and thermal properties by blending with the flexible and toughness materials.

Some literatures have studied to improve the mechanical, thermal and barrier properties, and degradation behavior of PLA by blending with rubber, plasticizers, thermoplastic starch (TPS) and other thermoplastic polymers [8-12]. For instance, Juntuek et al. [8] studied the properties of PLA/natural rubber (NR) blends at various ratios using an internal mixer. The impact strength and elongation at break of PLA/NR blends dramatically increased with increasing NR content up to 10% (w/w). Glycidyl methacrylate-grafted natural rubber (NR-g-GMA) was used as a compatibilizer for PLA/NR blends. The results showed that the addition of NR-g-GMA in PLA/NR blends significantly improved impact strength and elongation at break of PLA/NR blends when compared with that of neat PLA and PLA/NR blends without NR-g-GMA. Thermal property results indicated NR had higher thermal stability than PLA. The increase of thermal stability of PLA due to the good interaction between the blend components was reported in other PLA blends and composites. In the same way, Tanrattanakul and Bunkaew [9] prepared NR/PLA (60/40) blends by melt blending in an internal mixer and investigated the effect of plasticization of PLA on the mechanical and physical properties of the derived NR/PLA. Four plasticizers were used: tributyl acetyl citrate (TBAC), tributyl citrate (TBC), glycerol triacetate (GTA), and triethyl-2-acetyl citrate (TEAC). They reported that all plasticizers increased the strain at break. The presence of 4 phr of plasticizer provided the highest strength and tensile toughness, and the strain at break increased with the increasing plasticizer content.

Ferrarezi et al. [10] prepared PLA/TPS blends by using poly(ethylene glycol) (PEG), a non-toxic polymer, as a compatibilizer. They found that PLA/TPS blends presented large TPS phase size distribution and low adhesion between phases, which was responsible for the lower elastic modulus of this blend when compared to pure PLA. The addition of PEG resulted in the increase of PLA crystallization, due to its plasticizing effect, and improvement of the interfacial interaction between TPS and PLA matrix. The incorporation of PEG increased the impact strength of the ternary blend and that the elastic modulus remained similar to the PLA/TPS blends. Furthermore, Bee et al. [11] studied the properties of

[a] Corresponding author: sirirat.che@gmail.com

PLA and low density polyethylene (LDPE) blends. They reported that the tensile strength and Young's modulus of the PLA/LDPE blends reduced with an increasing percentage of LDPE due to the lower intermolecular bonding of LDPE than of PLA. Thus, they used the electron beam irradiation to improve PLA/LDPE properties. The result showed the increasing irradiation dosage enhanced the mechanical properties of PLA/LDPE blends at higher LDPE content.

From some researchers have studied using the rubber and flexible polymer to improve the mechanical properties of PLA, and using organoclay to improve the miscibility and properties of PLA blends for more applications. In stance, Hasook et al. [13] studied the mechanical properties of PLA/poly(ε-caprolactone) (PCL)/organoclay nanocomposites. They reported that Young's modulus of PLA/organoclay nanocomposite was more than that of neat PLA. The tensile strength of the PLA/PCL (molecular weight 10K and 40K)/organoclay nanocomposites was higher than that of neat PLA, and the elongation of the PLA/PCL/organoclay composites also improved. Moreover, the thermal stability of the PLA/organoclay nanocomposite increased with the addition of PCL. Bijarimi et al. [14] investigated the mechanical, thermal and morphological properties of PLA/liquid natural rubber/Cloisite C30B (C30B) nanocomposites. The results found that Young's modulus and flexural modulus increased with the addition of C30B to the PLA/liquid natural rubber blends. The elongation at break of PLA/liquid natural rubber increased significantly as compared to nanocomposite with 1% of C30B. Nevertheless, the elongation at break and impact strength decreased gradually when nanoclay content increased above 3%. This study indicated that PLA/liquid natural rubber-toughened nanocomposites with a higher modulus and that thermal stability could be produced.

From above the literature review showed that the organoclay can improve the mechanical properties of PLA blends, Thus, this work studied the influence of Clay-ODA on morphology and mechanical properties of PLA/EMAC1820 blends. The blends and composites with varying concentration of Clay-ODA were prepared by a melt mixing technique in an internal mixer, and the sample was molded by compression molding.

2 Experimental

2.1 Materials

PLA with the melt flow index (210°C/2.16 kg) of 6.0 g/10 min and specific gravity of 1.24 was produced by NatureWorks LLC, USA, under the trade name of "Ingeo™ Biopolymer 2003D". The ethylene-methyl acrylate copolymer (EMAC) has the melt flow index (190°C/2.16 kg) of 3.95 g/10 min and density of 0.926 g/cm3, and EMAC was produced by DuPont Company. EMAC with 19.60 wt% of methyl acrylate was called EMAC1820. The nanoclay surface modified with 25-30 wt% of octadecylamine (Clay-ODA) was purchased from Aldrich Company.

2.2 Sample Preparation

Before blending, PLA was dried in an oven at 80°C for 4 h, and Clay-ODA was dried at 80°C 24 h. EMAC1820 was not dried before blending process. PLA/EMAC1820 (80/20 by weight) were prepared by melt blending in an internal mixer at 180°C and a rotor speed of 80 rpm for 15 min. PLA/EMAC1820 (80/20 by weight) with Clay-ODA contents 1, 3, 5 and 7 phr was prepared in a same condition, and Clay-ODA was added after 5 min of mixing. After compounding, polymer blends without and with Clay-ODA were dried in an oven at 80°C for 1 h for reject moisture. Afterward, the blends and composites were compression molded at 180°C for 15 min for impact and tensile tests. DMA samples were prepared by compression molding at 180°C for 12 min. TGA samples were prepared as a film by compression molding at 180°C for 5 min.

2.3 Sample Characterization

The morphology of impact fractured surfaces of PLA/EMAC1820 blends without and with Clay-ODA was investigated by scanning electron microscope (SEM) instruments (Maxim 2000S, CamScan Analytical) at 15 kV accelerating voltage after gold coating.

The impact strength of PLA/EMAC1820 blends without and with Clay-ODA was measured using a noted Izod impact tester (Zwick/material testing August-Nagelstr.11.D-89079 Ulm) at room temperature. The impact test was performed according to ASTM D 256. Tensile test was conducted according to ASTM D 638 with a universal tensile testing machine (LR 50k from Lloyd instruments) at a crosshead speed of 50 mm/min. Each value obtained represented the average of five samples.

Dynamic mechanical analysis (DMA) for the thermomechanical properties (storage modulus) of PLA/EMAC1820 blends without and with Clay-ODA were measured in bending film mode at a constant vibration frequency of 1 Hz, a temperature range of 30-140°C, and a heating rate of 5°C/min in a nitrogen atmosphere with a dynamic mechanical analyzer (Pyris Diamond DMA, Perkin Elmer).

The degradation temperature of PLA/EMAC1820 blends without and with Clay-ODA was measured by thermogravimetric analysis (TGA) (Model SDT Q600, TA Instruments). The temperature was in the range of 30-600°C at a heating rate of 10°C/min under nitrogen atmosphere.

3 Results and Discussion

3.1. Morphology

Figure 1 shows the SEM micrographs of the impact fractured surfaces of pure PLA, PLA/Clay-ODA and PLA/EMAC1820 (80/20) blends, respectively. The fractured surface of pure PLA was quite smooth which characteristic of a brittle semicrystalline polymer. The fractured surface of PLA/Clay-ODA composites was more roughness when compared with pure PLA. The fractured surface of PLA/EMAC1820 blends clearly demonstrated a two-phase separation of dispersed EMAC1820 phase and the PLA matrix phase. The SEM image observed the minor phase of EMAC1820 dispersed as spherical domains in PLA matrix and the microvoids surrounding the droplets of dispersed EMAC1820 phase indicated a weak interfacial adhesion in the blends. Due to the immiscible nature of the polymers, high interfacial tension between the two polymers and the spherical morphology is anticipated because of the minimization of the interfacial area [15].

Figure 2 shows the SEM micrographs of the impact fractured surfaces of PLA/EMAC1820/Clay-ODA composites. The addition of Clay-ODA in the PLA/EMAC1820 blends reduced the spherical domain sizes of dispersed EMAC1820 phase significantly when compared with the blends without Clay-ODA. The spherical domain sizes of EMAC1820 phase decreased with increasing Clay-ODA content. This indicated that Clay-ODA played an important role in reducing the spherical domains of EMAC1820 of the polymer blends and improved adhesion interaction between PLA and EMAC1820 phases.

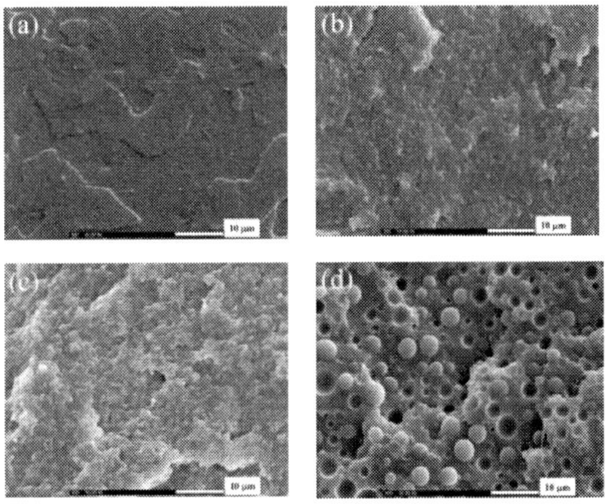

Figure 1. SEM micrographs of (a) pure PLA, (b) PLA/Clay-ODA (100/3), (c) PLA/Clay-ODA (100/7), and (d) PLA/EMAC1820 (80/20).

3.2 Mechanical Properties

Each value obtained of mechanical properties represented the average of five samples and the standard deviation displayed in error bar. Figure 3 shows the impact strength of PLA/EMAC1820/Clay-ODA composites with various Clay-ODA contents. The result found that the impact strength of PLA decreased with increasing Clay-ODA content. This implied that the addition of Clay-ODA did not improve the impact strength of PLA/EMAC blends due to the structure of Clay-ODA decreased the movement of entangled polymer chains, and as Clay-ODA loading increased the brittleness of the polymer composites increased.

Figure 4 shows Young's modulus of PLA/EMAC1820/Clay-ODA composites with various Clay-ODA contents. Young's modulus of PLA/EMAC1820/Clay-ODA composites increased with increasing Clay-ODA content. Young's modulus of PLA/EMAC1820/Clay-ODA composites was higher than that of PLA/EMAC1820 blends. Thus, the incorporation of Clay-ODA significant improved the Young's modulus of PLA/EMAC1820 blends due to the reinforcement effect of the rigid inorganic Clay-ODA induced a substantial increase in the modulus of the composites.

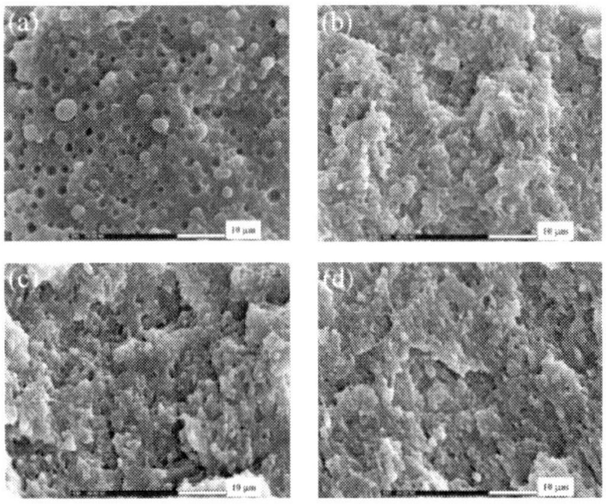

Figure 2. SEM micrographs of PLA/EMAC1820/Clay-ODA composites (a) 80/20/1, (b) 80/20/3, (c) 80/20/5, and (d) 80/20/7.

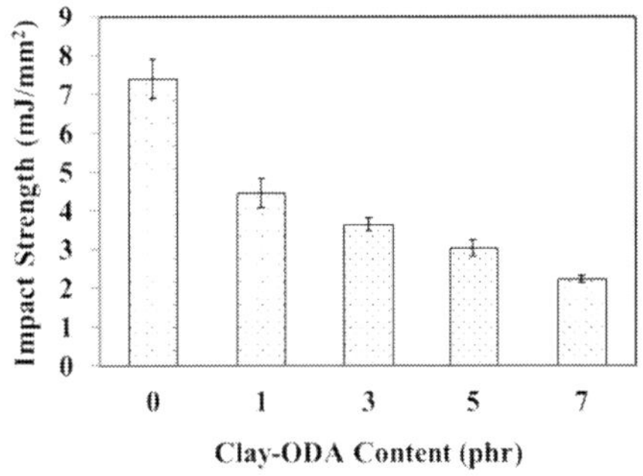

Figure 3. Impact strength of PLA/EMAC1820 blends without and with Clay-ODA.

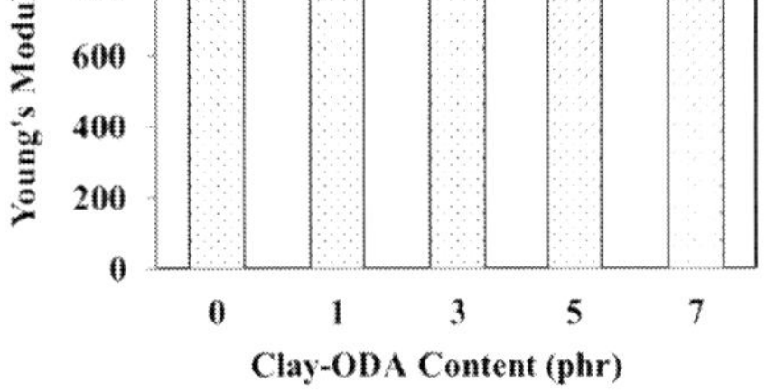

Figure 4. Young's modulus of PLA/EMAC1820 blends without and with Clay-ODA.

The tensile strength of PLA/EMAC1820 blends without and with Clay-ODA is presented in Figure 5. The tensile strength of PLA/EMAC1820 blends slightly decreased with increasing Clay-ODA content. This result

DOI: 10.1051/matecconf/201713007006

indicated that the addition of Clay-ODA did not improve the tensile strength of PLA/EMAC1820 blends.

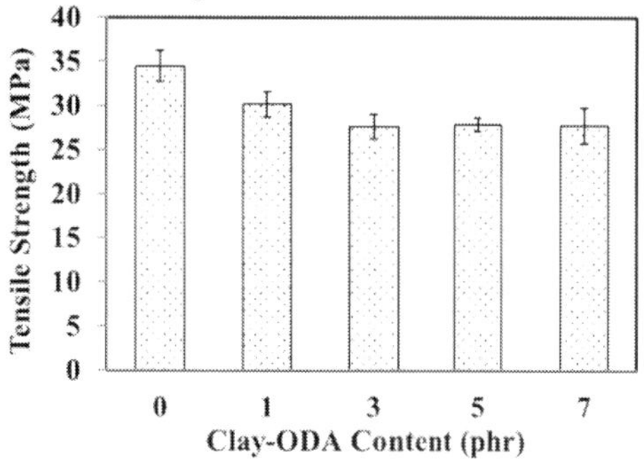

Figure 5. Tensile strength of PLA/EMAC1820 blends without and with Clay-ODA.

Figure 6 shows the stress at break of PLA/EMAC1820/Clay-ODA composites, the result found that the stress at break of PLA/EMAC1820/Clay-ODA composites increased at high Clay-ODA loading. The stress at break was lower than tensile strength due to the sample did not break at a yield point, but the sample could absorb the tensile stress and elongated before the breaking point. This can be explained that the addition of Clay-ODA at high content could improve the stress at break of the PLA/EMAC1820 blends due to Clay-ODA could absorb the tensile stress from the polymer matrix before the breaking point.

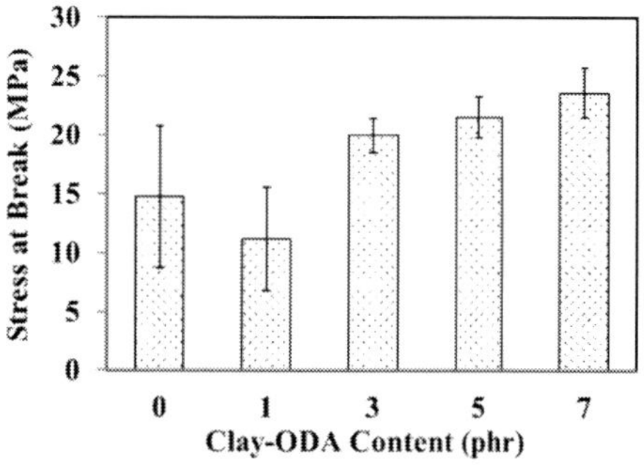

Figure 6. Stress at break of PLA/EMAC1820 blends without and with Clay-ODA.

Figure 7 shows the effect of Clay-ODA content on the strain at break of PLA/EMAC1820 blends. The result found that the strain at break of PLA/EMAC1820 blends increased when added Clay-ODA 1 phr and 3 phr and decreased at high Clay-ODA content. This can be explained that the layered Clay-ODA reduced the flexibility and elongation of a polymer chain after added high Clay-ODA loading.

The variation of storage modulus of PLA, PLA/EMAC1820 blends and composites is presented in Figure 8. The storage modulus PLA/EMAC1820 blends and PLA/EMAC1820/Clay-ODA composites tended to decrease with increasing temperature. The storage modulus of pure PLA showed the highest value, and it was higher than PLA/EMAC1820 blends and composites. Moreover, the addition of Clay-ODA could improve the storage modulus of PLA and PLA/EMAC1820 blends.

3.3 Thermal Properties

The results of thermal properties showed that the melting temperatures (T_m) of PLA and PLA/Clay-ODA (100/3) were 151.3°C and 153.4°C, while T_m of blends and composites were not significantly changed when EMAC1820 and Clay-ODA was incorporated in the polymer matrix. The glass transition temperature (T_g) of PLA was 66.4°C and the glass transition temperature of PLA/Clay-ODA composites was 67.7 ± 0.7°C. While the glass transition temperature of PLA blends and composites was 68.0 ± 1.0°C as shown in Table 1. Thus, the glass transition temperatures of the PLA and blends were not significantly changed when added Clay-ODA.

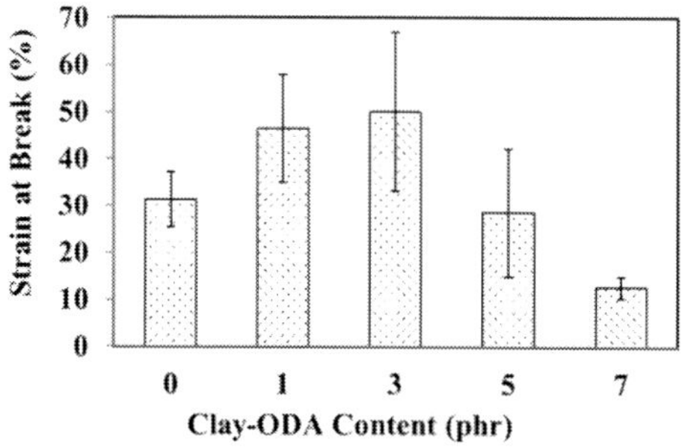

Figure 7. Strain at break of PLA/EMAC1820 blends without and with Clay-ODA.

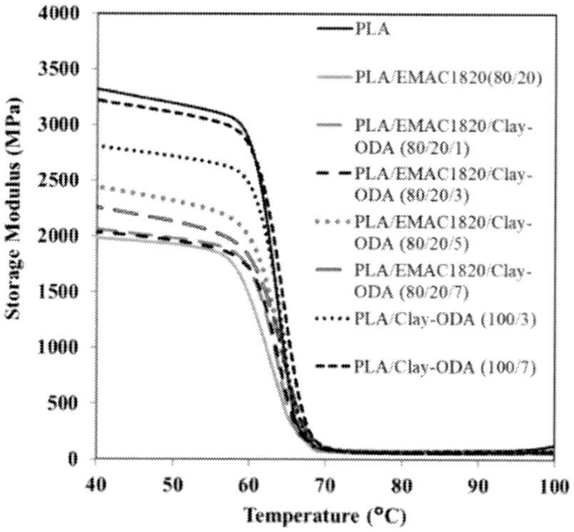

Figure 8. Storage modulus of PLA/EMAC1820 blends without and with Clay-ODA.

The degradation temperatures corresponding to the 10% weight loss (T_{d10}) were calculated as shown in Table 1. The degradation temperatures of PLA, EMAC1820 and Clay-ODA were 315.4, 426.2 and 358.6°C, respectively. This indicated the thermal stability of EMAC1820 was higher than that of PLA. The result showed the incorporation of EMAC1820 and Clay-ODA improved the degradation temperature of PLA. This indicates that an improvement in the thermal stability in the PLA/EMAC1820 blends due to EMAC1820 and Clay-ODA had the higher degradation temperature.

Figure 9 shows the thermogravimetric curves of thermal degradation of pure PLA, EMAC1820, Clay-ODA and PLA/EMAC1820 blends and composites. It is evident that the thermal degradation of PLA, EMAC1820, Clay-ODA showed only single step of weight loss, while the thermal degradation of PLA/EMAC1820 blends and composites proceeded by two degradation steps. It observed that the first step showed the large mass loss of thermal degradation of PLA. The second step showed a small mass loss of thermal degradation of EMAC1820.

Table 1. T_m, T_g and T_{d10} of PLA, PLA blends and composites.

Sample	T_m (°C)	T_g (°C)	T_{d10} (°C)
PLA	151.3	66.4	315.4
EMAC1820		-	426.2
Clay-ODA	-	-	358.6
PLA/Clay-ODA (100/1)	-	67.2	335.4
PLA/Clay-ODA (100/3)	153.4	67.8	331.8
PLA/Clay-ODA (100/5)	-	67.1	336.4
PLA/Clay-ODA (100/7)	-	68.7	335.8
PLA/EMAC1820 (80/20)	154.7	68.6	343.6
PLA/EMAC1820/Clay-ODA (80/20/1)	-	66.6	343.0
PLA/EMAC1820/Clay-ODA (80/20/3)	153.1	67.2	342.9
PLA/EMAC1820/Clay-ODA (80/20/5)	-	68.8	340.0
PLA/EMAC1820/Clay-ODA (80/20/7)	-	68.6	337.9

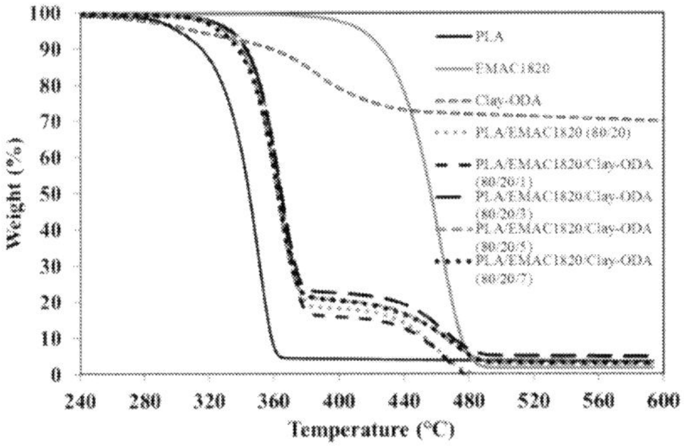

Figure 9. TGA thermograms of Clay-ODA, EMAC, PLA and PLA/EMAC1820 blends and composites.

4 Conclusions

The effects of Clay-ODA on mechanical, thermal and morphological properties of PLA/EMAC1820 blends with various Clay-ODA contents were investigated. The morphology analysis showed that the addition of Clay-ODA could improve the miscibility of PLA/EMAC1820 blends due to the decrease the domain sizes of EMAC1820 dispersed phase in the PLA matrix and improved the miscibility of blends. The mechanical properties showed that the impact strength of PLA/EMAC1820/Clay-ODA composites decreased with increasing Clay-ODA content. While Young's modulus, stress at break and storage modulus of PLA/EMAC1820 blends was improved after adding Clay-ODA. The addition Clay-ODA did not change the melting temperature and glass transition temperature of PLA and PLA blends. The addition of EMAC1820 and Clay-ODA could improve the degradation temperature of PLA.

5 Acknowledgements

The authors would like to thank Silpakorn University Research and Development Institute (SURDI) for the financial support of this project.

References

1. N.I. Akos, M.U. Wahit, R. Mohamed, A.A. Yussuf, Polym. Compos. **34**, 763 (2013)
2. I. Spiridon, K. Leluk, A.M. Resmerita, R.N. Darie, Composites: Part B **69**, 342 (2015)
3. R. Jaratrotkamjorn, C. Khaokong, V. Tanrattanakul, J. Appl. Polym. Sci. **124**, 5027 (2012)
4. K. Halász, L. Csóka, Journal of Engineering DOI: 10.1155/2013/329379: 1 (2013)
5. N.F.M. Rawi, K. Jayaraman, D. Bhattacharyya, J. Reinf. Plast. Comp. **32**, 1513 (2013)
6. R. Tokoro, D.M. Vu, K. Okubo, T. Tanaka, T. Fujii, T. Fujiura, J. Mater. Sci. **43**, 775 (2008)
7. S. Pilla, S. Gong, E. O'Neill, R.M. Rowell, A.M. Krzysik, Polym. Eng. Sci. **48**, 578 (2008)
8. P. Juntuek, C. Ruksakulpiwat, P. Chumsamrong, Y. Ruksakulpiwat, J. Appl. Polym. Sci. **125**, 745 (2012)
9. V. Tanrattanakul, P. Bunkaew, eXPRESS Polym. Lett. **8**, 387 (2014)
10. M.M.F. Ferrarezi, M. de Oliveira Taipina, L.C.E. da Silva, M. do Carmo Gonçalves, J. Polym. Environ. **21**, 151 (2013)
11. S.-T. Bee, C.T. Ratnam, L.T. Sin, T.-T. Tee, W.-K. Wong, J.-X. Lee, A.R. Rahmat, Nucl. Instr. Meth. Phys. Res. B **334**, 18 (2014)
12. J. Liu, H. Jiang, L. Chen, J. Polym. Environ. **20**, 810 (2012)
13. Hasook A, Tanoue S and Iemoto Y Polym. Eng. Sci. **46**, 1001 (2006)
14. M. Bijarimi, S. Ahmad, R. Rasid, J. Reinf. Plast. Comp. **32**, 1656 (2013)
15. S. Wacharawichanant, P. Amorncharoen, R. Wannasirichoke, Polym. Plast. Technol. Eng. **54**, 1349 (2015).

MATEC Web of Conferences 130, 07007 (2017) DOI: 10.1051/matecconf/201713007007
CMPSE2017

Synthesis and curing process of TMA-MDI polyamide-imides

Yuhong Qi[a], Bingnan Ji, Li Li and Zhanping Zhang

Department of Materials Science and Engineering, Dalian Maritime University, Dalian 116026, China

Abstract. The influence of molecular ratio of monomers, reaction time, reaction temperature and solid content on the molecular weight of polyamide imides(PAI), heating rate, temperature and time on tensile properties of PAI film was investigated in detail. The results showed that when monomer molar ratio (MDI: TMA)R >1, PAI is an isocyanate terminated, intrinsic viscosity increases with the increase of R, reaches maximum value at R=1.06, then it decreases obviously. Contrarily, when R <1, PAI is terminated with anhydride, intrinsic viscosity increases with the decrease of R, its maximum value at R= 0.97, then it decreases with the decrease of R. The relation between intrinsic viscosityof PAI and reaction temperature, reaction time, solid content all follows the parabolic law. The optimal reaction temperature,reaction time, solid content is 140℃, 120 min, 30%. Therefore, the best synthesis process of the PAI is reacting for 120 min at 140℃ when 30% solid content and R=1.06. Tensile strength and elongation increase with the increase of heating rate, curing temperature and curing time, after peak value, they decrease with the increase of heating rate, curing temperature andcuring time. The best curing process of the PAI is curing for 2h at 260℃and heating rate 100℃/h.

1 Background

Polyamide-imide (PAI) is a kind of high-performance thermoplastic resin.It isused for advanced applications, such as in the electrical, aerospace, automobile and energy fields in the form of adhesives, composite materials and coating materials, related to their combinationof high toughness, high thermal stability, high glass transition temperature and enhanced chemical resistance[1-5]. Polyamide-imide can be obtained from different strategies of synthesis. Commonly used synthesis routes are the acid chloride and the diisocyanate route. At the acid chloride route reaction between trimellitic acid chloride and an aromatic diamine yields in formation of a poly(amideamic acid). The formation of imide structures is obtained either due to dehydration of the acid intermediate, addition of tertiary amines or anhydrides [6-8]. But this route is long process flow and the storage stability of the product is poor. The diisocyanate route is based on the work of Hitachi Chemical Corp. [9] and today most common for preparation of PAI resins. Thisroute is simple, but the subsequent high temperature curing affects the properties of the material. Additionally obtained PAI product is very easy to cracking. The PAI structure obtained by diisocyanate route has been characterized extensively by NMR techniques [10-12]. Statistical distributions of imide–imide, amide–imide, and amide–amide linkages were observed [13]. In fact, the properties of PAI not only depend on its synthesis, but also curing process [14]. However there are little reports and publications about synthesis and curing process. It limits the application and development of PAI. Therefore, this work investigated indetail the influence of molecular ratio of monomers, reaction time, reaction temperature and solid content on the molecular weight of PAI resin; heating rate, curing temperature and curing time on tensile properties of PAIfilm, to understand and determine the best synthesis and curing process of PAI.

2 Experimental

2.1 Materials

1,2,4-Benzenetricarboxylic anhydride (TMA), 4,4'-Diphenylmethane Diisocyanate(MDI) and N-N-Diommethylacetamide (DMAc) were purchased from Aladdin. DMAc was dewatered over 4A molecular sieve before use. TMA and MDI were used as received.

2.2 Synthesis of PAIresin

The PAI from 1,2,4-Benzenetricarboxylic anhydride (TMA) and 4,4'-Diphenylmethane Diisocyanate(MDI) in N-N-Dimethylacetamide (DMAc) were synthesized in a four-neck, round bottom, Pyrex reaction flask equipped with a mechanical stirrer, reflux condenser, addition funnel and nitrogen inlet. Its reaction scheme is shown in Figure 1.

[a]Corresponding author: yuhong_qi@dlmu.edu.cn

 DOI: 10.1051/matecconf/201713007007

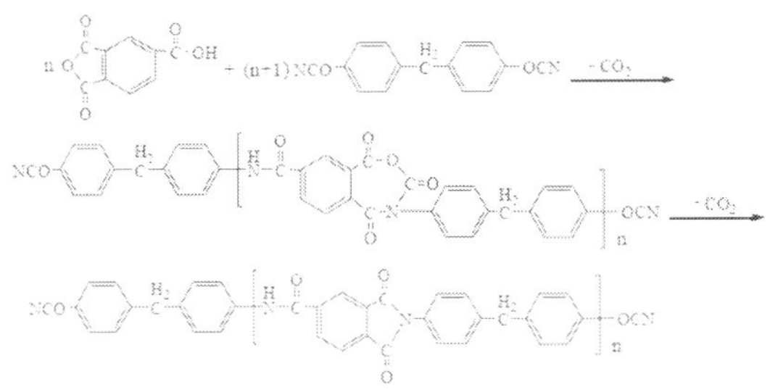

Figure 1. Reaction scheme of PAI

2.3 Curing of PAI

The curing test was carriedout in high temperature furnace. The curing samples were prepared by coating PAI resin on glass slide.

2.4 Characterizations

The molecular weight of PAI before curing were characterized with intrinsic viscosity η_r which tested with Ubbelohde Viscometer for 0.5 g/dL at 30℃.

Mechanical properties of PAI film were characterized with strength and elongation from tensile test which were carried out on XLW Tensile Tester (Labthink Instruments Co., Ltd) equipped with a 500 N load cell at a constant strain rate 50 mm/min for film dimensions of 4 mm width and 60 mm length and the gauge length was set as 20 mm. Three samples for each formulation were measured for statistical accuracy.

3 Results and Discussion

3.1 Synthesis

In the different stages of PAI synthesis, the intrinsic viscosity of the reaction products will change obviously with the reaction condition. Therefore, it is possible to evaluate the synthesis process of PAI by measuring the intrinsic viscosity of the product.

The results showed that the intrinsic viscosity (η_r) of PAI resin is closely related to reaction temperature, reaction time, solid content and monomer molar ratio (MDI: TMA) R. As shown in Figure 2, once the reaction temperature reached 120 ℃ , η_r began to increase significantly, and it increases with temperature; but once the temperature more than 140℃, the solution tends to be black and brown, it may be associated with oxidative crosslinking.

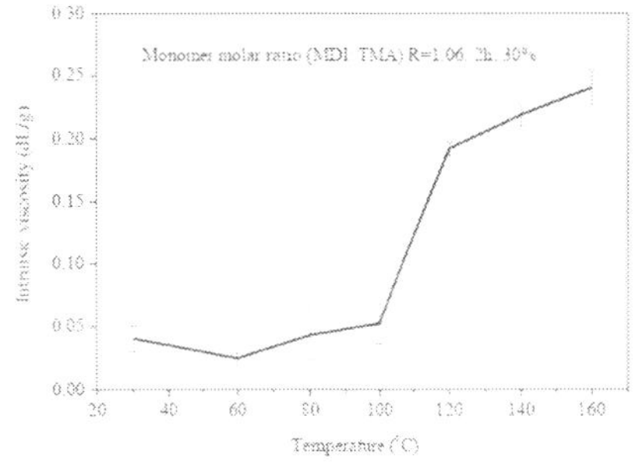

Figure 2. Variation of intrinsic viscosity of PAI with reaction temperature.

The reaction time is within 120 min, η_r increased with time, as shown in Figure 3, after 120 min, η_r decreases with time increase; η_r increases with solid content, as shown in Figure 4, it reaches the maximum value at solid content is 30%, then η_r decreases with increasing solid content.

As shown in Figure 5, when monomer molar ratio (MDI: TMA)R>1, PAI is an isocyanate terminated, η_r increases with the increase of R, η_r reaches maximum value at R=1.06, then η_r decreases obviously with the increase of R. Contrarily, when R <1, PAI is terminated with anhydride, η_r increases with the decrease of R, the maximum value of η_r at R= 0.97, then it decreases with the decrease of R. Therefore, the best synthesis process of the PAI is reacting for 120 min at 140℃ when 30% solid content and R=1.06.

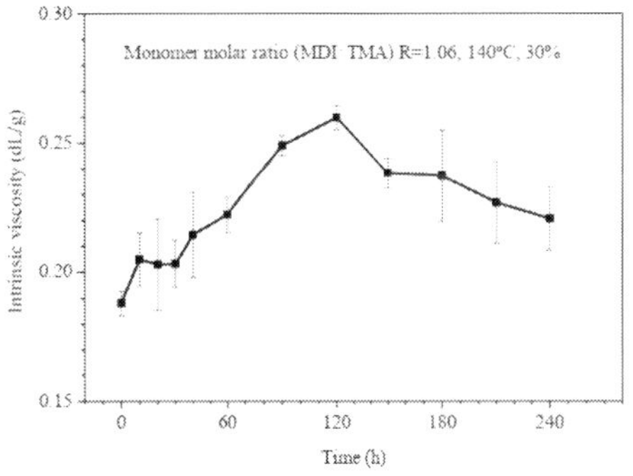

Figure 3. Variation of intrinsic viscosity of PAI with reaction time.

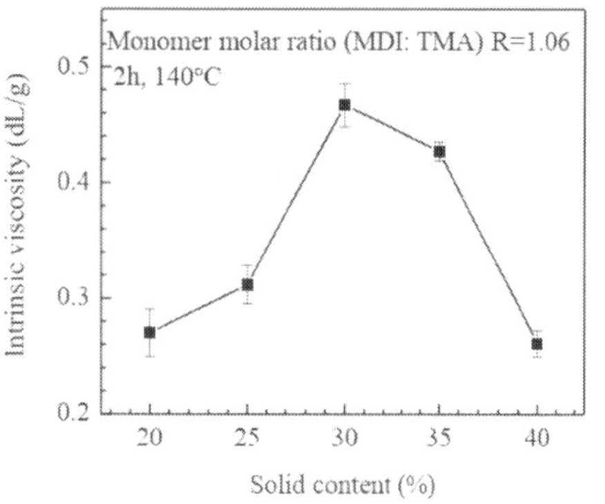

Figure 4. Variation of intrinsic viscosity of PAI with solid content.

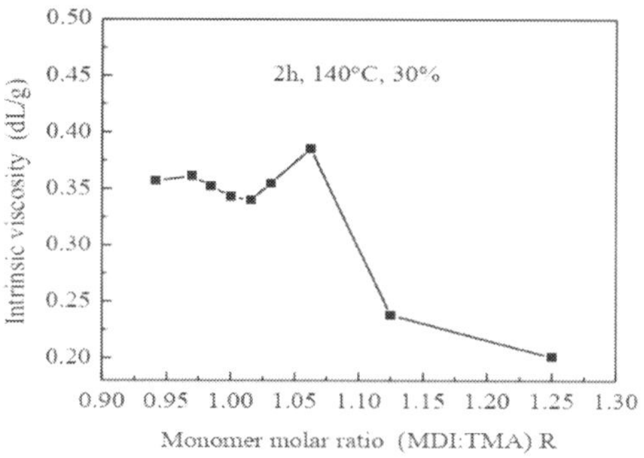

Figure 5. Variation of intrinsic viscosity of PAI with monomer molar ratio.

3.2Curing

The curing process parameters, such as heating rate, curing temperature and curing time, have important influence on the tensile properties of PAI film. The influence of curing parameters on the tensile strength and elongation is almost the same. As shown in Figure 6, Figure 7 and Figure 8, the results showed that tensile strength and elongation increase with the increase of heating rate, curing temperature and curing time, after peak value, they decrease with the increase of heating rate, curing temperature and curing time. For the TMA-MDI polyamide-imidesstudiedin this paper, proposed bestcuring process of the PAI is curing for 2h at 260℃and heating rate 100℃/h. Using this proposed best curing process, the highest tensile strength and elongation is respectively up to 117.03 MPa and 14.18%. PAI cured has good mechanical properties.

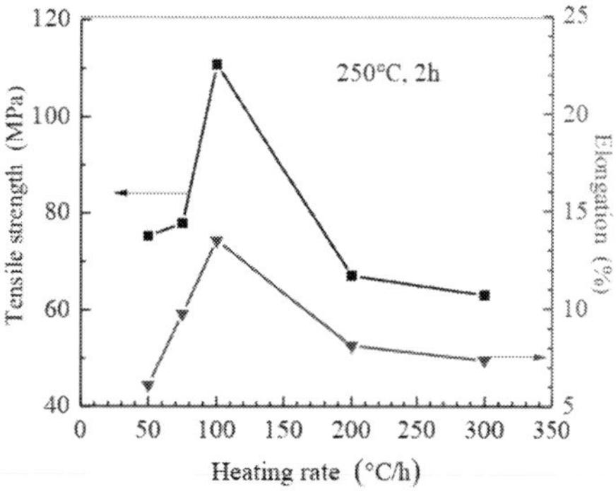

Figure 6. Variation of tensile strength and elongation of PAI films with heating rate.

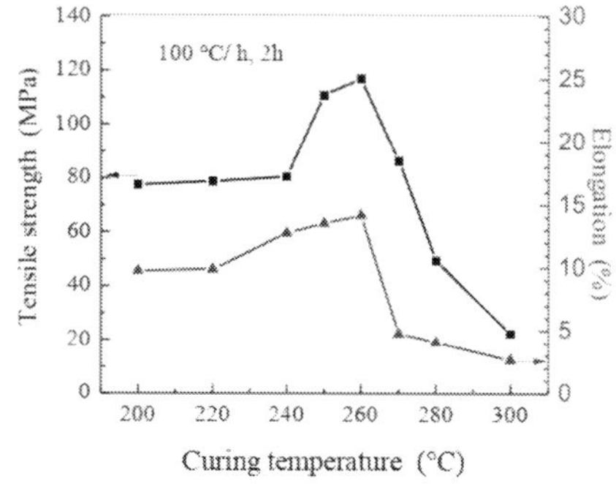

Figure 7. Variation of tensile strength and elongation of PAI films with curing temperature.

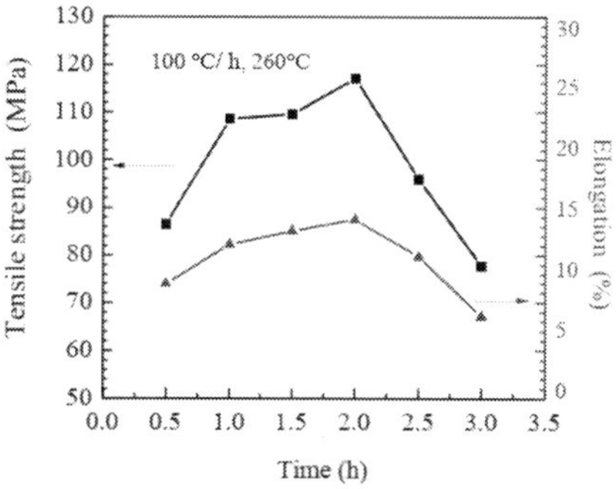

Figure 8. Variation of tensile strength and elongation of PAI films with curing time.

4 Conclusions

When monomer molar ratio (MDI: TMA) R>1, PAI is an isocyanate terminated, intrinsic viscosity increases with the increase of R, reaches maximum value at R=1.06, then it decreases obviously. Contrarily, when R <1, PAI is terminated with anhydride, intrinsic viscosity increases

 DOI: 10.1051/matecconf/201713007007

with the decrease of R, its maximum value at R= 0.97, then it decreases with the decrease of R. The relation between intrinsic viscosity of PAI and reaction temperature, reaction time, solid content all follows the parabolic law. The optimal reaction temperature, reaction time, solid content is 140℃, 120 min, 30%. Therefore, the best synthesis process of the PAI is reacting for 120min at 140℃ when 30% solid content and R=1.06. Tensile strength and elongation increase with the increase of heating rate, curing temperature and curing time, after peak value, they decrease with the increase of heating rate, curing temperature and curing time. The best curing process of the PAI is curing for 2h at 260℃ and heating rate 100℃/h.

Acknowledgments

This work was supported by National nature science foundation of China (51179011), defense fund. The authors gratefully acknowledge for financial support.

References

1. M. Barikani, S. M. Ataei, J. Polym. Sci. A: Polym. Chem. **37**, 2245(1999)
2. X. Ma, N.-H. Lee, H.-J. Oh, J.-S. Hwang, S.-J. Kim, Nanoscale Res. Lett. **5**, 1846 (2010)
3. D. Liaw, W. Chen, Polym. Degrad. Stab. **91**, 1731(2006)
4. M. Son, Y. Ha, M. C. Choi, T. Lee, D. Han, S. Han, C.S. Ha, Eur. Polym. J. **44**, 2236(2008)
5. Z. Nie, A. Petukhova, E. Kumacheva, Nat. Nanotechnol. **5**, 15(2010)
6. T. J. Murray, Macromol. Mater. Eng. **293**, 350–360 (2008)
7. Nemours P De, United States Patent 3179360 (1965)
8. A. L. Endrey, V. Nostnand, United States Patent 3179631 (1962)
9. A. L. Endrey, United States Patent 3179633 (1965)
10. J. Campa, la. de, Die Makromol. Chem. **578**, 571–578(1982)
11. J. Nieto, Die Makromol. Chem. **569**, 557–569(1982)
12. N. Sintès-Zydowicz, Die Makromol. Chem. **2307**, 2295–2307(1993)
13. D. Briesenick, W. Bremser, Progress in Organic Coatings, **82**, 26–32 (2015)
14. L. Li, Y. Qi. Polymer Mater. Sci. & Eng. **31**, 79(20

The Optimization Design of an Ionic Liquid Composite Material for Nitrogen Removal

Bao-you Liu[a] and Tao Wang

Key Laboratory of pollution prevention and control in Hebei Province, College of Environmental Science and Engineering, Hebei University of Science and Technology, Shijiazhuang Hebei 050018, China

Abstract. A kind of EMIMTA ionic liquid was prepared. The viscosity and the absorption amount of NO2 of EMIMTA-TEA composites were measured in different proportion and different temperature. Origin software was used to fit the mathematical equation of the relationship between the viscosity, the absorption amount of NO2 and the mole fraction of the ionic liquid. By the lingo software, the better proportion of Ionic liquid and triethanolamine is optimized. The results show that both the viscosity and the absorption amount of NO2 decrease with the increase of the mole fraction of EMIMTA, which are in accordance with the two order polynomial equation. In order to obtain better absorption amount of NO2 and lower viscosity of composites, taking the absorption amount of NO2 / viscosity maximization as the optimization goal, the viscosity and absorption amount equation NO2 as constraint conditions, the mathematical model of nonlinear programming is established and solved. The results show that when the molar ratio of the ionic liquid is 0.4091, the viscosity of the composite is 47.5 mPa / s, and the absorption amount of NO2 is 0.7744 g . g-1. This paper has some guiding significance for the optimization design of ionic liquid composites.

1 Introduction

Ionic liquid, as a kind of free combination salt of anion and cation, can be used as functional materials and play an important role in many fields. The ionic liquid showed different performance indexes, through physical and chemical reaction directly acting on the process of using, achieving the effect of the same with the traditional materials by its mechanical properties and chemical properties[1]. Moreover, under the same experimental conditions, the effect is better than the traditional materials, and it is an attractive alternative material. Compared to the traditional functional materials, ionic liquids have the characteristics of low vapor pressure, wide temperature range and high chemical stability, thus its effect is less affected by the environment, with a wide application range. In some extreme conditions, such as high pressure, vacuum, high temperature, low temperature, etc, the ionic liquid functional material can adapt to different temperature and pressure, and the effect is not affected by the change of temperature or pressure, and can carry out the stable and adequate relevant reaction. Based on this, ionic liquids can be used under different conditions, which is one of the important reasons for used in many fields as a functional material. It has a wide range of applications, and provide a feasibility study for the new field subject. Ionic liquid is composed of anions and cations, which can lead to a variety of chemical reactions, therefore, it has a good solubility for acid gas. Compared with traditional gas absorption materials, in the process of absorbing and

[a] Corresponding author: liuzhang544@163.com

desorbing gas, the thermal stability and low vapor pressure of Ionic liquid can avoid the problem of volatile. Ionic liquid is a new type absorbent of environmental protection, which shows great potential in the field of environmental protection. Up to now, there have been some reports on the absorption of acid gases in flue gas by ionic liquids. For example, Liu[2] prepared 5 kinds of low eutectic ionic liquids of acetamide-inorganic salts that have good absorption ability to SO2 gas, and the absorption capacity of SO2 can reach up to 0.554 g/g (20 ℃, 1.01 * 10^5 Pa), the recovered ionic liquid can be reused for at least 5 times. Duan reported that caprolactam / four Butyl Ammonium chloride ionic liquid absorbed NOx, NMR results show that the adsorption and desorption of NO and NO2 is reversible in ionic liquids, providing the possibility for the removal of NO2[3].

Lingo is a set of software packages for solving optimization problems, developed by Lindo systems company. Lingo is widely used in teaching, scientific research and industry because of its fast execution speed, easy input, simple solution process in analyzing mathematical programming problems[4]. Lingo is mainly used to solve linear programming, nonlinear programming, transportation planning, dynamic programming, goal programming and integer programming problem. And Lingo can also be used to solve the linear or nonlinear equations and the roots of algebraic equations. The biggest feature of the Lingo software is that it allows the variables in the optimization model is an integer, also known as integer programming. Actually, Lingo is a modeling language to solve optimization problems. In order to facilitate the user to establish the optimization model, Lingo also provides a number of commonly used functions for users to call. In addition, it also provides the interface with other data files, which makes it easier to input, solve and analyze large-scale optimization problems. On the basis of the existing research on the ionic liquid for desulfurization and denitration, in view of the advantages and disadvantages of ionic liquids denitrification in flue gas, preparation of ionic liquid composite absorbing material through a mixture of ionic liquid and organic alkali absorber-triethanolamine, the relationship between the viscosity of ionic liquid and organic alkali in different proportions and the absorption amount of NO2 were measured respectively, and the equation of the relationship between them was established by Origin in this paper. Considering that Lingo software has been successfully applied to the optimization design of composite materials[5], taking absorption amount / viscosity as the optimization objective, the optimum proportion of composite material was obtained by Lingo software to achieve the purpose of larger absorption capacity, lower viscosity and less energy consumption. This study can provide some theoretical guidance for the industrial application of ionic liquids, so as to realize NO2 treatment scheme of the energy saving and environmental protection.

2 Experimental

2.1 Instruments and reagents

The main instruments are Fu Liye transform infrared spectroscopy, Tencer 27, German Bruker company; Nuclear magnetic resonance spectromete, AVANCE II 400M, German Bruker company.The reagents used in this study were analysis pure (≥ 99%).

2.2 Experimental procedure

2.2.1 Preparation of ionic liquid EMIMTA

30mL 1-methylimidazolium and 90mL bromoethane are mixed in 250mL round bottom flask, heating for 8h, getting oily liquid, then cooling to obtain white solid. Acetonitrile is used for dissolving and filtering; filtrate is added with ethyl acetate to obtain white solid , dissolved in acetonitrile again,and then filtering. Recrystallization products were dried under vacuum for 36 hours,after that, 1- ethyl -3- methyl bromide (EMIMBr) was obtained. Take 19.1g EMIMBr and 22.4g three fluorine acetic acid silver, added to 200mL acetonitrile, after stirring, filtering to remove the silver bromide precipitation, rotary evaporation to remove acetonitrile. When the product was dried under vacuum at 60°C, 1- ethyl -3- methyl imidazole three acetic acid salt was obtained. The structure of EMIMTA ionic liquid was characterized by Fourier transform infrared (FT-IR) spectroscopy and nuclear magnetic resonance spectroscopy (1H NMR).

The results shown that the purity was more than 99%.

2.2.2 The viscosity of EMIMTA-TEA

In the calculation of the viscosity, the density value used can be measured by using GB4472-84 method-Specific gravity bottle method(10mL). The density was measured by the precision electronic balance (Precision is $\pm 10^{-4}$g), using the pure water bottle calibration before determination. Super thermostatic bath with accuracy of $\pm 10^{-2}$ K was used in temperature controlling. The viscosity was measured by Pinshi viscometer. Calibration liquid were glycerol, ethylene glycol and water respectively. Each experiment was repeated three times, each time interval of 10min, calculating viscosity by arithmetic mean.

2.2.3 EMIMTA-TEA absorption of NO2

Absorption device and process reference Literature 2. Temperature controlling accuracy is $\pm 10^{-1}$K, weighing accuracy is $\pm 10^{-2}$g. In the process of absorption, the quality of NO2 gas is measured by the electronic balance every 10 min, until the mass change range is less than 1%. Each experiment was repeated three times and the arithmetic mean was taken as the amount of absorption.

2.2.4 Optimization process

Taking the ratio of absorption and viscosity as objective function, taking the fitted equation of viscosity and X1 (molar fraction of ionic liquid) and the fitted equation between absorption amount and X1 (mole fraction of ionic liquid) as the main constraint condition, the mathematical model of nonlinear programming is established and solved by lingo11.0.

3 Results and Discussion

3.1 Viscosity of EMIMTA-TEA Composites

Table 1 Viscosity (mPa·s) of EMIMTA-TEA solution with different EMIMTA molar ratios (X_1) at different temperatures (T, K)

X_1 / T	303.15	313.15	323.15	333.15	343.15
0	405.92	207.24	108.69	60.25	36.29
0.1128	320.4	149.99	86.1	52.38	32.87
0.2006	234.79	117.78	65.22	40.93	27.13
0.3337	120.61	69.56	39.08	25.55	18.02
0.4299	67.35	41.65	26.19	17.46	12.69
0.5058	44.99	28.61	19.73	13.72	10.47
0.6051	37.59	23.83	17.74	12.32	9.27
0.6965	37.59	25.49	17.6	12.6	9.64
0.7994	35.58	26.28	17.44	12.36	9.74
0.9001	30.63	19.88	15.61	11.64	9.41
1.0000	26.56	17.24	13.04	10.39	9.34

The viscosity of EMIMTA-TEA mixture was determined at different temperatures (303.15 K-343.15 K). The viscosity of EMIMTA-TEA solution at different temperature and molar ratio is shown in Table 1. The viscosity of EMIMTA-TEA solution decreased with the increase of EMIMTA molar ratio,and the lower the molar ratio, the more significant the degree of decline. With 303.15 K as an example, when the molar ratio of X1 is less than 0.5, with the increase of the molar ratio, the viscosity of EMIMTA-TEA solution decreased rapidly from 405.92 mPa·s to 44.99 mPa·s, when the molar ratio is more than 0.5, the viscosity of EMIMTA-TEA decreased slowly from 37.59mPa·s to 26.56mPa·s. In addition, when the temperature is higher, such as 333.15 K and 343.15 K, the viscosity of EMIMTA-TEA solution decreases more gentle than 303.15 K. The results shown that the viscosity of EMIMTA-TEA solution with different EMIMTA molar ratios was lower at high temperature. Thus, the viscosity can be controlled by adjusting the temperature.

3.2 NO2 absorption amount of EMIMTA-TEA Composites

Under atmospheric pressure, at the same temperature point and different distribution ratio groups, ionic liquid-Triethanolamine solution (EMIMTA-TEA) has different absorption amount for NO2. When the temperature is the same, with the increase of the content of triethanolamine in the EMIMTA-TEA solution system, and the absorption amount of NO2 was gradually increased, which was higher than that of pure EMIMTA.

Table 2 NO2 absorption of EMIMTA-TEA solution ($g \cdot g^{-1}$) with different EMIMTA molar ratios(X_1) at different temperatures(T, K)

T / X_1	0.0000	0.4299	0.5006	0.6636	1.0000
303.15	1.0562	0.8570	0.8119	0.7687	0.6795
313.15	0.8846	0.7739	0.7409	0.7062	0.6233
323.15	0.7645	0.6757	0.6518	0.6261	0.5525
333.15	0.6506	0.5820	0.5683	0.5440	0.4933
343.15	0.5227	0.4753	0.4725	0.4594	0.4394

3.2 Model optimization better ratio

Under normal pressure, when temperature is the same, the absorption amount of NO2 by ionic liquids-triethanolamine solution (EMIMTA-TEA) of different groups was different. At the same temperature, with the increase of the triethanolamine content in the EMIMTA-TEA solution, the absorption amount of NO2 was gradually increased, which was higher than that of pure EMIMTA.

Table 3 NO2 absorption amount($g \cdot g^{-1}$) byEMIMTA-TEA Solution with different EMIMTA molar ratio(X_1) and temperature(T, K)

T / X_1	0.0000	0.4299	0.5006	0.6636	1.0000
303.15	1.0562	0.8570	0.8119	0.7687	0.6795
313.15	0.8846	0.7739	0.7409	0.7062	0.6233
323.15	0.7645	0.6757	0.6518	0.6261	0.5525
333.15	0.6506	0.5820	0.5683	0.5440	0.4933
343.15	0.5227	0.4753	0.4725	0.4594	0.4394

3.3 Model optimization

Taking 313.15K as an example, the absorption amount of NO2 increased with the increase of TEA molar ratio in EMIMTA-TEA solution. It is concluded that increasing TEA can increase the NO2 absorption amount of ionic liquids. However, the pure triethanolamine viscosity is larger, when the temperature is 313.15 K, the viscosity is 207.24 mPa/s, about 12 times of pure ionic liquid of the same conditions. With the increase of molar ratio of triethanolamine solution in EMIMTA-TEA , the viscosity of the solution gradually increases.When we are in the pursuit of higher absorption amount of EMIMTA-TEA solution, it is necessary to avoid the impact of greater viscosity. Therefore, through fitting the relationship between the molar ratio of EMIMTA and the absorption and viscosity of EMIMTA-TEA solution, this section establishes a mathematical model, using Origin and LINGO software to optimize. Then we can obtain EMIMTA-TEA molar ratio, under high absorption and low viscosity conditions. It can provide theoretical basis for practical production and application.

At 313.15 K, the viscosity and absorption of EMIMTA-TEA were fitted by polynomial equation. According to the fitting equation, the mathematical model is established and optimized. The results shown, when the molar ratio of EMIMTA in the EMIMTA-TEA solution was 0.4091, the absorption was better, and the corresponding viscosity was 47.50 mPa / s, and the absorption was about g .g-1. It also shows that the EMIMTA-TEA solution in this ratio can guarantee a higher absorption, but also reduce the viscosity. When EMIMTA-TEA is used to absorb NO2 in the industrial production, the cost is low and the operation is simple.

4 Conclusions

1- ethyl -3- methyl imidazolium based ionic liquid EMIMTA was prepared, the structure was characterized by Fourier transform infrared spectroscopy (FT-IR) and nuclear magnetic resonance spectroscopy (^1H NMR). The viscosity of EMIMTA-TEA ionic liquid composite was measured at different temperatures (303.15 K-343.15 K). At the same temperature, the viscosity of EMIMTA-TEA solution decreased with the increase of EMIMTA molar ratio. At same EMIMTA molar ratio, the viscosity of EMIMTA-TEA solution decreases with the increase of temperature. Under atmospheric pressure, when the temperature range is 303.15 K-343.15 K, NO2 gas absorption of EMIMTA-TEA ionic liquid composites is measured. At the same temperature, NO2 absorption amount gradually increased with the increase of EMIMTA-TEA (TEA) ratio of triethanolamine solution. The mathematical model is established and optimized by software calculation. The results showed that when the molar ratio of EMIMTA was 0.4091, the viscosity was 47.50 mPa / s, and the absorption was about g.g-1. At this time, the EMIMTA-TEA composite is the best absorbent with good absorption and low viscosity.

5 References

[1] C. Chiappe, C. S. Pomelli. Point-Functionalization of Ionic Liquids: An Overview of Synthesis and Applications. EUR J ORG CHEM, (2014), 2014(28): 6120-6139

[2] B. Y. Liu, F. X. Wei, J. J. Zhao, et al. Characterization of Amide–Thiocyanates Eutectic Ionic Liquids and Their Application in SO2 Absorption. RSC advance,(2013), 3(7): 2470-2476

[3] E.H. Duan, B. Guo, D. D. Zhang, et al. Absorption of NO and NO2 in Caprolactam Tetrabutyl Ammonium Halide Ionic Liquids. J Air Waste Manage Assoc, (2011), 61(**12**): 1393-1397

[4] J. X. Xie, Y..Xue, *optimization modeling and LINDO\LINGO software* [M], Beijing: Tsinghua University press, (2005.7)

[5] H. P. Han, Y. He, H. C. Lin, et al., the application of uniform design in rubber nano composite material preparation technology, Anhui Agricultural Sciences,(2008), 363064-3066.

Acknowledgments

The authors gratefully acknowledge Natural Science Foundation of Hebei Province (2015208122) and environment engineering key subject construction of Hebei Province for their financial supports.

 DOI: 10.1051/matecconf/201713007009

Surface Control of Thermosetting Fluorinated Polyurethane

Dongsheng Cai, Yangsheng Ren, Ying Li*, Xia Wang*

School of Material Science & Engineering, University of Shanghai for Science and Technology, Shanghai 200093, P. R. China

ABSTRACT: In this paper, thermosetting fluorine-containing polyurethane (S-FPU) was synthesized through a novel method with two-step process and its surface properties was studied. Firstly, fluorine-containing epoxy compound (FO) was obtained by epichlorohydrin (ECH) and tetrafluoropropanol (TFP). Further, the fluorinated polyether polyols (FPO) was synthesized by cationic copolymerization of tetrahydrofuran (THF) with the FO, the fluorine was located on the side chains of the FPO and the molecular weight of FPO was controllable, besides, the molecular weight distribution of FPO was narrow. Finally, the S-FPU was prepared by FPO as soft segment, MDI as a hard segment and TEA as chain extender. The influence of fluorine enrichment on the surface properties of S-FPU was studied by controlling curing temperature, curing time and structure of FPO. In addition, the surface fluorine content of S-FPU and the surface hydrophobicity were studied in detail. The results showed that with the curing temperature decreased, the surface fluorine content of S-FPU increased, and when the curing time was more than 10 hours, the fluorine element did not migrate to the surface. What's more, with the molecular weight of FPO increased, the surface fluorine content of S-FPU increased.

1. Introduction

Polyurethanes are used in a wide range of applications, mainly because their properties may be easily tailored to vastly different end results, ranging from stiff, thermoset materials to soft expanded elastomers[1]. It has been well-known that fluorine has the strongest electro negativity of all the elements ever known, and its atomic radius is merely larger than that of hydrogen [2-4]. Therefore, its atomic nucleus controls its electrons and electron clouds firmly. There is strong intermolecular force in the polymer containing C-F bond, which imparts remarkable properties including high thermal stability, low surface free energy, good hydrophobicity, good chemical-resistance to polymers and improved solvent and chemical resistance when compared with their analogs without fluorine. It is a good method to complement the disadvantages of PU by introducing fluorine-containing compounds into PU [5].The properties of fluorinated PU such as a low surface tension, and low coefficient of friction resulting in wide application in the areas of coating, leather decoration, textiles and medicine, especially anticorrosive coatings. So, extensive work has been done on the fluorinated polyurethanes, due to their unique properties [6-10].

There are also some problems with the synthesis of fluorinated polyurethane using fluorinated extenders. These extenders, because of their low molecular weights, do not present the improvement of properties [11]. Up to now, fluorinated polyurethanes (FPUs) with fluorinated chains in the hard-segment, soft-segment or fluorocarbon end-capped polyurethane has been reported. When fluorine segment in hard segment, the surface fluorine element is very low because of fluorocarbon chains are hindered. For FPU with fluorinated side-chain in the hard segment, fluorine can be enriched on the surface during processing resulting in the high fluorine content on the surface, but the mechanical properties of the material is poor. [12] FPU was mostly synthesized by fluorinated macromolecular as soft segment. Ameduri[13] proposed that the surface properties of fluorinated materials with fluorine-containing pendent groups were better than those with fluorinated groups in the main chains. But the fluorinated polyurethane was seldom synthesized by macromolecular diols as soft segment with fluorine-containing pendent groups. Tao Liu et al.[14] prepared side-chain fluorinated polyether diol with self-made flurorinated glycidyl ether by ring-opening polymerization, and a series of fluorinated thermoplastic polyurethane elastomer, the polyurethane that fluorinated side-chain in the soft-segment were prepared by two-step bulk, but the mechanical property of polyurethane was not high enough. Kim et al. reported that fluorinated polyurethane (PU) was blended with base PU prior to dispersion in water to modify the surface of UV cured WPU. Atomic force microscopy (AFM) showed that the surface was roughened with the addition of fluorinated PU. Jeong et al. modified conventional PU dispersion by

a Corresponding author: liying@usst.edu.cn

blending the base PU and fluorinated urethane oligomer prior to water dispersion. Surface roughness of the cast film increased over 13 times according to AFM tests, and the contact angles with water and methylene iodide were beyond 90 after the addition of 15% fluorinated urethane oligomer.[15,16]

In this work, S-FPU was synthesized and the surface properties of S-FPU was studied in detail. Firstly, FO was prepared by ECH and TFP. Further, the FPO with different molecular weight was synthesized by cationic copolymerization of tetrahydrofuran (THF) with the FO, and the structure of FPO was analyzed by FT-IR, ^{1}H NMR and ^{13}C NMR. Last, the S-FPU was prepared by FPO as soft segment, MDI as a hard segment and TEA as chain extender. The surface fluorine content of S-FPU was determined by X-ray photoelectron spectroscopy (XPS) and the hydrophobic of S-FPU was analyzed by contact angle testing. In addition the effects of curing temperature, curing time and structure of FPO on the surface of the S-FPU were studied.

2. Experimental Section

2.1. Materials

Tetrafluoropropanol (TFP) and diphenl-methane-diisocyante (MDI) were purchased from sigma(USA).The fluorinated epoxy compound(FO) and the fluorinated polyether polyol(FPO) were prepared according to the literature.

2.2. Synthesis of fluorinated epoxy compounds (FO) and fluorinated polyether polyol (FPO)

FO and FPO were prepared according to the reference[16,17].

2.3. Synthesis of S-FPU

First, the fluorinated polyether polyol was dehydrated in vacuum under 100℃ for 2 hours. Then, the calculated FPO was added into a three flask equipped with a mechanical stirrer and use nitrogen as protective gas. Next, the calculated MDI was added into the flask, heat up to 80℃ and react for 2 hours with stirring. Last, after the reaction, cool down to 40℃, the calculated TEA was added into the flask, stir fast and transfer polymer to the mould. The curing temperatures ranged from 40 ℃ to 100℃.

3. Results and discussion

The FT-IR spectra of FO and FPO were shown in Fig.1. For the FO, the stretching vibrations of the C–O groups occurred at 1109cm^{-1} and 761cm^{-1}, the stretching vibrations of the methyl groups occurred at 3019cm^{-1} and 2941cm^{-1}, the stretching vibrations of the three-ring groups and the C–F groups occurred at 1415cm^{-1} and 1210 cm^{-1} respectively. We could observe that the compound retained the three-ring structure and fluorine was introduced into the epoxide successfully. For the FPO, the stretching vibrations of the O-H groups occurred at 3442 cm^{-1}, the stretching vibrations of the C-H groups occurred at 2951cm^{-1} and 2843cm^{-1}, the bending vibrations of the C-H groups occurred at 1459 cm^{-1}. Above analysis proved that –OH and -CH_2 groups were in the compound. In addition the peaks at 1233cm^{-1} and 1370cm^{-1} indicated that the compound belonged to a kind of primary alcohol. The fluorine-containing groups was completely consumed in the reaction, which was consistent with the results of the ^{13}C NMR of FPO (Fig.2). It was evident that the peaks appear at 27 ppm and 70 ppm were attributed to THF. Two peaks at 65-80 ppm and 105-120 ppm were relative to the fluorinated epoxy compound. Thus, we concluded that the fluorinated polyether polyol was synthesized by cationic copolymerization of THF with the FO. Besides, the peaks of ^{13}C NMR assigned to the random copolymer of polyether polyol mainly appear at 70–80 ppm. This is the same as ^{13}C NMR of the random.

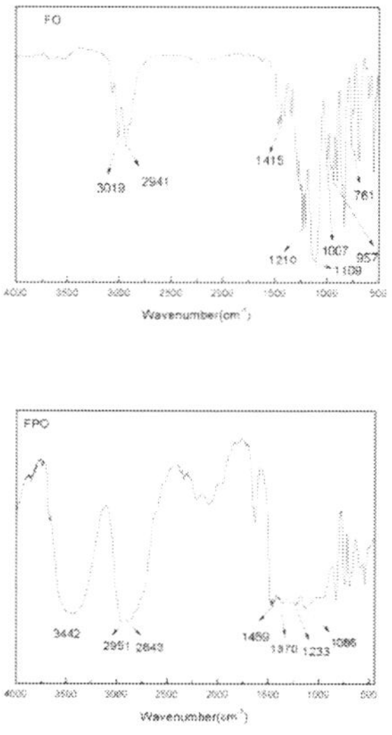

Fig.1 The FT-IR curves of FO and FPO

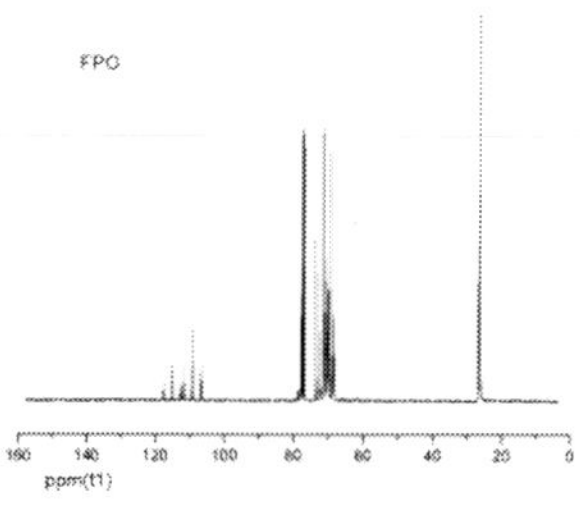

Fig.2 ^{13}C NMR curve of FPO

By controlling the curing temperature, thermosetting fluorinated polyurethane (S-FPU) was synthesized under different curing temperature. The surface element content was measured by XPS. The results were shown in table 1.

Table 1. The element content of S-FPU with different curing temperature

Samples	Curing temperature (℃)	Element content (mol%)			
		C	O	N	F
S-FPU_3-40-15-1530-1:2	40	61.56	26.22	3.68	8.54
S-FPU_3-60-15-1530-1:2	60	62.41	27.76	2.95	6.88
S-FPU_3-80-15-1530-1:2	80	65.92	26.66	2.17	5.25
S-FPU_3-100-15-1530-1:2	100	64.76	29.23	1.71	4.30

From table1, it was clearly observed that the surface fluorine content decreased with the increase of curing temperature, because that the formation of three-dimensional network structure was easier and faster when the curing temperature increased. The dimensional network would hinder the motion of the fluorine-containing side segments, which leaded to the decrease of surface fluorine content. Therefore, we concluded that lower curing temperature was advantaged for the migration of fluorine element to the surface.

Fig.3 showed the influence of curing temperature on surface fluorine content of S-FPU. When the curing time was more than 10h, the surface fluorine content was not changed, because that before the completely formation of three-dimensional network structure, it was easier for the side segments to migrate to the surface, but when the curing time got to a certain degree, it was harder for the fluorine-ontaining side segments to move. Thus, when the curing time was more than 10h, the three-dimensional network structure formed, so the side segments can't move, which leaded to decrease of surface fluorine.

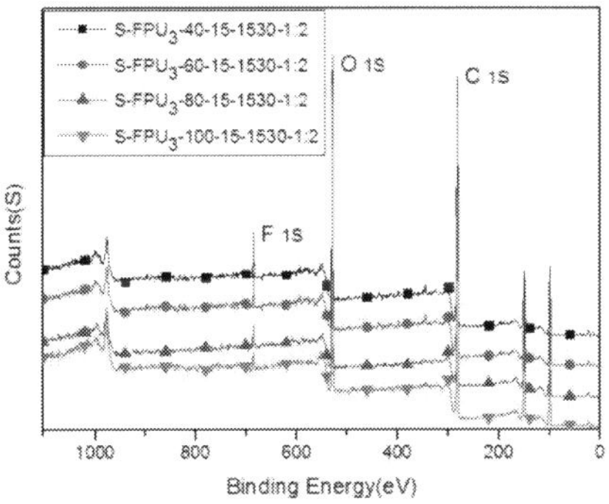

Fig.3 The XPS curves of S-FPU

From table.2, it could be clearly seen that the surface fluorine content of S-FPU_3-1600-1:2 was the highest, and the surface fluorine content of S-FPU_3-1600-1:1 was the least. On one hand, for the S-FPU_3-1600-1:1, the distance between fluorine-containing side segments was shorter, so the space steric hindrance was stronger, which hindered the motion of the segments resulting in the decrease of surface fluorine content. When S-FPU was synthesized, at the beginning, the fluorine-containing side chains could move to the surface, after the cross linked network structure formed, the position of the fluorine side chains was fixed. If the distance between fluorinated side chains was shorter, the entanglement between molecular chains would occur more easily, which increased the steric hindrance, so it was harder for the fluorine-containing side chains to move to the surface before the formation of three-dimensional network, which leaded to decrease of surface fluorine content. On the other hand, the noumenal fluorine content of S-FPU_3-1600-1:3 was lower than S-FPU_3-1600-1:2, and the amount of fluorine atom of S-FPU_3-1600-1:3 that could move to the surface was lower, so the surface fluorine content of S-FPU_3-1600-1:3 was lower than S-FPU_3-1600-1:2.

Table 2. The element content of S-FPU with different structure of FPO

Samples	Element content (mol/%)			
	C	O	N	F
S-FPU_3-30-10-1600-1:1	63.85	32.36	0	3.79
S-FPU_3-30-10-1600-1:2	62.41	27.76	2.95	6.88
S-FPU_3-30-10-1600-1:3	63.85	26.06	3.74	6.35

S-FPU_3-40-15-1530-1:2, S-FPU_3-60-15-1530-1:2, S-FPU_3-80-15-1530-1:2and S-FPU_3-100-15-1530-1:2 were synthesized under the 40 ℃ ,60 ℃ ,80 ℃ ,and 100℃.The Fig 4 and table 3 showed that with the curing temperature decreased, the surface contact angle increased. We concluded that with the increase of curing temperature, surface fluorine content gradually reduced resulting in the decrease of contact angle, which demonstrated that the hydrophobic of S-FPU was decided by the surface fluorine content.

S-FPU_3-40-15-1530-1:2

S-FPU_3-60-15-1530-1:2

S-FPU_3-80-15-1530-1:2

S-FPU_3-100-15-1530-1:2

Fig 4. The contact angle test of S-FPU

Table. 3 The contact angle and surface fluorine element content of S-FPU

samples	curing temperature (℃)	contact angle (°)	surface fluorine element content (%)
S-FPU3-40-15-1530-1:2	40	97.1	8.54
S-FPU3-60-15-1530-1:2	60	92.3	6.88
S-FPU3-80-15-1530-1:2	80	88.5	5.25
S-FPU3-100-15-1530-1:2	100	85.3	4.30

4. Conclusion

In Summary, we presented a new synthetic method of S-FPU by cationic copolymerization of THF with the FO and studied the surface properties of S-FPU. In this paper, we concluded that lower curing temperature was advantaged for the migration of fluorine element to the surface. After the formation of the three-dimensional network structure, it was hard for the fluorine-containing side segments to move to the surface. However, when the molecular weight increased, the flexibility of soft segment was improved so that it was easier for the side chains which contains fluorine to transfer to the surface. On one hand, the space steric hindrance was strong, which hindered the side chains to transfer to the surface, so the surface fluorine content became lower. On the other hand, the noumenal fluorine content was lower, the surface fluorine content was lower too. The surface hydrophobic would be better, if the surface fluorine content increased.

Acknowledgments

This work was supported by the National Natural Science Foundation of China (21101107, 51373100, 51173107, 5140030478), the innovation Project of the Shanghai Municipal Education Commission(No.15ZZ076), and the Hujiang Foundation of China(B14006).

References

1. Sheng, Y. P., Jiang,P.P., Zhang,D.Y., Journal of Applied Polymer Science. 2015, 132(2).41277.
2. Zheng, Z. J, Yu. Q, F. He, L. H, Ye, G. C, Xu, W. X, Liao, J. L, Liu,Y. Chem. Eng. J.2013, (5).810.
3. Zheng, F. L., Xiang, S. L., Lu, C. H., Zhao, S. L., Xu, Z. Z. PaintCoating Industry,2009, 39(7).45.
4. Zhang, F. C., Yang, J. J., Wu, M. Y., Wu, Q. Y., Zhang, J. A. New Chem. Mater.2011, 39(7).25.
5. Wang,S.J., Liu,W.Q., Tan,J.Q., Journal of Macromolecular Science, Part A: Pure and Applied Chemistry. 2016, 53(1):41.
6. A. Frick, A. Rochman, Polym. Test. 2004,23 (4). 413.
7. M. Rogulska, W. Podkoscielny, A. Kultys, S. Pikus, E. Pozdzik, Eur. Polym. J. 2006,42 (8). 1786.
8. H. Yeganeh, M. Barikani, F.N. Khodabadi, Eur. Polym. J. 2000,36 (10).2207.
9. Cao, Q., Li, Y., Jing, B., Liu, P.S., J. Appl. Polym. Sci. 2006,102 (6).5266.
10. T.M. Chapman,R. Benrashid,K. G.Marra,J.P.Keener,.Macromolecules.1995,28,331.
11. Tonelli, C. and Ajroldi, G.J. J. Appl. Polym. Sci. 2003, 87(14). 2279.
12. Zhao,X.W., Ding,J.J., Ye,L., Journal of Fluorine Chemistry.2014, 159,38..
13. Lahiouhel, D.Ameduri, B.Boutevin, B. J. FluorineChem.,2001,107, 81.
14. T. Liu, L. Ye, J. Fluorine Chem. 2010, 36, 131.
15. Lee MH, Jang MK, Kim BK. Eur Polym J .2007, 43(10),4271.
16. Jeong HY, Lee MH, Kim BK. Colloid Surface A. 2006, 290(1-3),178.
17. Wang,X., Hu,J.J., Li,Y., Zhang,J.R., Ding, Y.Y., Journal of Fluorine Chemistry. 2015, 176:14.
18. Wang,X.,Xu,J.C.,Li,L.L.,Liu,Y.,Li,Y.,Dong,Q.Z., Polymer.2016,98,311

Study on the preparation and properties of UV curable polyurethane materials modified by organic silicon

Xidan Luo [1] , Xuetao He[2], Pengcheng Xie[3], Jianyun He[4, a], Yumei Ding[5] and Bingqian Xia[6]
[1]Beijing , Chaoyang District North Third Ring Road No 15, Beijing University of Chemical Technology, China.

This study was supported by the National Natural Science Foundation of China (NO.51573017)

Abstract: In this paper, polyester polyol, toluene diisocyanate (TDI) and hydroxyethyl acrylate (HEA) were taken as the raw materials to synthesize polyurethane acrylate. UV curable organic silicon nano sol was synthesized by γ-methacryloxypropyltrime-thoxysilane. The UV curable polyurethane materials modified by organic silicon were prepared based on the UV curable organic silicon nano sol and synthesized polyurethane acrylate. This paper studies on the mechanical properties of organic silicon modified UV curable polyurethane materials and puts forward the micro model that inorganic phase is bonded to the resin in the form of chemical bonds. The results show that the heat resistance and mechanical properties of hybrid materials are improved with the increase of the nano silica sol's contents , the abrasion resistance of hybrid coatings is also improved and the silica nano particles are well dispersed in the organic phase.

1 Introduction

UV curable technology is a method to form a curable coating and products by taking ultraviolet light as curing energy. Photo-chemical reactions occurred in this process when liquid UV materials are excited by the UV light. UV curable technology has many advantages such as quick-curing, high-efficiency, energy saving, environmental protection, excellent performance of the cured product and so on. It is an environmentally friendly and energy efficient technology[1-2]. Polyurethane acrylate is an important light curing oligomer, which has high toughness and oil resistance after curing while the heat resistance is poor. Organic / inorganic hybrid material will have excellent performance because organic / inorganic hybrid material could combine the characteristics of high toughness, high elasticity, excellent machinability of the organic phase with the high hardness and high modulus performance of the inorganic phase. It has become a research focus due to its obvious advantages in many aspects[3-6].

The technology that prepares UV curable nano silica sol by sol-gel method has a lot of advantages.The mild reaction condition is a kind of this advantages ,it means the synthesis can realize at room temperature or slightly above the room temperature. In recent years, sol-gel technique has been rapidly developed in the preparation of organic / inorganic hybrid materials.

KeChangmei[7] prepared polyacrylate/TiO2 - SiO2 nano hybrid material with butyl titanate and ethyl orthosilicate as the metal alkoxide precursor, the appearance of inorganic particle in hybrid material is spherical, particle size is 90 ~ 110nm. Wei [8] finished the hydrolyzed ethyl silicatecondensation in the presence of hydroxy ethyl methacrylate and prepared hybrid materials that inorganic phase and the organic phase were combined though covalent bond. Yuan Jian prepared polyvinyl acetate/silica.

2 Experimental part

2.1 The main materials

Polyester polyol used in this experiment is provided by Qingdao Yu Tian New Material Co.LTD. Toluene diisocyanate (TDI) is bought from Tianjin chemical reagent factory, Hydroxyethyl acrylate (HEA) is bought from Beijing Dongfang acrylic Chemical Technology Co. LTD. Di-butyltin dilaurate (DBTDL) is bought from Beijing Zheng heng Chemical Co.LTD, Tripropylene glycol diacrylate(TPGDA) is bought from Tianjiao chemicalcompany of Tianjin city. The 2- hydroxy -2 methyl -1 phenyl -1 acetone (Darocur1173) is provided by Ciba Geigy, γ-methacryloxypropyltrimethoxysilane is bought from Nanjing Nengde chemical Co.LTD and the 37% hydrochloric acid, absolute alcohol are bought from Beijing chemical reagent factory.

2.2 The main instruments and equipments

The main instruments used in the tests are: UV-curing machine RW - UVA 201-20, Video based contact angle Measuring device OCAZO, thermogravimetric analyzer TG209C, Electronic Universal Material Testing Machine LR30KPlu, Differential scanning calorimetry DSC204F1, thermostat water bath HHS11-Ni1, Precision timing electric mixer JJ-1 and other major equipment.

2.3 Performance characterization

1.3.1 Infrared spectrum analysis

The infrared spectra of nano sol was tested through the KBr tabletting method by Fourier transform infrared spectrometer tensor27 type that is produced by Bruker Corporation of Germany, scanning range is 400 cm^{-1}~4000 cm^{-1}.

1.3.2 Thermogravimetric analysis (TGA)

The speed of heating is 20°C/min, temperature range from 30°C to 600°C and test the concentration of nitrogen.

1.3.3 Coating hardness test

The pencil hardness of the coating according to the national standard GB6739-86 is done by using a set of pencils whose hardness is known. The hardness of the coating is expressed as the depth that pencil core can penetrate the surface of the coating to substrate.

1.3.4 Coating adhesion test

For adhesion test, the QFH Paints-cross-cut tester is used, according to the determination of GB/T1720-1979. Test results are divided into six grades. Among them level 0 is for the best, while 5 is for the worst.

2.4 Synthesis of polyurethane acrylate

1)Put a certain amount of polyester polyol 2356 into three mouth flask equipped with a stirrer, a condenser pipe and a thermometer,then evacuate the flask for 1 hour at 110 °C.

2)Cool it until room temperature, then add toluene diisocyanate (TDI), add several drops of di-butyltin dilaurate (DBTDL) at the same time, stir it and warm it up to 30~35°C gradually,stop heating and let it self heating. When the temperature reached 68~70 °C, react for 1.5 hours under this temperature, then react further for 1.5 hours at 75°C.

3)Put a certain amount of Hydroxyethyl acrylate (HEA), then add several drops of di-butyltin dilaurate (DBTDL) and inhibitor (HEMQ) at the same time, react for 1.5 hours at 75°C, cool it down to 50°C ~ 60°C.

4)Add the diluent tripropylene glycol diacrylate (TPGDA), stir for a period of time and the resin of polyurethane acrylate was synthesized.

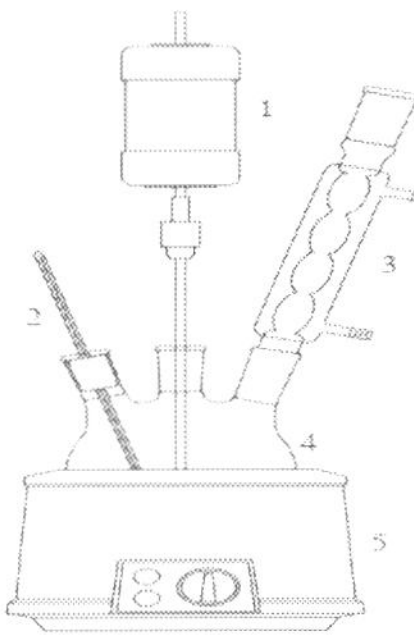

Fig. 1 Device for the synthesis of polyurethane acrylate (PUA)

1-stirrer 2-thermometer 3-condenser pipe 4-three mouth flask equipped 5-Heating jacket

2.5 Methods for synthesis of nano sol

A certain amount of anhydrous ethanol, γ-methacryloxypropyltrime-thoxysilane and 3.6 g of distilled water were put into three mouth flask equipped with a stirrer, a condenser pipe and a thermometer, drop in anhydrous ethanol and 4 ml of hydrochloric acid mixture, let the pH value of the solution become 2, and uniformly stir it at room temperature for 20 minutes, then heat it up to 65 °C, react for two hours,cool the solution and add the trimethylchlorosilane which is called capping agent, react half an hour. Finally, make the decompression and remove the solvent at 60 °C, cool it and get the UV-curable nano sol products.

2.6 The preparation of UV-curable hybrid coatings

A certain concentration of nano sol and photoinitiator1173 was put into the polyurethane acrylate, then mixed homogeneously to get UV-curable hybrid coating.

Table 1. Formula of UV-curable hybrid coating

Hybrid coating	Sol （wt%）	Light Initiator （wt%）
1#	0	4
2#	5	4
3#	10	4
4#	15	4
5#	20	4

3 Results and discussion

3.1 Infrared spectrum analysis of the synthesized polyurethane acrylate

Figure 2 is the FT-IR spectra of polyurethane acrylate, we can learn from the picture that the peak near 3379cm^{-1} and 1525 cm^{-1} is the stretching vibration band of —NH—, the peak near 1721 cm^{-1} is the stretching vibration band of C=O. The stretching vibration bands of —NH— and C=O indicate the formation of polyurethane. The peak near 2927 cm-[1] is the stretching vibration band of —CH_2—. The behavior that most of the stretching vibration band of —NCO near 2275~2240 cm^{-1} disappear shows the group of — NCO has been reacted basically. The stretching vibration bands of —C=C— near 1636 cm^{-1} and 810 cm^{-1} indicate that double bond of acrylate has been linked on the polyurethane molecular chain. The peak near 1460 cm^{-1} is the deformation vibration peak of —CH_2—and the peak near 1409 cm^{-1} is the stretching vibration band of

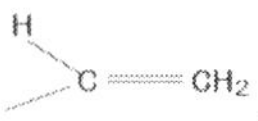

, the peak near 776 cm^{-1} is the stretching vibration band of benzene ring.

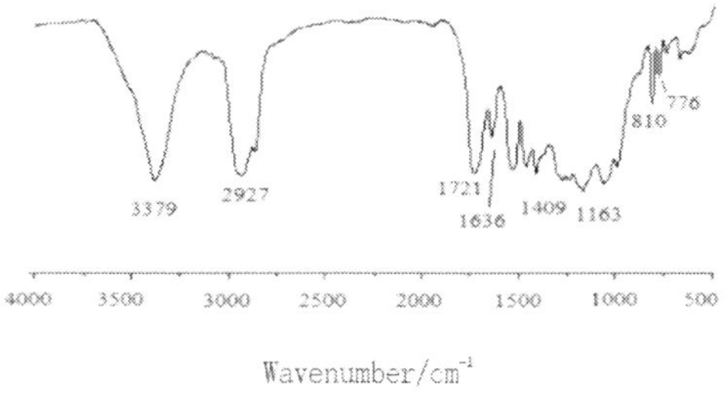

Fig. 2 FT-IR spectrum of PUA

3.2 The analysis of the infrared spectrum of hybrid coating

Figure 3 is the FT-IR spectra of UV curable nano sol /PUA hybrid coatings. The picture shows that the peaks near 3372 cm^{-1} and 1530 cm^{-1} are the stretching vibration band of —NH—, the peaks near 2951cm^{-1} and 1722 cm^{-1} are the stretching vibration band of —C—H and—COO—, the stretching vibration band of —Si—CH_3 near 1255cm^{-1} and 755cm^{-1} indicate the existence of $(CH_3)_3Si$—group.The stretching vibration band of —Si—CH_3 near 3000cm^{-1} disappear.It shows that the trimethylchlorosilane which is called capping agent combined with polymers by hydrolysis-condensation reaction.The stretching vibration band of -Si-O-Si- near

1046cm^{-1} indicates that the coating is an organic / inorganic hybrid system. The disappear of stretching vibration band of —C=C— near 1630cm^{-1} and 810 cm^{-1}indicates that the polymerization reaction occurred in the double bond.

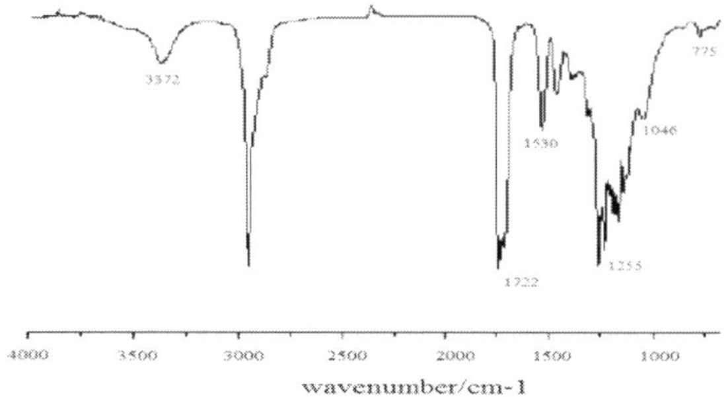

Fig. 3 FT-IR spectrum of hybrid coating

3.3 Thermogravimetric analysis of hybrid coating

The thermal stability of nano sol /PUA hybrid coating was analyzed by thermogravimetric analyzer. Nano sol and temperature have influence on the weightlessness of coating,as shown in Figure 4.The weightlessness of the coatings at 300°C~450°Cmainly due to the thermal decomposition of the polyurethane acrylic polymer.Thermal decomposition would not happen when the temperature was below 300 °C, thermal decomposition rate increased with temperature at 300°C~450°C,and the thermal decomposition was completed basically before 450°C.Because of the decomposition of organic compounds in nano sol,the content of nano sol in coating were reduced from 15% and 20% to 12.4% and 17.2% after reaction.

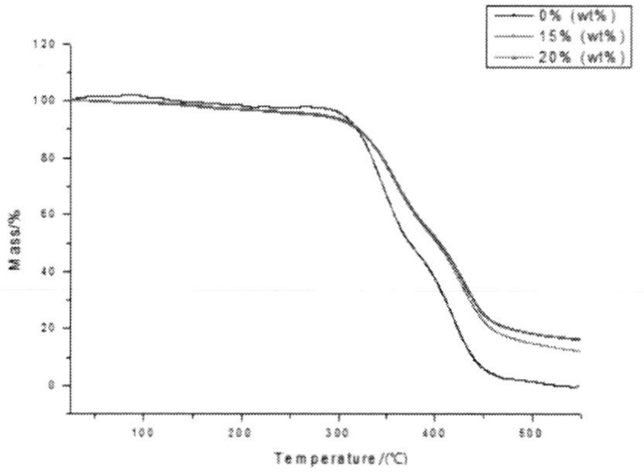

Fig. 4 The TGA analysis of nano-sol /PUA hybrid coating

We also learn from the figure 4 that below 300°C the adding of the nano-sol has few effect on thethermal stability. The nano-sol plays an role in the effect of thermal stability when the temperature is over 300℃. At 350℃, the weightlessness of the hybrid films without adding nano sol has reached 35%, and the weightlessness of hybrid films containing nano sol has just reached 19%. It can be seen from the results that the addition of nano - sol can improve the thermal stability and heat resistance of PUA materials. This is mainly due to the nano particles in the sol and the organic polymers connected together though chemical bond to form interpenetrating crosslinked network, which increases the resistance to rotation, bending and moving of molecular chain and hinders the thermal decomposition of molecular chain.

3.4 Pencil hardness of the filmlight cured nano hybrid materials

PUA/nano sol hybrid film is a kind of organic / inorganic hybrid coating, and it is cured by photochemical reaction under UV irradiation. The UV curable organic / inorganic nano hybrid films were prepared by mixing the synthetic polyurethane acrylate resin and syntheticnano sol and joining the active monomer whose photoinitiator is type 1173.The pencil hardness of hybrid coating is 1H when the content of the nano sol is 10%. The pencil hardness of hybrid coating has increased to 2H when the content of the nano sol is 15%. The pencil hardness of hybrid coating increased to 3H when the content of the nano sol is 20%. The organic/ inorganic phase structure in hybrid coating is shown in figure 5, it is easy to learn from the figure that pencil hardness of hybrid coating increase with the increase of the content of nano sol. Sol molecules are combined with each other through the chemical bond. Inorganic phase and organic phase form the cross-linked network under the irradiation of the ultraviolet light. With the increase of the nano sol content, the proportion of inorganic phase in the films is increased , the reinforcing effect of inorganic phase in hybrid coating is also improved. As a result, the hardness of the hybrid films is increased with the increase of the content of the nano sol. Figure 6 is the microscopic model of hybrid coating .

DOI: 10.1051/matecconf/201713007010

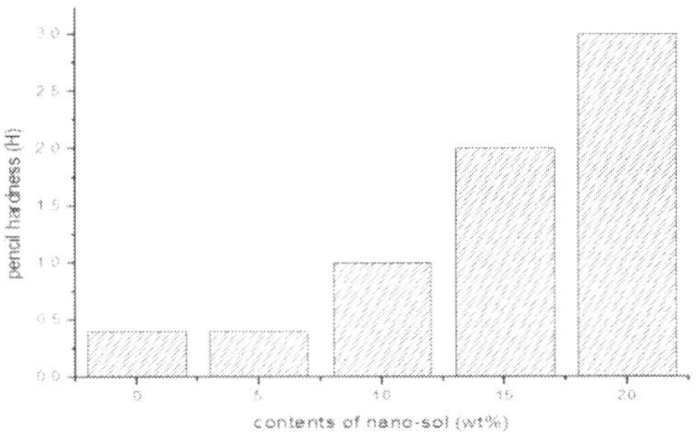

Fig. 5 Relationship between content of nano sol and pencil hardness of hybrid coating

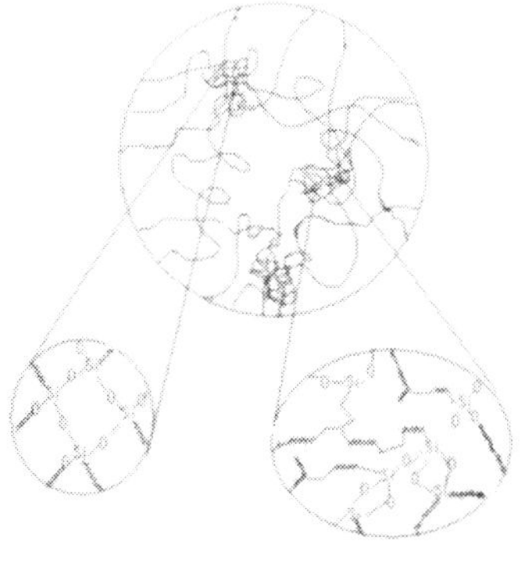

Fig. 6 The microscopic model of hybrid material

Figure 6 is the microscopic model of hybrid materials, inorganic phase of material is bonded to the resin in the form of chemical bond, chain in the inorganic phase containing photosensitive group and chain in the organic phase form interpenetrating network. The cross-linked network structure can improve the hardness of the material and change the mechanical properties of the material.

4 Conclusions

Polyester polyol, toluene diisocyanate (TDI) and Hydroxyethyl acrylate (HEA) were taken as raw material to synthesize polyurethane acrylate. UV curable organo silicon nano sol was synthesized by γ-methacryloxypropyltrime-thoxysilane. UV curable organic / inorganic nano hybrid materials were prepared by synthesized nano sol and polyurethane acrylate. Inorganic phase and organic phase form the cross-linked network under the irradiation of the ultraviolet light. The results showed that the heat resistance of the hybrid materials was increased with the increase of the amount of nano - sol and the hardness of the hybrid film was also increased.

References

1. A.Endruweit; M.S,Johnson.A.C,Long Polym,Compos [J].2006,27,119
2. E.Lackey;K.Inamdar,L.Worrel;W.Al-Akhdar;D.A.Wostrazky,RadTechReport[J]. 2001,15,36
3. Jin Shuping , Liu Mingzhu , Chen Shilan , et al.Synthesis, characterization and the rapid response property of the temperature responsive PVP-g-PNIPAM hydrogel[J].European Polymer Journal , 2008, 44: 2162-2170.
4. Zhang Jie, Chu Liangyin, Li Yuanke, et al.Dual thermo-and pH-sensitive poly(N-isopropylacrylamide-co-acrylic acid)hydrogels with rapid response behaviors[J].Polymer, 2007, 48: 1718-1728.
5. Liewen Liao, Zhengtang Liu, Xinming,Li. Solution properties of co-polymer hydrogel of AA/AMPS[J]. Advances in Chemical Engineering,2008,27(5):729-732
6. Abd El-Rehim H A, Hegazy E A, Khalil F H, et al.Radiation preparation of drug carriers based polyacrylic acid(PAAc)using poly(vinyl pyrrolidone)(PVP)as a template polymer[J].Nuclear
7. ChangmeiKe, HouzhiWang, LilinLu, Preparation and properties of Polyacrylate /TiO2-SiO2 nanohybrid materials [J]. Glass just / composite, 2005 (4):23-26.
8. Wei Y,JinD, YangC.J Sol-Gel Sci Tech.1996, 7:191-196

Thermoelectric Properties of Composite Nanoparticle Array with Protoporphyrin Molecule Junctions

I-Chih Ni[1], and Shien-Der Tzeng[1]

[1]Department of Physics, National Dong Hwa University, 97401 Shoufeng, Hualien, Taiwan, R.O.C.

Abstract. Composite nano-structures have many interesting properties and potential applications. For a molecule-nanoparticle (NP) composite array, its properties may strongly depend on the molecules on/between NPs. For instance, the resistance of NP composite film sensitively change with the length of surface molecules,[1,2] and could be applied to vapor sensors.[3] In this report, we show a homemade system to measure thermoelectric properties of protoporphyrin molecule junctions in molecule-NP composite array. Many kinds of metal-protoporphyrin molecules (Heme, Hematin, Zinc-protoporphyrin, Cobaltic-protoporphyrin, Protoporphyrin IX) and dl-cysteine/cysteamine molecules were studied. We found metal ion in protoporphyrin molecule is matter for the Seebeck coefficient of molecule-NP composite array.

1 Introduction

Recent years, thermoelectric properties of molecule junctions were reported in theories [4-6] and in experiments [7-9]. For instance, Yu-Shen Liu *et al.* presented the thermoelectricity in molecular junctions formed by a single molecule contact through first-principle calculation.[4] The Seebeck coefficient of molecule-metal junction is related to the slope of the transmission function near Fermi level, and could be described by eq. (1).

$$S = -\frac{\pi^2 K_B{}^2 T}{3e}\frac{\partial \ln \tau(E)}{\partial E}\Big|_{E=E_F} \quad . \tag{1}$$

Here $\tau(E)$ is transmission function of electron with energy E passing through molecule junction between electrodes. The Seebeck coefficient would be equal to zero when peak position of LUMO or HOMO aligns with the Fermi level (E_F). In addition, Seebeck coefficients are related to the structure and functional group of molecules.[9] For instance, amine-terminated molecule junctions may have positive Seebeck coefficient because of HOMO-conducting, while pyridine-terminated molecule junctions have negative Seebeck coefficient because of LUMO-conducting. B. Wang *et al.* reported calculated thermoelectric property of different length of alkane molecules between Au electrodes.[6] They found the thermoelectric voltages are positive when the number of carbon is odd, while the voltages is negative when the number of carbon is even.

Recent years, scanning probe microscopy had been used to measure thermoelectric properties of molecule junctions.[4-5,7-8] In a scanning tunneling microscope (STM), Au tip is kept at room temperature with a large thermal reservoir as a cool side; Au substrate is heated to a controlled temperature, then the temperature difference (dT) is formed between substrate and tip. The Au substrate surface has some thiol-terminated molecules, then Au STM tip slowly moves to approach these molecules by monitoring tunneling current between tip and substrate. After tip traps molecules, tip slowly withdraw to a sufficiently large distance. Meanwhile, thermoelectric voltage is measured. Although, STM measurement could be capable of measuring the Seebeck coefficient of single molecule, its operation is relatively difficult and STM is expensive.

In this report, we use gold nanoparticle (AuNP) array to form abundant molecule junctions, and measure their thermoelectric properties. Our measurement method is simple, inexpensive, convenient, stable, and can provide rich molecule-NP nano-composite structures for future applications.

2 Basic Principles

Thermoelectric voltage of molecule junctions could come from differential Fermi levels of left and right molecule-electrode junctions.[4-6,10-12] In our experiment, electrodes are AuNPs, and metal-protoporphyrin molecules are modified on AuNP surfaces. When thermoelectric voltage is created, the current of molecule junction could be described with Landauer function:

$$I = \frac{2q}{h}\int \overline{T(E)}(f_1 - f_2)dE \tag{2}$$

Here $\overline{T(E)}$ is the transmission function, and f_1 and f_2 are equilibrium Fermi-Dirac distributions for heating and

a Corresponding author: author@e-mail.org

cooling contacts. Fermi level $f_1 - f_2$ could be described by

$$f_1 - f_2 \approx \frac{\partial f}{\partial \mu}\Delta\mu + \frac{\partial f}{\partial T}\Delta T \tag{3}$$

$$f(\mu, T) = \frac{1}{e^{(E-\mu)/K_B T}+1} \tag{4}$$

Here $\Delta T = T_1 - T_2$ is the temperature difference, and $\Delta\mu$ is the Fermi level difference between contacts. The current of molecule junction could be described by

$$\mathrm{I} = \frac{2q}{h}\int \overline{T(E)}\left(\frac{\partial f}{\partial \mu}\Delta\mu + \frac{\partial f}{\partial T}\Delta T\right)dE \tag{5}$$

$$\frac{\partial f}{\partial \mu} = e^{\frac{E-\mu}{K_B T}} \cdot \frac{1}{K_B T}\left(e^{\frac{E-\mu}{K_B T}} + 1\right)^{-2} \tag{6}$$

$$\frac{\partial f}{\partial T} = e^{\frac{E-\mu}{K_B T}} \cdot \frac{(E-\mu)}{K_B T^2}\left(e^{\frac{E-\mu}{K_B T}} + 1\right)^{-2} = \frac{E-\mu}{T}\left(\frac{\partial f}{\partial E}\right) \tag{7}$$

For linear response regime, heating and cooling rates of contacts are equal, the current of molecule junction could be described by

$$\mathrm{I} \approx <\mathrm{G}> \Delta\mathrm{V} + <\mathrm{SG}> \Delta T \tag{8}$$

Here S is Seebeck coefficient, G is conductance, and

$$<\mathrm{G}> \Delta\mathrm{V} = \frac{2q}{h}\int \overline{T(E)}\left(\frac{\partial f}{\partial \mu}\Delta\mu\right) dE \tag{9}$$

According to $\Delta\mu = -q \times \Delta V, \Delta V = V_1 - V_2$,

$$<\mathrm{G}> \Delta\mathrm{V} = \frac{2q^2}{h}\int \overline{T(E)}\left(-\frac{\partial f}{\partial E}dE\right)\Delta V \tag{10}$$

$$<\mathrm{G}> = \frac{2q^2}{h}\int \overline{T(E)}\left(-\frac{\partial f}{\partial E}dE\right) \tag{11}$$

Another part,

$$<\mathrm{SG}> \Delta T = \frac{2q}{h}\int \overline{T(E)}\left(\frac{\partial f}{\partial T}\Delta T\right) dE$$

$$= -\frac{2q}{hT}\int \overline{T(E)}(E-\mu)\left(-\frac{\partial f}{\partial E}\right)\cdot dE(\Delta T) \tag{12}$$

$$<\mathrm{SG}> = -\frac{2q}{hT}\int \overline{T(E)}(E-\mu)\left(-\frac{\partial f}{\partial E}\right)\cdot dE \tag{13}$$

Then, Seebeck coefficient could be described by

$$\mathrm{S} = \frac{<\mathrm{SG}>}{<\mathrm{G}>} = \frac{-\frac{2q}{hT}\int \overline{T(E)}(E-\mu)\left(-\frac{\partial f}{\partial E}\right)\cdot dE}{\frac{2q^2}{h}\int \overline{T(E)}\left(-\frac{\partial f}{\partial E}dE\right)} = \frac{K_B}{-q}\frac{\int \overline{T(E)}\frac{(E-E_F)}{K_B T}\left(-\frac{\partial f}{\partial E}\right)\cdot dE}{\int \overline{T(E)}\left(-\frac{\partial f}{\partial E}dE\right)} \tag{14}$$

Because

$$\frac{\partial f}{\partial T} = \frac{E-\mu}{T}\left(\frac{\partial f}{\partial E}\right), \frac{\partial f}{\partial T} = -\left(\frac{\partial f}{\partial E}\right)\delta(E-\mu) \tag{15}$$

Then

$$\frac{\Delta\mu}{\Delta T} = \frac{1}{\overline{T(E)}}\int \overline{T(E)}\frac{(E-\mu)}{T}\left(\frac{\partial f}{\partial E}\right)dE|_{E=u} = \frac{\pi^2 K_B{}^2 T}{3}\frac{1}{\overline{T(E)}}\frac{\partial\overline{T(E)}}{\partial E}|_{E=\mu} \tag{16}$$

Finally, Seebeck coefficient could be described by

$$\mathrm{S} = -\frac{\Delta V}{\Delta T} = -\frac{\pi^2 K_B{}^2 T}{3e}\frac{1}{\overline{T(E)}}\frac{\partial\overline{T(E)}}{\partial E}|_{E=\mu} \tag{17}$$

3 Experiment method and measurement system

3.1 Metal-protoporphyrin-AuNP film

AuNP surfaces are modified with metal-protoporphyrin molecules (such as Heme, Hematin, Zinc-protoporphyrin, Protoporphyrin IX, and Cobaltic-protoporphyrin) or dl-cysteine/cysteamine molecules, then deposited on glass substrate surface with symmetrical electrodes by a centrifuge deposition method.[1] The symmetrical electrodes were 20nm/50nm Cr/Au deposited on glass substrate (Corning). High-purity Cu wires were adhered respectively on two pairs of electrodes by silver filled electrically conductive adhesives.

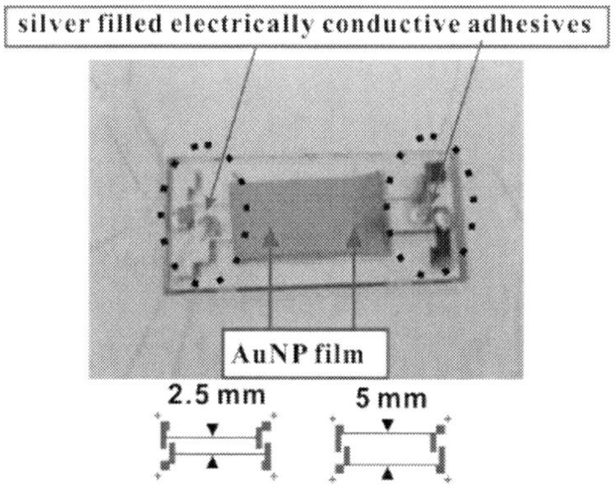

Figure 1. Sketch images of electrodes, and a picture of molecule-AuNP film on a 2.5 mm-gap electrodes.

3.2 Measurement system

We built up a system for thermoelectric voltage measurements with high sensitivity (0.1 μV) and low thermal noise for high resistance samples (to 300 kΩ). The design of homemade measurement system is shown in figure 2. Measurement instruments and data acquisition are connected and controlled by LabVIEW program. The vacuum pressure of system chamber is about 10^{-5} torr. Sample is placed on a metal bar connected between heater and cooler stages. In experiment, the initial temperatures of hot and cold stage are kept at about 150 °C and 4 °C respectively, and then slowly return to room temperature through shutdown the heating and cooling power simultaneously. Thermoelectric voltage of molecule-AuNP film is measured by HP 34420A micro-voltage meter when temperatures slowly return to room temperature. Thermal radiation from heat source is shielded by aluminium foils, as shown in figure 3. Voltage of thermocouples (T_{cool} and T_{hot}) on both sides of the chip are measured by keithley 2001 and Agilent 34410A voltage meter. All valves, pumps, gauges, mass

DOI: 10.1051/matecconf/201713008001

flows, heater, and cooler are controlled or measured through digital I/O and analog I/O by a data acquisition board (NI Elvis II).

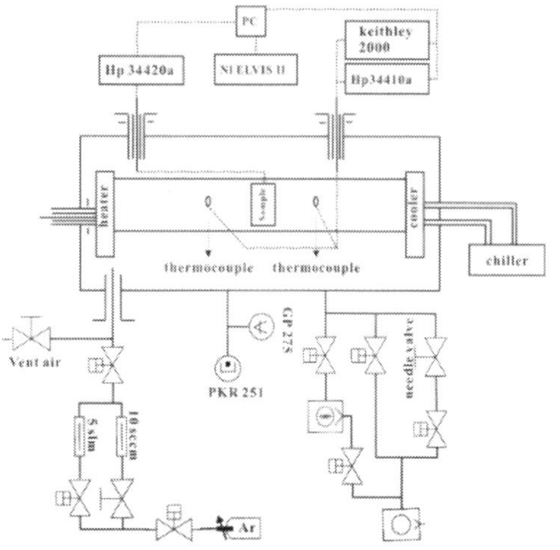

Figure 2. The design of homemade system for molecule thermoelectric measurements.

Figure 3. Picture of the measurement system.

4 Result and Discussion

Six kinds of molecule-AuNP composite films are studied. Heme, Hematin, Zinc-protoporphyrin, Cobaltic-protoporphyrin or Protoporphyrin IX molecules are linked on NH_2 function groups of dl-cysteine$_{0.75}$/cysteamine$_{0.25}$-AuNPs by ENC/NHS crosslinking reaction. Figure 4 (a)(c) shows thermoelectric voltage-temperature curves of two Heme-AuNP films on symmetrical electrode of 5 mm and 2.5 mm gap. Three voltage curves are corresponding to original sample direction(dV), sample rotated with 180 degree (dV_R), and sample lift up from heater bar (dV_L). The curve of dV-dT_{film} is averaged from repeated measurements for 2~3 times. We found the thermoelectric voltage of dV_L is near to zero (because dT_{film}~0). The inverse symmetrical behavior of dV and dV_R curves were due to opposite temperture gradient. This displays the acquired voltage are correct thermoelectric voltage from Heme-AuNPs film. Figure 4 (b)(d) show thermoelectric voltage of the AuNP films by eq. (18).

$$dV_{film} = (dV - dV_R)/2 \tag{18}$$

We found the relationship between dV_{film} and dT_{film} is nearly linear, and the slopes for different gap of symmetrical electrodes are similar. We also calculate the Seebeck coefficient (S) of Heme-AuNP films through eq. (19), and get S are about 2.5 and 2.4 μV/K for 5 mm and 2.5 mm gap electrodes, respectively.

$$S = -\frac{dV_{film}}{dT_{film}} = -\text{slope} \tag{19}$$

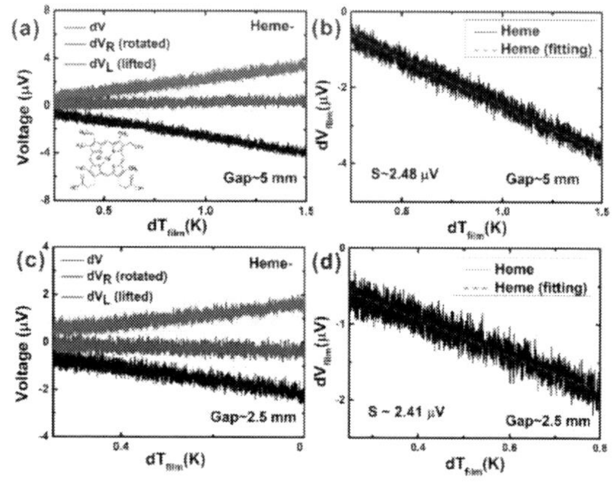

Figure 4. Thermoelectric voltage of Heme-AuNPs films for (a) 5 mm and (b) 2.5 mm gap electrodes. (b)(d) The dV_{film}- dT_{film} curve. Inset sketch is molecule structure of Heme.

We also found Seebeck coefficients of cysteine$_{0.75}$/cysteamine$_{0.25}$-AuNP films are negative (≈ -2.2 μV/K). But after linking protoporphyrin molecules on surfaces, its Seebeck coefficient is changed to positive. Table 1 shows Seebeck coefficients of some molecule-AuNPs films (each S was averaged from 4 ~ 6 sample). We found Proto-porphrin (no metal ion)-AuNPs films have a Seebeck coefficient (S ≈ 1.0 μV/K) smaller than other metal-Protoporphrin-AuNPs films, and the Seebeck coefficient would increase with electronic configuration of the ceter-matal elements in 3d orbital. The Zinc($3d_{10}4s_2$)-protoporphyrin samples have largest Seebeck coefficient (S ≈ 9.6 μV/K) than Cobaltic($3d_74s_2$)-, Heme($3d_64s_2$)-, and Hematin($3d_64s_2$)-protoporphyrin samples. This may due to its filled 3d orbitals. Besides, Heme-AuNP and Hematin-AuNP films have similar Seebeck coefficient. These results show center-metal ion has important influence to the Seebeck coefficient of protoporphyrin molecules.

5 Conclusion

We use novel nano-assembly method to fabricate composite molecule-AuNP films and build up a measurement system to study their thermoelectric properties. The composite molecule-AuNP arrays have lots of molecule junctions in the films. We got the Seebeck coefficients of metal-protoporphyrin-AuNP films and cysteine/cysteamine-AuNPs films, and found the Seebeck coefficients of metal-protoporphyrin-AuNPs films were positive, but was negative for cysteine/cysteamine-AuNP film. We also found Heme-AuNP and Hematin-AuNP films have similar Seebeck coefficient. This may due to the same ceter-metal ion (Fe). Zinc-protoporphyrin have largest Seebeck coefficient than others that may due to its filled 3d orbitals.

Table 1. Seebeck coefficient of molecule-AuNP films.

	Molecule Structure	Seebeck coefficient (μV/K)
Heme	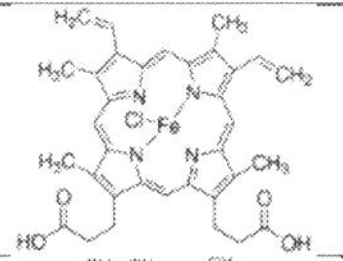	3.1 ± 1.0
Hematin	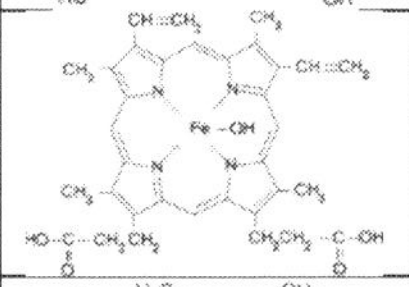	3.8 ± 0.5
Cobaltic-protoporphyrin	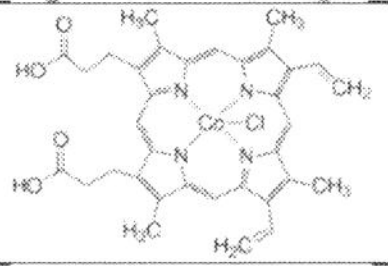	5.1 ± 0.5
Zinc-protoporphyrin	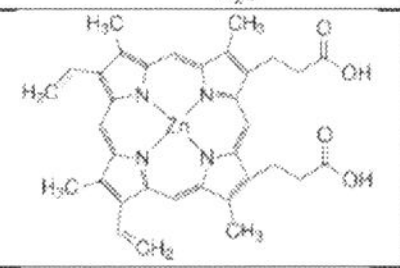	9.6 ± 0.9
Protoporphyrin IX	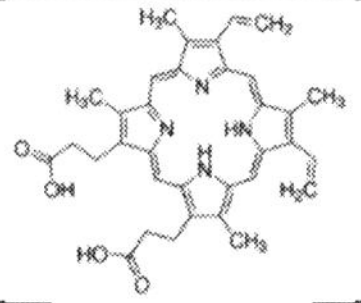	1.1 ± 0.2
cysteine$_{0.75}$/ cysteamine$_{0.25}$	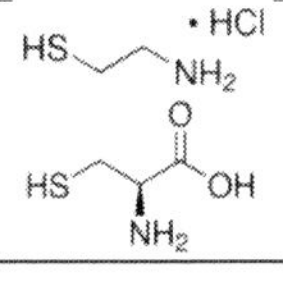	-2.3 ± 0.2

References

1. C.-W. Jiang, I-C. Ni, S.-D. Tzeng, C.-S. Wu and W. Kuo, Nanoscale **6**, 5887-5893 (2014).
2. Amir Z.-K., P.-E. Trudeau, Y. Suganuma, A.-A. Dhirani, and B. Statt, Phys. Rev. Lett. **96**, 156403 (2006).
3. Y. Joseph, A. Peić, X. Chen, J. Michl, T. Vossmeyer, and A. Yasuda, J. Phys. Chem. C **111**, 12855-12859 (2007).
4. Y.-S. Liu and Y.-C. Chen, Phys. Rev. B **79**, 193101 (2009).
5. Y.-S. Liu, Y.-R. Chen and Y.-C. Chen, ACS Nano **3**, 3497-2504 (2009).
6. Wang B, Xing YX, Wei YD, Wan LH, Wang J, Carbon **43**, 2786-2791 (2005).
7. J. R. Widawsky, W. Chen, H. Vázquez, T. Kim, R. Breslow, M. S. Hybertsen, and L. Venkataraman, Nano Lett. **13**, 2889-2894 (2013).
8. P. Reddy, S.-Y. Jang, R. A. Segalman, and A. Majumdar, Science **315**, 1568-1571 (2007).
9. J. R. Widawsky, P. Darancet, J. B. Neaton, and L. Venkataraman, Nano Lett. **12**, 354-358 (2012).
10. J. P. Bergfield, M. A. Solis, and C. A. Stafford, ACS Nano **4**, 5314-5320 (2010).
11. Y. Dubi and M. D. Ventra, Nano Lett. **9**, 97-101 (2008).
12. M. Tsutsui and M. Taniguchi, Sensors **12**, 7259-7298 (2012).

Preparation and properties of silicone fouling release coatings with long-life afterglow fluorescent

Zhanping Zhang[a], Hao Li and Yuhong Qi

Department of Materials Science and Engineering, Dalian Maritime University, Dalian 116026, China

Abstract. Based on polydimethylsiloxane, three-component coatings were prepared with different content of luminescence powder. The results showed that the illuminance of coatings increases with the content of luminescence powder, decays exponentially with the afterglow time, increases exponentially with the increase of exposure time. The afterglow illuminance augments with irradiated light illuminance. All coatings are hydrophobic and oleophilic. Surface free energy decreases with the increase of luminescence powder. They have highest impact-resistance and bend flexibility. The luminescence powder does not change obviously the shore hardness, tensile breaking strength, breaking elongation rate, elastic modular and roughness of coatings. The static test panels in sea generally could be covered obviously by biofouling including sponges, bryophytes and mussels, hydra, kelp, green algae after 2 months of immersion during growing season. But it never found that the barnacle attached on the coating surface during 4 years of immersion test. The static anti-fouling ability of the coatings is very limited. In addition, the sea creatures attached on the coating surface can be easily removed; even attached organisms will fall off and expose again the smooth coating surface. Consequently, all coatings with long-life afterglow fluorescent have a significant effect on preventing adhesion of barnacle and fouling-release performance.

1 Introduction

All man-made structures in marine environments suffer from the problem of biofouling that the unwanted growth of bacteria, algae and invertebrates. It causes extra financial burdens, excessive consumption of energy, and contamination of the environment. Current self-polishing antifouling coating technologies derived from copper-based compounds, could induce detrimental effects on marine environment [1]. Hence there is an exigent need for developing efficient and non-toxic marine coatings that would possess no ecological concern.

During past years, more researches were focused on the non-toxic antifouling coatings, especially on fouling release coatings. Fouling release coatings present a surface which can reduce the adhesion strength of fouling organisms enabling cleaning by pressurized water jets or in some cases by hydrodynamic forces generated as the ship sailing [2]. These include silicone and fluoropolymer coatings. Amount of studies has been conducted on fouling-release coating systems based on polydimethylsiloxane (PDMS), appear to out-perform fluoropolymers and be a viable fouling-release coating system. Surface properties of the PDMS coatings including surface free energy, elastic modulus and thickness are considered be important to the fouling release performance [3]. In addition, some researchers pointed out that incorporation of low molecular weight silicone oils into silicone coatings improves their fouling release properties [4].

Biofouling process occurs on the interface between seawater and coating surface. Previous researches showed that the colour and brightness of coating surface can affect the attachment behaviour of marine biofouling [1]. Furthermore, invertebrates like barnacle, adhere to all kinds of surfaces in the worldwide ocean and the animals settle as larval cyprids, prefer live in dark environment [2]. Considering there is sunlight undersea in the depth of 200m, we therefore conceived to develop a novel technique that long afterglow luminous coating will be used to antifouling in marine environment in previous study. The results have proved that phosphorescent materials in coating can be activated by sunlight during day even the coating was immersed in seawater [5]. In addition, Cao's investigation has proven that the luminescence of coating during night will disturb the normal physiological activity and metabolism of marine organism, and provoke larval cyprids to keep away from the coating [6].

In this paper, based on PDMS, five fouling release coatings were prepared consisting of 107 silicone gum, dimethyl silicone oil, tetraethylorthosilicate as crosslinked curing agent, heavy calcium carbonate as filler, alkali soils silicon aluminate as luminescence powder, dibutyltin dilaurate as catalyst. The aim of this study is to evaluate the influence of luminescence powder

[a] Corresponding author: zzp@dlmu.edu.cn

 DOI: 10.1051/matecconf/201713008002

content on the properties of silicone fouling release coatings with long-life afterglow fluorescent.

2 Materials and methods

2.1 Preparation of coatings

Based on polydimethylsiloxane (PDMS), five three-component paints were prepared with controlling the content of luminescence powder (Alkali soils silicon aluminate, body colour is light-yellow; glowing colour is blue) respectively 2.5, 5, 7.5, 10, 12.5 wt. %. Component A included PDMS, silicone oil, fillers, luminescence powder and additives. TEOS mixed with xylene to make curing agent (Component B), the mixture of DBTDL and acetylacetonate was prepared into catalytic agent (Component C). The dimethyl silicone oil (DY-201-100), which formula is $(CH_3)_3SiO[(CH_3)_2SiO]n\text{-}Si(CH_3)_3$. The viscosity is 100±5(25°C) mPa·s, was bought from Shandong Dayi Chemical Co. Ltd. The 107 silicone gum (DY-OH501), which formula is $HO\text{-}Si(CH_3)_2O(Si(CH_3)_2O)_nSi(CH_3)_2\text{-}OH$, possesses the reactivity of hydroxyl group that can be made into some crosslinked as elastomer coating. The tetraethylorthosilicate (TEOS) is crosslinked curing agent, the dibutyltin dilaurate (DBTDL) is catalyst. Both of them and xylene were bought from Tianjin Kemiou Chemical Reagent Co., Ltd. BYK-161 and BYK-066N is respectively dispersing agent and defoamer produced by BYK-Chemie GmbH. The composition of studied paints is listed in Table 1. The component A is first mixed at 4000rpm in high-speed stirrer, then grinded to 40 μm by cone mill. Three components were mixed for 15 minutes at A:B:C=18.3:4:1 before painting, then cured in ambient for at least 7 days before tests.

Paint was brushed on glass slides with dimensions of 75mm×25mm×1mm, and also cast into a Teflon mould with dimensions of 150mm×150mm×3mm. The slide samples were for contact angle and surface characteristics measurement of coatings. Cast specimen was cut strip with 40mm×250mm for mechanical properties tests. Each coating was prepared three pieces of panels for antifouling immersion test in ocean. The steel plate samples with 300mm×200mm×3mm were in turn coated with epoxy anti-rust primer, intermediate anti-corrosive paint and prepared silicone fouling release coatings with long-life afterglow fluorescent.

2.2 Mechanical properties

Tensile samples were tested on a Labthink XLM auto tensile tester. Elastic modulus was fitted with the data which the strain is less than 0.02 mm/mm. HT220 shore hardness tester was applied to test the hardness of coating.

2.3 Fluorescence of the coatings

After some period exposure of the panels under artificial illumination in HPG-280B container, lamp-house was turned off and the fluorescence of the coatings was measured with illuminometer ST-900 at regular intervals.

Table 1. Composition of studied paint.

Component		Paint				
Composition / wt.%		P5	P10	P15	P20	P25
A	Heavy calcium carbonate	10	7.5	5	2.5	0
	Luminescence powder (PLO-7C)	2.5	5	7.5	10	12.5
	PDMS (DY-107)	50				
	Xylene	10				
	Fillers	7.5				
	Additive	2.5				
	Silicone Oil (DY-201)	2.5				
B		12				
C		3				
Total		100				

2.4 Surface morphology and topography

The morphology was investigated by confocal laser scanning microscopy (CLSM) OLS4000 from Olympus. Roughness was measured with software LEXT. The objective lens is used respectively BF plan semi-apochromat and LEXT-Dedicated plan apochromat. The coating was observed respectively after curing 7 days, 30 days, 180 days and 360 days.

2.5 Surface characteristics of the coatings

Contact angle measurements were conducted using the sessile drop method on a JC2000C (Zhongchen, P. R. China) contact angle measuring system. The surface free energy was calculated based on the contact angle measurements of deionized water and diiodomethane.

2.6 Antifouling performances of the coatings

The antifouling property of the samples in the marine environment was also investigated. Panel samples for antifouling test in sea were spread onto 300mm×200mm×3mm steel plate, pre-coated with epoxy anti-rust primer and iron oxide red epoxy intermediate coating. Then panel samples were hung on a floating dock in Lvshun Port in China, and the panels were immersed in the Yellow Sea at depths of 1–2 m form April 2012. The samples were checked regularly to monitor the progress of the fouling process.

3 Results and discussion

3.1 Illuminance

The results showed that the illuminance of coatings increases with the content of luminescence powder, the illuminance of every coating decays exponentially with the increase of afterglow time. With the increase of exposure time, the illuminance of coating increases exponentially. When the coating was irradiated under higher light illuminance, the afterglow illuminance of coating augments.

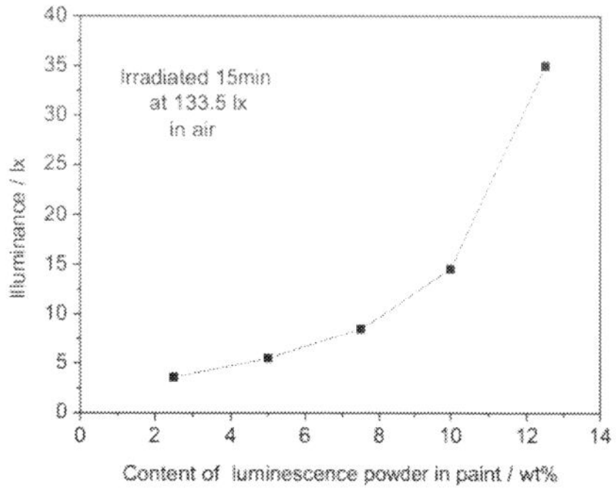

Figure 1. Illuminance of coatings vs. content of luminescence powder.

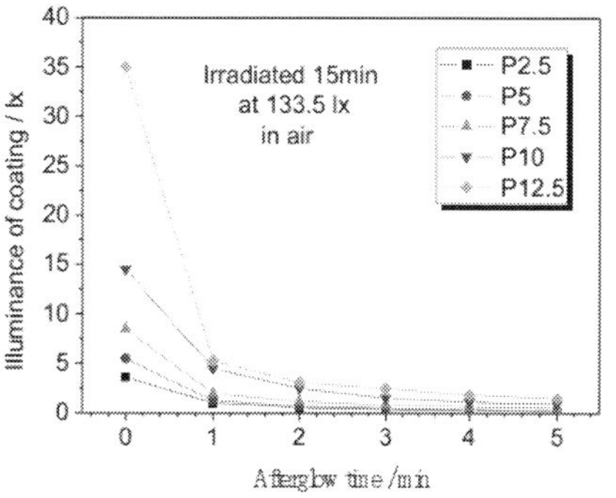

Figure 2. Illuminance of coatings vs. afterglow time.

3.2 Surface morphology and topography

Because both of PDMS and silicone oil have very low surface tension, studied paints have excellent levelling property. Cured coating looks very smooth. Morphology of studied coating showed in Figure 3. Measured roughness Ra of coatings changes from 0.13μm to 0.17μm, it is not related to the content of luminescence powder.

3.3 Surface characteristics

The surface free energies and components were calculated from contact angle measurements, using the acid-base theory for solids according to the van Oss-Chaudhury-Good approach [6]. The test liquids were distilled water and diiodomethane. The contact angles, the surface free energies and their corresponding polar and dispersive components of the samples were listed in Table 2.The results showed that the water contact angle of coatings is between 103° and 110°, they are hydrophobic. CH_2I_2 contact angle of coatings increases, it showed be oleophilic. Surface free energy of coatings, ranged from 29 mJ/m^2 to 27 mJ/m^2, decrease with the increase of luminescence powder. As shown in Figure 4, it was found that the surface free energy not only results from its dispersive component, but also is related obviously to the dispersive component.

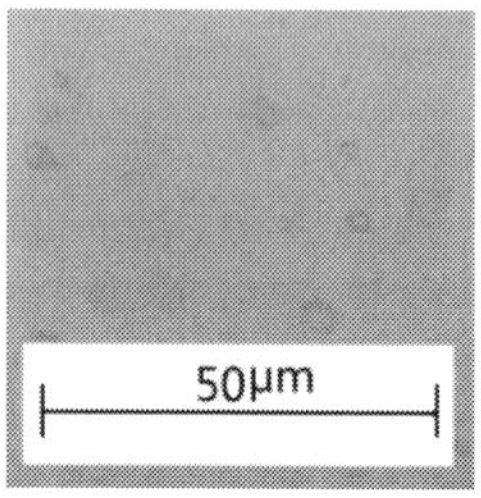

Figure 3. Morphology of studied coating.

Table 2. Contact angle and surface free energy of coatings.

	Contact angle / °		Surface energy / mJ/m^2		
Coating	H_2O	CH_2I_2	Polar	Dispersive	Total
P2.5	103.6	59.9	0.16	28.70	28.85
P5	109.6	62.4	0.01	28.01	28.02
P7.5	108.2	62.9	0.01	27.48	27.49
P10	108.0	63.0	0.01	27.39	27.40
P12.5	110.8	64.3	0.01	26.98	27.00

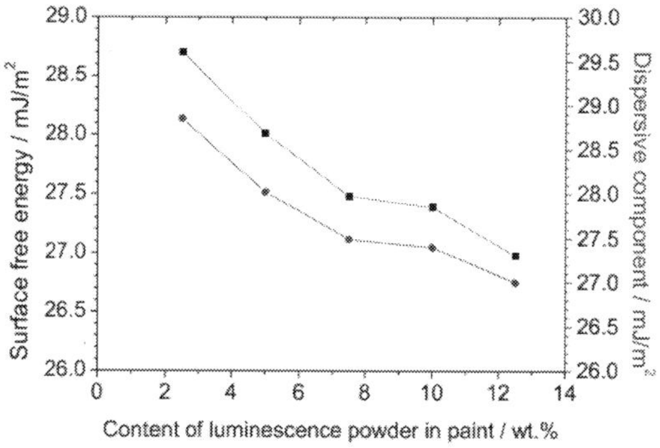

Figure 4. Surface free energy and its dispersive component.

3.4 Mechanical properties

All coatings have highest impact-resistance and bend flexibility, they are respectively 50kg·cm and 1mm. The shore hardness, tensile breaking strength, breaking elongation rate and elastic modular of coatings were listed in Table 3. They are not affected by the content of luminescence powder. The shore hardness is about 27

 DOI: 10.1051/matecconf/201713008002

HA. The breaking elongation rates of coatings are range from 180% to 240%. The elastic modular is much low, it is about 0.2MPa. Therefore the coatings showed good elastic deformation capacity.

Table 3. Mechanical properties of coatings.

Coating	Breaking strength (MPa)	Elongation rate (%)	Elastic modular (MPa)	Shore's hardness HA (%)
P2.5	1.17	237.88	0.23	28.8
P5	1.08	213.33	0.20	25.8
P7.5	1.07	235.33	0.20	27.3
P10	0.96	179.00	0.26	28.7
P12.5	1.21	238.70	0.23	24.9

3.5 Antifouling performance

Because the coatings are non-toxic, as shown in Figure 5, the coating surfaces of static test panels generally could be covered obviously by biofouling after 2 months of immersion in sea during growing season for sea creatures. The marine organisms found on the surface of the coating were sponges, bryophytes and mussels, Hydra, kelp, green algae. But it never found that the barnacle attached on the coating surface during 4 years of immersion test in sea. The static anti-fouling ability of the coatings is very limited. In addition, the sea creatures attached on the coating surface can be easily removed, even attached organisms will fall off and expose again the smooth coating surface. As shown in Figure 6, taking coating P12.5 as an example, the coatings repeatedly undergo organism adhering and fouling, fouling-released and self-cleaning. Consequently, all coatings with long-life afterglow fluorescent have a significant effect on preventing adhesion of barnacle and good fouling-release performance.

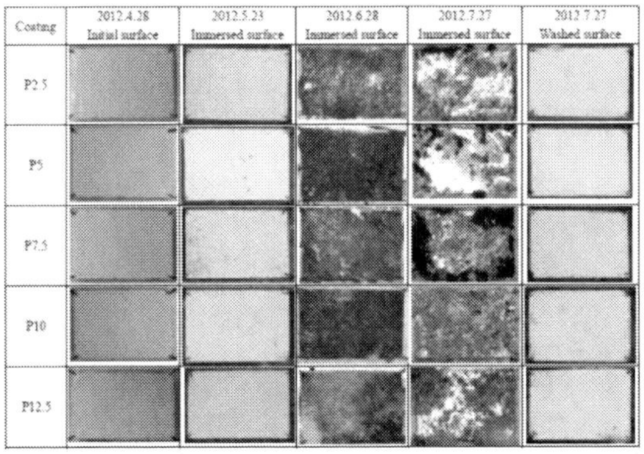

Figure 5. Photos of the coatings for static test panels.

3.6 Discussion

Above-mentioned results appear that the antifouling property of coatings is neither related to illuminance nor surface free energy of coatings. This is consistent with the results of previous studies on silicone fouling-release coatings. In other words, the attachment of fouling organisms to such coatings may be related with surface free energy, but not positively correlated [7]. It can also be related to the elastic modulus of the coatings and other factors (e.g., polarity, electronegativity, thickness and smoothness). Specially, it was experimentally confirmed that adhesion is proportional to the square root of the product of elastic modulus (E, MPa) and surface energy (γ, mJ/m^2) [8,9]. Accordingly, the square root of the product can be defined as the relative adhesion factor. This factor used to as a key reference for laboratory design and for selecting fouling-release coatings. Calculated values of relative adhesion factor for studied coatings were listed in Table 4. Clearly, the antifouling performance of studied coatings is not well proportional to the relative adhesion factor. It means that the antifouling performance of the coatings is not only related to single physical parameter, but influenced by some physical and structure factors.

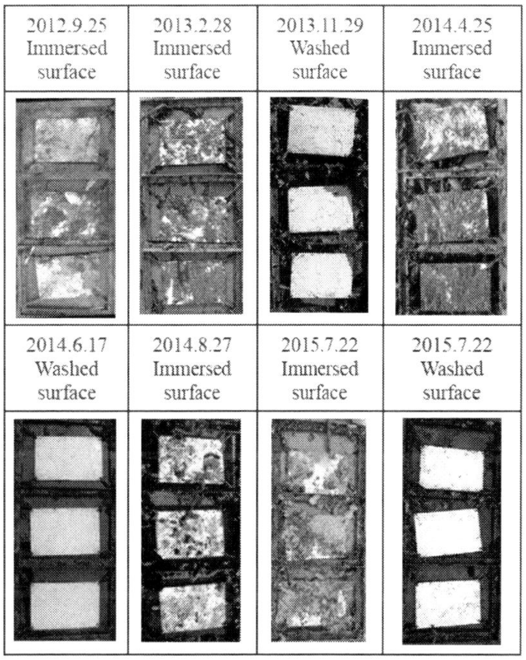

Figure 6. Photos of coating P12.5 for static test panels.

Table 4. Relative adhesion of coatings.

Coating	P2.5	P5	P7.5	P10	P12.5
Relative adhesion factor	2.60	2.39	2.34	2.64	2.50

In addition, non-reactive silicone oil was added to the silicone coating in this study. Observation results under confocal laser scanning microscopy showed that no silicone oil appears on the surface of coating samples exposed in air for 180 days. Leached silicone oil was observed on the coating samples exposed in air for 360

DOI: 10.1051/matecconf/201713008002

days, as shown in Figure 7. It's clearly that the amount of leached silicone oil increases with the content of luminescence powder. This can be attributed to the effect of oil absorption of luminescence powder and fillers.

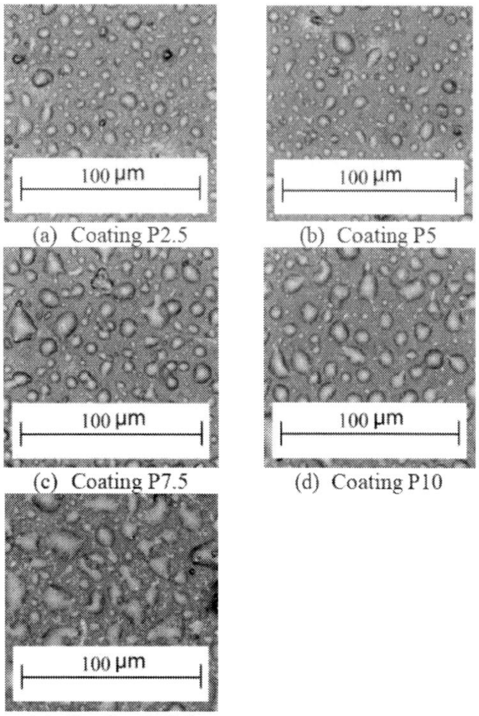

(a) Coating P2.5 (b) Coating P5 (c) Coating P7.5 (d) Coating P10 (e) Coating P12.5

Figure 7. The coating surfaces exposed in air for 360 days.

Above-mentioned results showed that dimethyl silicone oil leach enough slowly from studied coating in air. However leaching rate of silicone oil will accelerate once the coating was immersed in seawater. This has been proven in previous study, relative details can be found in reference [10]. Leached silicone oil will weaken the adhesion between the attached organism and the coating surface. As shown in Figure 5, when the attached organisms accumulate and silicone oil leach to a certain extent, the attached organism will spontaneously fall off. Consequently, studied coatings have good self-cleanly and fouling-release performance.

4 Conclusions

The illuminance of coatings increases with the content of luminescence powder, decays exponentially with the afterglow time, increases exponentially with the increase of exposure time. The afterglow illuminance augments with irradiated light illuminance. All coatings are hydrophobic and oleophilic. Surface free energy decreases with the increase of luminescence powder. They have highest impact-resistance and bend flexibility. The luminescence powder does not change obviously the shore hardness, tensile breaking strength, breaking elongation rate, elastic modular and roughness of coatings. The static test panels in sea generally could be covered obviously by biofouling including sponges, bryophytes and mussels, hydra, kelp, green algae after 2 months of immersion during growing season. But it never found the barnacle attached on the coating surface during 4 years of immersion test. The dimethyl silicone oil can leach slowly from studied coating. It will weaken the adhesion between the attached organism and the coating surface. Through the static anti-fouling ability of the coatings is very limited, however, the marine organism attached on the coating surface can be easily removed, even will fall off and expose again the smooth coating surface. Consequently, all coatings with long-life afterglow fluorescent have a significant effect on preventing adhesion of photophobic barnacle and fouling-release performance.

Acknowledgements

This work was supported by National nature science foundation of China (51179018), defense fund and the application basic research fund of the Ministry of Transport (2013329225330). The authors gratefully acknowledge for financial support.

References

1. Elisabete A., Teresa C. D., Orlando de S, Prog. Org. Coat., **59**, 2-20(2007)
2. D. M. Yebra, S. Kiil, K. Dam-Johansen. Prog. Org. Coat., **50**, 75 (2004)
3. T. Vladkova, Journal of the University of Chemical Technology and Metallurgy, **42**,239(2007)
4. Chambers L.D., Stokes K.R., Walsh F.C., et al. Surface & Coatings Technology, **201**, 3642-3652(2006)
5. Yu-Hong Qi, Zhan-Ping Zhang, Hui Gao, *Proc. of SPIE*, **7493**,74934M(2009)
6. Shan Cao, Jiadao Wang, Darong Chen, Microbial Ecology, **62**(4): 931-940(2011)
7. R.E. Baier, E.G. Shafrin, W.A. Zisman, Science, **162**, 1360–1368 (1968)
8. R.F. Brady, Prog. Org. Coat. **43**,188–192 (2001)
9. T.J. Gui, Ocean University of China, Tsingtao, (2008)
10. Zhanping ZHANG, Yuhong QI, Miao BA, et al. Advanced Materials Research, **842**, 737-741(2014)

Study the Influence of the Anodizing Process Parameters on the Anodized Copper Hardness

M.H. Mahmood [1a], Suryanto [1,b], Muataz H. F. Al Hazza [1,c] and Farag I Haider [1,d]

[1]*Department of Manufacturing and Material Engineering, International Islamic University Malaysia, Kuala Lumpur, Malaysia*

Abstract. The metals anodization process used to enhance the surface hardness and corrosion resistance. This study developed a durable hard Nano copper oxide coating on copper using anodization technique in solutions of 0.1 to 0.5 M oxalate concentrations and 0 to 24°C operating temperature. The settings of the process parameters determined by using Taguchi's experimental design method. The EDX and XRD results confirm the formation of cupric oxide coating with monoclinic lattice crystalline structures. The FESEM results for the coated samples showed that the grain size was in the range between 25 to 68 nm. Microhardness tests for the anodized copper samples characterized by microhardness tester. Analysis of Variance for the orthogonal arrays of Taguchi identified that the most affecting parameter on the microhardness of the coating was the anodizing temperature. The results show that the hardness of the anodized coating was decreased with the anodizing temperature, where maximum hardness, with smaller grain size, were produced at lower anodizing temperatures.

1 Introduction

Anodization is an electrochemical process used to create a uniform anodic metal oxide coating on the metal surface. In this electrochemical process, the sample set as the anode part of the electrical circuit. The metal surface after the anodization process converted to a thick and hard oxide layer, which provides some degree of corrosion protection for the underlying metal. Recently Nano copper oxide coating is being developed for various applications as, gas detection sensors, superhydrophobic surfaces, specialized solar cells [1, 2, 3]. Other researchers study the increasing of the surface hardness and the corrosion resistance of metals by anodization process [4, 5]. The hardness of the copper oxide is superior to that of the pure copper, and it is a highly non-conductive material. Therefore, it improves the hardness of the anodized coated surface and its corrosion protection. According to Mindat.org, the hardness of the pure copper surface is between 77 and 99, (about 80 HV) (2.5-3 Mohs) [6], while copper oxide Hardness is (190–300 HV) (3.5-4 Mohs) [7, 8]. The latest study on copper anodization [9] showed that in anodization the copper electrode passivized and the dissolution of copper reduced. It recommends for further research studies for developing the surface properties to improve copper corrosion resistance [9]. The present work, investigate the influence of the most affecting parameters of electrolyte concentration, temperature and applied voltage on coating hardness.

M.H. Mahmood: mahmoodfattah@yahoo.com

2 Experimental Procedure

2.1 Copper anodization

The experimental work starts with the copper anodization process in oxalate containing solutions. Copper foil samples, purity 99.5 % ASTM B280, of 0.12 mm thickness and 1cm x 2cm dimensions. The samples polished and cleaned according to the standard of ASTM, 2003, then it washed with tap water, distilled water and acetone. The anodization process parameters have been chosen in the ranges of, oxalate concentrations 0.1 - 0.5 M, electrolyte temperatures 0 - 24°C and applied voltage 7.5 - 9 V. Experiments were performed using Ag/AgCl (SSE) as a reference electrode and Platinum foil as the counter electrode. Anodization measurements performed by Auto Lab Potentiostat/Galvanostat, PGSTAT-AUT86037, supported (ANOVA Software). The anodization was carried out using Linear Sweep Voltammetry (LSV) at scan rate 0.02 V/S.

2.2 Taguchi's Design of Experiment

Taguchi DOE is the most efficient and powerful over other traditional analytical methods. It is a systematic approach to investigate system or process. A series of experimental tests designed with a planned input variables change for a process or system. The effects of these changes on detailed output responses then evaluated. Four levels for each input variable are selected. These levels represent the desired range for the effect of the variable that wants to be studied within. The response for each run are measured, then results analyzed by comparing the differences between response output readings for different variables input groups. These differences attributed to the input variables acting alone as a single effect. The software used for generating the Taguchi's orthogonal array is Minitab 17. The main advantage of Taguchi DOE is to reduce the experiments number that has to test as compared with the full factorial method. The total tests to be conducted for three factors and four levels combinations in full factorial method is 64, as compared with Taguchi method that only 16 operations must be tested. These experiments investigate the effects of the main anodization factors on the anodized copper hardness results, an experimental strategy Design of Experiments, DOE, for the running tests at various levels of the factors was used to obtain the optimal anodizing condition with highest coating hardness.

For studying the influence of the process parameters during copper anodization process in oxalate containing solutions, three factors and four levels were used. The four levels parameters of the anodizing experiments were 0.1, 0.23, 0.36, 0.5M oxalate concentrations, 0, 8, 16, 24 °C anodizing temperatures, and 7.5, 8, 8.5, 9 V applied voltage, as shown in Table 1.

Table 1. Anodization parameters and their levels

Parameters	Levels and values				Response
	Level 1	Level 2	Level 3	Level 4	
Electrolyte Concentration (M)	0.1	0.23	0.36	0.5	Micro hardness
Temperature (°C)	0	8	16	24	
Applied Voltage (V)	7.5	8.0	8.5	9.0	

In this research, sixteen experiments conducted at different parameter levels. Orthogonal array L16 was used, which has sixteen rows corresponding to the tests numbers. For monitoring the effects of anodization parameters on the hardness of the anodized copper coating, this study considered the surface microhardness as the primary output response.

3 Results and discussions

3.1 Coating Hardness Characterization

To investigate copper anodization parameters affecting on anodized copper surface hardness. The microhardness of the anodized copper samples tested on Automatic microhardness tester, MHT- Smart by Omnitech. The hardness tests were carried out by pyramid indenter with an applied load of 100 gm.f. The indentation length determined using optical microscopy and image-j software used for analyzing the resulted image. The tests carried out at five different locations on all the samples surfaces to avoid the possibility of any error readings. The indentation length determined using optical microscopy and image-j software used for analyzing the resulted image. The average measured microhardness was170 Hv.

3.2 Morphological analysis

The surface morphology and composition of the fabricated anodized copper coating, characterized using energy dispersive X-ray spectroscopy, EDX, X-ray diffraction (XRD) and Field Emission Scanning Electron Microscope, FESEM. The EDX results of coated and uncoated copper samples show the increase of the oxygen content in the coated sample, as illustrated in figure 1.

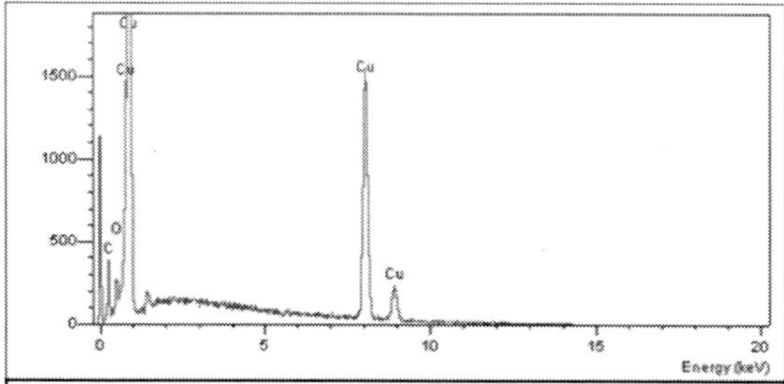

Uncoated sample		
Element	Wt.%	Atom %
O_2	6.33	21.17
Cu	93.67	78.83

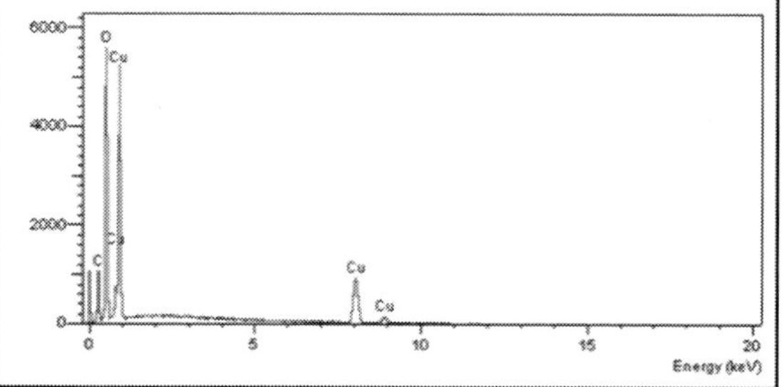

Coated sample		
Element	Wt.%	Atom %
O_2	69.20	79.60
Cu	30.8	20.4

Figure 1. EDX spectra distribution for anodized and non-anodized coating

The X-ray diffraction pattern graphs produced from the reflected spectrum on the anodized copper samples show diffraction peak identical with the standard card of cupric oxide spectrum peaks JCPDS 05-0661. The prominent peaks in the XRD diffraction pattern for the anodized copper samples at 38 and 65 °, refers to the formation of cupric oxide CuO on the coated surface with a preferred orientation plans (111) and (022), as illustrated in figure 2.

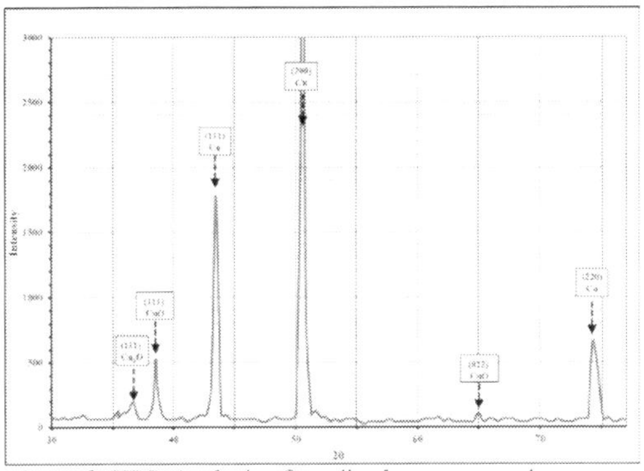

Figure 2. XRD Analysis of anodized copper samples

FESEM, micrographic results of anodized specimens, show a uniformly distributed Nanoparticles of different grain sizes on the surface as illustrated in figure 3. The analyzed FESEM results using, Image J, software produced that the coating grain sizes were in the range between 25 to 68 nm.

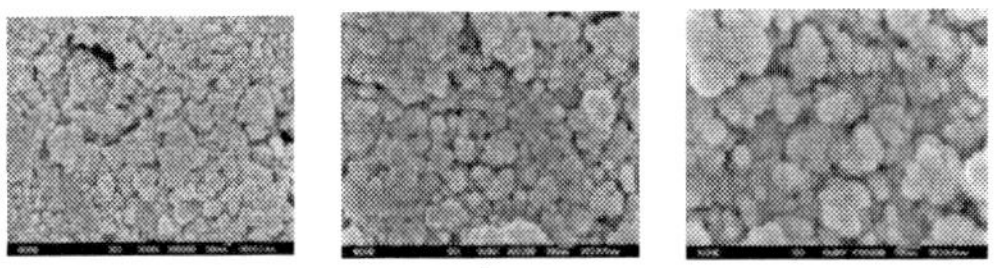

Figure 3. FESEM images for copper oxide Nanoparticles with magnifications of a; 30,000, b; 60,000 and c; 100,000

3.3 Effects of anodization parameters on the Coating hardness

After the copper anodizing experiments completed, the anodized coating microhardness measured by fully automatic microhardness tester MH Smart by Omnitech. The anodized coating experimental results were analyzed using Design Expert, Minitab 17, and JMP11.0.0 software to evaluate the effects of the anodization parameters of oxalate concentration, temperatures, and applied voltage on the microhardness of the anodized coating. According to Taguchi analysis method, the results were analyzed based on the signal - to – noise, S/N, ratio response variation, to minimize the change of the characterization quality due to the uncontrollable parameter. The coating hardness response considered as a quality characteristic with the concept of "larger-the-better":

$$S/N = -10 \times \log \left[1/n \sum (1/Y_i^2)\right] \quad (1)$$

Where n is the number of observations, and Yi is the measured output value. The resulted output values of the surface coating hardness and their corresponding S/N ratio listed in Table 2.

Table 2. Average hardness values along with their corresponding S/N ratio values

Exp.	Concentration (M)	Temperature (°C)	Voltage (V)	Hardness (Hv)	S/N
1	0.1	0	7.5	192	45.67
2	0.1	8	8	190	45.58
3	0.1	16	8.5	188	45.48
4	0.1	24	9	186	45.39
5	0.23	0	8	179	45.06
6	0.23	8	7.5	177	44.96
7	0.23	16	9	174	44.81
8	0.23	24	8.5	172	44.71
9	0.36	0	8.5	168	44.51
10	0.36	8	9	165	44.35
11	0.36	16	7.5	163	44.24
12	0.36	4	8	160	44.08
13	0.5	0	9	155	43.81
14	0.5	8	8.5	156	43.86
15	0.5	16	8	153	43.69
16	0.5	24	7.5	145	43.23

The values of S/N ratio of the anodized coating hardness by factor levels shown in Table 3.

Table 3. S/N ratio values by levels according to the coating hardness

Level	Temperature	Concentration	Voltage
1	45.53	44.76	44.52
2	44.88	44.69	44.60
3	44.30	44.56	44.64
4	43.65	44.35	44.59
Delta	1.88	0.41	0.12
Rank	1	2	3

From this result, it was concluded that the anodization parameters effect on hardness as shown in Figure 4. This plot illustrates the Signal to noise S/N ratio as a function of the anodization parameters.

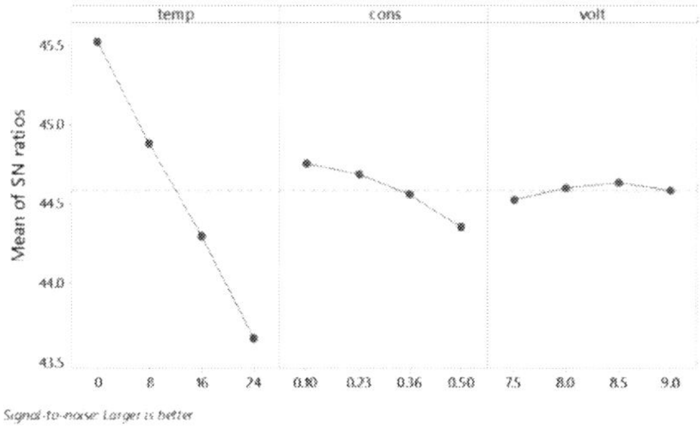

Figure 4. Signal to noise S/N plot showing the effects of anodization parameters on coating hardness

These results evaluate the effect of process parameters on the surface hardness by using Signal to noise S/N ratio. The electrolyte temperature was the most affecting parameter on the anodized coating hardness. The oxalate concentration was the next affecting parameter, while the applied voltage was the less affecting parameter on the surface hardness

3.3.1 Effect of anodizing temperature on the surface hardness

The anodization temperature was the most affecting parameter on the hardness of the coated surface. The analysis of the results revealed that the microhardness of the coating was decreased with the increase of the anodization temperature, as shown in figure 5. The maximum surface hardness of 190 HV resulted at the anodized coating at 0 °C, and minimum surface hardness of 150 HV produced at 24 °C, as shown in figure 5.

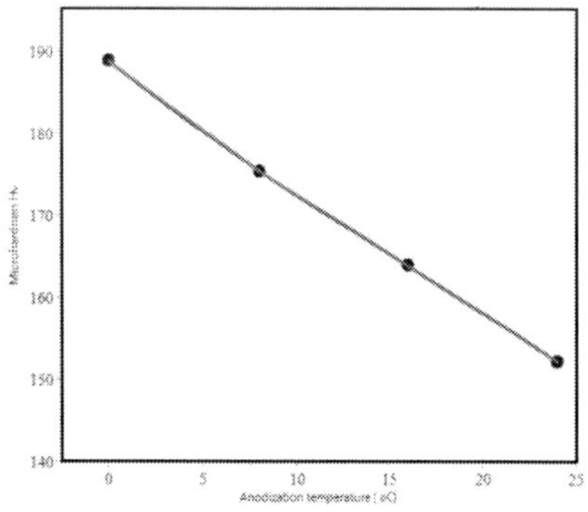

Figure 5. Anodized coating micro-hardness as a function of temperature

3.3.2 Effects of oxalate concentration on the surface hardness

The oxalate concentration was the second affecting parameter on the hardness of the coated surface. The analysis of the results revealed that the microhardness of the coating was slightly decreased with the increase of the anodizing oxalate concentration, as shown in figure 6.

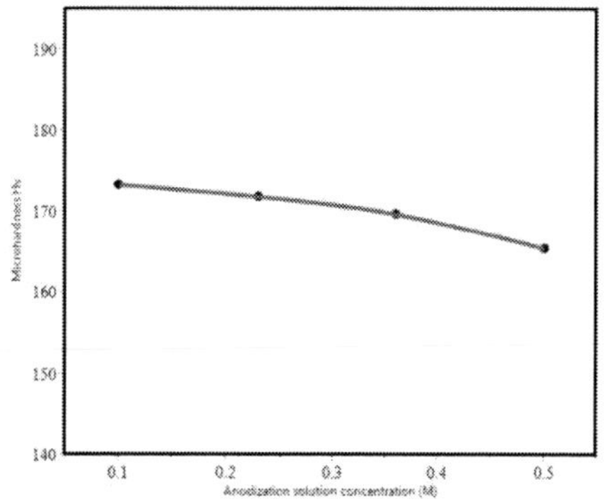

Figure 6. Microhardness of coated surface as a function of the anodizing oxalate concentration

The analysis of the results also revealed that the coated surface hardness was decreased with the increase of the anodizing temperature at the same oxalate concentration. The coated surface hardness also was reduced with the rising of the oxalate concentration at the same anodizing temperature, as shown in Figure 7. Therefore, the harder coating was produced at a lower temperature and higher oxalate concentration.

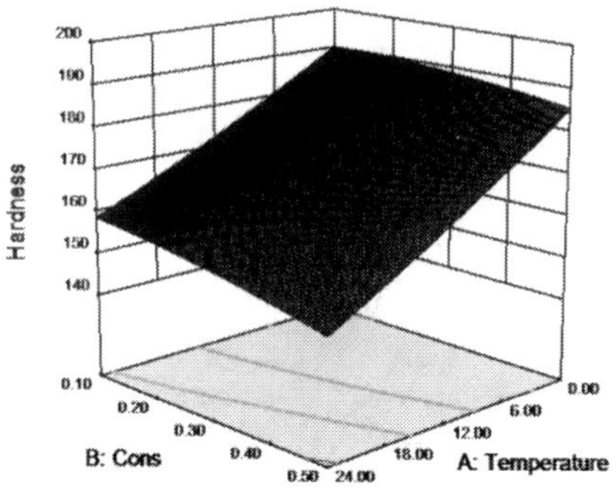

Figure 7. average hardness of coated surface as a function of (A) anodization temperature and (B) electrolyte concentration

3.3.3 Effects of applied voltage on the coated surface hardness

The analysis of the experimental results shows that the applied voltage was the less affecting parameter on the surface hardness. The relation between the applied voltage and the coated surface hardness was reversible. Whereas, the increase of the applied voltage was slightly decreased the surface hardness as shown in figure 8.

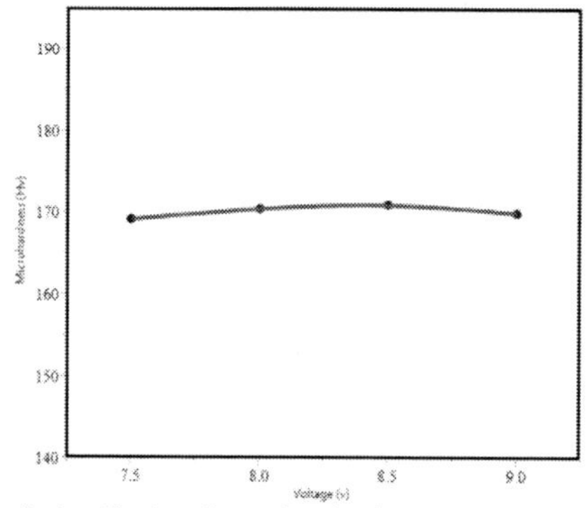

Figure 8. Anodized surface micro-hardness as a function of the anodization applied voltage

4 Conclusions

From this results was concluded that the anodized coating increased the surface hardness of copper and protecting it from the surrounding environment. Results show that the temperature was the most affecting parameter on the hardness of the anodized coating and the oxalate concentration was the second affecting parameter, while the applied voltage is the less affecting parameter on the coated surface hardness.

Results show that the microhardness of the anodized coating decreased with the increase of the anodizing temperature. Moreover, it also reduced with the rising of the oxalate concentration at the same anodizing temperature. So that harder anodized copper coating produced at a lower temperature and higher oxalate concentration.

References

1. L. Xu, "Novel urchin-like CuO synthesized by a facile reflux method with efficient olefin epoxidation catalytic performance," Chem. Mater., vol. **21**, no. 7, pp. 1253–1259, (2009).
2. M.S. Yong, "Natural and Biomimetic Artificial Surfaces for Superhydrophobicity, Self-Cleaning, Low Adhesion, and Drag Reduction Dissertation," The Ohio State University, (2009).
3. H. Singh, "Fabrication and Characterization of Copper Nanowires," J. Nanosci. Nanoeng. Appl., vol. **1**, no. 1, (2011).
4. N. Fredj and T. D. Burleigh, "Transpassive Dissolution of Copper and Rapid Formation of Brilliant Colored Copper Oxide Films," J. Electrochem. Soc., vol. **158**, no. 4, p. C104, (2011).
5. H. Chang, "Fabrication of CuO Nanofilm preparation by Electrophoretic Deposition," vol. **4**, pp. 370–373, 9, (2007).
6. P. Ivan, "Planarization composition for metal surfaces comprising an alumina hydrate abrasive," PCT/US2006/046225, (2008).
7. K. Rajiv, "Chemical-Mechanical polishing slurry for polishing of copper or silver films," US 6,821,309 B2, (2004).
8. H. Battez, "CuO, ZrO2 and ZnO nanoparticles as an anti-wear additive in oil lubricants," Wear, vol. **265**, no. 3–4, pp. 422–428, (2008).
9. A. Romeiro, "Polyphenazine films as inhibitors of copper corrosion," J. Electroanal. Chem., vol. **688**, pp. 282–288, (2013).

Si doping superlattice structure on 6H-SiC(0001)

Lianbi Li[1, a], Yuan Zang[2] and Jichao Hu[2]

[1]School of Science, Xi' an Polytechnic University, Xi' an 710048, China
[2]Department of Electronic Engineering, Xi' an University of Technology, Xi' an 710048, China

Abstract. Si-DSL structures multilayers are prepared on 6H-SiC(0001) successfully. The energy offsets of the n-Si/n-6H-SiC heterojunction in the conduction band and valance band are 0.21eV and 1.65eV, respectively. TEM characterizations of the p/n-Si DSL confirms the epitaxial growth of the Si films with [1-11] preferred orientation and the misfit dislocations with a Burgers vector of 1/3 <21-1> at the p-Si/n-Si interface. J-V measurements indicate that the heterostructure has apparent rectifying behavior. Under visible illumination with light intensity of $0.6W/cm^2$, the heterostructure demonstrates significant photoelectric response, and the photocurrent density is $2.1mA/cm^2$. Non-UV operation of the SiC-based photoelectric device is realized.

1 Introduction

SiC is a desirable material for power devices due to its superior physical properties such as wide bandgap, high thermal conductivity, and high critical electric field, etc..[1-3] However, because of the wide bandgap, SiC-based photoelectric devices can only be driven by ultraviolet (UV) light, which essentially limit its application for a detection of visible and infrared light. Moreover, UV light is harmful to the human body and is not the conventional light source in optical communications. In order to realize non-UV operation, Si film grown on SiC is used as a non-UV light-absorption layer[4, 5], so as to develop the SiC-based photoelectric devices applied in high temperature and high power regions.

At present, the SiC-based Si/SiC heterojunction is comparatively less studied[6-9], and the studies just focused on using Si/SiC heterostructure to improve the performance of the SiC SBD[7], or using Si/SiC heterostructure to solve the problem of SiC/SiO2 interface defect states in SiC MOSFET[8, 9], the non-UV photoelectric applications of the Si/SiC heterostructure are rarely reported. In our previous work, it was found that the Si/SiC heterojunctions are sensitive to the long-wave length lights ranging from visible to infrared region of the optical spectrum[4, 5]. However, because of the large lattice mismatch between Si and 6H-SiC (~19.6%), there are high density of structural defects such as edge misfit dislocations at the hetero-interface[10], stacking faults and twins in Si films[11], which led to a small carrier lifetime and poor device properties[4]. The doping superlattice (DSL) is a structure composed of the periodic p/n doped layers. In doping superlattice structures, the space charge potential of ionized impurities modulates the bandgaps of the materials, separates the electrons and holes spatially and thus enhances the carrier lifetime and the photoelectric properties. The DSL structures based on amorphous hydrogenated Silicon (a-Si:H) are investigated[12, 13]. Compared to the unstructured a-Si:H, up to tenfold increase of the photoconductivity is observed.

In this paper, p/n-Si DSL structures on 6H-SiC substrates were adopted to promote the photoelectric properties of the Si/SiC heterojunctions. The non-UV photodetector with the DSL structure was fabricated on 6H-SiC by low-pressure chemical vapor deposition (LPCVD). Transmission electron microscopy (TEM) and selected area electron diffraction (SAED) were employed to investigate the interface structure of Si-DSL/SiC heterojunction, and the photoelectric characteristics were investigated.

2 Experimental

p/n-Si DSL structures are grown on n-type 6H-SiC(0001) Si-face by LPCVD. An n-type doped (impurity concentration of $\sim 10^{17}cm^{-3}$) 6H-SiC(0001) wafer with a thickness of 300 μ m was purchased from II-VI Inc.. Silane (SiH_4), diborane (B_2H_6) and hydrogen (H_2) are used as a silicon source, a p-type doping source and a carrier, respectively. After the standard RCA cleaning processes[15], 6H-SiC substrates were treated in high-purity H_2 at 1050℃ for 10min. And then the p/n-Si DSL structures were prepared at 900 ℃. To prepare the ohmic contacts, Al electrodes on Si and Ni electrodes on 6H-SiC are prepared by magnetron sputtering followed by annealing at 900℃ and 1050℃, respectively.

[a] Corresponding author: xpu_lilianbi@163.com

3 Results and discussion

Figure 1(a) shows the schematic diagram of Si-DSL/6H-SiC heterostructure. Because 6H-SiC can be used as the penetration window of the visible and near-infrared lights, the light irradiated from the 6H-SiC side in photoelectric test. As is well known, when light is from the high energy bandgap side (6H-SiC), photons of energy between the two energy band gaps (wavelength range from 400nm to 1100nm), pass freely through the wide gap side and are absorbed in the low-energy bandgap material side (Si). The photoelectric characteristics of Si-DSL/6H-SiC heterostructure are simulated by Silvaco-TCAD. In simulation, the bandgap and electron affinity of 6H-SiC are set as 3.0eV and 3.85eV[14], respectively. The impurity concentrations of the n-Si and p-Si layers are set as $10^{15}cm^{-3}$ and $10^{16}cm^{-3}$. The impurity concentration of n-type 6H-SiC with a thickness of 300 μm is $10^{17}cm^{-3}$, which are consistent with the experimental parameters. Figure 1(b) shows the

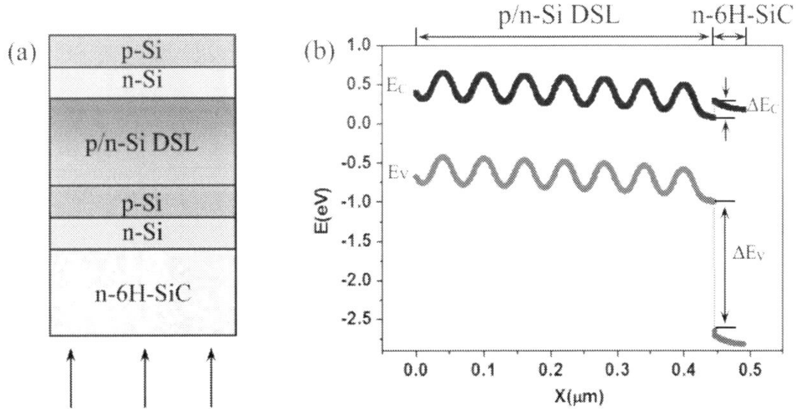

Fig.1 Schematic diagram (a) and energy band diagram (b) of Si doping superlattice structure on 6H-SiC in a thermal equilibrium state. ΔE_C=0.21eV, ΔE_V=1.65eV. The arrows indicate the incident direction of the light.

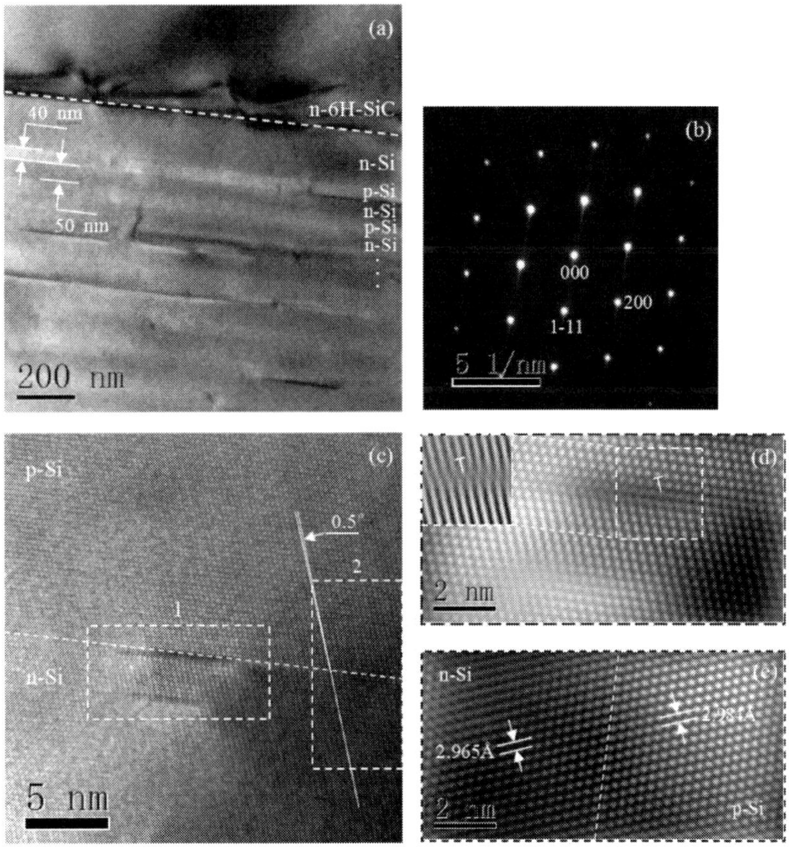

Figure 2 TEM and SAED images of the Si-DSL/6H-SiC heterostructure. (a) Cross-sectional low magnification TEM image, (b) SAED patterns of the Si films, (c) HRTEM image of the p/n-Si doping superlattice interface, and the processed HRTEM images of region 1(d) and region 2 (e) by using FFT and Fourier mask filtering technique.

energy band diagram of p/n-Si DSL structures on 6H-SiC in a thermal equilibrium state. It is shown that there are energy offsets of the n-type isotype Si/6H-SiC heterostructure in the conduction band ΔE_C and valance band ΔE_V, which are about 0.21eV and 1.65eV, respectively. The electron affinity of poly-Si (~4.05eV) is close to that of 6H-SiC, which possibly makes the electron barrier height of the Si/6H-SiC heterostructure very low. The electrons are the dominated current carriers because of the lower electron barrier.

The low magnification cross-sectional TEM bright-field image of the Si-DSL/6H-SiC heterostructure is shown in Fig. 2(a). In this image, the upper part belongs to the 6H-SiC substrate, while the lower part represents the p/n-Si DSL structures. The Si-DSL structures with obvious contrast differences show 40nm-p-Si/50nm-n-Si multilayers are prepared on 6H-SiC(0001) Si-face successfully. The interface of the p/n-Si DSL is abrupt in structure. The SAED patterns of the p/n-Si DSL corresponding to Si[011] zone axes are shown in Fig. 2(b), confirming the epitaxial growth of the Si films with [1-11] preferred orientation. Figure 2(c) shows a high-resolution TEM image of the p-Si/n-Si interface. There are some structural defects at the interface. The [111] direction was tilted by 0.5 ° at the interface. Fourier filtering technique (FFT) is applied to reveal the periodic atomic arrangement of the p/n-Si DSL interface. Figure 2(d) shows the FFT image of region 1 in Fig. 2(c). It is shown that relatively small amounts of misfit dislocations (MD, indicated by the arrows) appear at the p-Si/n-Si interface, which can be easily identified by extra lattice fringes in p-Si layer. The MDs are of the pure edge type with a Burgers vector of 1/3 <21-1 > parallel to the interface, which are labeled in Fig. 2(d). Moreover, the crystal plane spacing of the Si film has relatively obvious change at the p/n-Si DSL interface, as shown in Fig. 2(e). The crystal plane spacing of p-Si is 0.64% lower than n-Si layer. It demonstrates that the occurrence of the interfacial MDs can not accommodate the most of lattice mismatch strain and makes the lattice change at the p/n-Si DSL interface.

Figure 3(a) presents the J-V characteristics of the Si-DSL/6H-SiC heterostructure under dark condition. It is shown that the heterostructure has obvious rectifying characteristics with rectifying ratio up to 40 at ±5V. The turn-on voltage is about 1.2V. J-V characteristics of the fabricated Si-DSL/6H-SiC heterostructure on 6H-SiC under visible illumination of 0.6 W/cm^2 are shown in Fig. 3(b). The solid and the dashed curves represent the J-V characteristics measured under dark and illuminated conditions, respectively. The heterostructure demonstrates apparent photoelectric behavior, the short-circuit current density J_{SC} is about 2.1mA/cm^2. After stopping the visible illumination, the J-V curve immediately returned to the identical one under dark. Non-UV operation of the SiC-based photoelectric device is realized.

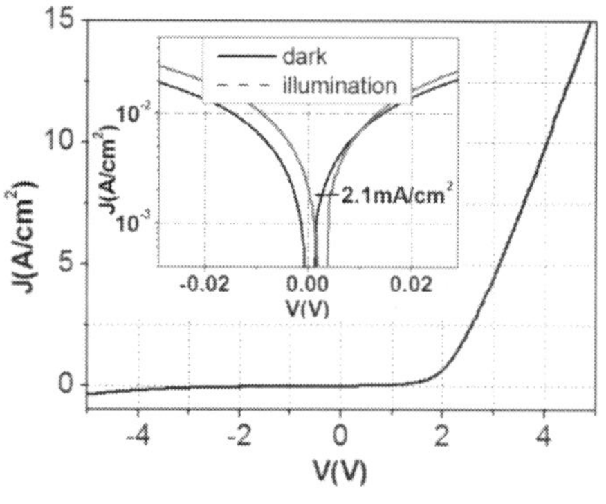

Figure 3 The electrical and photoelectrical characteristics of the p/n-Si DSL structures on 6H-SiC

4 Conclusion

In this paper, Si-DSL structures with 40nm-p-Si/50nm-n-Si multilayers are prepared on 6H-SiC(0001) Si-face successfully. The energy offsets of the n-Si/n-6H-SiC heterojunction in the conduction band and valance band are found to be 0.21eV and 1.65eV, respectively. The electrons are the dominated current carriers because of the low electron barrier. TEM characterizations of the p/n-Si DSL confirms the epitaxial growth of the Si films with [1-11] preferred orientation and the misfit dislocations with a Burgers vector of 1/3 <21-1> at the p-Si/n-Si interface. Under visible illumination with light intensity of 0.6W/cm^2, the heterostructure demonstrates significant photoelectric response, and the photocurrent density is 2.1mA/cm2 Non-UV operation of the SiC-based photoelectric device is realized.

Acknowledgements

This work was Supported financially by the National Natural Science Foundation of China (Grant No. 51402230, 21503153), the Project Supported by Natural Science Basic Research Plan in Shaanxi Province of China (Grant No. 2015JM6282), Scientific Research Program Funded by Shaanxi Provincial Education Department (Grant No. 14JK1302) and China Postdoctoral Science Foundation funded project (Grant No. 2013M532072).

References

1. J. Hu, X. Xin, J.H. Zhao, F. Yan, B. Guan, J. Seely, B. Kjornrattanawanich, Opt. Lett. **31**, 1591 (2006).
2. W.T. Liu, Z.M. Chen, Power Electronics **45**, 89 (2011).
3. L.B. Li, Z.M. Chen, Z.Q. Ren, Z.J. Gao, Chin. Phys. lett. **30,** 097304 (2013).
4. L.B. Li, Z.M. Chen, W.T. Liu, W.C. Li, Electron. Lett. **48**, 1227 (2012).

5. H. Tetsuya, T. Hideaki, S. Yoshio, T. Satoshi, H. Masakatsu, Mater. Sci. Forum, **483 – 485**, 953 (2005).
6. A. Pérez-Tomás, M.R. Jennings, M. Davis, J.A. Covington, P.A. Mawby, V. Shah, T. Grasby, J. Appl. Phys. **102**, 014505 (2007).
7. A. Pérez-Tomás, M.R. Jennings, M. Davis, V. Shah, T. Grasby, J.A. Covington, P.A. Mawby, Microelectron. J. **38**:1233 (2007).
8. O.J. Guy, T.E. Jenkins, M. Lodzinski, A. Castaing, S.P. Wilks, P. Bailey, T.C.Q. Noakes, Mater. Sci. Forum **556 – 557**, 509 (2007).
9. L. B. Li, Z. M. Chen, Y. Zang, S. Feng, Mater. Lett. **163**, 47 (2016).
10. L. B. Li, Z. M. Chen, Y. Zang, CrystEngComm **18**, 5681 (2016).
11. J. Kakalios, H. Fritzsche, Phys. Rev. Lett. **53**, 1602 (1984).
12. M. Hundhausen, L. Ley, R. Carius, Phys. Rev. Lett. **53**, 1598 (1984).
13. Tokuyuki Teraji, Shiro Hara. Control of interface states at metal/6H-SiC(0001) interfaces, Phys. Rev. B **70**, 035312 (2004).
14. W. Kern and D. A. Puotinen, RCA Rev. **31**, 187 (1970).

Effect of SiO_2 nanoparticles on the hydrophobic properties of waterborne fluorine-containing epoxy coatings

Xiaofang Liu[1, a], Rongwei Wang[1], Yongkang Fan[1], Ming Wei[1]

[1]School of Chemistry, Chemical Engineering and Life Sciences, Wuhan University of Technology, 430070 Wuhan, Hubei, China

Abstract. A nano-composite coating was developed by adding nano-SiO_2 to a home-made Waterborne fluorine-containing epoxy resin. The surface hydrophlicity of the coating was investigated by the contact angle with water. As the particle sizes of SiO_2 changed from 7-18nm, the water contact angle of the films changed from 62.8° to 85.6°. The surface energy of the coating was calculated out through measuring the contact angles of the coating with water and 1-bromonaphthalene. The presence of SiO2 significantly enhanced the corrosion resistance of the composite coatings, which showed two times than that of the neat waterborne fluorine-containing epoxy coatings. And the thermos dynamic stability of the composite coatings has been significantly improved. When the silica content is 2.0%, the coating has the largest contact angle. Other performance of the coating was investigated in terms of their thickness, flexibility, pencil hardness, impact resistance.

1 Introduction

Hydrophobic coatings create an attractive interest both in industry and academia due to their unique self-cleaning properties as a result of their water repelling character[1]. Based on the lotus leaf effect and other natural phenomena[2-6], highly hydrophobic surfaces have been prepared by roughening low surface energy materials. The chemical composition of a coating determines its surface free energy. The C-F bond is a kind of the strong organic bonds[7].The introduction of fluorinated component to the coating is an effective method to improve water impermeability and reduce surface free energy on the architecture of the molecular structure, which will increase the hydrophobic properties of coatings because of their small dipole and the low polarizability of the C–F bond together with the large free volume[8-9]. Various techniques have been applied to increase the surface roughness, including plasma etching [10-12], electrochemical deposition[13-14], chemical vapor deposition[15-16], nanoparticles deposited[17-18], etc. Meanwhile the addition of nanoparticles such as nano-Ti [19], nano-SiO_2[20], nano-ZnO[21-22] increase of the corrosion resistance.

In the present work, we modified waterborne fluorine-containing epoxy coatings with different contents of SiO_2 nanoparticles and different particle sizes. This study aims to make a hydrophobic coating from the SiO_2 nanoparticles as deposit sediment to improve the roughness. And anticorrosion properties of modified waterborne fluorine-containing epoxy coatings were also studied.

2 Experimental

2.1 Materials

SiO_2 nanoparticles(A380, A300, A200, A150, S103583) was purchased from Evonik and Aladdin company, with the particle sizes of 7, 9, 12, 14, and 18nm. 1-bromonaphthalene and NaCl were purchased from Aladdin Industrial Corporation. Waterborne fluorine-containing epoxy resin was home-made its structure is shown in Fig.1 and waterborne epoxy curing agent 8545 was supplied by Momentive Performance Materials Company. Defoaming agent, leveling agent, corrosion inhibitor agent, wetting and dispersing agent were obtained from Elementis-specialties.

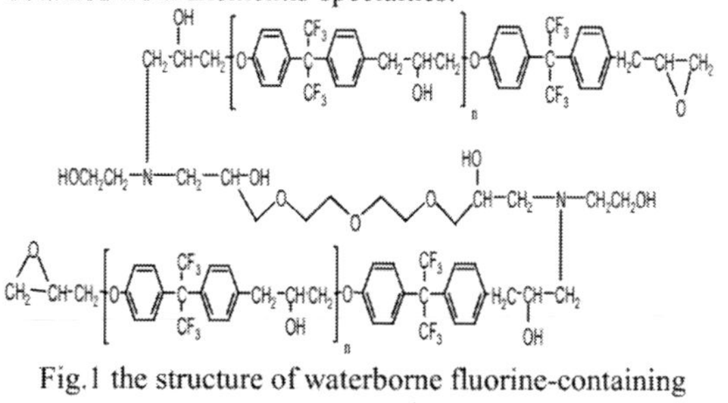

Fig.1 the structure of waterborne fluorine-containing epoxy resin

2.2 Preparation of waterborne fluorine-containing epoxy coatings

a Corresponding author: Xiaofang Liu@sdjclxf@sina.com

The epoxy emulsion was prepared by the waterborne fluorine-containing epoxy resin, SiO_2 nanoparticles and deionized water via phase inversion method (75℃, 3000 rpm, high speed dispersers.) with a solid content of 50% and average particle size of 75.8 nm. The deionized water dropped into the system at a speed of 1 mL/min. The waterborne fluorine-containing epoxy coating was obtained by mixed waterborne fluorine-containing epoxy emulsion, waterborne epoxy curing agent and defoaming agent, leveling agent, corrosion inhibitor agent, wetting and dispersing agent evenly. In the waterborne epoxy coating, the ratio of epoxy and amine hydrogen is controlled as 1:0.5, and the total content of leveling agent, corrosion inhibitor agent and wetting and dispersing agent are 0.2wt%, together with the defoaming agent content of 0.4wt %.

2.3 Measurement and Characterization

Surface water contact angles of coatings were studied by contact angle measuring instrument (JC2000C3, Chengdu best instrument Co., China). The surface energy of the coating is calculated through the Owens-Wendt-Kaelble theory[23-24] . The corrosion resistance of coating was studied by EIS (AUTOLAB PGSTAT30, Metrohm Co., Switzerland). The thermal gravimetric analysis of coating was studied by TG-DSC (STA449F3, Netzsch Co., Germany). Coatings were applied on previously sanded tinplates. After drying, the coating properties such as: bending test[25], adhesion test[26], impact[27], scratch hardness by pencil test[28], were determined by standard methods. The thickness of the film was measured by coating thickness gauges (QNIX4500, QNIX Co., Germany).

3 Results and Discussion

3.1 Effect of the particle sizes of SiO_2 on the water contact angles of coatings

The influences of the particle sizes of SiO_2 on the water contact angles of coatings are illustrated in Fig 2, with the constant of 1 wt % SiO_2 nanoparticles.
The SiO_2 addition with different particle sizes affected the water contact angles of coatings. The minimum contact angle is 62.8°while the largest is 85.6°. The water resistance of the paint film is mainly determined by two factors: the structure and the solid surface energy of the paint film. The different particle sizes of SiO_2 could make different scales of structure.

3.2 Effect of the contents of SiO2 nanoparticl

-es on the water contact angles of coatings

With the choose of 18 nm SiO_2, the effect of SiO_2 (S103583) contents on water contact angles for coatings was investigated, as shown in Fig. 3 The water contact angles of the composite coatings gradually increased with the content of SiO_2 increased from 0.5 to 2.0wt %. Specifically, the introduction of 1.5wt% and 2.0wt% SiO_2 nanoparticles significantly advanced the hydrophobicity of the composite coatings, which induced the water contact angles of coatings higher than 90°. However, an excess of SiO_2 addition caused the decrease of the water contact angles, because a small amount of SiO_2 were exposed to the outer edge of the paint film. The exposed hydrophilic SiO_2 results in the decrease of the hydrophobic of the paint film .When the content of SiO_2 is higher than 1.5wt%, the viscosity of emulsion has increased dramatically, and presents a paste. When the content of SiO_2 was 2.0wt % the advantage of the hydrophobic effect is not obvious compared with 1.5 wt%. Taking the coating process and the hydrophobic effect into consideration, we chose 1.5wt% as an appropriate content.

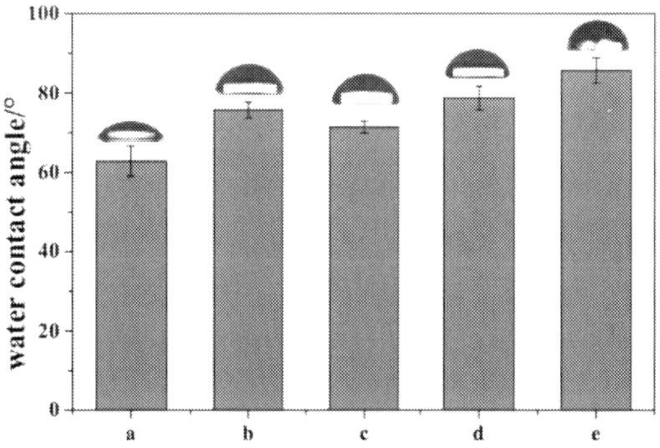

Fig.2 Water contact angles of coatings with different particle sizes of SiO_2

Table 1. Water contact angles of coatings with different particle sizes of SiO_2

Code	a	b	c	d	e
Particle sizes of SiO_2/nm	7	9	12	14	18
Contact angle(°)	62.8 ±3.8	75.7 ±2.0	71.4 ±1.5	78.7 ±3.0	85.6 ±3.2

Fig.3 Water contact angles of coatings with different contents of SiO_2

3.3 Evaluation of comprehensive performanc-e of the coating

3.3.1 Surface energy of the coating

The surface energy of solids is an important index to evaluate its' hydrophobic properties. Deionized water and

1-bromonaphthalene were used for the contact angles measurement of the composite coatings. The surface tension and its components for water and 1-Bromonaphthalene are listed in Table 2. The γ_{lv}, γ_{plv} and γ_{dlv} represent the surface tension the polar and nonpolar component of the liquid. The contact angles of coating with the water and 1-bromonaphthalene were shown in Fig. 4, which exhibited the values of 94.06° (water) and 53.5° (1-bromonaphthalene). According to Eq. (2), the surface energy of the coating was calculated as 29.93 mN/m.

$$\gamma_{lv}(1+\cos\theta) = 2\sqrt{\gamma_{sv}^{d}\gamma_{lv}^{d}} + 2\sqrt{\gamma_{sv}^{p}\gamma_{lv}^{p}} \quad (1)$$

γ_{sv}^{p}=1.57 γ_{sv}^{d}=28.36

$$\gamma_{sv} = \gamma_{sv}^{p} + \gamma_{sv}^{d} = 1.57+28.36=29.93 \quad (2)$$

Table 2. Surface tension and its components for water and 1-Bromonaphthalene

Liquid	γ_{lv}(mN/m)	γ_{lv}^{p}(mN/m)	γ_{lv}^{d}(mN/m)
Water	72.8	51.0	21.8
1-Bromonaphthalene	44.6	0	44.6

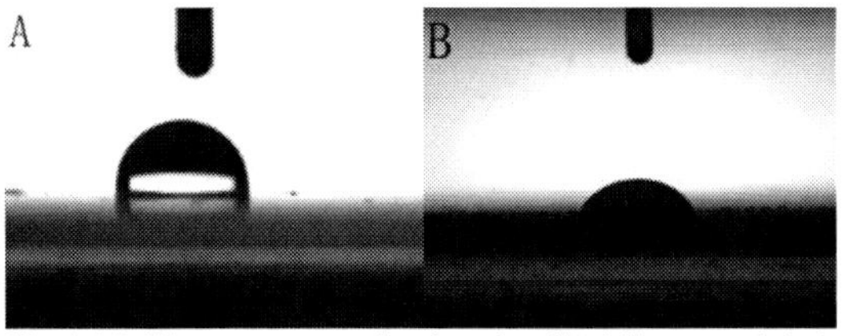

Fig.4 (A) Water contact angle of coating (B) 1-Bromonaphthal-Ene contact angle of the coating

3.3.2 The corrosion resistance of the coatings

Electrochemical Impedance Spectroscopy (EIS) measurement was used to investigate the corrosion resistance of the samples. Electrochemical measurements were conducted using a three electrode system. The fluorine-containing epoxy-coated steel coupon served as the working electrode, while the counter electrode and the reference electrode used were a platinum grid and a saturated calomel electrode respectively. The EIS experiment was performed at room temperature in aqueous 3.5 wt% NaCl solution prepared by dissolving sodium chloride in deionized water. The steel was polarized at ±10 mV around its open circuit potential by an alternating current signal with its frequency ranging from 10^{5}Hz to 10^{-2}Hz. Fig. 5 was the nyquist diagram of EIS test. Z′ and Z″ represents the real and imaginary parts of the impedance, respectively. The result of the coating contained SiO2 was curve A and curve B was the fitted value by using the equivalent circuit shown in Fig.6. The result of coating without SiO2 was curve C and curve D was the fitted value by using the equivalent circuit shown in Fig.7. The measured curves basically are consistent with the calculated curves, and thus the fitted value can better reflect the real situation. It could be seen that the incorporation of SiO_2 increased the coating resistance obviously and improved the quality of the cured epoxy coating. The calculated resistance of the coating contained SiO_2 was 64909 Ω·cm². The calculated resistance of the coating without SiO_2 was 35990 Ω·cm². The addition of SiO_2 improved nearly twice corrosion resistance of the coating.

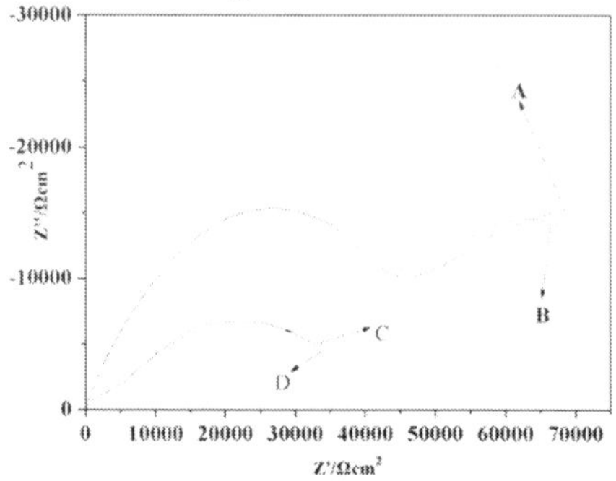

Fig.5 EIS Nyquist diagram for fluorine-containing epoxy-coated steel after 7 days (A) measured value with SiO2 in the coating (B) fitted value(C) measured value without SiO2 in the coating (D) fitted value

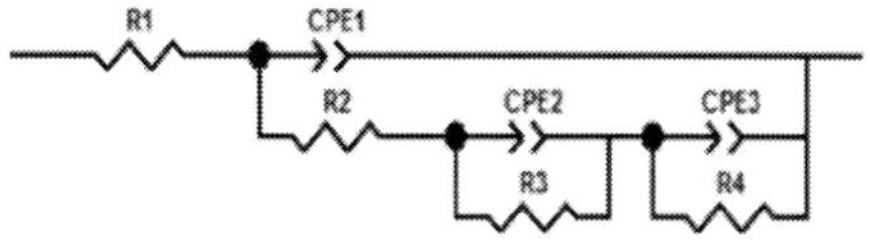

Fig.6 Schematic drawing of the equivalent circuit. R1 is associated with the electrolyte resistance. R2 and CPE1 are the resistance and capacitance of Outer layer including the passivation film and corrosion products etc, respectively. R3 and CPE2 are the resistance and capacitance of coating porosities, respectively. CPE3 is the capacitance of the double layer. R4 is the charge transfer resistance.

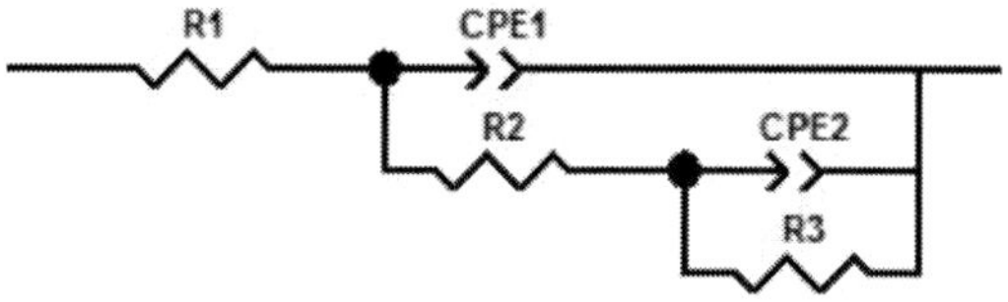

Fig.7 Schematic drawing of the equivalent circuit . R1 is associated with the electrolyte resistance. R2 and CPE1 are the resistance and capacitance of Outer layer including the passivation film and corrosion products etc, respectively. CPE2 is the capacitance of the double layer. R3 is the charge transfer resistance.

3.3.3 The thermal gravimetric (TG) analysis of coating

The thermal properties of the coating were studied by the TG-DSC, which is used to analyze the degradation steps and thermal stability of epoxy resin coatings. Fig. 8 A, B displays TG weight loss and weight loss derivative curves in air. The DTG curve clearly indicated that the coating underwent four main degradation stages. The first stage was the volatilization of residuary water, in the temperature range of 20-250 ℃. The second stage from 250 to 380℃ was the thermal degradation process of the low crosslinking density of curing products and some monomer resin, curing agent not involved in the curing reaction. The third stage, between 380 and 470 ℃, was the thermal degradation process of side chain and end group of the skeleton structure. The final process was the degradation of the skeleton structure in the temperature range of 470-670℃.

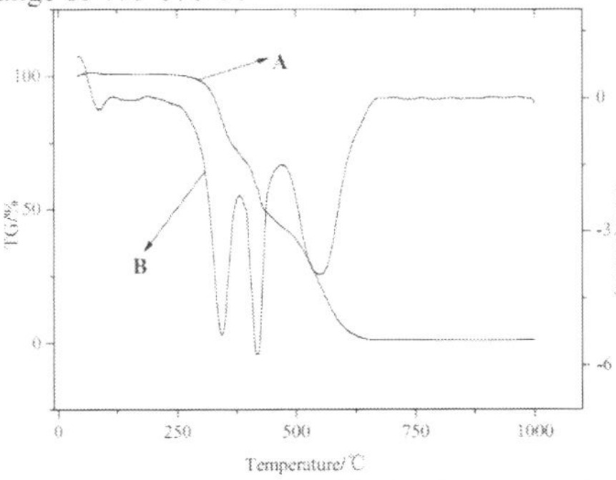

Fig.8 Thermogravimetric and Derivative curves of coating (A) TG (B) DTG

The comprehensive performances of the waterborne fluorine-containing epoxy coating loaded with 1.5 wt% SiO_2 nanoparticles (18 nm) are listed in Table 3, which shows the promising surface hydrophobicity, high corrosion resistance, and excellent.

Table.3 Comprehensive performances of waterborne fluorine-containing epoxy coating.

Performances	Results
Resistance/Ω·cm²	64909
Contact angel/°	94.1±2.1
Thickness/(μm)	48.7
Flexibility/(mm)	2
Adhesion/(grade)	1
Pencil hardness/H	3
Impact resistance/(kg.cm)	50

4 Conclusions

The waterborne fluorine-containing epoxy coatings modified by SiO_2 nanoparticles showed a higher contact angle. With the presence of 18 nm SiO_2 nanoparticles at the loading level of 1.5wt%, the coating exhibited the hydrophobic property from and the water contact angle of 94.1°. The surface energy of the coating is calculated as 29.9 mN/m. The hydrophilic-to-hydrophobic transition of fluorine-containing epoxy coating derived from the interaction of low surface energy and SiO_2 nanoparticles. The addition of SiO_2 improved corrosion resistance of the coating obviously, and the coating showed a good thermodynamic stability.

References

1. Liu X, Liang Y, Zhou F, et al. Extreme wettability and tunable adhesion: biomimicking beyond nature [J]. Soft Matter, 2012, 8(7): 2070-2086.
2. Sun, T.; Feng, L.; Gao, X; et al. Bioinspired surfaces with special wettability [J]. Acc. Chem. Res. 2005, 38, 644-652.
3. Hosono, E.; Fujihara, S.; Honma, I.; et al. Superhydrophobic perpendicular nanopin film by the bottom-up process [J]. Am. Chem. Soc. 2005, 127, 13458-13459.
4. Zhang, X.; Shi, F.; Niu, J.; et al. Superhydrophobic surfaces: From structural control to functional application [J]. Mater. Chem.2008, 18, 621-633.
5. Wenzel R N. Surface Roughness and Contact Angle [J]. The Journal of Physical Chemistry, 1949, 53(9): 1466-1467.
6. Nishino T, Meguro M, Nakamae K, et al. The lowest surface free energy based on-CF3 alignment [J]. Langmuir, 1999, 15(13): 4321-4323.
7. Lee J R, Jin F L, Park S J, et al. Study of new fluorine-containing epoxy resin for low dielectric constant [J]. Surface and Coatings Technology, 2004, 180: 650-654.
8. Merkel T C, Toy L G. Comparison of hydrogen sulphide transport properties in fluorinated and nonfluorinated polymers [J]. Macromolecules, 2006, 39(22): 7591-7600.
9. Tao Z, Yang S, Ge Z, et al. Synthesis and properties of novel fluorinated epoxy resins based on 1, 1-bis (4-glycidylesterphenyl)-1-(3′-trifluoromethylphenyl)-2, 2, 2-trifluoroethane[J]. European polymer journal, 2007, 43(2): 550-560.
10. Jin M, Feng X, Xi J, et al. Super-hydrophobic PDMS surface with ultra-low adhesive force[J]. Macromolecular rapid communications, 2005, 26(22): 1805-1809.
11. Balu B, Breedveld V, Hess D W. Fabrication of "roll-off" and "sticky" superhydrophobic cellulose surfaces via plasma processing [J]. Langmuir, 2008, 24(9): 4785-4790.
12. Fernández-Blázquez J P, Fell D, Bonaccurso E, et al. Superhydrophilic and superhydrophobic nanostructured surfaces via plasma treatment[J]. Journal of colloid and interface science, 2011, 357(1): 234-238.
13. Wang M F, Raghunathan N, Ziaie B. A nonlithographic top-down electrochemical approach for creating hierarchical (micro-nano) superhydrophobic silicon surfaces [J]. Langmuir, 2007, 23(5): 2300-2303.
14. Jiang Y, Wang Z, Yu X, et al. Self-assembled monolayers of dendron thiols for electrodeposition of gold nanostructures: toward fabrication of superhydrophobic/superhydrophilic surfaces and pH-responsive surfaces[J]. Langmuir, 2005, 21(5): 1986-1990.
15. Intranuovo F, Sardella E, Rossini P, et al. PECVD of fluorocarbon coatings from hexafluoropropylene oxide: glow vs. afterglow [J]. Chemical Vapor Deposition, 2009, 15(4-6): 95-100.
16. Li S, Li H, Wang X, et al. Super-hydrophobicity of large-area honeycomb-like aligned carbon nanotubes[J]. The journal of physical chemistry B, 2002, 106(36): 9274-9276.
17. Kylián O, Polonskyi O, Kratochvíl J, et al. Control of Wettability of Plasma Polymers by Application of Ti Nano-Clusters [J]. Plasma Processes and Polymers, 2012, 9(2): 180-187.

18. Solař P, Kylián O, Polonskyi O, et al. Nanocomposite coatings of Ti/C: H plasma polymer particles providing a surface with variable nanoroughness [J]. Surface and Coatings Technology, 2012, 206(21): 4335-4342.
19. Zhang X Z, Li Y J. Effects of Nano-sized Titanium Powder on the Anti-corrosion Property of Epoxy Coatings on Stee l[J]. Kemija u industriji, 2014, 63(9-10): 317-322.
20. Sadreddini S, Salehi Z, Rassaie H. Characterization of Ni–P–SiO2 nano-composite coating on magnesium [J]. Applied Surface Science, 2015, 324: 393-398.
21. Ramezanzadeh B, Attar M M. Studying the effects of micro and nano sized ZnO particles on the corrosion resistance and deterioration behavior of an epoxy-polyamide coating on hot-dip galvanized steel [J]. Progress in Organic Coatings, 2011, 71(3): 314-328.
22. Duraibabu D, Ganeshbabu T, Manjumeena R, et al. Unique coating formulation for corrosion and microbial prevention of mild steel[J]. Progress in Organic Coatings, 2014, 77(3): 657-664.
23. Owens D K, Wendt R C. Estimation of the surface free energy of polymers[J]. Journal of applied polymer science, 1969, 13(8): 1741-1747.
24. Kaelble D H. Dispersion-polar surface tension properties of organic solids[J]. 1970, 2(2): 66-81.
25. Standard Test Method for Bend Test, China Standard Specification, GB1731-79.
26. Standard Test Method for Adhesion, China Standard Specification, GB 1720-79.
27. Standard Test Method for Impact, China Standard Specification, GB/T 1732-1993.
28. Standard Test Method for Coating Hardness by Pencil Test, China Standard Specification, GB/T 6739-2006.

Fabrication and dynamics of two layers vanadium carbide reinforced iron matrix surface composite

Fan Bai[1], Li-sheng Zhong[a,1], Xin Wang[1], Yun-hua Xu[1], Zi-yuan Zhao[1]
[1]School of Materials Science and Engineering, Xi'an University of Technology, Xi'an 710048, China

Abstract. Vanadium carbide particulates-reinforced iron matrix composites were prepared by in situ solid-phase diffusion method with different heat treatment times (30, 60, 120 and 240 min) at 1223 K, 1273 K and 1323 K after high temperature treatment at 1373 K for 5 min. The surface vanadium carbide layers (V_2C layer and V_8C_7 layer) were characterized by means of scanning electron microscopy and X-ray diffraction. The kinetics of vanadium carbide layers showed that a parabolic relationship between carbide layer thickness and treatment time, which indicating that the composite layer's growth is controlled by diffusion. The apparent activation energy of V_2C layer and V_8C_7 layer were 47.27 kJ/mol and 32.26kJ/mol, respectively.

1 Introduction

It is well known that vanadium carbide is one of the most appropriate reinforcement for iron matrix surface composite, which due to its high hardness, good thermal stability, and well wettability with iron matrix [1]. Vanadium carbide reinforced iron matrix surface composite shows excellent wear resistance and high temperature properties, which can be widely used in metallurgy, mining, power, machinery, and water conservancy [2]. The diffusion dynamics based on the Arrhenius equation of carbide coating [3], boride coating [4], and nitrogen coating [5] on the surface of steel or iron matrix have been reported elsewhere. S. Barzilai and N. Frage [3] have developed an effective method, which thin films of niobium (~8 μm) were deposited on an ATJ graphite substrate by radio-frequency magnetron sputtering at low temperatures (420±50 K) under argon pressure of 0.67 Pa. It was found that the activation energies are 190 kJ/mol for Nb_2C, 164 kJ/mol for NbC, and 169.4 kJ/mol for the combined carbide layers, respectively.
The main goal of this study is to study the kinetics of two layers vanadium carbide (V_2C and V_8C_7) reinforced iron matrix surface composite, which were prepared by solid phase diffusion reaction.

2 Experimental procedures

The main raw materials for preparing two layers vanadium carbide are vanadium plate (wt.%, 99.97 V) and gray cast iron (wt.%, 2.57C–1.04Mn–1.03Si–0.046P–0.018S). The two layers vanadium carbide reinforced iron matrix surface composite were prepared by solid-phase diffusion reaction process. The details of the process are presented in the author's prior work [6]. Each specimen was heated to 1373 K and maintained for 5 min, followed by incubation at 1223, 1273, and 1323 K for 30-240 min in a horizontal tube furnace with a modest flow of argon (5 mL/min).

After being polished with diamond paste and etched with a 4% Nital solution, the microstructures of specimens were examined using a JSM-5800 scanning electron microscope (JEOL, Japan) equipped with an energy dispersive X-ray spectroscopy (EDS). The X-ray diffraction (XRD) data were recorded on a PW 1730 X-ray diffractometer (Philips, The Netherlands) with monochromated Cu Kα radiation at 40 kV and 40 mA in the 2θ range of 20-90°.

3 Results and discussion

3.1 Microstructure of the vanadium carbide reinforced iron matrix surface composite.

Fig.1 is the microstructure of the vanadium carbide reinforced iron matrix surface composites by incubation at 1223 K for 30 min, 60 min, 120 min, and 240 min after high temperature treatment at 1373 K for 5 min, respectively. It is clear that, there are two dense and compact layers between vanadium plate and the matrix, which have good bonding to the gray cast iron. In detail, the longitudinal cross-section microstructure of the composite can be divided into two layers: light gray layer (I) and dark gray layer (II). The location of layer I is close to the vanadium plate, and its thickness are only 6~15 μm with dense carbide. Layer II shows less density and carbide distribution with 21~61 μm in thickness. Besides, the thicknesses of the layers gradually increase

[a] Corresponding author: zhonglisheng@xaut.edu.cn

with the heat treatment times increased from 30 min to 240 min.

Fig.2 is the XRD results of the vanadium carbide reinforced iron matrix surface composites by incubation at 1223 K for 30 min, 60 min, 120 min, and 240 min after high temperature treatment at 1373 K for 5 min, respectively. The XRD results showed that V, V_8C_7, and V_2C are the main phases of the composites. By prolonging heat treatment time from 30 min to 240 min, the intensities of the diffraction peaks corresponding to metallic vanadium decrease along with increasing the intensities of the diffraction peaks of V_8C_7 and V_2C. The author's previous work [7] shows that layer Ⅰ and layer Ⅱ are V_2C layer and V_8C_7 dense layer, respectively.

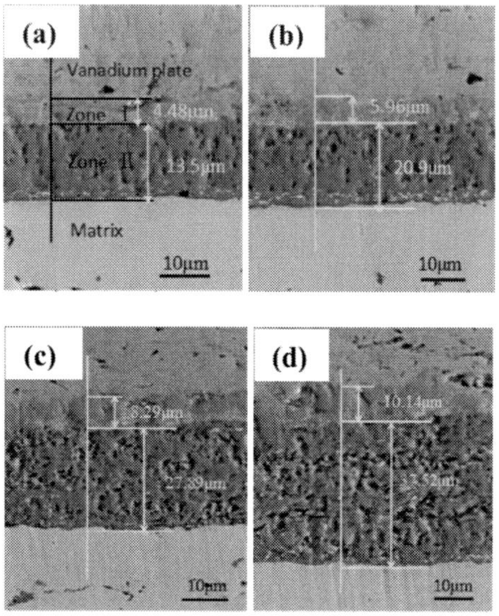

Figure.1. Microstructure of the vanadium carbide reinforced iron matrix composite by incubation at 1223 K for 30 min (a), 60 min (b), 120min (c), and 240 min (d), after high temperature treatment at 1373 K for 5 min, respectively.

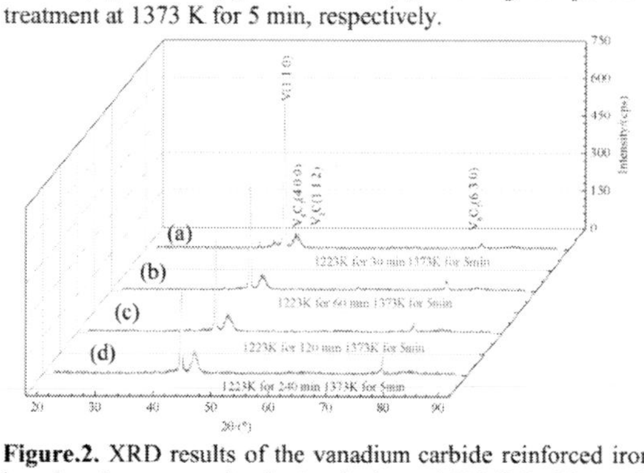

Figure.2. XRD results of the vanadium carbide reinforced iron based surface composites by incubation at 1223 K for 30 (a), 60 (b), 120 (c), and 240 min (d) after high temperature treatment at 1373 K for 5 min.

To obtain more information on the layer Ⅰ and layer Ⅱ in the composite, the specimen by incubation at 1223 K for 240 min were examined. According to the EDS results obtained from the crystalline particles present in each layer, the ratio of vanadium to carbon is 1:0.55 and 1:0.93 for the layers of Ⅰ and Ⅱ, respectively, in the specimen heat treated at 1223 K for 240 min. Based on the XRD results, these values are considered to be close to the atomic ratio of V_2C and V_8C_7, respectively.

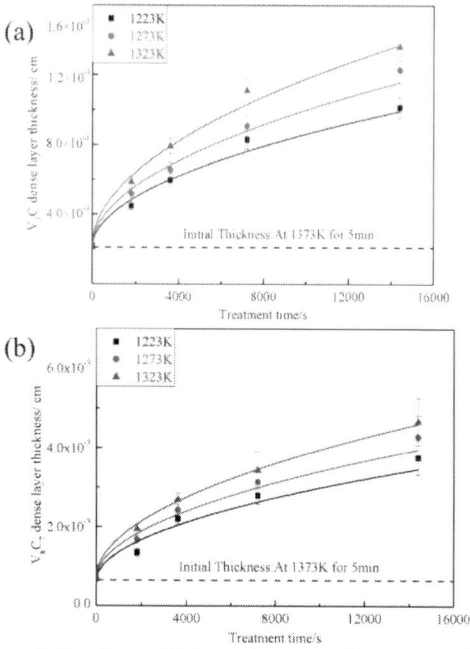

Figure.3. Two layers thickness increases with treatment time for each process temperature for V_2C layer (a) and V_8C_7 layer (b).

3.2 Kinetics of two layers vanadium carbide reinforced iron matrix surface composites.

The experiments can measure the V_2C and V_8C_7 layers thickness. These measured values can be used to calculate the kinetics parameters of the vanadium carbide reinforced layer. The graphical representation in Fig.3 shows that two layers thickness increases with treatment time for each process temperature. It was observed that the longer the treatment time and the higher the process temperature, the thicker carbide layer became. On the basis of the classical kinetic theory.

Fig.4 shows the square of the layer thickness versus treating time at different temperatures. It is clear that the square of the layer thickness changes linearly with the treating time. The fitted six straight lines in Fig.4 correlation coefficients R^2 are higher than0.98, indicating that it is credible that the growth rate constant K at the same temperature conditions is a constant.

In addition, the relationships between the growth rate constant K, activation energy Q, and the process temperature in Kelvin T, can be expressed as an Arrhenius equation [8]:

$$\ln K = \ln(K_0) - \frac{Q}{RT} \qquad (1)$$

where K_0 is the pre-exponential constant and R is the gas constant(8.314 J/mol K). Q and K_0 can be calculated from a plot of lnK versus 1/T. The activation energy, Q is the slope of line in Fig.5. It can be seen the value of Q for V_2C layer and V_8C_7 layer are 47.27 kJ/mol and 32.26 kJ/mol, and the value of K_0 for V_2C layer and V_8C_7 layer are 7.26×10^{-3} cm^2/s and $6.31\times10^{-3}cm^2/s$, respectively.

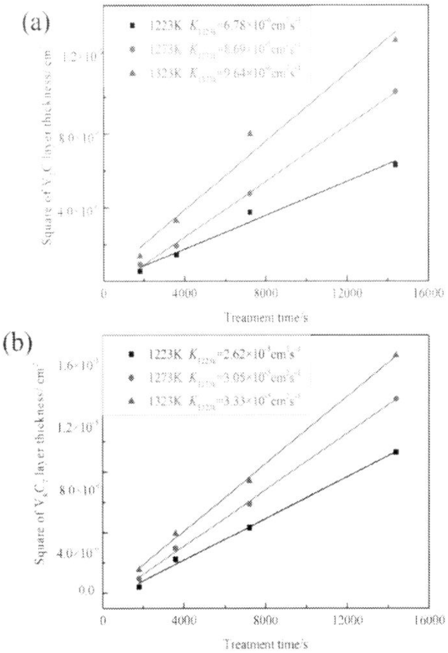

Figure.4. The square of the layer thickness versus treating time at different temperatures for V_2C layer (a) and V_8C_7 layer (b).

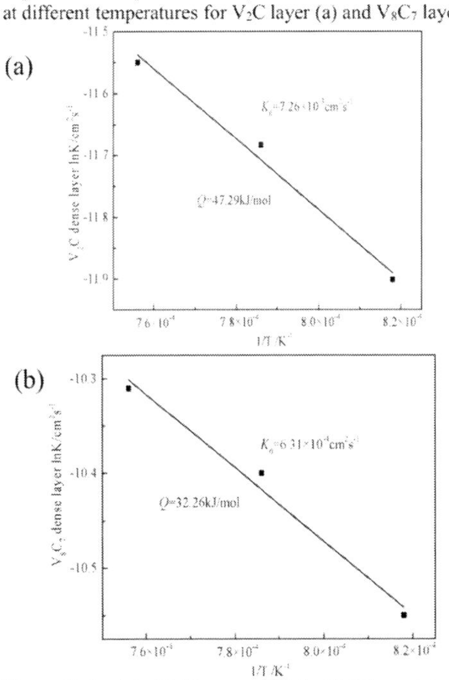

Figure.5. A plot of lnK versus 1/T for V_2C layer (a) and V_8C_7 layer (b).

Inserting the values of Q and K_0 in Eq. (1), Eq. (2) are obtained by inserting the calculated value of Q and K_0. For V_2C layer and V_8C_7 layer, the K are,

$$K = 7.26\times10^{-3}\exp\left(-\frac{3930.52}{T}\right) \quad (2)$$

$$K = 6.31\times10^{-3}\exp\left(-\frac{3880.03}{T}\right) \quad (3)$$

Then, the relationship among V_2C layer or V_8C_7 layer thickness, treating time and temperature can be obtained.

Conclusions

The kinetics of composite layer showed that a parabolic relationship between carbide layer thickness and treatment time, which indicating that the composite layer growth kinetics is controlled by diffusion. The relationship between growth rate constant and reaction temperature meet the classical Arrhenius Equation. The apparent activation energy of V_2C layer and V_8C_7 layer were 47.27 kJ/mol and 32.26kJ/mol, respectively.

Acknowledgements

The project was supported by the Shaanxi Natural Science Basic Research Project (2016JQ5044) and the Scientific Research Plan Projects of Shaanxi Education Department (16JK1561).

References

1. X.J. Di, M. Li, Z.W. Yang. Mater.Des. 96 (2016)
2. Y.P. Ji, S.J. Wu , L.J. Xu. Wear. 295, 31 (2012)
3. S. Barzilai, N. Frage, A. Raveh. Surf. Coat. Technol. 200 (2006)
4. V. Ataibis, S. Taktak. Surf. Coat. Technol. 279 (2015)
5. H. Liu , J.C. Li, Y.T. Chai. Surf. Coat. Technol. 302 (2016)
6. S.Z. Fan. L.S. Zhong, Y.H Xu, Y.H. Fu, L.L. Wang. Adv. Eng. Mater. 17, 11 (2015)
7. F.X. Ye, M. Hojamberdiev, Y.H. Xu. Appl. Surf. Sci. 280, 8 (2013)
8. M. Aghaie-Khafri , F. Fazlalipour, J. Phys. Chem. Solids. 69, 10(2008)EPL, 84 (2008)

Electrodeposition of Adherent Polypyrrole Film on Titanium Surface with Enhanced Anti-corrosion Performance

Jilin Tan[1], Zhong Zhang[2,a] and Dongtao Ge[1,a]

[1]*Department of Biomaterials, College of Materials, Xiamen University, Xiamen 361005, China*
[2]*Department of Stomatology, The Affiliated Zhongshan Hospital of Xiamen University, Xiamen 361004, China*

Abstract. A method of producing extremely adhesive polypyrrole film on Titanium (Ti) substrate was investigated. The Ti substrate was chemically pretreated and then modified by polydopamine (PDA), the polypyrrole film synthesized by electrochemical method on the treated Ti substrate displayed good adhesion and enhanced anti-corrosion performance. The study of the corrosion process was conducted through open circuit potential, tafel polarization and alternating current impedance test. The adhesive polypyrrole film coated titanium showed higher positive shift in corrosion potential and lower corrosion rate, indicating the great enhanced anti-corrosion performance.

1 Introduction

Polypyrrole (PPy), one of the typical conducting polymers (CPs), has been widely investigated in metal anti-corrosion applications due to its good air and water stability, high conductivity and ease of synthesis at room temperature[1, 2]. However, the poor adhesion force between the PPy film and substrate material seriously inhibit its further application[3]. Although lots of significant progress has been made to improve its adhesion problem, there still has the desire to find a mild and effective method[4, 5].

Dopamine (DA), a biological neurotransmitter, which has a similar structure to the essential adhesive component of mussel protein, was famous for its good adhesion and easy synthesis[6, 7]. The good adhesion property and the ability to be secondary modified by other molecules make it a suitable material for improving the adhesion force between films and substrate by being used as an interlayer[8].

In the present study, we investigate an effective method to produce adherent PPy film on Ti substrate. Before electrosynthesis of PPy film, the Ti substrate was chemically pre-treated and modified by PDA interlayer. The obtained PPy film showed good adhesion force and enhanced anti-corrosion performance.

2 Experimental Section

Pyrrole (Py, 98 %), Sodium dodecyl benzene sulfonate (SDBS, 95 %) and DA was purchased from J & K Chemical Technology. Titanium foil (99.7 %, 1.5×2.5 cm) was purchased from Alfa Aesar. All other chemicals were of analytical grade and were used as received.

Ti surface was chemically pretreated by dipping the Ti pieces into the solution contained 0.5 M NaOH and 1.0 M H_2O_2 for 10 minutes at room temperature. Polydapomine film was deposited on Ti surface via the solution oxidation method reported by Lee et al[6]. Ti pieces were immersed into a dopamine solution (2 mg/ml DA in 10 mM Tris, pH 8.5) for 24 hours, and then washed with ultrapure water and dried. The electrodeposition of PPy film was performed from an electrolyte solution containing 0.2 M SDBS and 0.2 M Py monomer via three electrode system. Platinum wire electrode, saturated calomel electrode (SCE), and Ti electrode was used as counter electrode, reference electrode, and working electrode, respectively. A constant potential of 0.75 V was applied to the system for 600 s, the obtained PPy film was removed from the electrolyte and rinsed with ultrapure water after electrodeposition.

The morphology of the PDA and PDA-PPy films were analyzed using SU-70 scanning electron microscopy (SEM). Adhesion force between the films and Ti substrate was tested by the Scotch™ Magic™ Tape 810 (3M). The electrochemical impedance spectra were measured in 3.5 % NaCl solution at 25 °C with a CHI 660E Electrochemical Workstation using 5 mV (rms) AC sinusoid signal at a frequency range from 1000 kHz to 0.01 Hz. The Tafel polarisation curves were performed by potentiodynamic polarisation in 3.5 % NaCl solution with a rate of 0.333 mV s^{-1}.

3 Results and Discussion

Fig. 1 illustrated the potentiostatic method deposition of PPy films from SDBS and Py monomer solution on

[a] Corresponding author: zhongz@xmu.edu.cn; gedt@xmu.edu.cn

untreated Ti substrate, PDA modified Ti substrate, and pretreated plus PDA modified Ti substrate, the obtained samples were denoted as Ti/PPy, Ti/PDA–PPy and SC2Ti/PDA–PPy, respectively. As shown in Fig. 1, the current of all the electropolymerization curves initially decreases at the beginning and then increases at a fast rate, which correspond to Ti surface oxidation and PPy nucleation and growth, respectively[9]. The electropolymerization current of Ti/PPy was apparently larger than Ti/PDA–PPy and SC2Ti/PDA–PPy in the whole process, we deduced the current change was due to the pretreated process and the modification of PDA, which changed the surface state and electrical conductivity of Ti electrode, leading to the PPy formation more difficult.

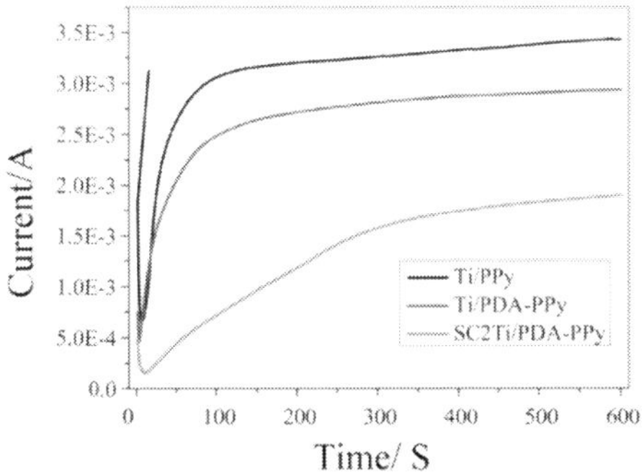

Figure 1. Chronoamperometric curve of PPy films

The morphology of the PDA and PPy films were characterized by SEM and shown in Fig. 2. From Fig. 2a and Fig. 2c, we can see that after PDA modification, lots of isolated islands were formed and the polishing scratch of Ti surface was coated by uniform thin film. It was consistent with the PDA layer formation process reported by Jiang et al[10], nanoaggregates of PDA and uniform PDA film were formed in the suspension and substrate simultaneously, and the particle of PDA will deposit onto substrate after a period of time. Fig. 2b and Fig. 2d shows the cauliflower-like morphology of PPy film, which means the PDA modification or chemically pretreating for Ti substrate did not change the morphology of PPy films.

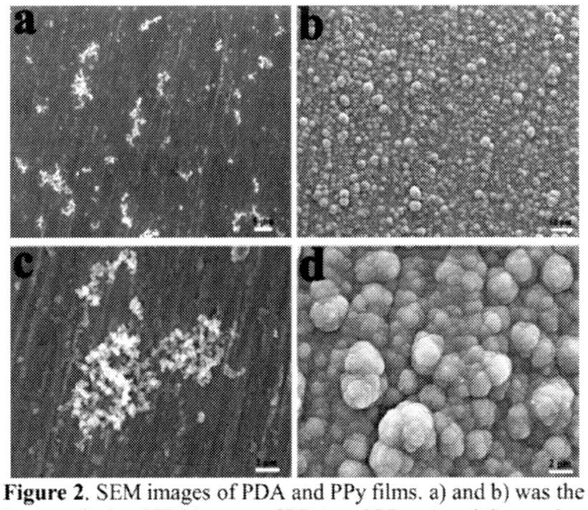

Figure 2. SEM images of PDA and PPy films. a) and b) was the low-resolution SEM images of PDA and PPy. c) and d) was the high-resolution SEM images of PDA and PPy.

Adhesion force of the PPy and PDA-PPy films was examined by peel-off test after the electrosythesis, the results were shown in Table 1. For the Scotch tape peel-off test, a piece of tape was firstly taped on PPy films, the air bubbles was removed by pressure to make sure good contact between the tape and the film, then the tape was peeled off from the bottom upwards at a quick speed. If the film was strongly bonded with the substrate, the delamination will occur at the tape/film interface, otherwise the delamination will occur at the film/substrate interface. In the cases of PPy on untreated Ti, PPy films were completely removed from Ti surface for all 6 samples, indicating that PPy films were barely bonded to the substrate. In the cases of PPy on PDA modified Ti substrate, 20% of PPy films was removed from the Ti surface for 2 samples and 100% adhesion for other 4 samples, indicating that the cohesive strength between PPy film and Ti substrate was greatly enhanced. DA was a small biomolecule bearing similar structure to the essential adhesive component of mussel protein, so we supposed that the existence of PDA improved its adhesive force to Ti substrate. The best adhesive force between PPy and Ti substrate was in the cases of PPy on pretreated plus PDA modified Ti, 100% adhesion was observed in all 6 samples, it was impossible to remove any PPy from the Ti substrate. We attribute the good adhesion to the pretreated process and the PDA modification, which change the topography and chemical composition of Ti substrate, leading to a more adhesive PPy film formation.

Table 1
Scotch tape test of PPy, and PDA-PPy films after the electrosythesis.

Sample	Number of samples	Result
PPy on Ti (without any treatment)	6	100% film off (6 cases)
PDA-PPy on Ti (without any treatment)	6	0% film off (4 cases) 20% film off (2 cases)
PDA-PPy on SC2 treated Ti	6	0% film off (6 cases)

The corrosion resistance of PPy and PDA-PPy coated Ti in 3.5% NaCl solution was analyzed by the open circuit potential, alternating current impedance test and tafel polarization test, the results was shown in Fig. 3. The open circuit potential (OCP, Fig. 3a) for the PPy coated Ti was positive shift to above than 0 V compared to pure Ti (-0.33 V), and the OCP of SC2Ti/PDA-PPy was more positive than Ti/PDA-PPy and Ti/PPy. Typical Nyquist plot of pure Ti, Ti/PPy, Ti/PDA-PPy, and SC2Ti/PDA-PPy was shown in Fig. 3b, all the curves showed only one capacitive loop in the measured frequency region. The enhancement of corrosion resistance can be reflected by the increasing radius of

Nyquist plot[11]. It was observed that the radius of Ti/PPy was much higher than pure Ti, indicating the anti-corrosion performance of Ti/PPy. The radius of SC2Ti/PDA-PPy was higher than Ti/PPy and Ti/PDA-PPy, indicating the best corrosion resistance. The tafel polarization test (Fig. 3c) further confirmed the enhanced corrosion resistance of Ti/PDA-PPy and SC2Ti/PDA-PPy. The positive shift in corrosion potential of Ti/PPy, Ti/PDA-PPy and SC2Ti/PDA-PPy (compared to pure Ti) was 0.289 V, 0.355 V, and 0.372 V, respectively.

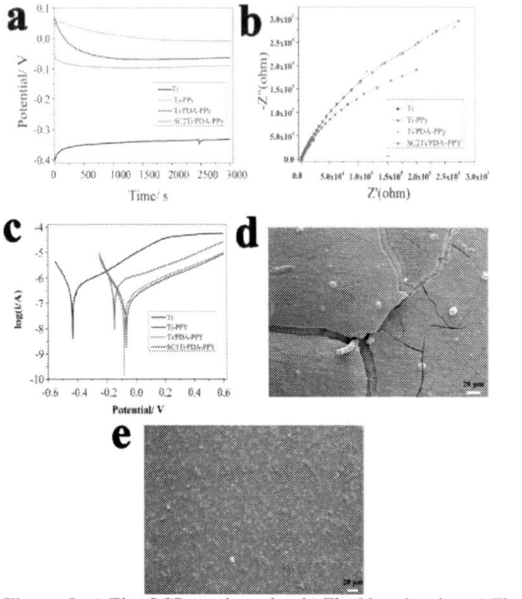

Figure 3. a) The OCP vs. time plot. b) The Nyquist plot. c) The tafel polarization plot of Ti and PPy coated Ti in 3.5% NaCl solution. The SEM images of d) Ti/PPy and e) SC2Ti/PDA-PPy after polarization.

The corrosion rate for Ti, Ti/PPy, Ti/PDA-PPy, and SC2Ti/PDA-PPy was 4.29×10^{-7} m/s, 3.18×10^{-7} m/s, 1.41×10^{-7} m/s, and 1.34×10^{-7} m/s, respectively. The much more positive shift in corrosion potential and the lower corrosion rate indicate the greatly enhanced anti-corrosion performance of SC2Ti/PDA-PPy. The SEM images of Ti/PPy and SC2Ti/PDA-PPy after polarization was shown in Fig. 3d and Fig. 3e. It was observed that Ti/PPy film was badly damaged while the SC2Ti/PDA-PPy film surface seems no change, which further verified the enhanced anti-corrosion performance of SC2Ti/PDA-PPy.

4 Conclusions

Adherent SC2Ti/PDA-PPy film was electrosynthesized on chemical pretreat and PDA modified Ti substrate, the adhesion force between PPy film and substrate was evaluated by peel-off test. The adherent SC2Ti/PDA-PPy film showed improved protective behavior against the corrosion of Ti compared to pure Ti and Ti/PPy, which was reflected by open circuit potential, the polarization curves and the electrochemical impedance spectroscopy test. The results were also in agreement with the observed morphology of Ti/PPy film and SC2Ti/PDA-PPy film after polarization in 3.5% NaCl solution.

Acknowledgements

This work was supported by the National Nature Science Foundation of China (31271009, 81271689), the Fundamental Research Funds for the Central Universities (No. 2011121001), the Program for New Century Excellent Talents in University, and the Program for New Century Excellent Talents in Fujian Province University and the Xiamen Municipal Science and Technology project (3502Z20144026).

References

1. N. Sheng, Y. Lei, A. Hyonoo, M. Ueda, T. Ohtsuka, Prog. Org. Coat **77**, 1724 (2014).
2. M. Torabi, S.K. Sadrnezhaad, Mater. Corros **62**, 252 (2011).
3. A.J.A. F. Faverolle, B. Bioch, P. Audebert, C. P. Andrieux, Chem. Mater **10**, 13 (1998).
4. O.I. Katrin Idla, Marek Strandberg, Electrochim. Acta **45**, 10 (2000).
5. S. Popescu, C. Ungureanu, A.M. Albu, C. Pirvu, Prog. Org. Coat **77**, 1890 (2014).
6. H. Lee, S.M. Dellatore, W.M. Miller, P.B. Messersmith, Science **318**, 426 (2007).
7. Q. Ye, F. Zhou, W. Liu, Chem. Soc. Rev **40**, 4244 (2011).
8. Y. Liu, K. Ai, L. Lu, CHEM REV **114**, 5057 (2014).
9. K. Cysewska, J. Karczewski, P. Jasiński, Mater. Lett **183**, 397 (2016).
10. J. Jiang, L. Zhu, L. Zhu, B. Zhu, Y. Xu, Langmuir **27**, 14180 (2011).
11. Z. Zhao, Y.W. Shao, T.M. Wang, D.P. Feng, W.M. Liu, Corros. Eng. Sci. Techn **46**, 330 (2013).

Investigation of subsurface fatigue crack in PEEK shaft under one-point rolling contact by using 2.5D layer observation method

Hitonobu KOIKE[1a], Genya YAMAGUCHI[2], Koshiro MIZOBE[2], Yuji KASHIMA[3], and Katsuyuki KIDA[2]
[1] *Kyushu Sangyo University, 2-3-1, Matsukadai, Higashi-ku, Fukuoka, 813-8503, Japan*
[2] *University of Toyama Gofuku 3190, Toyama, 930-8555, Japan*
[3] *Kashima bearings Inc. 2-9-21 Himesato Nishi-Yodogawa-ku Osaka, 555-0025,Japan*

Abstract. Subsurface fatigue cracks under rolling contact area of the PEEK shaft against an alumina bearing's ball were investigated for application of frictional part in mechanical element in special situations such as chemical environments. In order to explore the flaking process of the PEEK shaft, the rolling contact fatigue tests were carried out by using a one-point radial loading rolling contact machine. The flaking occurred on the rolling track of the PEEK shaft at approximate 4×10^5 fatigue cycles. The subsurface fatigue crack propagation was investigated by using 2.5-Dimension layer observation method. The flaking was caused by the propagations of surface cracks and subsurface shear cracks, and the flaking shape was half-ellipse. Moreover, beach marks as fatigue crack propagation in the flaking were observed.

1 Introduction

Lightweight function parts using premium plastic are extended in various mechanical elements under special situation such as chemical, medical, space or flying devices [1-2]. Poly-ether-ether-ketone (PEEK), which is semi-crystalline polymer containing phenyl and ketone groups, has high corrosion resistance, self-lubrication ability, low specific gravity, high impact durability, and high melting temperature of 340°C. Therefore PEEK is one of the most promising materials for precision-machined custom bearing or gear etc. Researchers investigated tribological failures of PEEK materials for frictional parts [3-6]. Dearn et al. reported that the location of the maximum shear stress depends on the friction condition and is related with the fracture of PEEK polymer gears [3]. Hoskins et al. stated that in the PEEK twin disc test, failure including melting was observed on the contact surfaces in severe high slip-ratio, high load conditions [4]. Harras et al. stated that in rolling wear test, the heat accumulated inside of the material due to cyclic revolutions is too high because of the polymer's low thermal conductivity. As a result, the material starts softening [5]. Fridrich et al. investigated various failure patterns through their works. They stated that Mode I fatigue was not the dominant mechanism from results of the brock on ring wear tests of PEEKs with different molecular weight against hard steels [6]. The PEEK's tribological fatigue flacture mechanism is not well understood.

Meanwhile, the practical application research on PEEK's tribological fatigue fracture of the combination between PEEK plastic and ceramic part is only few studies [7-9]. In the previous study, we reported that the subsurface crack propagated along the PEEK shaft rolling direction in terms of fatigue failure mechanism when cyclic compressive stress was applied to the PEEK shaft [9].

In this study, the PEEK shafts were manufactured by machining for practical applications. The one-point loading rolling contact fatigue (one-point RCF) tests under an oil lubricant condition were performed, in order to explore the subsurface fatigue crack and flaking failure mechanism of the PEEK shaft against an alumina ball.

2 Experimental Procedure

2.1 PEEK shaft

Figure 1 shows the PEEK shaft specimen for the one-point radial loading rolling contact fatigue (RCF) test. Table 1 shows the grade and dimensions of the machined PEEK shaft for the one-point RCF test. The PEEK material grade was 450G in VICTREX™. Young's modulus and Poisson's ratio of PEEK were 3.5 GPa and 0.4, respectively. The PEEK shaft specimens were made through a lathe maching and a griding. After the grinding process on the PEEK shafts, the average roughness (R_a)

[a] Corresponding author: koike@ip.kyusan-u.ac.jp

DOI: 10.1051/matecconf/201713009001

on the surface of the shaft were 0.69~1.10 μm (Maximum height roughness R_z= 4.81~5.74 μm).

Table 1. The machined PEEK shaft specimen.

Grade	Young's modulus, GPa	Length, mm	Diameter, mm	R_a, μm	R_z, μm
450G	3.5	290	17	0.69-1.10	4.81-5.74

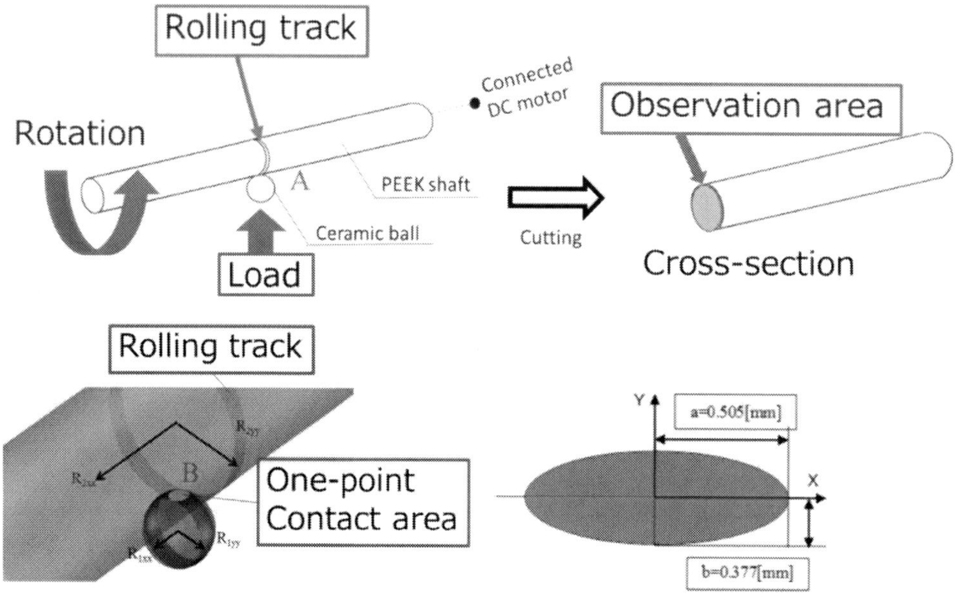

Detail A. Rolling track **Detail B.** Hertzian contact area

Fig. 1. Dimension and testing position of PEEK shaft specimen in one-point loading RCF. Note that X axis represents axial direction.

2.2 One-point rolling contact fatigue test

The RCF tests were carried out with the one-point radial loading machine (Yosinori industry Ltd.) under oil lubricant condition [10]. The VG46 lubricant oil was used. A shown in Figure 1, one alumina ball with 9.52 mm diameter contacted the PEEK shaft surface. 120 N radial load was applied to the PEEK shaft. The advantage of the one-point RCF testing is efficient observation. Whole subsurface under the rolling track of the shaft specimen can be observed continuously. Thus, fatigue fracture mechanism by rolling contact can be investigated easily. The contact was Hertzian point contact as shown in Detail B of Figure 1. The size of the contact area was 'a'= 0.505 mm and 'b'= 0.377 mm.The maximum Hertzian contact stress on the contact area between the PEEK shaft and the alumina ball was 301 MPa from Hertzian theory. The RCF test speed was 1200 rpm. The rolling track can be concentrated only one line under a radial load as shown in Detail A of Figure 1. The one-point RCF tests were continued until fatigue failure. The PEEK shaft after the test was cut along the rolling track, in order to observe the subsurface of the PEEK shaft. The cross-sections of the PEEK shaft were polished by diamond suspension. The subsurface fatigue cracks on the cross-sections of the PEEK shaft were observed by laser confocal microscopy (LCM). The maximum crack depths of the subsurface fatigue cracks were measured by the LCM.

3 Experimental results and discussion

3.1 Fatigue fracture of PEEK shaft under rolling contact

Figure 2 shows the microscope image of the fatigue fracture on the rolling track of the PEEK shaft after the tests. A flaking as fatigue fracture occurred on the rolling track at 4.1×10^5 fatigue cycles. The maximum lengths of the flaking toward rolling direction and axial direction were approximately 2000 and 1400 μm, respectively. The rolling track of the PEEK shaft was little worn. Figures 3 shows the 2D-contour line of maximum crack depths of the flaking of the PEEK shaft. The dot-break line at 0 μm represents the CTR line in Figure 2. The X axis, the axial direction of the PEEK shaft, represents the distance from the CTR line. The points of black solid circle with numbers correspond to the layer numbers of the LCM images in Figures 4. At measurement point CTR (Layer #8), the maximum crack depth was 352 μm as the 1st flaking bottom. The maximum crack depth depended on the distribution of maximum shear stress (τ_{yz}) under Hertzian contact. In the axial direction, the maximum crack depths around 0 ~ 375 μm distance were approximately 350 μm. The internal shear crack occurred at this maximum crack depth of plateau zone due to τ_{yz}. The maximum crack depths around 375 ~ 825 μm distance in the axial direction was deeper than that of the 1st flaking. This observation results show the multi-flaking occurrence.

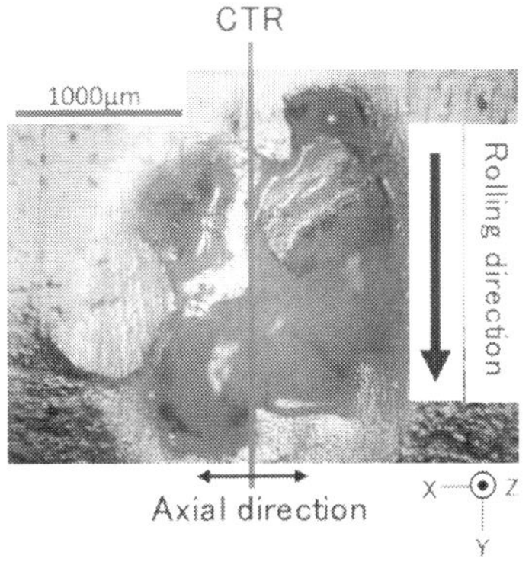

Fig. 2. Microscope image of flaking failure.

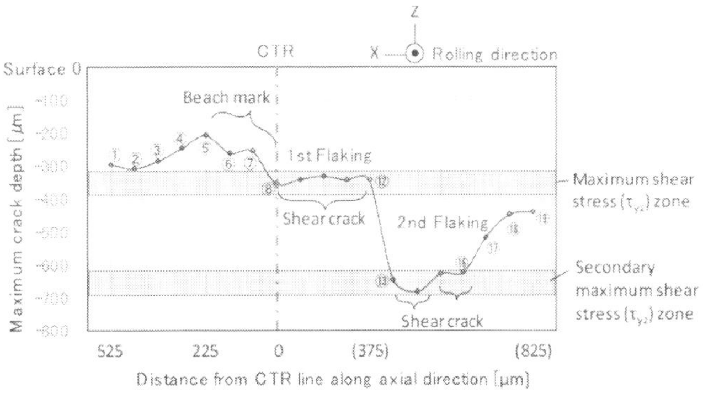

Fig. 3. Maximum crack depth from surface in the flaking area.

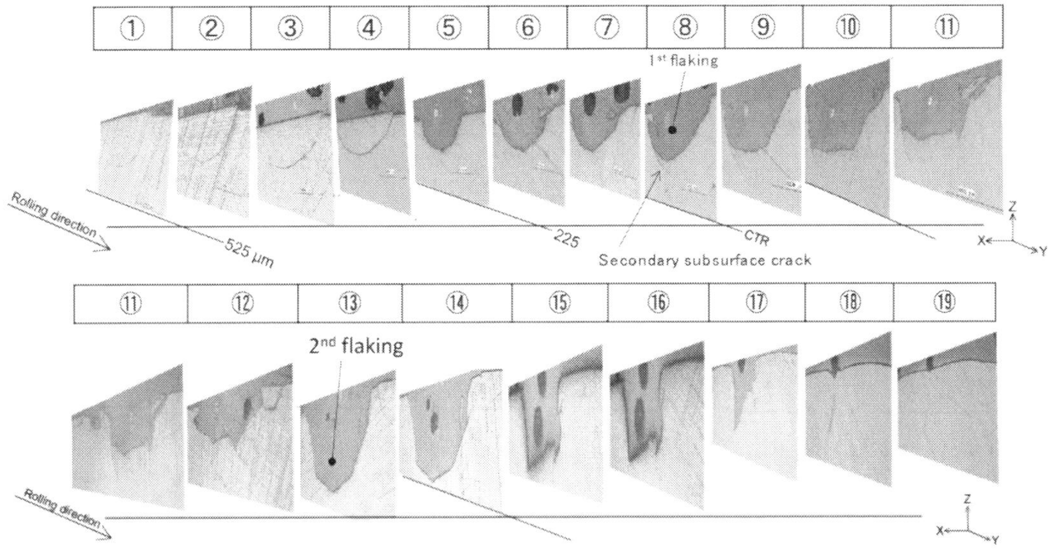

Fig. 4. Microscope images of 2.5D layer of the PEEK shaft subsurface.

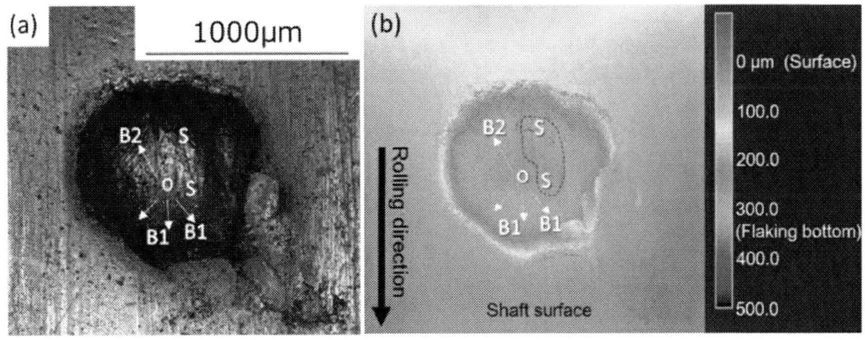

Fig. 5. LCM image of the flaking ; (a) laser image and (b) contour image. Note that ' O' represents crack initiation point, 'B' means beachmark and crack's direction, and 'S' means shear crack.

3.2 Subsurface crack propagation

Figure 4 shows the LCM images of 2.5D layer of the shaft subsurface. The distance between layers was 75 µm. The layer image (Layer #8) corresponds to the cross-section at CTR position in Fig. 2. The maximum crack depth in layer #8 was 352 µm. In the previous work, we reported that the subsurface crack propagated along the PEEK shaft rolling direction [9]. After that, the subsurface crack direction changed toward the shaft surface. From this work, it was found that the subsurface crack also propagated toward the axial direction and shaft surface. Moreover, in the layer images (Layer #9-12), the secondary subsurface crack was observed at the bottom of the 1st flaking. It was thought that the secondary subsurface fatigue crack influenced the secondary flaking occurance as shown in Layer #13. The flakings grew gradually due to complex subsurface fatigue crack propagation. Figures 5a and 5b show the LCM image and the contour image of the flaking after 4.6×10^5 fatigue cycles, respectively. The fracture with plastic deformation was localy observed on the S-S area in Figure 5a. The depth at the S-S area was measured approximately 300 µm from the shaft surface as shown in Figure 5b. This means that the shear crack occurred in the subsurface around 300 µm depth due to compressive and shear stresses. Beach mark indicating internal fatigue cracks was observed at 'B' point in Fig. 5a. The beach mark was located in the blue-green color area in Figure 5b. The subsurface crack direction changed in this area. The beach mark was influenced by cyclic tensile crack rather than shear crack. This means that the subsurface fatigue crack propageted toward radial direction from bottom ('O' point) to the shaft surface. In the beach marks at B1 and B2 points, the subsurface fatigue crack propagation rates in the rolling direction (B1) and the opposite direction (B2) were different as seen in the beach mark stripe's width. These resluts show that the fatigue crack propagation rate difference influenced the flaking occurrance time and the flaking shape. The mechanism of subsurface fatigue crack is discussed following; Firstly, the subsurface fatigue crack ocuured due to shear stress. Then, the subsurface fatigue crack propageted toward radial direction, and changed crack direction. The flaking shape was half-ellipse.

This research work is financial supported by JSPS KAKENHI Grant Number JP16K06042. A part of this work was supported by JKA and its promotion funds from KEIRIN RACE (No. 27-124, 28-138).

Conclusions

In order to explore the flaking process of the PEEK shaft against an alumina bearing's ball, the one-point radial loading rolling contact fatigue tests were carried out. The important results obtained from this work can be summarized as follows:

(1) Subsurface shear crack as Mode II fatigue was caused by maximam shear stress at 352 µm depth from the shaft surface.

(2) Beach mark as Mode I fatigue crack propagation was observed locally in the flaking. This area was corresponded to the zone where the subsurface crack propagation direction changed.

(3) From the reasons above (1) and (2), the subsurface fatigue crack propagated toward the rolling and axial direction.

References

1. T. A. Stolarski, *Adv. Comp. Trib.edited by K. Friedrich*, Elservier, Chapter 17, (1993).
2. K. Friedrich, G. Theiler and P. Klein, *In: SinhaSK, BriscoeBJ, editors. Polymer Tribology*, London UK: Imperial College Press, 375-415, (2009).
3. K. D. Dearn, S. N. Kukureka, and D. Walton, *In: SinhaSK, BriscoeBJ, editors. Polymer Tribology.* London UK: Imperial College Press; 470-505, (2009).
4. T. J. Hoskins, K. D. Dearn, Y. K. Chen, and S. N. Kukureka, Wear, **309**, 35–42, (2014).
5. K. Friedrich, Z. Lu and A. M. Häger, Theoretical and Appl. Frac. Mech., **19**, 1-11, (1993).
6. M. Harrass, K. Friedrich, and A. A. Almajid, Trib. Int'l, **43**, 635-646, (2010).
7. H. Koike, K. Kida, T. Honda, K. Mizobe, S. Oyama, J. Rozwadowska, Y. Kashima, and K. Kanemasu, Adv. Mat. Res., **566**, 109-114, (2012).
8. H. Koike, K. Kida, E. C. Santos, J. Rozwadowska, Y. Kashima and K. Kanemasu, Trib. Int'l, **49**, 30-38, (2012).
9. H. Koike, G. Yamaguchi, K. Mizobe, Y. Kashima, and K. Kida, Mat. Sci. Forum, **878**, 137-141, (2017).
10. S. Hazeyama, J. Rozwadowska, K. Kida, E. C. Santos, T. Honda, K. Kanemasu, and T. Shibukawa, Adv. Mat. Res., **566**, 182-186, (2012).

Investigation of wear, groove shape and load capacity of PPS-PTFE hybrid radial ball bearings

Hitonobu KOIKE[1], Katsuyuki KIDA[2a], Toshihiko MATSUMURA[2], Koshiro MIZOBE[2], and Yuji KASHIMA[3]
[1] *Kyushu Sangyo University, 2-3-1, Matsukadai, Higashi-ku, Fukuoka, 813-8503, Japan*
[2] *University of Toyama Gofuku 3190, Toyama, 930-8555, Japan*
[3] *Kashima bearings Inc. 2-9-21 Himesato Nishi-Yodogawa-ku Osaka, 555-0025,Japan*

Abstract. In order to improve the radial load capacity and wear loss of the PPS-PTFE hybrid bearings, the groove shape of the PPS inner ring in the PPS-PTFE hybrid bearing was designed based on ceramic bearing's ball radius (r). The rolling contact fatigue tests were performed by using the PPS-PTFE hybrid bearings with the original (r*1.0) and modified (r*1.1) inner rings. The radial load capacity of the bearing using the r1.1 inner ring was improved 150% over the capacity of the bearing using the r1.0 inner ring. While the operation temperature of the bearing with the r1.1 inner ring was stable at 40 °C, that of the bearing with the r1.0 inner ring incresed rapidly over 100 °C and thermal failure occurred. Moreover, the wear loss of the bearings with the r1.1 inner ring was lower than that of the r1.0. Thus the groove shape influenced the tribology performance of the PPS-PTFE hybrid bearings.

1 Introduction

Premium plastic bearings of lightweight and custom-made for low-volume production are expected for assemblies or equipments in special stuations such as medicine, sky/space, underwater etc. [1-4]. Poly-phenylene-sulfide (PPS), which is semi-cystalline thermoplastic polymer, has a glass transition temperature of 90 °C and a melted tenperature of 278 °C [5]. Specific density and bending strength are 1.35 g/cm^3 and 142 MPa, respectively. With lightweight, good machinability, and corrosion resistance, PPS is promising material for precision-machined custom bearings.

Concerning tribological failures in plastic ball bearings, Marshek stated that overheating, rapid wear or plastic flow occurs on a raceway of bearing's ring [6] when bearing's operation conditions are heavy loads and high rotational speeds. Overheating softens or melts ring, to cause a disassembled bearing. In previous polymer bearing's study, Koike et al stated that the lifespan and wear resistance of the hybrid polymer bearing with the PTFE composite retainer was superior to that without the PTFE composite retainer [7,8]. Also, Mizobe et al. reported that the wear resistance of the PPS bearing with the PTFE composite retainer could be improved under dry condition [9,10].

In this study, the groove shapes of the PPS inner rings of the PPS-PTFE hybrid bearings were designed and then manufactured by lathe machining, in order to improve the load capacity, wear loss and operation temperature for practical applications. Rolling contact fatigue (RCF) tests were performed by using the original and modified PPS-PTFE hybrid bearings under a dry condition.

2 Experimental Procedure

2.1 PPS-PTFE Hybrid Bearing Specimens

Figure 1 shows the bearing specimen. The PPS-PTFE hybrid bearings were designed based on JIS-B-1512 #6205. The groove shapes on the bearing's inner rings were designed as the 103% and 114% curvature radiuses based on alumina ball radius. Hereafter, the inner rings of 103% and 114% curvature radiuses are called as 'r1.0' inner ring and 'r1.1' inner ring (Fig. 2). The bearing's inner ring and outer ring were produced of PPS from KUREHA EXTRON CO.,LTD. The retainers were produced of PTFE composite [5]. 9 alumina balls were used for each bearing.

2.2 Rolling Contact Fatigue (RCF)

Figure 3 shows the illustration of the RCF test machine. The RCF tests were carried out by using the radial-type rolling contact fatigue machine under test conditions of the radial loads ranging from 113 N to 333 N at room temperature. Rotation speeds were 600 and 3000 rpm. The RCF tests were continued until 1.0×10^6 fatigue cycles or bearing's failure. The contact surface of each tested bearing's specimen was observed by a Laser Confocal Microscope (LCM). The operation temperature

[a] Corresponding author: kida@eng-u.toyama.ac.jp

 DOI: 10.1051/matecconf/201713009002

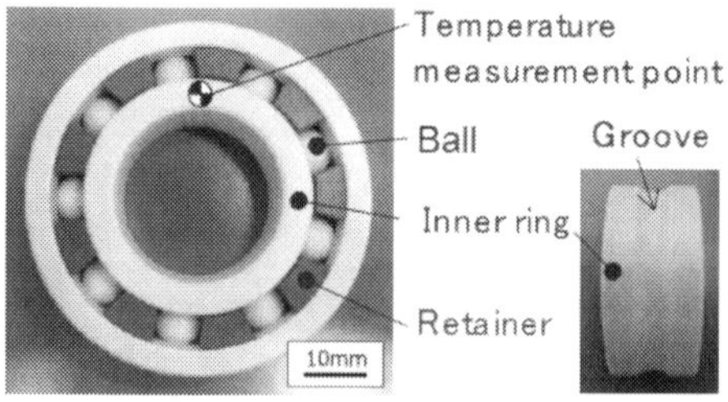

Fig. 1. PPS-PTFE radial bearing.

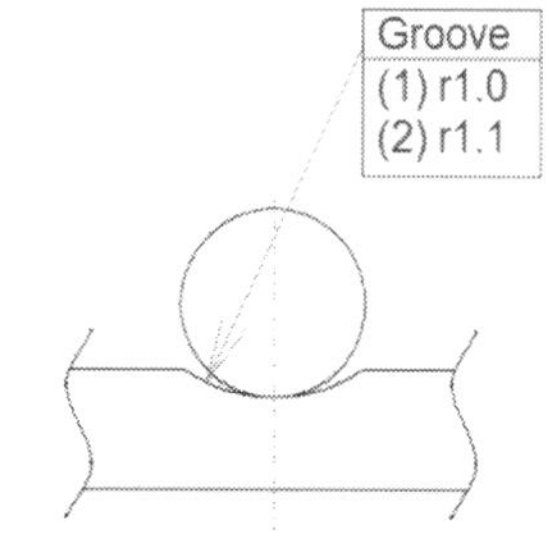

Fig. 2. Truck curvature radius of inner rings of bearings.

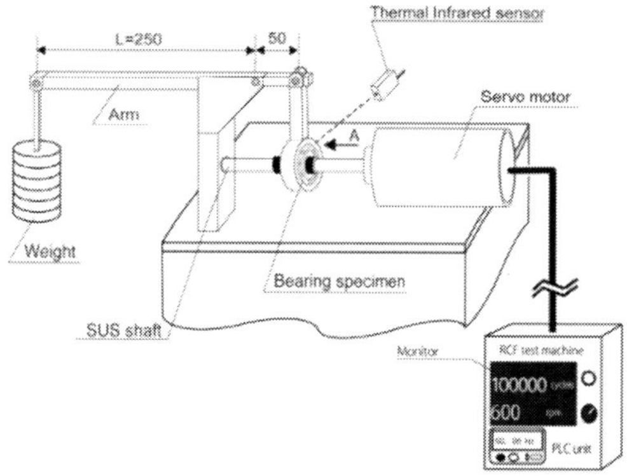

Fig. 3. Radial-type rolling contact fatigue machine.

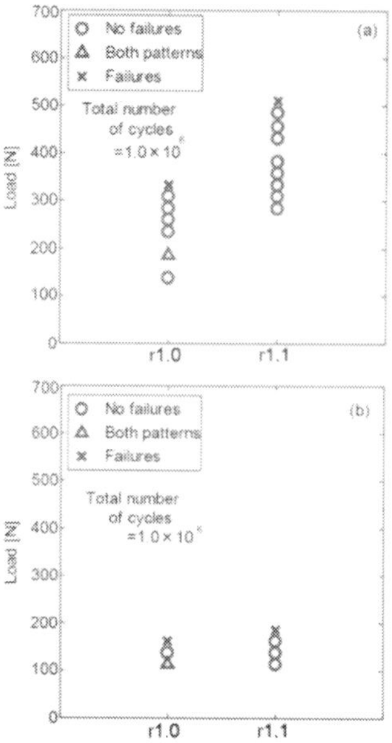

Fig. 4. Load capacity of r1.0 and r1.1 bearings ; (a) at 600rpm and (b) at 3000rpm.

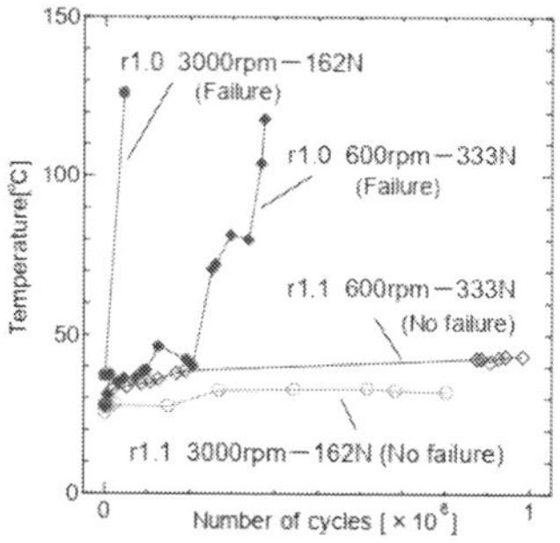

Fig. 5. Transition of operation temperature of r1.0 and r1.1 bearings at 600 rpm.

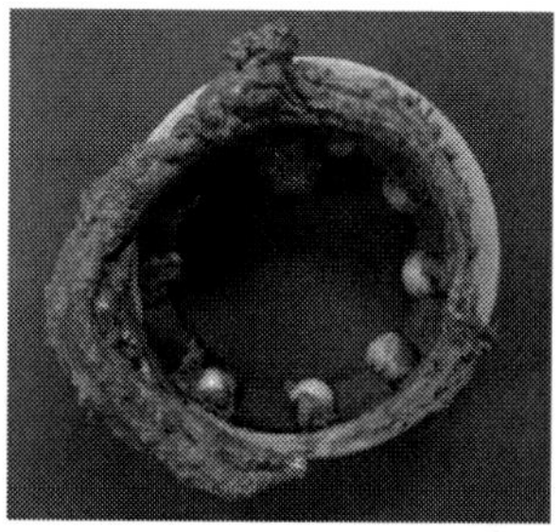

Fig. 6. Thermal failure at 3000 rpm and 162 N load.

of the bearing's inner ring was measured by a thermal infrared sensor. The temperature was measured on the inner ring surface from rotating axis direction as shown in Figure 3. In order to calculate wear loss of the bearing's part, the weights of the inner ring, outer ring and retainer of PPS-PTFE hybrid radial bearing were measured by a Shimadzu electric balance.

3 Experimental results and discussion

3.1 RCF test results

Figure 4 shows the load capacities of the bearings with the r1.0 and r1.1 inner rings. The red circle (○) represents no failure at 1.0×10^6 fatigue cycles. The cross mark (×) represents the bearing's rotation stop due to thermal deformation during RCF testing. When the test conditions were 600 rpm - 333 N or 3000 rpm - 162 N, the thermal deformation of the hybrid bearing with the r1.0 inner ring occurred. The radial load capacity of the PPS-PTFE bearings with the r1.1 inner ring during the RCF test at 600 rpm was improved 150% of the PPS-PTFE bearings with the r1.0 inner ring. It was found that the load capacity of the bearing with the r1.1 inner ring was higher than that of the r1.0 ring when rotational speeds were 600 rpm and 3000 rpm. Figure 5 shows the transition of operation temperature of the r1.0 and the r1.1 inner rings during the RCF tests at 600 rpm.

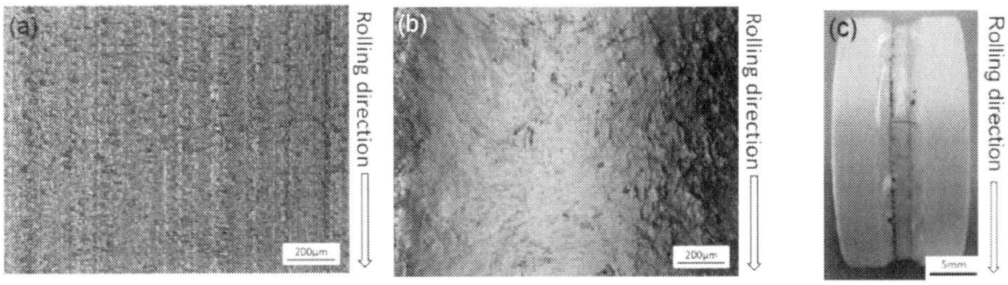

Fig. 7. Specimen surface of r1.0; (a) before the test, (b) after the test at 600 rpm-284 N, and (c) macro image of (b). Note that the red rectangle in (c) means observation area.

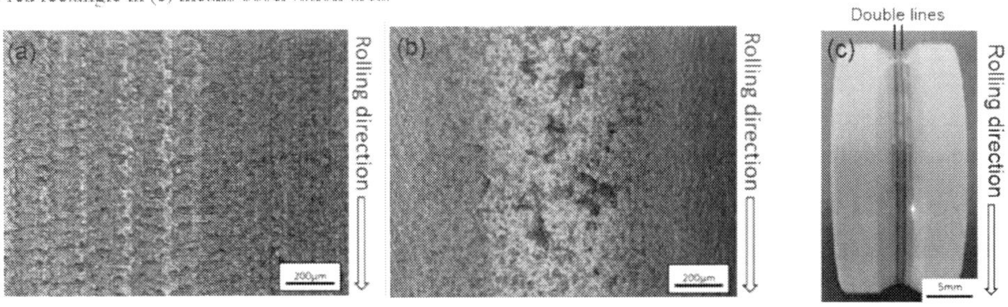

Fig. 8. Specimen surface of r1.1 ; (a) before the test and (b) after the test at 600 rpm-284 N, and (c) macro image of (b). Note that the red rectangle in (c) means observation area.

While the temperature of the PPS-PTFE bearing with the r1.1 inner ring was approximately 40°C during the test at 600 rpm under 333 N load, the temperature of the PPS-PTFE bearing with the r1.0 inner ring increased rapidly during the early operation stage. The rotation of the PPS-PTFE bearing with the r1.0 inner ring was stopped due to thermal deformation as shown in Figure 6.

3.2 Contact Surface

Figures 7 show the LCM images of the r1.0 inner ring raceway of the PPS-PTFE bearing before and after 1.0×10^6 fatigue cycles under 284 N radial load at 600 rpm. The r1.0 inner ring raceway surface was worn and flattened during the tests due to the rolling contact. Figures 8 show the LCM images of the r1.1 inner ring raceway of the PPS-PTFE bearing before and after 1.0×10^6 fatigue cycles. The dark gray double lines were observed on the r1.1 inner ring raceway (Fig. 8c). Because of the tribo-film accumulated along this double lines, the bearing rotated smoothly during the RCF test.

3.3 Wear Loss

Figure 9 shows the wear loss of the each part in the PPS-PTFE bearing after 1.0×10^6 fatigue cycles. The wear losses of the inner ing (○) and the outer ring (□) in the r1.1 PPS-PTFE bearing were lower than that of r1.0. The tribo-film was formed on the raceway of r1.1 inner ring rather than on the raceway of r1.0 inner ring. Therefore, the deformation and friction heat of the inner ring

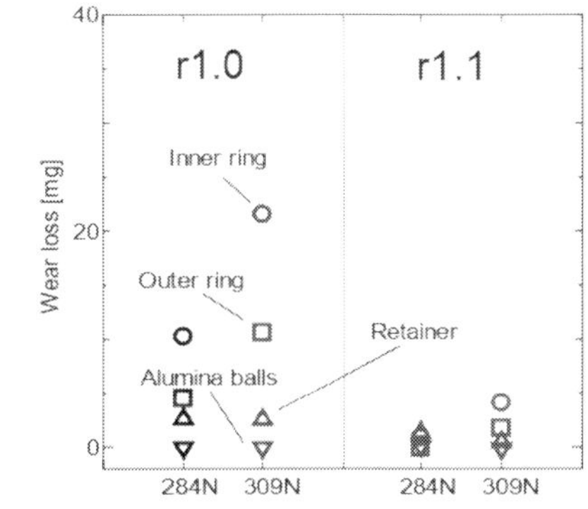

Fig. 9. Wear loss of the PPS-PTFE bearings (600 rpm).

raceway surface decreased. Thus, it is found that the difference of the groove radius values between the r1.1 and the r1.0 influenced tribo-film formation.

Conclusions

In order to improve the load capacity and wear loss of the radial PPS-PTFE hybrid alumina ball bearings, the r1.0 and r1.1 groove shapes of the PPS inner rings were designed. The rolling contact fatigue tests were carried out under a dry condition at 600 and 3000 rpm rotation speeds. The important results obtained from this work can be summarized as follows:

(1) The radial load capacity of the PPS-PTFE hybrid bearings with the r1.1 inner ring during the RCF test at 600 rpm could be improved 150% compared to the PPS-PTFE hybrid bearings with the r1.0 inner ring.

(2) While the PPS inner ring with the r1.0 groove shape was deformed at 333 N radial load, the PPS inner ring with the r1.1 radius raceway was achieved at 1.0×10^6 fatigue cycles.

(3) The wear loss and operation temperature of the PPS-PTFE hybrid bearings with the r1.1 inner ring were lower than those of the r1.0 when the test conditions were 113-333N radial load and 600 rpm rotational speed.

This work was supported by JKA and its promotion funds from KEIRIN RACE (No. 27-124, 28-138).

References

1. T. A. Stolarski, *Advanced in Composite Tribology edited by K. Friedrich* (Elsevier Science Publishers B.V., 1993), Chapter 17.
2. K. Friedrich, G. Theiler and P. Klein, *In: SinhaSK, BriscoeBJ, editors. Polymer Tribology* (Imperial College Press, London UK, 2009), 375-415.
3. V. Buck, Trib. Int'l, **19**, 1, 25-28, (1986).
4. M. Harrass, K. Friedrich, A. A. Almajid, Trib. Int'l, **43**, 635–646, (2010).
5. K. Mizobe, T. Honda, H. Koike, E. C. Santos, Y. Kashima and K. Kida, Adv. Mat. Res., **683**, 90-93, (2013).
6. K. M. Marshek, Wear, **52**, 141-146, (1979).
7. H. Koike, K. Mizobe, S. Oyama, Y. Kashima, and K. Kanemasu, Adv. Mat. Res., **683**, 385-390, (2013).
8. H. Koike, K. Kida, K. Mizobe, X. Shi, S. Oyama, and Y. Kashima, Trib. Int'l, **90**, 77–83, (2015).
9. K. Mizobe, T. Matsumura, Y. Kashima, and K. Kida, Key Eng. Mat. , **703**, 197-201, (2016).
10. K. Mizobe, H, Honda, H, Koike, E. C. Santos, K. Kida, Y. Kashima, Adv. Mat. Res. **566**, 157-161, (2012).

Rolling Contact Fatigue and Static Compression Deformation of UHMWPE thrust bearing in water

Masahiro Inagaki[1], Naoki Fujimura[1], Koushiro Mizobe[1], Katsuyuki Kida[1,a] and Yuji Kashima[2]

[1]University of Toyama, Gofuku 3190, Toyama, 930-8555, Japan
[2]Kashima Plastic Bearings Corporation, 2-9-21 Himesato, Nishiyodogawa-ku, Osaka, 555-0025, Japan

Abstract. In this study, the rolling contact fatigue tests were performed in order to investigate the effect of thrust load on life of UHMWPE bearings. Furthermore, the static compression tests were performed in order to investigate the effect of static load on deformation. It was found that thrust load controls limitations of the bearings, and the deformation by the static load is not the main reason of the adhesion failure.

1 Introduction

The UHMWPE (Ultra-High-Molecular-Weight-Polyethylene) has some good mechanical features such as wear resistance and corrosion resistance. In our previous studies on UHMWPE thrust bearings [1,2], we investigated the effect of thrust load on bearing life in dry air. The failure pattern of these bearings was the thermal deformation. The thermal deformation depended on the thrust load and rotation speed. Furthermore, we studied PEEK, PPS, and PTFE bearings in water [3-6].

The compressive strength of the UHMWPE is lower than that of other engineering plastics. The constant load deforms the UHMWPE samples, and the effect of static load on the deformation of UHMWPE was investigated [7,8]. However, the effect of the static load on the life of UHMWPE bearing is unclear.

In this study, the rolling contact fatigue tests were performed in order to investigate the effect of thrust load on life of UHMWPE bearings. Furthermore, the static compression tests were performed in order to investigate the effect of static load on deformation.

2 Test method

2.1 Specimens

The bearing specimen (two UHMWPE races - one UHMWPE retainer with nine soda glass balls) was designed based on the Japanese Industrial Standards (JIS) B 1512-2 2011 (Rolling bearing - Boundary dimensions - Part2: Thrust bearings) and 1513 1995 (Rolling Bearing Designation, #51305). Figure 1 is a photograph of the bearing specimen. The outer diameter, the inner diameter and pitch circle diameter were 52 mm, 25 mm and 38.5 mm, respectively. The ball diameter was 9.525 mm (3/8 inch). Table 1 shows the material properties of UHMWPE.

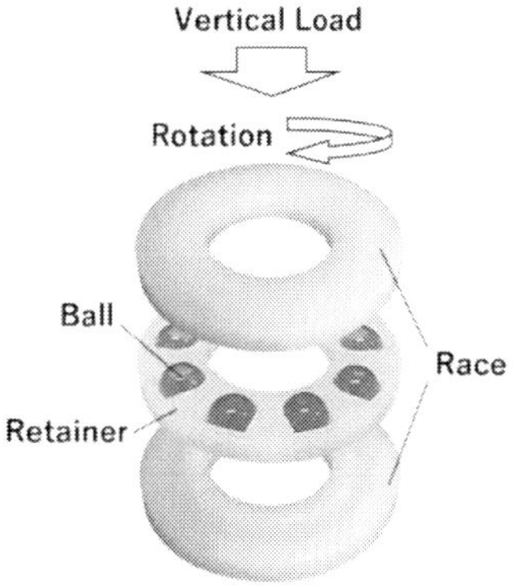

Figure 1. Photograph of UHMWPE thrust bearing.

Table 1. Material properties of UHMWPE [9, 10].

Melting point [°C]	Glass transition point [°C]	Water absorption [%]	Specific gravity	Compression strength [MPa]
136	-20	<0.01	0.94	19

2.2 Rolling contact fatigue (RCF) test

The RCF tests of UHMWPE thrust bearings were done using a RCF test machine which was developed by the Kida's group [3-6]. Figure 2 shows a schematic illustration of the RCF testing machine. All tests were performed in water and total number of cycles was 1.0×10^6. The RCF tests were stopped when the rotating resistance increased rapidly.

[a] Corresponding author: kida@eng.u-toyama.ac.jp

 DOI: 10.1051/matecconf/201713009003

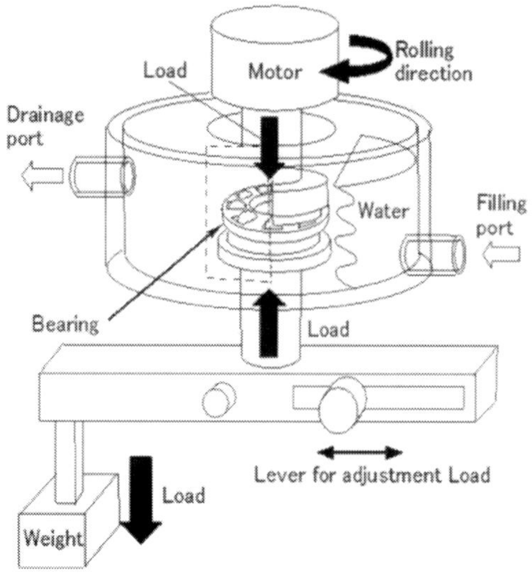

Figure 2. Schematic illustration of RCF testing machine.

2.3 Static compression (SC) test

Figure 3 shows a schematic illustration of the static compression testing machine. The ball was pushed by the static load. The displacement is defined as "the deformation length of the specimen's surface". A laser analogue sensor (KEYENCE CORPORATION, IA-030) measured the displacement. All tests were performed for 14 hours. (This value means the operation time during 1.0×10^{6} cycles at 1200 rpm). The tests were done in water and air. The static loads were 50 N/ball, 100 N/ball, 150 N/ball, and 200 N/ball.

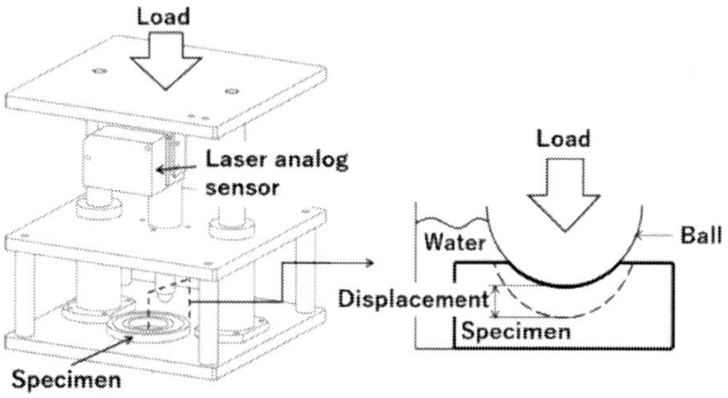

Figure 3. Schematic illustration of RCF testing machine.

3 Results and discussion

3.1. Relation between rotation speed and thrust load

Figure 4 shows the failure conditions of the specimen. We found two failure patterns: adhesion and indentation. These failures are plotted as "×". In the case of the adhesion failure, the race adhered to the retainer. This type of failure occurred at 2000 N. In the case of the indentation failure, the specimen did not rotate. The race surface deformed into the ball shape. This type of failure occurred at 3500 N. In addition, no failure was observed on a contact area between the race and the retainer under a load of 2000 N. This indicates the thrust load controls the limitations of the bearings.

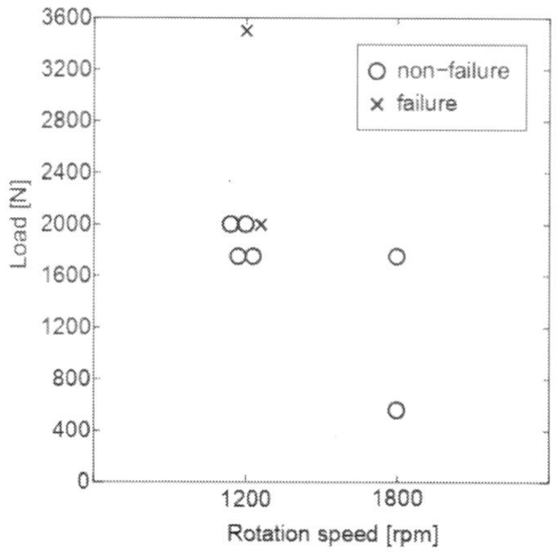

Figure 4. Relation between rotation speed and thrust load.

3.2 Wear loss

Figure 5 shows the relation between the wear loss and the thrust load. The wear loss ranged from -2.8 mg to 20.6 mg. Some values of wear loss are lower than "zero" because the amount of water absorption is higher than the wear loss. This indicates that we can ignore the wear loss in this test.

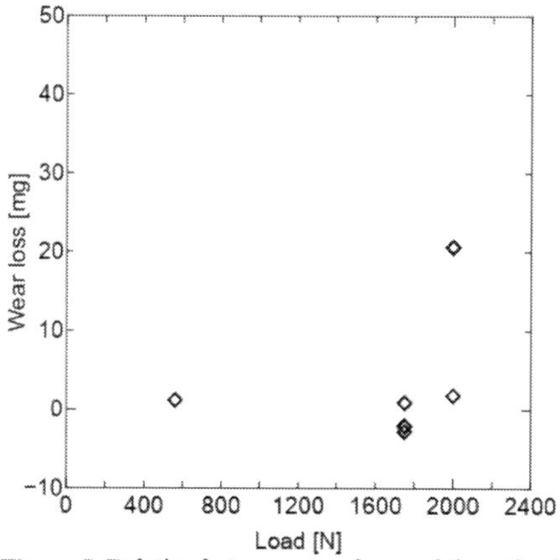

Figure 5. Relation between wear loss and thrust load.

3.3 Relation between displacement and time

Figures 6 show the relation between the displacement and the time. Five static compression tests were performed under a load of 50 N/ball. The average values are plotted as "bold line". The deformation behavior in air was similar to that in water. After 14 hours, the difference of the average of displacement was about 0.01 mm between air and water conditions. These results indicate the maximum displacement in air is similar to that in water.

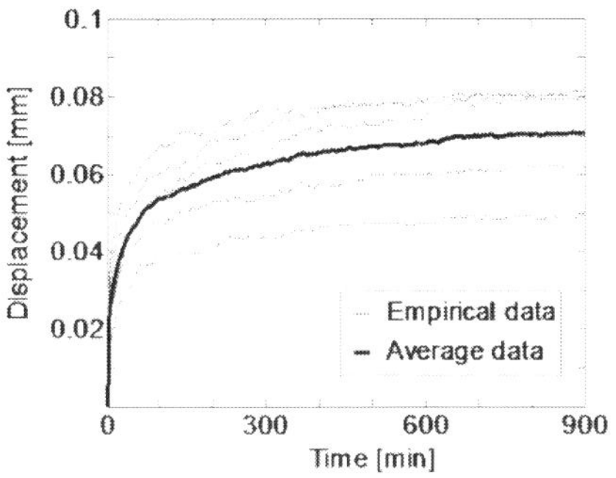

(a) In air.

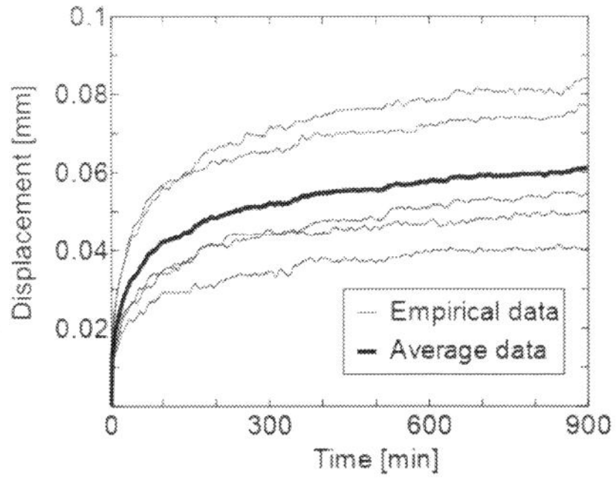

(b) In water.

Figure 6. Relation between displacement and time at 50 N/ball.

3.4 Relation between maximum displacements and static loads

Figure 7 shows the effect of the static load on the maximum displacement in air and water after 14 hours. The SC tests were performed under 50 N/ball, 100 N/ball, 150 N/ball, and 200 N/ball. The maximum displacements increase with the load when the load is over 100 N/ball. The maximum displacement of 200 N/ball was 0.1269

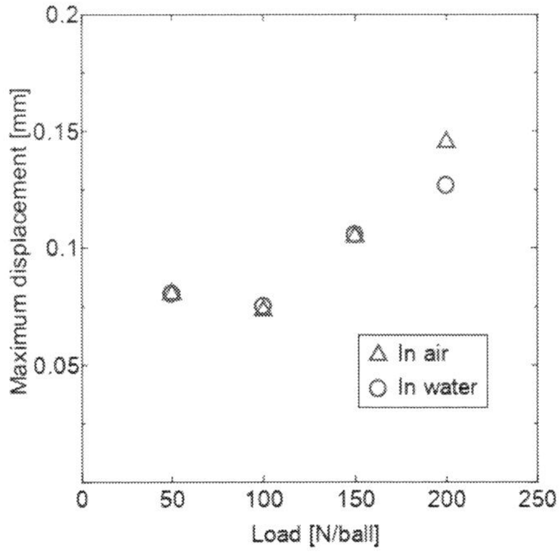

Figure 7. Relation between maximum displacement and static load in air and water.

mm in water. Even in the condition, the race does not contact the retainer. This indicates the static load is not the main cause of the adhesion failure.

4 Conclusions

In the present study, RCF tests and SC tests of UHMWPE thrust bearings were performed. The effect of thrust load on life was examined. Furthermore, the effect of static load on the deformation was studied. The following conclusions were obtained:
(1) The thrust load controls the limitations of the bearings.
(2) The failure is not governed by the wear loss.
(3) The maximum displacement after 14 hours in air is similar to that in water.
(4) The static load is not the main cause of the adhesion failure.

References

1. K. Mizobe, N. Fujimura, Y. Kashima, K. Kida, Key Engineering Materials, **703**, 192-196, (2016)
2. S. Hazeyama, S. Oyama, K. Kida, T. Honda, K. Mizobe, H. Koike, K. Itakura, Y. Kashima, K. Kanemasu, Advanced Materials Research, **683**, 77-81, (2013)
3. M. Orito, K. Kida, Y. Hashizume, M. Ishida, K. Mizobe, Y. Kashima, Materials Science Forum, **878**, 117-121, (2016)
4. T. Honda, K. Kida, E. C. Santos, Y. Kashima, K. Kanemasu, *Proc. World Tribology Congress*, **488**, (2009)
5. X. Shi, Y. Kashima, K. Kida, Applied Mechanics and Materials, 563, 270-274, (2014)
6. K. Itakura, T. Honda, S. Oyama, K. Kida, S. hazeyama, Y. Kashima, Advanced Materials Research, **683**, 391-395, (2013)
7. F. Metin, M. Cengil, Journal of Achievements in Materials and Manufacturing Engineering, **78(2)**, 65-70, (2016)
8. F. Quinci, M. Dressler, A. M. Strickland, A. Metcalfe, M. Taylor, G. Limbert, *Multifunctional Materials for Tribological Applications*, 339-364, (2015)
9. Mitsuboshi Belting Ltd., http://www.mitsuboshi.co.jp/english/
10. NSK Ltd., technical report, http://www.nsk.com/jp

Observation of fracture behavior of 3-D printed specimens under rolling contact fatigue in water

Koshiro Mizobe [1,a], Masahiro Inagaki[1] and Katsuyuki Kida[1]

[1]*University of Toyama, Department of mechanical engineering, 3190 Gofuku, Toyama, Japan*

Abstract. Polymer bearing was widely used in the corrosive conditions because of its high corrosion durability. The polymer bearing had been formed using molding and machining until the new 3-D printing method was developed. In this study, we performed the rolling contact fatigue tests of the 3-D printed specimens in water and observed the fracture behaviour of the specimens. We found that the surface cracks are related to both the rolling direction and the lamination directions.

1 Introduction

When engineers use bearings, they require various properties from the materials such as high hardness, low friction coefficient and formability. Bearing steel is popular because of its high hardness. However, it cannot be used in oil-less conditions; for example, in dry conditions which result in direct contact between the materials, in solutions which include some corrosive materials. In these special conditions we use the polymer bearings.

Some of the good properties of polymer are self-lubrication ability; light weight; and high corrosion resistance. In particular, corrosion resistance is important because wet and corrosive lubricating conditions often occur in industrial situations. Therefore researchers have studied polymer bearings in wet conditions[1-3].

Two major methods of making polymer bearing's formation are the injection molding and machining. Recently, a new 3-D printing method was developed and it had become popular because we can easily make complex and seamless parts using the method. Along with the expansion of the use of 3-D printing, researchers had focused on the effect of the laminated layers joins on the mechanical properties of the 3-D printed parts[4-5].

In this study, we performed rolling contact fatigue tests on the 3-D printed specimens in water and observed these fracture behaviour of the specimens

2 Experimental procedure

2.1 Specimens

The specimens were printed by 3-D printer (Projet 3510 HD plus, 3-D systems, Inc.). Figure 1 shows the photograph of a donut-shaped specimen. The thickness was 5 mm, the outer diameter of each bearing was 52 mm, the inner diameter was 25 mm, and the pitch circle diameter was 38.5 mm. All lamination directions were X direction and the lamination pitch was 16 μm. The material was ABS-like polymer (VisiJet M3X, 3-D systems, Inc.).

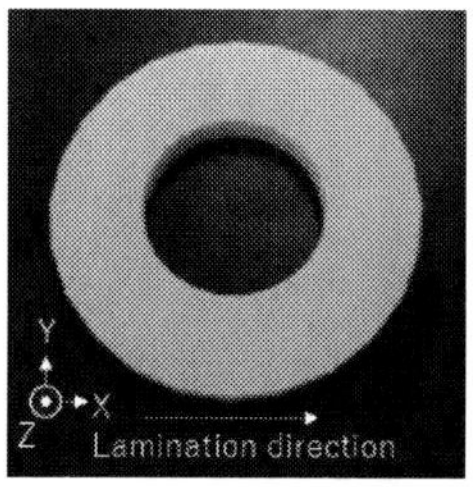

Fig. 1 3-D printed specimen

2.2 Rolling contact fatigue (RCF) test

Before testing we cleaned all specimens in ethanol and dried by the blower. After that, we performed RCF tests using a thrust type test machine which is shown in Fig. 2. The donut-shaped specimen set under the retainer with ball. The retainer and races were designed based on the JIS B 1513 #51305 standard. The race and retainer material were PEEK (Dupont-Mitsui). The retainer has nine 9.525 mm (3/8 inch) diameter Al2O3 balls. The outer, inner and pitch circle diameter were similar to the donut-shaped specimen

The tests were done in water conditions. The specimens set-up was similar to our previous work[6]. The rotation speed (rpm) was 1200 rpm and the total number of rotations was 2.88×10^5 cycles. This number of rotations simulated one day's working time (8 hours at 600 rpm). After testing the specimens were washed in ethanol and measured the weights. We observed the

[a] Corresponding author: kmizobe@eng.u-toyama.ac.jp

specimens using Laser Confocal Microscope (LCM, VK9710, Keyence).

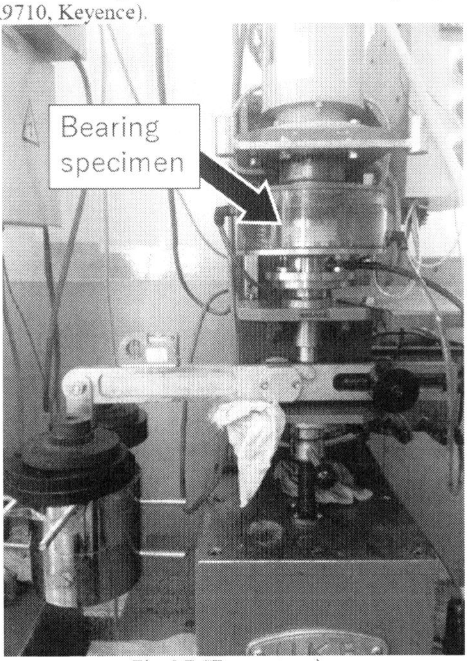

Fig. 2 RCF test set up image

3 Results and discussion

3.1 Specimen failure

Figure 3 shows the failure conditions of the specimens in water. In the case of the 500 N thrust load, the rotation of the bearing stopped because the donut-shaped specimen started flaking. Under the 250 N thrust load, the specimen showed no signs of flaking. However, the weight increased by 13.7 mg and this means that the water absorption of the specimen was higher than the wear weight loss. The wear loss was low which indicates that the fracture was strongly related to the flaking.

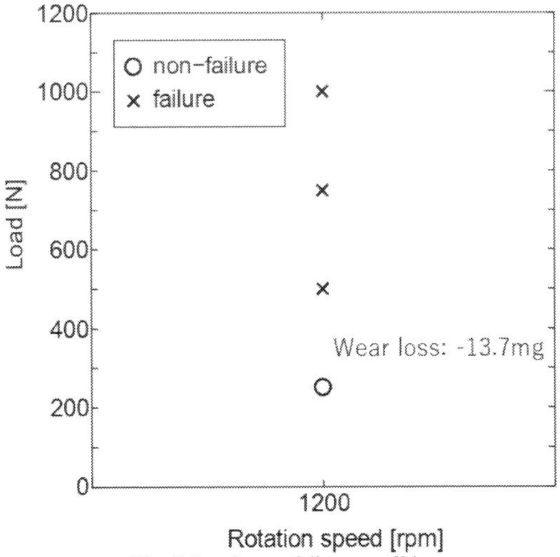

Fig. 3 Specimens failure conditions

3.2 Specimen surface observation

We observed the specimen surfaces using a LCM. Figure 4 shows the donut-shaped specimen's surface whose lamination direction was perpendicular to the rolling direction. Figure 4(a) shows the surface before the test. We observed the lamination gaps, and the roughness of the surface was about R_a = 4.6 μm (the measurement direction parallel to rolling direction). Figure 4(b) shows the donut-shaped specimen's surface after the 500 N thrust load test. The circular cracks and the deep peeling can be seen. The circular cracks are evenly distributed at increments between 50 and 100 μm. Figure 4(c) shows a 3-D image of Fig. 4(b). The peeling depth was over 350 μm after the 500 N thrust load test.

Figure 5 shows the donut-shaped specimen's surface whose lamination direction was the same as the rolling direction. Figure 5(a) shows the surface before the test. The roughness of the surface was about R_a = 4.0 μm (the measurement direction parallel to rolling direction). Figure 5(b) shows the donut-shaped specimen's surface after the 500 N thrust load test. We observed a deep hole which was the result of flacking. We also observed two types of cracks, surface circular cracks which are perpendicular to the rolling direction, and cracks which are parallel to the rolling direction at the bottom of the flaking failure. Figure 5(c) shows a 3-D image of Fig. 5(b) and the two types of cracks can be clearly observed.

Figure 6 shows the donut-shaped specimen's surface whose lamination direction was 45 degrees to the rolling direction. Figure 6(a) shows the surface before the test. The roughness of the surface was about R_a = 4.7 μm (the measurement direction parallel to rolling direction and rolling direction opposite to that in other pictures). Figure 6(b) shows the donut-shaped specimen's surface after the 500 N thrust load test. Some of the cracks were in the same direction as the lamination direction. Figure 6(c) shows a 3-D image of Fig. 6(b) with cracks in various directions.

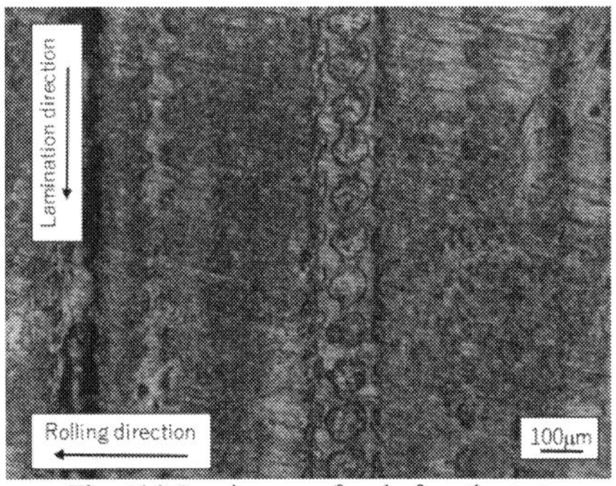

Fig. 4(a) Specimen surface before the test

DOI: 10.1051/matecconf/201713009004

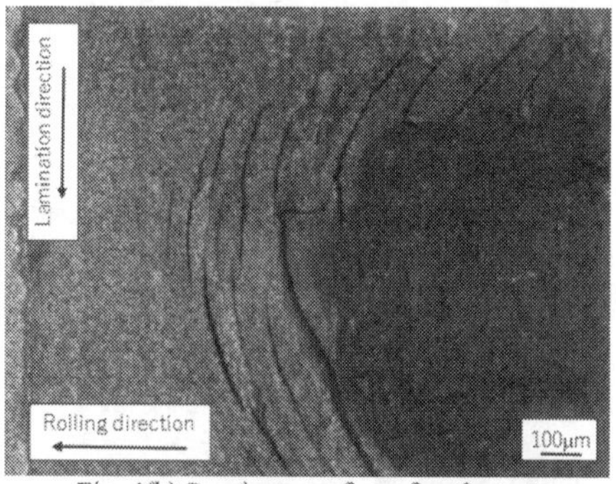

Fig. 4(b) Specimen surface after the test

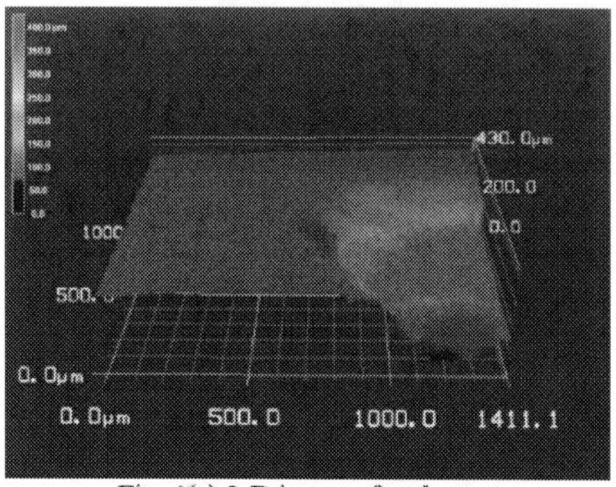

Fig. 4(c) 3-D image after the test

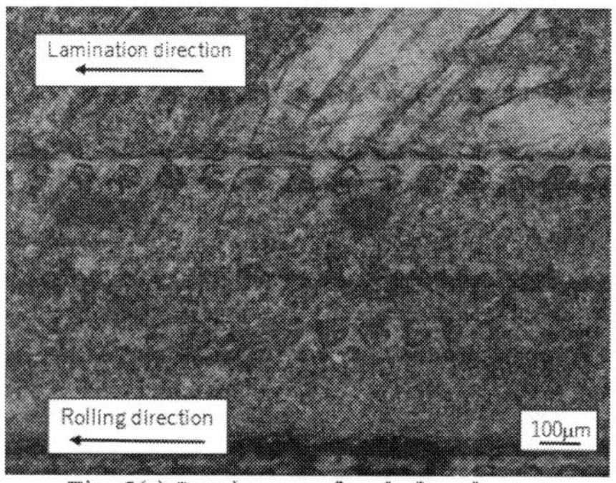

Fig. 5(a) Specimen surface before the test

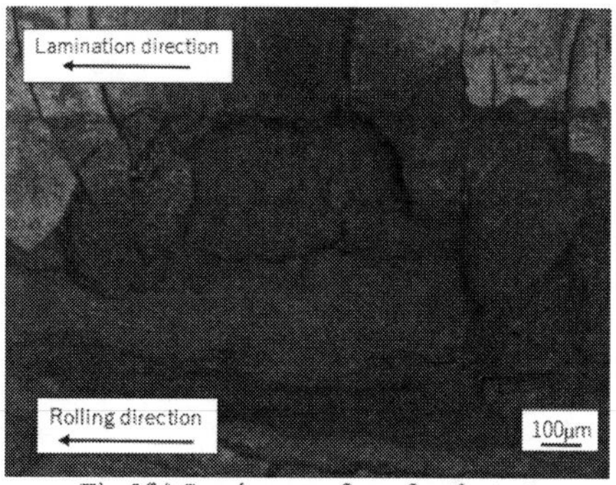

Fig.5(b) Specimen surface after the test

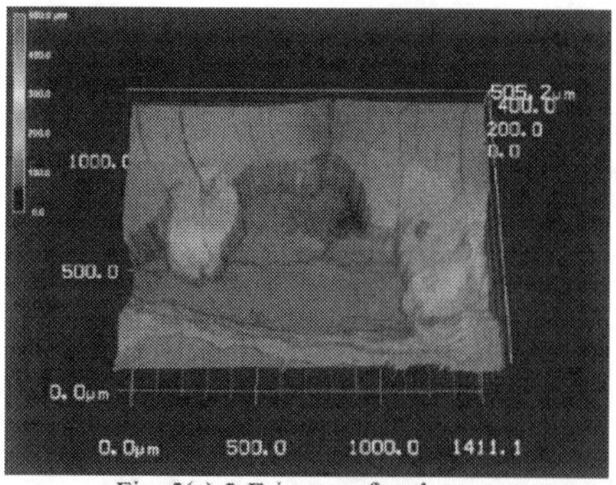

Fig. 5(c) 3-D image after the test

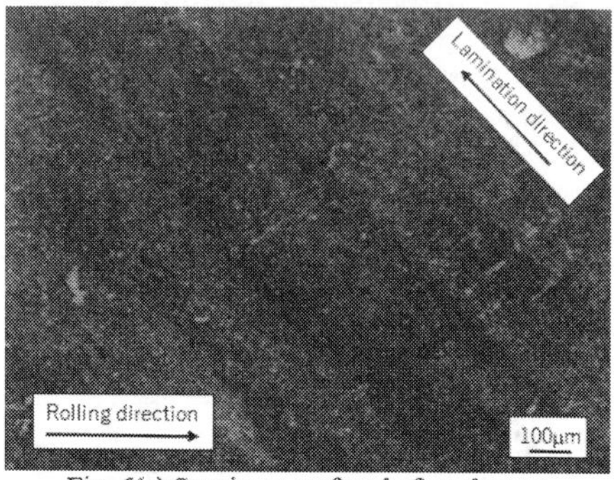

Fig. 6(a) Specimen surface before the test

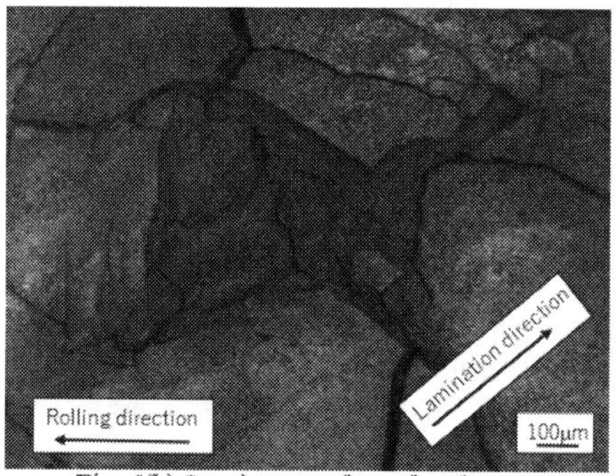

Fig. 6(b) Specimen surface after the test

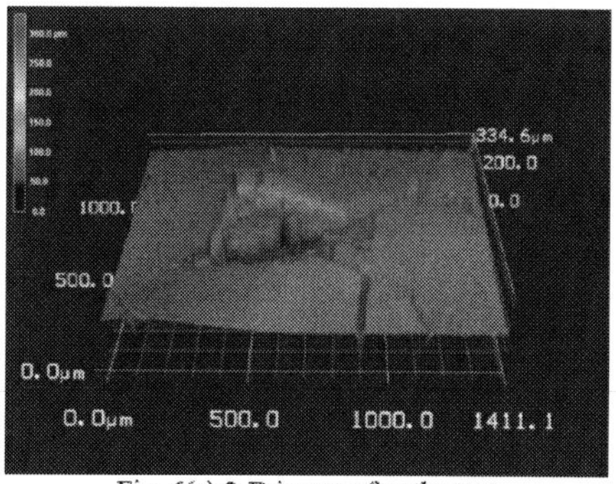

Fig. 6(c) 3-D image after the test

4 Conclusions

In this study, we performed rolling contact fatigue tests on 3-D printed specimens in water and observed the fracture behaviour of the specimens. Our findings are as follows:

1. The flaking failure occurred in the 3-D printed donut-shape specimen in water.
2. When the lamination direction was parallel to the rolling direction, circular cracks perpendicular to the rolling direction were observed.
3. When the lamination direction was perpendicular to the rolling direction, two types of cracks occurred: circular cracks at the surface, and long parallel cracks at the bottom of the flaking failure.
4. When the lamination direction was 45 degrees to the rolling direction, cracks occurred in various directions.

These facts indicate that the surface cracks are related to both the rolling and lamination directions.

References

1. X. Shi, K. Kida and Y. kashima, Mat. Res. Inov. **18**, 3, S42-S47 (2014)
2. R. Prehn, F. Haupert and K. Fridrich, Wear, **259**, 1-6, 693-696 (2005)
3. J. Jia, J. Chen, H. Zhou, L. Hu and L. Chen, Com. Sci. Tech. **65**, 7-8, 1139-1147 (2004)
4. B. M. Tymrak, M. Kreiger and J. M. Pearce, Mat. Des. **58,** 242-246 (2014)
5. J. M. G. Farstad, and O. Netland and T. Welo, Procedia CIRP, 60, 247-252 (2017)
6. X. C. Shi, M. Orito, Y. Kashima, K. Mizobe and K. Kida, Key Eng. Mat. **703,** 172-177 (2016)

The Study of the Composite Material Go/CF/PTFE Tribological Property

Li-hu Wang[1], Shou-ren Wang[1,a],Dao-sheng Wen[1] ,Gao-qi Wang[1] and Li-ying Yang[1]
[1]School of Mechanical Engineering, Jinan University, Jinan 250022, China

Abstract: In this paper, the composite material Go/CF/PTFE tribological property was studied. The test of its mechanical property, and the fabrication of the filled PTEE composite material sample which is based on the technology of cold press molding and sinter molding proved that adding Go and CF moderately to the composite material was an efficient way to improve its mechanical property. Meanwhile the process of friction and wear trial and SEM analysis results of the micro-structure of wear pattern proved that the addition of the Go and CF tremendously improved the anti-wear property and that after the addition the plowing effect which took place on the material surface would turn into a kind of mixed wear effect that includes plowing effect and fatigue wear. Working as pinning and bridging, the Go which distributing uniformly in the matrix was able to improve the resistance and substantially resisted the crack propagation, therefore to a certain degree enhanced the intensity of composite material and prolong its lifespan.

1Introduction

The PTFE is a kind of polymer with the great comprehensive property, such as the high tensile strength(range from 10MPa to 35MPa), bending strength(range from 18MPa to 20MPa), compressive strength(range from 18MPa to 27MPa), impact strength(range from 24MPa to 31MPa) and low friction coefficient (range from 0.13 to 0.16), which are attributed to its special supramolecular helical structure. But the insufficient property of wear resistance limited the its application of production, especially in the high speed and heavy load conditions which demanded the great property of wear resistance[1]. Recently, although many researchers have improved the property of wear resistance of PTFE by adding the compounds such as fiberglass, SiO_2, tin bronze, CF and Go etc. to PTFE, few researchers have studied the mechanical property and wear resistance property which were affected by Go and CF at the same time. This paper includes the study where we added the some kinds of padding to PTFE to change the property, then investigated the affect of content of GO and CF on the mechanical property by several trials. We also have tested the friction coefficient by MMG-10 friction wear testing machine, then in order to analyze the wear mechanism we analyzed the Micro-structure of crack and fracture area deeply.

2The material of experiment and method

2.1The manufacturing of composite material sample

The manufactured composite material is consisted of PTFE used as matrix and Go and CF used as reinforcement. According to the different quantity of addition, the sample was divided into 10 groups which were distinguished by the mass fraction: Go(1%, 2%, 3%), CF(1%, 2%, 3%). The specific quantity of adding units arrange is shown in Table 1. The original material specification and the manufacture factory are shown in Table 2. Before the trial, the Go and CF must be disposed by KH-550 silane coupling agent, and the procedure was that the solution with 8% water+70% alcohol+22% silane coupling agentwas prepared, then Go and CF were added to the solution according to Table 1. When they were dissolved completely,the new solution should be placed into the ultrasonic oscillator for 2 minutes, then it was dried by the air-dry oven. Finally, the material was collected whose size is smaller than 300 microns by standard sieve[2]. The manufacturing procedure of composite material is shown as follows. First of all, the material which was weighed and took in a certain amount with the calculated mass fraction should be placed into mill pot in order to mix them and then corundum balls should be added to the sample with the proportion 7:1 (sample: balls) and sealed it. Secondly,the mill pot need to be evacuated to vacuum and filled it with argon gas whose concentration is 0.075MPa to mill the sample with 180r/min revolving speed for 5 hours. Subsequently, finished sample was placed into the self-made mold (figure 1)for cold compression with 10MPa/min compression rate. When the stress arrived 130MPa, it

[a] Corresponding author: sherman0158@163.com

should be kept for 2 minutes. Then the sample was placed into electric furnace for sintering with the 60°C/h heating rate. When the temperature rose to 327°C, it should be kept for 0.5 hours, then the temperature rose to 375°C with the same rate and it was soaked for 2 hours. Finally, the sample was furnace cooled to room temperature. The technological parameters of sintering are shown in Figure 2. The sample after manufacturing should be individually ground coarsely, finely and accurately based on different test aims. .

Table 1. Preparation of GO / CF / PTFE composites sample

Sample number	1	2	3	4	5	6	7	8	9	10
PTFE/%	100	89	79	69	88	78	68	87	77	67
CF/%	0	10	20	30	10	20	30	10	20	30
Go/%	0	1	1	1	2	2	2	3	3	3

Table 2. Raw materials of Go/CF/PTFE composites

Material Name	Specification	Manufacturer
PTFE	Diameter of about 48μm	Dongguan prosperous plastic materials Co., Ltd
Go	7 layers	Global Technology
CF	Diameter of about 50μm	Zibo Kun Mao Da New Materials Co., Ltd
Silane coupler	KH-550	Nanjing Creation Chemical Additives Co., Ltd

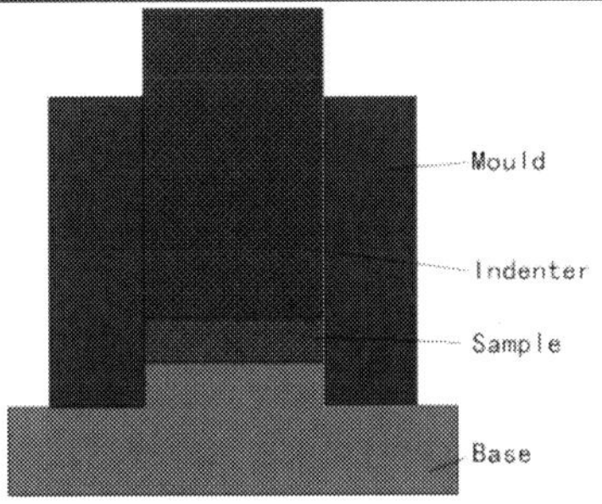

Figure 1. Schematic diagram of cold forming for sample in the mould

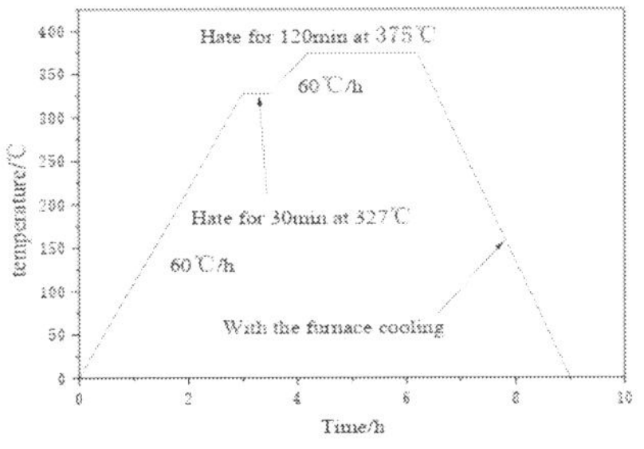

Figure 2. Sintering processes

2.2 The mechanical property trial

2.2.1 The test of tensile property

We tested the tensile strength of the sample we gained by electronic universal testing machine. Before the test, the sample should be molded into hexahedron structure with dimension of 40mm×4mm×4mm. Stretching rate was 2mm/min, and the pre-streching rate was 1mm/min. After the test of all 10 groups, the maximum and minimum results should be ignored, then calculated the average of the rest results.

2.2.2 The test of bending property

The bending strength of the samples in every group was tested by three point bending method. In this experiment, the span between two fulcrums on the clamp is L=30mm, and the load rate is 2mm/min. There are 10 samples in each group to be tested then calculate the arithmetic average of the results.

2.2.3The test of compressive property

We tested the compressive property by computer-controlled high-temperature universal testing machine. Before the test, the sample should be molded into hexahedron structure with dimension of 10mm×4mm×4mm. compressing rate was 2mm/min, and the pre-compressing rate was 1mm/min. After the test of all 10 groups, the maximum and minimum results should be ignored, then calculated the average of the rest of the results.

2.2.4 The test of impact toughness

We tested the impact toughness by the XJJ-5 impact testing machine. Before the test, the sample should be molded into an incomplete hexahedron structure with dimension of 10mm×4mm×4mm. The energy of impact is set as 4J. After the test of all 10 groups, the maximum and minimum result should be ignored, then calculated the average of the rest of the results.

2.3 The friction-wear test

We tested 9 groups of sample by friction-wear testing machine. The other component is stainless steel rim with the size that the outer diameter was 15mm and the inner diameter was 10mm and lubricated by 5201 silicone grease. The primitive revolving rate was set as 100r/min. The vertical stress is set as 100N. The rubbing length of each sample was 1km in this trial then the group with the best performance was selected to have a test. In the later trial we set the stress as 100N and revolving rate as 50r/min, 100r/min, 150r/min and 200r/min orderly. Then we selected the group with the best performance to be tested. Controlling the revolving rate, the stress was set as 100N, 200N,300N and 400N orderly and record each results which would be the standard for evaluating the affection on wear property from the vertical stress[3-6].

2.4 Micro- structure

We analyzed the micro-structure of the composite material Go/CF/PTFE crack and fracture area, and therefore we could analyze the the wear mechanism deeply.

3 Results and discussion

3.1 Mechanical property

The mechanical property test result of composite material Go/CF/PTFE is shown in Table 3.

Table 3. Mechanical properties of composite material Go / CF / PTFE

Sample number	1	2	3	4	5	6	7	8	9	10
Tensile strength /MPa	15.96	19.39	15.97	9.22	19.86	13.15	9.61	17.72	9.58	7.45
Bending strength /MPa	20.67	21.54	18.20	16.07	21.47	18.01	14.52	20.11	17.32	10.70
Compressive strength /MPa	28.87	33.73	25.51	21.24	32.99	21.04	18.32	30.11	18.95	14.54
Impact strength /MPa	24.85	28.94	21.04	16.09	28.86	26.19	14.67	23.64	18.49	14.29

What is shown in Figure 3 is the strength value of composite Go/CF/PTFE when the mass fraction of Go was set as 1% and we could conclude that the addition of CF was able to improve the tensile strength, bending strength, compressive strength and impact strength. But if the CF content is too high, the strength will decrease. When the CF content was about 10%, the tensile strength, bending strength, compressive strength and impact strength could be increased by 21.5%, 4.2%, 48% and 16.46%, respectively. In the test of tensile property, the CF would work as skeleton to bear and deliver the load because of its high strength and elasticity modulus, and so when the CF content is about 10%, the tensile strength will increase. As the CF content increased, the tensile strength would decrease considerably. The reasons are that the PTFE is a kind of organic material with stable property, and the interface issue between CF and PTFE is stubborn. Although CF have been disposed by KH-550 silane coupling agent, when the additional content is low the interface issue of CF won't be apparent, then the CF will perform as enhancement; when the CF content is more than 20%, the interface issues will perform dominantly, so the tensile strength will decrease.

Figure 4 shows the strength value of composite material Go/CF/PTFE when the CF content is 10%. The reasons are that the increasing of CF content will extrude the interface issue between CF and PTFE. It would lead to amounts of defect inside the sample, and therefore the strength would decrease slightly.

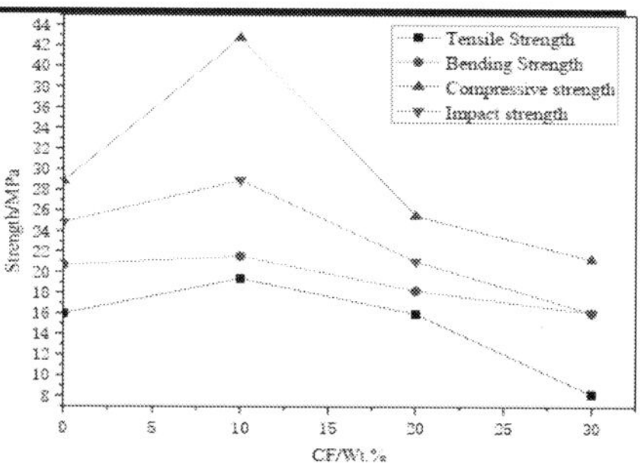

Figure 3. Dependence of mechanical properties on CF content of the Go/CF/PTFE composites with the graphene mass fraction of 1%

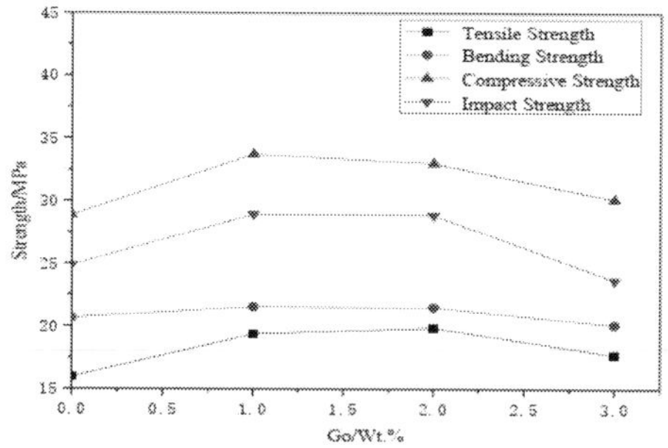

Figure 4. Dependence of strength on Go content of the Go/CF/PTFE composites with the carbon fiber mass fraction of 10%

3.2 The study of wear mechanism

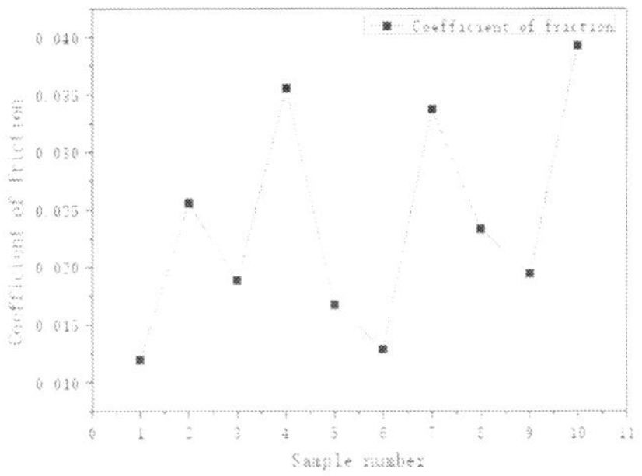

Figure 5. Friction coefficients of the Go/CF/PTFE composites of different samples tested at a pressure of 100 N, with a rotation speed of 100 r/min

Figure 5 has shown all stable friction coefficient value of each trial groups. It can be concluded that the addition of CF and Go could increase the value of stable friction coefficient. The tribological property of the composite material is affected by the original property of additional phase and the interface issue between additional phase and matrix. The high strength and hardness of CF would increase the hardness of composite material, and then increase the friction torque and friction coefficient. CF is a kind of new material with very low friction coefficient and the self-fabrication property will diminish the friction coefficient when CF is used as additional phase. When the content of CF and Go is low, the interface issue of the additional phase would not be obvious and the self-fabrication of CF will neutralize a part of the increase of friction coefficient which is attributed to CF content. So the increase of friction coefficient is not remarkable. But with increasing the additional phase content, the interface issue between additional phase and matrix will be more and more obvious. Meanwhile, the increase of friction coefficient will depend largely on the interface issue. When the Go content is not more than 1%, the friction coefficient will be large when the CF content is 10% and 30%, and when CF content is 20% the friction coefficient will decrease 26.6% and 50.1% compared with the friction coefficient when CF content is 10% and 30%. The former trends are same when Go content is 2% and 3%. If the CF content is not more than 20%, you can find out the law that how Go affected the friction coefficient just like what CF content did. When the Go content is 2%, the stable friction coefficient will decrease by 31.4% and 33.5% compared with stable friction coefficient when Go content is 1% and 3%.

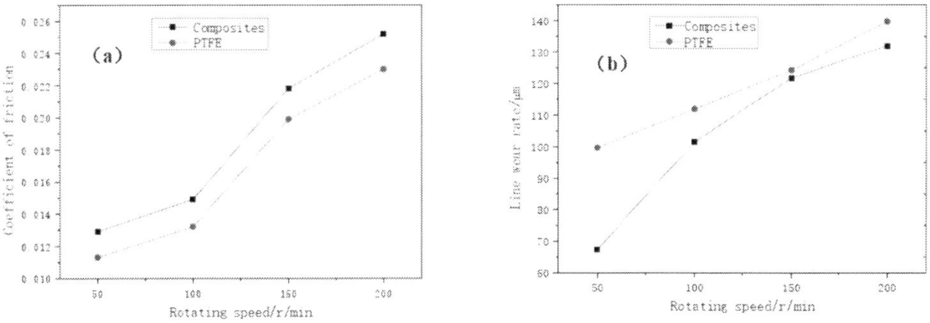

Figure 6. Coefficient of friction (a) and wear weight (b) for composite sample at pressure of 100N

Figure 6 shows the composite wear property (Figure 8(a)) in various revolving rate when the stress is 100N, and also shows that as revolving rate(Figure 8(b)) increases, the friction coefficient and Line wear rate will increase as well. The friction coefficient of composite material is larger than PTFE, but the line wear rate of composite material is lower. It proves that the friction coefficient of composite material is as low as the friction coefficient of PTFE. The additional phase improves the wear resistance property of PTFE, because the addition of CF can harden the surface of composite material, and the hard surface is able to prevent efficiently matrix from forming the completed wear film on the other rubbing component. Then the wearing capacity would decrease which is attributed to the stripping of wear film. Especially with the 50r/min revolving rate, the line wear rate decreased by 50.8%. The friction coefficient and line wear rate did not increase tremendously with the relatively low revolving rate. However, when the revolving rate was larger than 100r/min, the friction coefficient and line wear rate would increase considerably. That was because the increase of revolving rate led to the shear stress increase of CF, and CF shed from the matrix will increase largely. Meanwhile, the increase of revolving rate would increase the wear surface chafe heat which could soften the matrix and then weaken the fixation between the matrix and CF. Under the conditions that the interface issue was not obvious, the chafe heat was generated by high revolving rate change the interface issue into an essential reason that affected the CF drop. The dropped CF acted as hard particle would intensify the wear phenomenon and increase the friction coefficient and line wear rate[7-9].

Figure 7 shows the composite material tribological property according to different stress value with with the revolving rate of 50r/min,. From Figure 7(a), it can be seen that the friction coefficient of composite material is higher than that of PTFE, and it would increase as the stress increases. From Figure 7(b) , the line wear rate of composite material was lower than that of PTFE, and it would increase as the stress increases. It proves that the friction coefficient of composite material is as low as that of PTFE. Because of the addition of CF the surface got harder it prevented the wear film from forming. Therefore the high line wear rate was generated by dropped wear film would be diminished, and also the increased hardness would prevent abrasive wear happening. Consequently, the line wear rate of composite material was lower than that of PTFE.

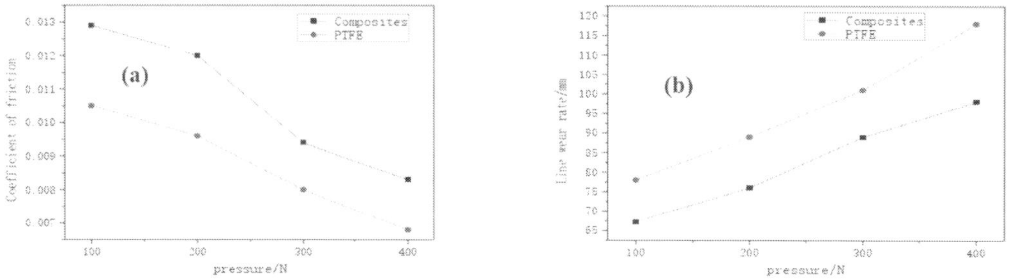

Figure 7. Coefficient of friction (a) and wear weight (b) for composite sample at 50r / min

3.3 Micro- structure analysis

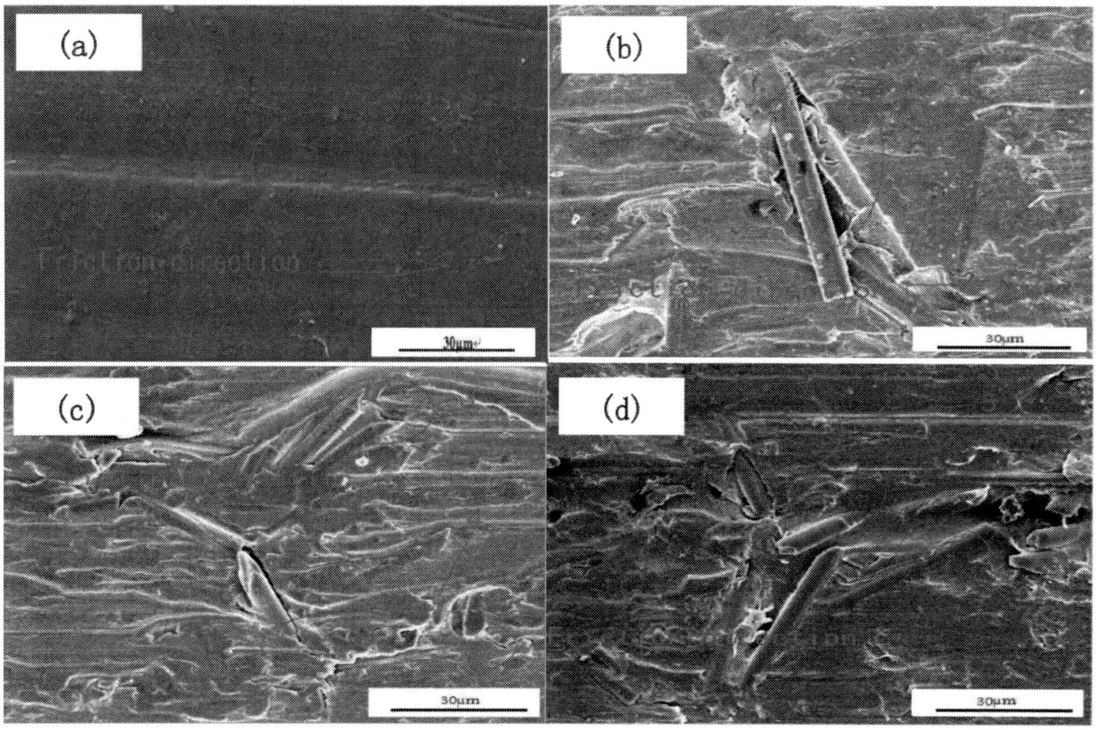

Figure 8. SEM wear morphology of PTFE and its composite material
(a)PTFE, 100r / min, 200N; (b)composite sample, 100r / min, 200N; (c) composite sample, 100r / min, 300N;(d) composite sample, 200r / min, 200N

Figure 8 shows the microstructure of different composite material. The wear form was changedby the addition of CF and Go. Figure 8(a) demonstrated the wear surface of PTFE sample, and it can be found that the scratch on the area 1 was caused by the uneven wear facet of the stainless steel component in the micro level. The protruding part acted as the cutting edge scratched the surface, the procedure which is called plowing effect[10-12]. After disposing by KH-550 silane coupling agent, the reinforcement CF prevented the matrix from wearing and the CF of the composite material sample(b) on area 2 would shed nearly. It have been supposed that it was caused by two reasons: firstly, after disposing by KH-550 silane coupling agent, the interface issue between CF and matrix was not good and the CF could not inset the matrix tightly; secondly, the protruding part of the rubbing component kept cutting the rubbing surface during the trial, and then it brought a shear force to the rubbing surface. So the CF in the matrix would undertake the load and diminish the wear and the CF whose form was vertical to the rubbing orientation would suffer from the shear force. After being loaded for a long time, the crack would take place first on the loosest combination part between CF and matrix(shown in area 3 in Figure 10 (c)) and the crack would keep expanding during the rubbing procedure till the loosest part shed completely. The addition of CF would change the wear form and the plowing effect which took place on the material surface would turn into a kind of mixed wear effect that includes plowing effect and fatigue wear. Area 4 in Figure 10 (c) shown that the semi-enveloped crack had took place, because the wear surface was forced by the shear force for a long time and the crack source took place on the weak under area of CF. The shear force would keep the crack source expanding and would cause the crater on the material surface. The CF which was non-paralleled to the rubbing orientation would prevent the crack happening because of the original high strength. Area 5 on Figure 8 (d) has shown the CF could prevent the matrix from plowing effect and then the wear resistance of composite material would be improved. Area 6 on Figure 8 (d) shows that CF was higher than the surrounding matrix, because the surrounding matrix would be wore firstly. The load-bearing function would be more dominant than wear resistance function. When the matrix of CF upper part was wore off, the CF was exposed on the rubbing surface. At the same time, CF and rubbing component contacted directly,so CF which has great wear resistance property kept matrix from wearing. When CF was wore off completely, the matrix and rubbing component would contact directly. The circulation of the former process would form the entire rubbing and wear process[13-16].

Figure 9. Carbon fiber at high magnification

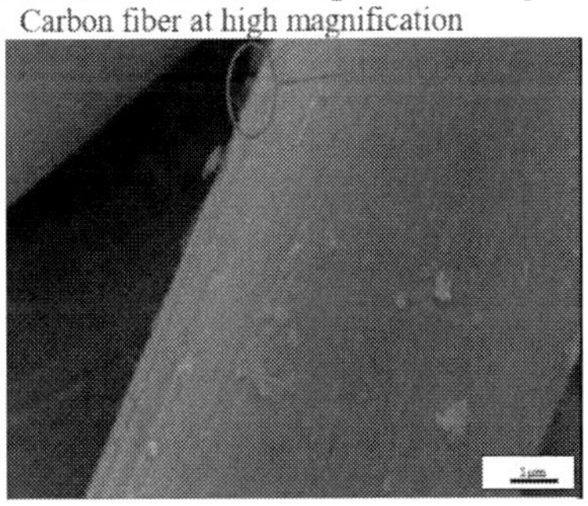

As a kind of self-fabricated material, CF would resist the wear when it was added into the composite material. The interface issue would be stubborn when inorganic Go was added into organic PTFE. This paper disposed the Go by silane coupling agent and improved the interface issue on the contact facet among Go, CF and PTFE. The combination between Go and CF has been shown in area 1 in Figure 9. It can be found that the CF which has plicated lamellar structure combined CF tightly. During the trial, the wear would be resisted by the excursion between layers, which is attributed to CF plicated lamellar structure.

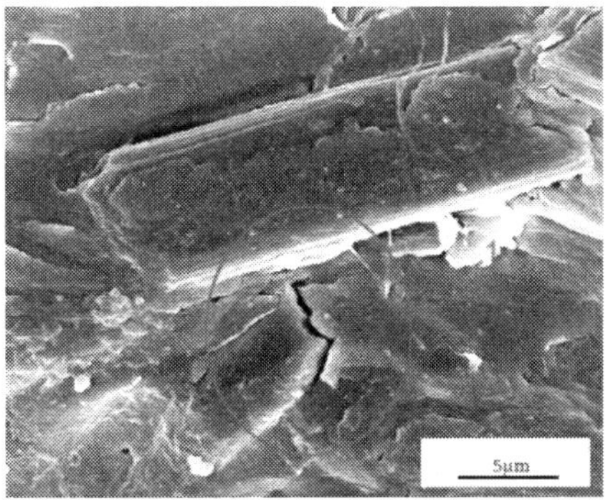

Figure 10. Pinning of carbon fibers

Figure 10 shows the affection form of CF among the matrix. Because the composite material sample was imposed by a shear force for a long time, the crack source would occur in the weak area of the material and then develop in a certain direction until it arrived at one side of CF which worked as pinning and prevented the material from shedding. When there were more than two cracks in different orientations expanding in one certain direction to one side of CF but did not intersect with each other, CF would worked as bridge to prevent the expansion efficiently as well as spinning. When force was accumulated to a certain extent, the cracks expanding path would be diverged from the original one by CF spinning and bridge affection. The new path would be parallel to the outer profile of CF. Finally, CF would shed completely, and therefore the wear would be more and more intensive. So the conclusion is that the interface issue between CF and PTFE would affect the tribological property of composite material considerably[17].

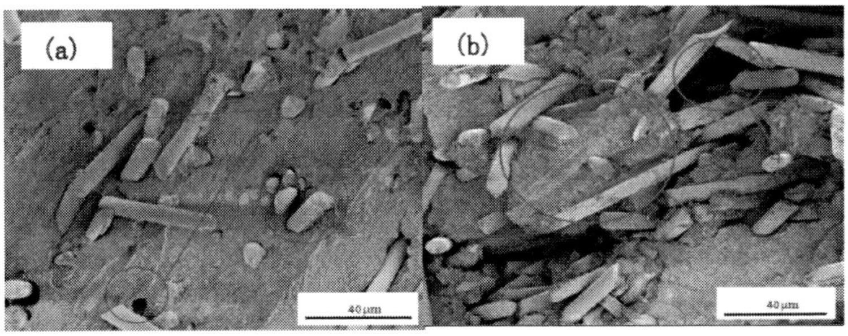

Figure 11. Composites Specimen Impact Section: (a) CF losening; (b) CF aggregate

Figure 11 shows the impact section micro-structure of composite sample. We can conclude that the interface issue was not good from area 1 in Figure 11(a). CF was extracted out from the material when it was impacted. The combination between CF and matrix would turn to be loose. Maybe it was caused by the incomplete disposition by silane coupling agent , or the silane coupling agent disposition couldn't improve the interface issue between CF and PTFE. The tight combination between CF and matrix which is shown on area 3 in Figure 11 (b), which proved that the interface issue was solved seemingly. What is shown on area 2 in Figure 11 (a) is that the mid area of CF has been broken, which proved that CF could strengthen the matrix. Area 4 in Figure 11 (b) has shown the gaps in composite material sample. Maybe it was caused by the fast compressing rate of cold-press molding, or the liquid evaporation during sintering process which was attributed to the halfway drying of sample .

4 Conclusion

(1) The addition of CF and Go improved the strength property of composite material. When the content of CF and Go was 10% and 1%, the strength of sample would be improved, but the excessive addition of CF and Go would cause the sudden decrease of strength.
(2) .The addition of CF and Go has changed the wear form of composite material, and it would turn into a kind of mixed wear effect that includes plowing effect and fatigue wear from the single plowing effect.
(3) Working as pinning and bridging, the Go which distributed uniformly in the matrix was able to improve the resistance and substantially resist the propagation of crack efficiently, and therefore to a certain degree the intensity of composite material was enhanced and its life span was prolonged.

Acknowledgments

This work was supported by the National Natural Science Foundation of P.R. China for the financial support (ID: 51372101, 51405195), Distinguished Middle-Aged and Young Scientist Encourage and Reward Foundation of Shandong Province (ID: ZR2016EMB01) and Taishan Scholar Engineering Special Funding (2016-2020).

References

1.Cai Xian-bin, Liu Xiu-bo.Study on Friction and Wear Properties of Polytetrafluoroethylene Composites [J]. Chemical Industry Management, 2015 (30): 15-15.
2.Wang Yong-hu, Wang Ling-juan, SONG Liming, et al.Studies and properties of graphene-polytetrafluoroethylene nanocomposites [J].
3.Chemical Production and Technology, 2014, 21 (2): 12-14.
4.Zhu En-bo.Study on the preparation and properties of polytetrafluoroethylene composites modified by fiber and powder filling [J]. Jiangsu University, 2010.
5.Yan Ling, Wang Wen-dong.Application and Application of Carbon Fiber Filled Polytetrafluoroethylene [J]. Chinese Journal of Organic Chemistry, 2005 (3): 23-25.
6.Shi Guo-jun, LI Cui, Yuan Yue.Preparation and tribological properties of polytetrafluoroethylene composites co-filled with mullite and carbon fiber [J]. Journal of Materials Science, 2016, 33 (9): 1886-1898.
7.Jiang Bo,Zhou Qian-qian,Zhu Ai-ping, et al. The thermal expansion and mechanical properties of erilite-glass fiber / polytetrafluoroethylene composites under friction and wear [J]. Chinese Journal of Composites, 2016, 33 (9): 1999-2005.
8.Shang Guan -qian .Study on tribological properties of rare earth modified carbon fiber reinforced polytetrafluoroethylene composites [D]. Shanghai Jiaotong University, 2007.
9.Wen Qin-shi, Huang Ping. Tribology [M]. Tsinghua University Press, 2012.
10.Gu Na. Study on Friction and Wear Properties of Polytetrafluoroethylene Composites [D]. Shandong University, 2009.
11.Chen Zhong-hua, Wang Jian-chuan, Yu Fei, et al.Preparation and thermal properties of graphene / melamine resin composites [J]. Functional Materials, 2015, 46 (1): 1125-1128.
12.Yang Wen-bin, Zhang Li, Liu Jing-wei, et al. Progress in the preparation and application of graphene composites [J]. Materials Engineering, 2015, 43 (3): 91-97.
13.Kuzharov A S. Polytetrafluoroethylene fiber-based composite antifriction coatings[J]. Industrial & Engineering Chemistry Research, 1993, 32(5):1072-1078.
14.Conte M, Igartua A. Study of PTFE composites tribological behavior[J]. Wear, 2012, 296(s 1–2):568-574.
15.Zhao Zhu,Qun Ji Xue.Friction and wear properties of metal powder filled PTFE composites under oil lubricated conditions[J].wear,1997,210:151～156.
16.Beckford S, Mathurin L, Chen J, et al. The effects of polydopamine coated Cu nanoparticles on the tribological properties of polydopamine/PTFE coatings[J]. Tribology International, 2016, 103:87-94.
17.Jotaki K, Miyatake M, Stolarski T, et al. Tribological performance of natural resin urushi containing PTFE[J]. Tribology International, 2016.

AUTHOR INDEX

AUTHOR INDEX